RY/07

MEDICAL RADIOLOGY
Radiation Oncology

Editors:
L. W. Brady, Philadelphia
H.-P. Heilmann, Hamburg
M. Molls, Munich

W. Schlegel · T. Bortfeld · A.-L. Grosu (Eds.)

New Technologies in Radiation Oncology

With Contributions by

J. R. Adler · N. Agazaryan · D. Baltas · Y. Belkacémi · R. Bendl · T. Bortfeld · L. G. Bouchet
G. T. Y. Chen · K. Eilertsen · M. Fippel · K. H. Grosser · A.-L. Grosu · P. Häring
J.-M. Hannoun-Lévi · G. H. Hartmann · K. K. Herfarth · J. Hesser · R. Hinderer
G. D. Hugo · O. Jäkel · M. Kachelriess · C. P. Karger · M. L. Kessler · C. Kirisits
P. Kneschaurek · S. Kriminski · E. Lartigau · S.-L. Meeks · E. Minar · M. Molls · S. Mutic
S. Nill · U. Oelfke · D. R. Olsen · N. P. Orton · H. Paganetti · A. Pirzkall · R. Pötter
B. Pokrajac · A. Pommert · B. Rhein · E. Rietzel · M. A. Ritter · M. Roberson · G. Sakas
L. R. Schad · W. Schlegel · C. Sholz · A. Schweikard · T. D. Solberg · L. Sprague
D. Stsepankou · S.E. Tenn · C. Thieke · W. A. Tomé · N. M. Wink · D. Yan · M. Zaider
N. Zamboglou

Foreword by

L. W. Brady, H.-P. Heilmann and M. Molls

With 299 Figures in 416 Separate Illustrations, 246 in Color and 39 Tables

 Springer

Wolfgang Schlegel, PhD
Professor, Abteilung Medizinische Physik
in der Strahlentherapie
Deutsches Krebsforschungszentrum
Im Neuenheimer Feld 280
69120 Heidelberg
Germany

Thomas Bortfeld, PhD
Professor, Department of Radiation Oncology
Massachusetts General Hospital
30, Fruit Street
Boston, MA 02114
USA

Anca-Ligia Grosu, MD
Privatdozent, Department of Radiation Oncology
Klinikum rechts der Isar
Technical University Munich
Ismaningerstrasse 22
81675 München
Germany

Medical Radiology · Diagnostic Imaging and Radiation Oncology
Series Editors: A. L. Baert · L. W. Brady · H.-P. Heilmann · M. Molls · K. Sartor

Continuation of Handbuch der medizinischen Radiologie
Encyclopedia of Medical Radiology

Library of Congress Control Number: 2004116561

ISBN 3-540-00321-5 Springer Berlin Heidelberg New York
ISBN 978-3-540-00321-2 Springer Berlin Heidelberg New York

Springer is part of Springer Science+Business Media

http//www.springeronline.com
© Springer-Verlag Berlin Heidelberg 2006
Printed in Germany

Medical Editor: Dr. Ute Heilmann, Heidelberg
Desk Editor: Ursula N. Davis, Heidelberg
Production Editor: Kurt Teichmann, Mauer
Cover-Design and Typesetting: Verlagsservice Teichmann, Mauer

Printed on acid-free paper – 21/3151xq – 5 4 3 2 1 0

Foreword

Radiation oncology is one of the most important treatment facilities in the management of malignant tumors. Although this specialty is in the first line a physician's task, a variety of technical equipment and technical know-how is necessary to treat patients in the most effective way possible today.

The book by Schlegel et al., "New Technologies in Radiation Oncology," provides an overview of recent advances in radiation oncology, many of which have originated from physics and engineering sciences. 3D treatment planning, conformal radiotherapy, with consideration of both external radiotherapy and brachytherapy, stereotactic radiotherapy, intensity-modulated radiation therapy, image-guided and adaptive radiotherapy, and radiotherapy with charged particles are described meticulously . Because radiotherapy is a doctor's task, clinically orientated chapters explore the use of therapeutic radiology in different oncologic situations. A chapter on quality assurance concludes this timely publication.

The book will be very helpful for doctors in treating patients as well as for physicists and other individuals interested in oncology.

Philadelphia   Luther W. Brady

Hamburg   Hans-Peter Heilmann

Munich   Michael Molls

Preface

In the 1960s radiation therapy was considered an empirical, clinical discipline with a relatively low probability of success. This situation has changed considerably during the past 40 years.

Radiation therapy is based heavily on fields such as physics, mathematics, computer science and radiation biology as well as electrical and mechanical engineering, making it a truly interdisciplinary field, unparalleled by any other clinical discipline. Now radiation therapy can be applied so safely, precisely and efficiently that the previously feared side effects no longer play a role. At the same time, tumour control, and the probability of cure, has significantly increased for many tumour patients. This change from an empirical and qualitative discipline to a scientifically based, precise clinical science has been accompanied by groundbreaking innovations in physics and technology (Fig. 1).

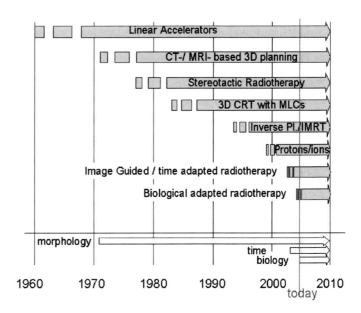

- The first important step was the replacement of cobalt-60 and betatrons as irradiation sources by electron-linear accelerators (also known as "linacs") between 1960 and 1980. Modern computer-controlled linacs are comparatively compact and reliable, have a high mechanical accuracy and deliver sufficiently high dose rates. Having become the "work-horses" of radiation oncology, they have been introduced in nearly every radiotherapy department in the world, providing the basis of modern precision radiotherapy.
- The next important milestone, which sparked a revolution not only in radiological diagnostics but also in radiotherapy, was the invention of X-ray computed tomography (CT). Computed tomography was introduced to the radiotherapy process at the end of the 1970s, and this resulted in 3D computerized treatment planning, now a standard tool in all radiotherapy departments.

- The CT-based treatment planning was later supplemented with medical resonance imaging (MRI). By combining CT and MRI, and using registered images for radiotherapy planning, it is now possible to assess tumour morphology more precisely, and thus achieve improved definition of planning target volumes (PTV), improving both percutaneous radiotherapy and brachytherapy.
- The computer revolution, characterized by the development of small, powerful and inexpensive desktop computers, had tremendous impact on radiation therapy. With new tools from 3D computer graphics, implemented in parallel with 3D treatment planning, it was possible to establish "virtual radiotherapy planning", a method to plan and simulate 3D irradiation techniques. New 3D dose calculation algorithms (e.g. "pencil-beam algorithms") made it possible to precalculate the 3D dose distributions with sufficient accuracy and with acceptable computing times.
- With the aforementioned advent of 3D imaging, 3D virtual therapy simulation and 3D dose calculation, the preconditions for introducing an individualized, effective local radiation treatment of tumours were fulfilled. What was still missing was the possibility to transfer the computer plans to the patient with high accuracy. This gap was filled by the introduction of stereotaxy into radiotherapy in the early 1980s. Prior to this development, stereotaxy was used in neurosurgery as a tool to precalculate target points in the brain and to precisely guide probes to these target points within the tumour in order to take biopsies or implant radioactive seeds. The transfer of this technique to radiotherapy resulted in significantly enhanced accuracy in patient positioning and adjustment of radiation beams. Stereotactic treatment techniques were first developed for single-dose irradiations (called "radiosurgery"), then for fractionated treatments in the brain and the head and neck region ("stereotactic radiotherapy"). Later, it became possible to transfer stereotactic positioning to extracranial tumour locations ("extracranial stereotactic radiotherapy") as well. This opened up the possibility for high-precision treatments of tumours in nearly all organs and locations.
- The next important step which revolutionized radiotherapy came again from the field of engineering. The development of computerized multi-leaf collimators (MLCs) in the middle of the 1980s ensured the clinical breakthrough of 3D conformal radiotherapy. With the advent of MLCs, the time-consuming fabrication of irregularly shaped beams with cerrobend blocks could be abandoned. Conformal treatments became less expensive and considerably faster, and were applied with increasing frequency. The combination of 3D treatment planning and 3D conformal beam delivery resulted in safe and efficient treatment techniques, which allowed therapists to escalate tumour doses while at the same time lowering the dose in organs at risk and normal tissues.
- By the mid 1990s, 3D conformal radiotherapy was supplemented by a new treatment technique, which is currently becoming a standard tool in modern clinics: intensity-modulated radiotherapy (IMRT) using MLC-beam delivery or tomotherapy, in combination with inverse treatment planning. In IMRT the combination of hardware and software techniques solves the problem of irradiating complex target volumes with concave parts in the close vicinity of critical structures, a problem with which radio-oncologists have had to struggle from the very beginning of radiotherapy. In many modern clinics around the world, IMRT is successfully applied, e.g. in the head and neck and in prostate cancer. It has the potential to improve results in many other cancer treatments as well.
- The IMRT with photon beams can achieve a level of conformity of the dose distribution within the target volume which can, from a physical point of view, not be improved further; however, the absolute dose which can be delivered to the target volume is still limited by the unavoidable irradiation exposure of the surrounding normal tissue. A further improvement of this situation is possible by using particle radiation. Compared with photon beams, the interaction of particle beams (like protons or heavier charged particles) with tissue is completely different. For a single beam, the dose delivered to

the patient has a maximum shortly before the end of the range of the particles. This is much more favourable compared with photons, where the dose maximum is located just 2–3 cm below the surface of the patient's body. By selecting an appropriate energy for the particle beams and by scanning particle pencil beams over the whole target volume, highly conformal dose distributions can be reached, with a very steep dose fall-off to surrounding tissue, and a much lower "dose bath" to the whole irradiated normal tissue volume. Furthermore, from the use of heavier charged particles, such as carbon-12 or oxygen-16, an increase in RBE can be observed shortly before the end of the range of the particles. It is expected that this radiobiological advantage over photons and protons will result in a further improvement in local control, especially for radioresistant tumours. However, particle therapy, both with protons and heavier charged particles, is still in the early stages of clinical application and evaluation on a broad scale. Ongoing and future clinical trials must demonstrate the benefit of these promising, but costly, particle-beam treatments.

At the beginning of the new millennium, the field of adaptive radiotherapy evolved from radio-oncology:

- After 3D CT and MRI enabled a much better understanding of tumour morphology, and thus spatial delineation of target volumes, the time has arrived where the temporal alterations of the target volume can also be assessed and taken into account. Image-guided and time-adapted radiotherapy (IGRT and ART) are characterized by the integration of 2D and 3D imaging modalities into the radiotherapy work flow. The vision is to detect deformations and motion between fractions (inter-fractional IGRT) and during irradiation (intra-fractional IGRT), and to correct for these changes either by gating or tracking of the irradiation beam. Several companies in medical engineering are currently addressing this technical challenge, with the goal of implementing IGRT and ART in radiotherapy as a fast, safe and efficient treatment technique.

- Another innovation which is currently on the horizon is biological adaptive radiotherapy. The old hypothesis that the tumour consists of homogeneous tissue, and therefore a homogeneous dose distribution is sufficient, can no longer be sustained. We now know that a tumour may consist of different subvolumes with varying radiobiological properties. We are trying to characterize these properties more appropriately by functional and molecular imaging using new tracers in PET and SPECT imaging and by functional MRI (fMRI) and MR spectroscopy, for example. We now have to develop concepts to include and integrate this information into radiotherapy planning and beam delivery, firstly by complementing the morphological gross tumour volume (GTV) by a biological target volume (BTV) consisting of subvolumes of varying radioresistance, and secondly by delivering appropriate inhomogeneous dose distributions with the new tools of photon- and particle-IMRT techniques ("dose painting"). Furthermore, biological imaging can give additional information concerning tumour extension and tumour response to radiotherapy or radiochemotherapy.

Currently, we have reached a point where, besides the 3D tumour morphology, time variations and biological variability within the tumour can also be taken into account. The repertoire of radiation oncology has thus been expanded tremendously. Tools and methods applied to radiotherapy are increasing in number and complexity. The speed of these developments is sometimes breathtaking, as radiation oncologists are faced more and more with the problem of following and understanding these modern innovations in their profession, and putting the new developments into practice. This book gives an introduction into the aforementioned areas. The authors of the various chapters are specialists from the involved disciplines, either working in research and development or in integrating and using the new methods in clinical application. The authors endeavoured to explain the very often complicated and complex subject matter in an understandable manner. Naturally, such a collection of contributions from a heterogeneous board of authors cannot completely cover the whole field of innovations. Some overlap, and variations in the depth of descriptions and explanations were unavoidable. We hope that the book will be particularly helpful for physicians and medical physicists who are working in radiation oncology or just entering the field, and who are trying to achieve an overview and a better understanding of the new technologies in radiation oncology.

The motivation to compile this book can be traced back to the editors of the book series Medical Radiology/Radiation Oncology, by Michael Molls, Munich, Luther Brady, Philadelphia, and Hans-Peter Heilmann, Hamburg. We thank them for continuous encouragement and for not losing the belief that the work will eventually be finished. We extend thanks to Alan Bellinger, Ursula Davis, Karin Teichmann and Kurt Teichmann, who did such an excellent job in preparing the book. Most of all, thanks to all the authors, who wrote their chapters according to our suggestions, and a very special thanks to those who did this work within the short period of time before the deadline.

Heidelberg
Boston
Munich

WOLFGANG SCHLEGEL
THOMAS BORTFELD
ANCA-LIGIA GROSU

Contents

1 New Technologies in 3D Conformal Radiation Therapy: Introduction and Overview

Wolfgang Schlegel

CONTENTS

1.1
Clinical Demand for New Technologies in Radiotherapy

Radiotherapy is, after surgery, the most successfully and most frequently used treatment modality for cancer. It is applied in more than 50% of all cancer patients.

Radiotherapy aims to deliver a radiation dose to the tumor which is high enough to kill all tumor cells. That is from the physical and technical point of view a difficult task, because malignant tumors often are located close to radiosensitive organs such as the eyes, optic nerves and brain stem, spinal cord, bowels, or lung tissue. These so-called organs at risk must not be damaged during radiotherapy. The situation is even more complicated when the tumor itself is radioresistant and very high doses are needed to reach a therapeutic effect.

W. Schlegel, PhD
Professor, Abteilung Medizinische Physik in der Strahlentherapie, Deutsches Krebsforschungszentrum, Im Neuenheimer Feld 280, 69120 Heidelberg, Germany

At the time of being diagnosed, about 60% of all tumor patients are suffering from a malignant localized tumor which has not yet disseminated, i.e., no metastatic disease has yet occurred; thus, these patients can be considered to be potentially curable. Nevertheless, about one-third of these patients (18% of all cancer patients) cannot be cured, because therapy fails to stop tumor growth.

This is the point where new technologies in radiation oncology, especially in 3D conformal radiotherapy, come into play: it is expected that they will enhance local tumor control. In conformal radiotherapy, the dose distribution in tissue is shaped in such a way that the high-dose region is located in the target volume, with a maximal therapeutic effect throughout the whole volume. In the neighboring healthy tissue, the radiation dose has to be kept under the limit for radiation damage. This means a steep dose falloff has to be reached between the target volume and the surroundings; thus, in radiotherapy there is a rule stating that with a decrease of dose to healthy tissue, the dose delivered to the target volume can be increased; moreover, an increase in dose will also result in better tumor control (tumor control probability, TCP), whereas a decrease in dose to healthy tissue will be connected with a decrease in side effects (normal tissue complication probability, NTCP). Increase in tumor control and a simultaneous decrease in side effects means a higher probability of patient cure.

In the past two decades, new technologies in radiation oncology have initiated a significant increase in the quality of conformal treatment techniques. The development of new technologies for conformal radiation therapy is the answer to the wishes and guidelines of the radiation oncologists. The question of whether clinical improvements are driven by new technical developments, or vice versa, should be answered in the following way: the development of new technologies should be motivated by clinical constraints.

In this regard the physicists, engineers, computer scientists, and technicians are service providers to the radiologists and radiotherapists, and in this context new technologies in conformal radiation therapy are technical answers to a clinical challenge.

An improvement in one field automatically entails the necessity of an improvement in other fields. Conformal radiation therapy combines, in the best case, the advantages of all new developments.

1.2
Basic Principles of Conformal Radiotherapy

The basic idea of conformal radiation therapy is easy to understand and it is close to being trivial (Fig. 1.1). The problem is that depth dose of a homogeneous photon field is described by an exponentially decreasing function of depth. Dose deposition is normally higher close to the surface than at the depth of the tumor.

To improve this situation normally more than one beam is used. Within the overlapping region of the beams a higher dose is deposited. If the apertures of the beams are tailored (three dimensionally) to the shape of the planning target volume (PTV) masking the organs at risk (OAR), then the overlapping region should fit the PTV. In the case of an OAR close to the PTV, this is not true for such a simple beam configuration such as the one shown in Fig. 1.1. In such cases one needs a more complex beam configuration to achieve an acceptable dose distribution; however, the general thesis is that, using enough beams, it should be possible to a certain extent to fit a homogeneous

dose distribution to the PTV while sparing the OARs. One hopes that the conformity of dose distributions can be increased using individually tailored beams compared with, for example, a beam arrangement using simple rectangular-shaped beams, and for the majority of cases this is true. Nevertheless, there are cases, especially with concave-shaped target volumes and moving targets, where conventional conformal treatment planning and delivery techniques fail. It is hoped that these problems will be solved using intensity-modulated radiotherapy (IMRT; see Chap. 23) and adaptive radiotherapy (ART; see Chaps. 24–26), respectively.

1.3
Clinical Workflow in Conformal Radiotherapy

The physical and technical basis of the radiation therapy covers different aspects of all links in the "chain of radiotherapy" (Fig. 1.2) procedure. All parts of the "chain of radiation therapy" are discussed in other chapters separately and in great detail, and an overview on the structure of this book within this context is given in Table 1.1.

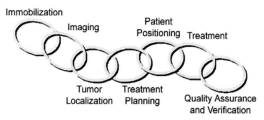

Fig. 1.2. Chain of radiotherapy. (From SCHLEGEL and MAHR 2001)

Table 1.1 Structure of the book with respect to the chain of radiotherapy

Part of the radiotherapy chain	Described in chapter
Patient immobilization	21, 22
Imaging and information processing	2–6
Tumor localization	7–11
Treatment planning	13, 14
Patient positioning	12, 21, 22, 26
Treatment	20–28
Quality assurance and verification	32, 33

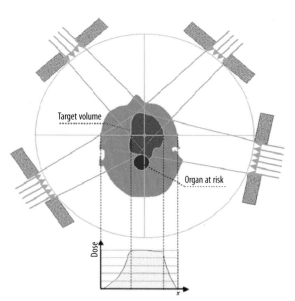

Fig. 1.1. Basic idea of conformal radiotherapy

1.3.1
Patient Immobilization

It is obvious that a reliable and exact fixation of the irradiated body area or organ is a substantial prerequisite for conformal radiation therapy. Modifications of the position of the patient relative to the treatment machine can lead to dangerous dose errors.

Numerous immobilization devices and techniques have been developed for radiotherapy, most of them using casts and moulds.

The highest immobilization accuracy is required when patients have to be treated with single-dose irradiation under stereotactical boundary conditions (see Chaps. 21, 22).

1.3.2
Imaging and Tumor Localization

Presently tumors and organs at risk are localized with 3D imaging techniques. One of the most important implicit constraints facing a technician developing new radiation therapy treatment techniques is that the segmented OAR and PTV are identical with the true organs and the true tumor. This is not a trivial statement, and it is still the subject of controversy. It is well known that different observers, and even the same observer, may create slightly different outlines at sequential attempts to define OARs and the PTV.

Without doubt, advances in the field of medical imaging, especially in the use of computed tomography (CT), magnetic resonance imaging (MRI), ultrasound (US), and positron emission tomography (PET), have led to improved precision in tumor localization. In particular, the gross tumor volume can be reconstructed in three dimensions from tomographic slices, and taken in the tumor region, thus forming the basis for 3D treatment planning (see Chaps. 2–5, 7, 8). Computed tomography is an ideal basis for 3D treatment planning, as it has the potential to quantitatively characterize the physical properties of heterogeneous tissue in terms of electron densities which is essential for dose calculation (see Chaps. 15, 16). On the other hand, MRI is very often superior to CT, especially for the task of differentiating between healthy tissue and tumor tissue (see Chap. 9). In addition, MRS and PET imaging have the potential to include information on tumor metabolism and heteronegeity (see Chaps. 10, 11, 13).

1.3.3
Treatment Planning

Computer-assisted 3D treatment planning can be considered as state of the art in most modern hospitals. The planning process can be divided into the following steps:

1. Determination of the target volume and organs at risk
2. Virtual therapy simulation
3. Dose calculation
4. Visualization and evaluation of dose distributions

The goal of treatment planning is the determination of a suitable and practicable irradiation technique which results in a conformal dose distribution; thus, treatment planning is a typical optimization problem. Whereas in conventional "forward planning" a trial-and-error method is applied for interactive plan optimization (see Chap. 14), the new method of "inverse planning" is able to automatically calculate a treatment technique which leads to the best coverage of the target volume and sufficient sparing of healthy tissue (see Chap. 17).

1.3.3.1
Defining Target Volumes and Organs at Risk

The best way of determining the PTV and OAR is on the basis of multiple-modality 3D image data sets such as X-ray computed tomography (CT), magnetic resonance (MRI and MRS), and PET. Routinely, X-ray CT is the most common tomographic imaging method (see Chaps. 7–8, 13).

The registration of all these imaging modalities for the purpose of defining target volumes and organs at risk is highly desirable. Three-dimensional image registration is a computer tool which is able to match the 3D spatial information of the different imaging modalities by use of either external or internal anatomic landmarks. The involved methods are described in Chap. 5. The problem of target-volume definition using multi-modal imaging techniques from the radio-oncologist's point of view are described in Chaps. 10 and 13.

Another problem that is more important, if highly conformal dose distributions are delivered, is organ movement. It is clear that in conformal therapy organ movements cannot be ignored.

Time-adapted radiotherapy is a field which is trying to solve this problem, and some chapters of this book report the approaches which are currently being investigated (see Chaps. 8, 24–26).

1.3.3.2
Definition of the Treatment Technique

Conformal radiation therapy basically requires 3D-treatment planning. After delineating the therapy-relevant structures, various therapy concepts are simulated as part of an iterative process. The search for "optimal" geometrical irradiation parameters – the "irradiation configuration" – is very complex. The beam directions and the respective field shapes must be selected. The various possibilities of volume visualization, such as Beams Eye View, Observers View, or Spherical View, are tools which support the radiotherapist with this process (Chap. 14).

1.3.3.3
Dose Calculation

The quality of treatment planning depends naturally on the accuracy of the dose calculation. An error in the dose calculation corresponds to an incorrect adjustment of the dose distribution to the target volume and the organs at risk. The calculation of dose distributions has therefore always been a special challenge for the developers of treatment-planning systems.

The problem which has to be solved in this context is the implementation of an algorithm which is fast enough to fulfill the requirements of daily clinical use, and which has sufficient accuracy. Most treatment-planning systems work with so-called pencil-beam algorithms, which are semi-empiric and meet the requirements in speed and accuracy (see Chap. 15). If too many heterogeneities, such as air cavities, lung tissue, or bony structures, are close to the target volume, the use of Monte Carlo calculations is preferred. Monte Carlo calculations simulate the physical rules of interaction of radiation with matter in a realistic way (see Chap. 16). In the case of heterogeneous tissue they are much more precise, but also much slower, than pencil-beam algorithms.

1.3.3.4
Evaluation of the 3D Dose Distribution

The 3D-treatment planning leads to 3D-dose distributions, which must be evaluated in an appropriate way. In particular this concerns the occurrence of hot and cold spots, as well as the homogeneity and conformity of the dose distribution. Numerous computer graphics and mathematical tools have been developed to support the evaluation of dose distribution.

The weak point in treatment planning still is that the evaluation of dose distributions occurs usually on the basis of the physical dose distribution and not on the basis of quantified radiobiological or clinical effects. Physical dose is only a surrogate for the effects that radiation induces in healthy tissue as well as in the target volume. In the context of radiotherapy planning, it has always been emphasized that radiobiological models to predict normal tissue complication probabilities (NTCP) and tumor control probabilities (TCP) would be much better suited to treatment planning than the sole consideration of physical dose distributions. Unfortunately, lack of clinical data still hinders the development of adequate biological planning algorithms.

Some planning programs permit the calculation of values for the TCP and for the NTCP as yardsticks for the biological effect of the dose distribution; however, radiobiological models are still the topic of much controversial discussion (see Chap. 18).

1.4
Patient Positioning

The fifth link in the chain of radiotherapy is the link between treatment planning and the irradiation, the so-called problem of patient positioning. The problem here is to accurately transfer the planned irradiation technique to the patient. In practice, this means that the patient first of all has to be placed in exactly the same position as during 3D imaging. This is performed with a suitable immobilization device which can be used during imaging and treatment. Secondly, the treatment couch with the patient has to be adjusted until the isocenter position matches the pre-calculated coordinates. A variety of different techniques are currently used to reach this goal: conventionally, X-ray simulators are used to control the patient's position with the use of radiographic imaging. In connection with 3D-treatment planning, the use of digital reconstructed radiographs (DRRs) has been established (see Chaps. 4, 14). More recently, stereotactic patient positioning techniques have been introduced, techniques which initially could only be applied to target volumes in the brain but presently can also be applied to extracranial targets (see Chaps. 21, 22). The most modern approach to patient positioning is the use of navigational techniques, which offer not only tools to position the patient at the beginning of each fraction but also can monitor movement during irradiation (see Chap. 26). A promising approach to the problem of patient positioning is image-guided therapy (IGRT), where a 2D or even 3D X-ray imaging procedure is integrated in the irradiation unit, thus offering the possibility of controlling and monitoring the position of the target volume under treatment (Chaps. 24–26). This approach, which has the advantage that the treatment conditions could be dynamically matched to a moving and changing target volume, is also called adaptive radiotherapy (ART).

1.5
Treatment

The next and most essential link in the chain of radiotherapy, of course, is treatment, itself characterized by radiation delivery. Modern radiotherapy, especially when there is a curative intention, is practiced as 3D conformal radiotherapy. Most conformal radiotherapy treatments are performed by external radiation with photons, but the obvious physical and probably also biological advantages of charged particle therapy with proton or carbon beams are currently leading to a worldwide increasing number of particle-therapy installations. On the other hand, 3D conformal therapy with internal sources (brachytherapy) has also been established and proven to be very efficient for special indications.

1.5.1
Photons

The most common treatment modality presently is the use of a high-energy X-ray machine (Linac) in conjunction with a conformal irradiation technique realized by multiple irregular-shaped fixed beams. Beam shaping is often still performed with blocks, but computer-controlled multi-leaf collimators (MLCs) are increasingly replacing them. The MLCs also have the potential for intensity-modulated radiotherapy (IMRT), which can be considered as the most advanced treatment technique of 3D conformal radiotherapy with photons. Whereas the general aspects of conventional conformal radiation therapy are described briefly in Chap. 20, IMRT techniques are described in Chap. 23 and ART in Chaps. 24–26.

1.5.2
Charged-Particle Therapy

The use of heavier charged particles, such as protons and ^{12}C, is still restricted to a very limited number of centers worldwide. The physical advantages of heavier charged particles are obvious: due to the Bragg peak, they result in a favorable dose distribution in healthy tissue. ^{12}C beams, which is high linear energy transfer radiation, also seem to have radiobiological advantages (see Chaps. 27, 28). The high costs of charged-particle irradiation units are the major hindrance to broader introduction.

1.5.3
Brachytherapy

The implantation of radioactive sources into tumors has the potential to produce conformal dose distributions with a very steep dose gradient to neighboring structures and shows an excellent sparing effect for normal tissue. Depending on the tumor type, it can be applied either in short-term irradiations (with high dose rates) or in a long-term irradiation with radio-emitters at low dose rates. In Chaps. 19, the 2D- and 3D-planning techniques which are presently applied in modern brachytherapy are described, and Chaps. 29–31 describe clinical applications in vascular and prostate brachytherapy.

1.6
Quality Management in Radiotherapy

All steps and links of the chain of radiotherapy are subject to errors and inaccuracies, which may lead to treatment failure or injury of the patient. A careful network of quality assurance and verification has to be established in a radiotherapy unit in order to minimize these risks. A quality management system has to cover all the components involved and all aspects of the chain, e.g., dosimetry, software and hardware testing, standardization, documentation, archiving, etc. Three-dimensional conformal radiotherapy includes medical, biological, mechanical and electronic engineering, and computer science, as well as mathematical and physical aspects and components, and can be considered to be among the most complex and critical medical treatment techniques currently available. For this reason, efficient quality management is an indispensable requirement for modern 3D conformal radiotherapy (see Chaps. 32, 33).

References

Schlegel W, Mahr A (2001) 3D conformal radiotherapy: introduction to methods and techniques. Springer, Berlin Heidelberg NewYork

Basics of 3D Imaging

2 3D Reconstruction

Jürgen Hesser and Dzmitry Stsepankou

CONTENTS

Röntgen's contribution to X-ray imaging is a little over 100 years old. Although there was early interest in the extension of the technique to three dimensions, it took more than 70 years for a first prototype. Since then, both the scanner technology, reconstruction algorithm, and computer performance have been developed so significantly that in the near future we can expect new techniques with suppressed artefacts in the images. Current developments of such methods are discussed in this chapter.

2.1
Problem Setting

We begin with a presentation of the inverse problem: Given a set of X-ray projections p of an object, try to compute the distribution of attenuation coefficients

J. Hesser, PhD, Professor
D. Stsepankou
Department of ICM, Universitäten Mannheim und Heidelberg, B6, 23–29, C, 68131 Mannheim, Germany

μ in this object. We can define the projection function P: $p=P(\mu)$. Formally, reconstruction is therefore the calculation of an inverse of P, i.e. $\mu=P^{-1}(p)$. This inverse need not necessarily exist or, even if it does exist, it may be unstable in the sense that small errors in the projection data can lead to large deviations in the reconstruction.

In order to determine the projection function P we have to look at the physical imaging process more closely. X-ray projection follows the Lambert-Beer law. It describes the attenuation of X-rays penetrating an object with absorption coefficient μ and thickness d. A ray with initial intensity I_0 is attenuated to $I=I_0 e^{-\mu d}$. If the object's attenuation varies along the ray, one can write $I=I_0 e^{-\int \mu(x)dx}$, where the integral is along the path the ray passes through the object. We divide by I_0 and take the negative logarithm: $p=-\ln(I/I_0)=\int \mu(x)dx$. Let $u(x)$ describe the contribution of each volume element (infinitely small cube) to the absorption of the considered ray, the integral can be written as $p = \int_{volume} \mu(x)u(x)dx$, where the integration is now over the full volume. Discretizing the volume, we consider homogenous small cubes with a non-vanishing size, each having an absorption coefficient μ_i; thus, the integral is replaced by a sum reading: $p=\sum_i \mu_i u_i$, where u_i is a weighting factor. Solving for μ_i requires a set of such equations being independent of each other; therefore, n projections of size m^2 result in $n \cdot m^2$ different rays. For each of them $p_j=\sum_i \mu_i u_{ij}$ which can be rewritten in vector format $\vec{p} = U\vec{\mu}$. $\vec{p}$ describes all $n \cdot m^2$ different rays, $\vec{\mu}$ represents the absorption coefficients for all discrete volume elements and matrix element $(U)_{ij}=u_{ij}$ says how much a volume element i contributes to the projection result of ray j, i.e. all what we have to do is to invert this set of equations, namely $\vec{\mu}=U^{-1}\vec{p}$.

Under certain restrictions imposed on the projection directions this set of equations is solvable, i.e. U^{-1} exists. A direct calculation of the inverse of the matrix U, however, is impossible for present-day computers due to the extraordinary large number of more than 10^7 variables for typical problems. The following methods are, therefore, techniques that allow solving such systems more efficiently.

2.2
Parallel-Beam Reconstruction Using the Fourier-Slice Theorem

We start with the simplest case, parallel-beam reconstruction in two dimensions. To understand how the reconstruction is realized, let us assume a circle-like object in a rectangular 2D region (Fig. 2.1). The negative logarithm of the projection data is backprojected (or distributed) along the ray direction. This results, as seen in Fig. 2.2, in a rough approximation of the object. The projection data are smoothed and processed by a special filter (Fig. 2.3). It has the property that all contributions accumulate where the object is located and around the object negative and positive contributions cancel out.

Parallel-beam reconstruction belongs to the first approaches used for CT (CORMACK 1963; HOUNSFIELD 1973). A significant increase in scanning time was subsequently achieved by fan-beam configurations.

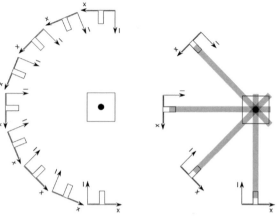

Fig. 2.1. *Left:* projection data from point like object. *Right:* backprojection of $-\log(I/I_0)$

2.3
Fan-Beam Reconstruction

Fan-beam CT reduces scanning times since it captures a full scanline in parallel. Nevertheless, the parallel-beam reconstruction technique can be used as well, since individual parallel rays can be selected from each fan and considered as a parallel projection (as shown in Fig. 2.5). In order to find enough paral-

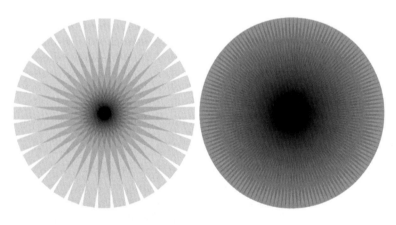

Fig. 2.2. Simulated backprojection for 18 projections (*left*) and 72 projections (*right*)

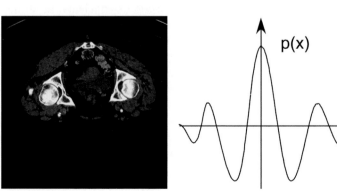

Fig. 2.3. Result for backprojection with filtering (360 projections); *right:* the filter. (From CHANG and HERMAN 1980)

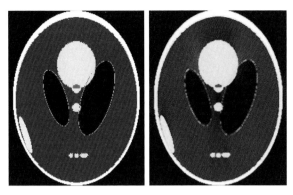

Fig. 2.4. Results on simulated data. Shepp-Logan phantom (Shepp and Logan 1974; *left*), filtered backprojection result for 360 projections (*right*)

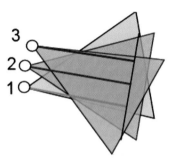

Fig. 2.5. Three different fan-beam projections are shown [shown as *triangles* with different colours; *circles* denote the X-ray sources (*1–3*)]. We identify three approximately parallel rays in the three projections shown as *green*, *violet*, and *blue*. These could be interpreted as a selection of three rays of a parallel projection arrangement.

lel rays, a sufficient number of projections covering 180°+ fan-beam angle is required.

Fan-beam CT was used for a long time until in the 1990s Kalender invented the spiral CT, i.e. a fan-beam CT where the table is shifted with constant speed through the gantry so that the X-ray source follows a screw line around the patient (Kalender et al. 1990).

2.4
Reconstruction Methods of Spiral CT

For spiral CT the projection lines do not lie in parallel planes as for fan-beam CT. Linear interpolation of the captured data on parallel planes allows use of the fan-beam reconstruction method. Data required for interpolation may come from either projections being 360 or 180° apart where the latter yields a better resolution.

In recent years, CT exams with several detector lines have been introduced reducing the overall scanning time for whole-body scans. Despite the divergence of the rays in different detector rows, standard reconstruction techniques are used. Recently, further developments have reconstructed optimally adapted oblique reconstruction planes that are later interpolated into a set of parallel slices (Kachelriess et al. 2000).

2.5
Cone-Beam Reconstruction

For cone-beam reconstruction we differentiate between exact methods, direct approximations, and iterative approximations.

2.5.1
Exact Methods

Exact reconstruction algorithms have been developed (Grangeat 1991; Defrise and Clack 1994; Kudo and Saito 1994), but they currently do not play a role in practice due to the long reconstruction time and high memory consumption. Cone-beam reconstruction imposes constraints on the motion of the source-detector combination around the patient. As shown in Fig. 2.6, a pure rotation around the patient does not deliver information about all regions and therefore a correct reconstruction is not possible; however, trajectories, as shown in Fig. 2.7, for example, solve this problem (Tam et al. 1998).

2.5.2
Filtered Backprojection for Cone Beams

In filtered backprojection (FBP) the projection data is filtered with an appropriate filter mask, backprojected and finally accumulated (Feldkamp et al. 1984; Yan and Leahy 1992; Schaller et al. 1997). For small cone-beam angles fairly good results are obtained, but for larger angles the conditions for parallel beams are violated leading to typical artefacts. Nevertheless, filtered backprojection is currently the standard for commercial cone-beam systems (Euler et al. 2000). From the computational point of view, FBP is more time-consuming compared with the parallel beam, fan beam, or spiral reconstruction since the latter use the Fourier transform to solve backprojection

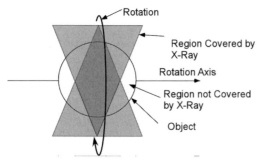

Fig. 2.6. Using only a circular path about the object, one is not able to reconstruct parts of the volume.

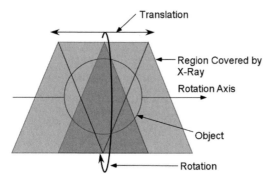

Fig. 2.7. Source detector motion that allows reconstructing the volume for cone beam configurations

fast. Although fast FBP implementations exist (Toft 1996), they do not reach a similar reconstruction quality.

2.6
Iterative Approaches

Filtered backprojection can be characterized as an example of direct inversion techniques. For particularly good reconstruction quality FBP is not the method of choice since it can produce severe artefacts especially in the case of the presence of high-contrast objects or a small number of projection data. Iterative techniques promise better reconstructions.

2.6.1
Algebraic Reconstruction Techniques

The algebraic reconstruction technique (ART; GORDON et al. 1970) is one of the first approaches to solve the reconstruction problem using an iterative method. The basic idea is to start with an a priori guess about the density distribution. The density is often assumed to be zero everywhere. Next, the simu-

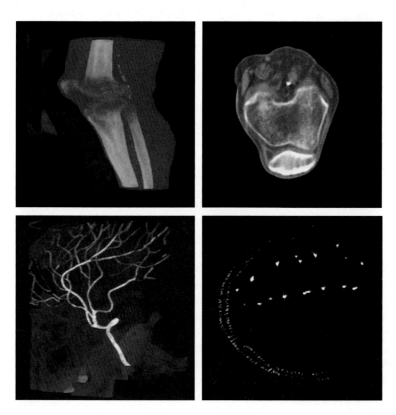

Fig. 2.8 Result of a reconstruction using filtered backprojection. One hundred twenty projections of size 512×512

lated projection is calculated and compared with the acquired projection data. The error is backprojected (without filtering) and then accumulated from each projection leading to an improved guess. Iteratively applying this course of simulated projection, error calculation and error backprojection, the algorithm converges towards the most likely solution.

There are different implementations. The original ART operates on a ray-to-ray basis, i.e. the projection image of a single ray is calculated, then the error is determined and is finally backprojected to correct the volume. In the simultaneous iterative reconstruction technique (SIRT) the error is considered in all projections simultaneously. SART (ANDERSEN and

KAK 1984) can be seen as a combination of ART and SIRT. It uses a more accurate algorithm for the simulated projection and a heuristic to emphasize corrections near the center of a ray.

2.6.2
Expectation Maximization Technique

Expectation maximization (EM; LANGE and CARSON 1984) is based on a statistical description of the imaging process. It considers the noise distribution in the image and tries to find a solution that generates the observed result with maximum probability. Ordered

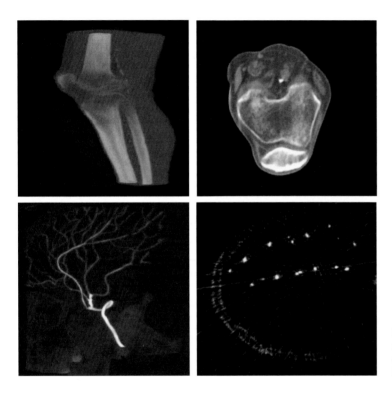

Fig. 2.9. Example of simultaneous iterative reconstruction technique reconstructions on a phantom, 36 projections (512×512), 30 iterations

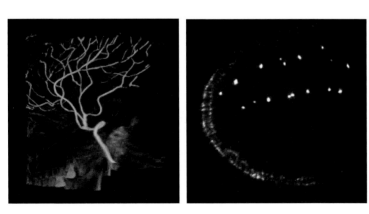

Fig. 2.10. Ordered-subsets expectation maximization result: 36 projections (512×512); three subsets (12 projections in subset); and 3 iterations

subset methods where a subset of all projections for each iteration step is chosen (HUDSON and LARKIN 1994; HSIAO et al. 2002) can speed up convergence by a factor of approximately 10, although this is not as fast as SIRT or SART.

2.7
Regularization Techniques

Iterative approaches, at best, yield a maximum-likelihood solution. This is a solution, given only the information about the physical process of imaging and the data, which describes the most probable density distribution of physical parameters. In many cases, however, one has additional knowledge about the object to be imaged. For example, restrictions can be imposed on the size of the object and the maximal, minimal or average X-ray density; there may also be an a priori data set, etc. Including this information in the reconstruction process can substantially increase the reconstruction quality and accuracy.

Let us refer to the reconstruction problem as described above: Given the function (matrix) that describes the physical imaging process, U, find the density distribution $\vec{\mu}$ so that the error $E(\vec{\mu}) = (\vec{\mu} - U\vec{p})^2$ is minimal. This is the least-square approximation. A regularization modifies this function by adding a regularization function $R(\vec{\mu})$: $E(\vec{\mu}) = (\vec{p} - U\vec{\mu})^2 + \alpha R(\vec{\mu})$, where α is a coupling constant that has to be chosen manually.

Finding suited regularization functions is a difficult task (YU and FESSLER 2002). Simple examples are TIKHONOV regularizations (TIKHONOV et al. 1997), where $R(\vec{\mu})$ has been chosen as the scalar product $[L(\mu-\mu^\star)] \cdot [L(\mu-\mu^\star)]$, where L is typically either the identity matrix or the discrete approximation of the derivative operator and $\mu^\star$ is the a priori distribution of the absorption coefficient. More advanced and recent techniques are impulse-noise priors (DONOHO et al. 1992; QI and LEAHY 2000), Markov random-fields priors (GEMAN and YANG 1995; VILLAIN et al. 2003) and total variation regularization (RUDIN et al. 1992; PERSSON et al. 2001).

2.8
Hardware Acceleration

2.8.1
Parallelization

One of the significant disadvantages of the cone-beam reconstruction technique is the high computational demands preventing use in daily practice; therefore, parallelization has been a natural means for acceleration. The naive approach is to subdivide the volume into small subvolumes, assign each subvolume to one processor and then compute simulated projection and backprojection locally. Since the projection result is combined by the partial projection results from all subvolumes, processors have to send their intermediate results to a central node where the final projection is generated and then distributed to all other processors again. Current parallel computers are generally limited for this sort of processing by their network bandwidth. In other words, the processors process the data faster than the network can transmit the results to other processors; therefore, parallelization is not very attractive. There are, however, two new upcoming technologies that promise a solution.

2.8.2
Field Programmable Gate Arrays

Field programmable gate arrays are chips where the internal structure can be configured to any hardware logic, e.g. some sort of CPU or, which is interesting in our application, to a special-purpose processor. The main advantage is that recent chips have hundreds of multipliers and several million logic elements that can be configured as switches, adders or memory. Recently, it has been demonstrated that these systems can be more than ten times faster than a normal PC for backprojection, and this factor will grow over the next few years since the amount of computing resources grows faster than the performance of CPUs (STSEPANKOU et al. 2003).

2.8.3
Graphics Accelerators

Graphics cards contain special-purpose processors optimized for 3D graphics and that internally have the processing power which is much higher than that of normal PCs. Internal graphics can now be programmed by a C-like language. This opens the opportunity to implement projection (using so-called texture mapping) and backprojection on them. While projection is relatively fast (since it is similar to 3D graphics algorithms), backprojection is still the limitation where currently the performance of standard PCs is achieved (CABRAL et al. 1994; MUELLER 1998; CHIDLOW and MÖLLER 2003). Since the internal performance of graphics chips grows much faster than for normal CPUs, it is foreseeable that in the future these accelerators will be a good candidate for implementation of the advanced reconstruction algorithms and, therefore, will allow their use in practice.

2.9
Outlook

In the past few years, with the advent of high-performance PCs, the aspect of reconstruction has become increasingly important in medicine. We have seen that from the current state of the art of cone-beam CT there exist already reconstruction algorithms that promise fewer visible artefacts and a reduction of the required dose for obtaining a given image quality. Most of these algorithms are of iterative nature, partially including regularization techniques based on a priori knowledge. Due to the current hardware developments and the ever-increasing speed of normal CPUs in PCs, these techniques will find their way into practice in the next few years and may revolutionize the way 3D volumes are generated. Normal X-ray systems may therefore be considered for configuration as CT devices, as they are much more cost-effective and more flexible in practice.

References

Andersen AH, Kak AC (1984) Simultaneous algebraic reconstruction technique (SART): a superior implementation of the ART algorithm. Ultrasound Imaging 6:81–94

Cabral B, Cam N, Foran J (1994) Accelerated volume rendering and tomographic reconstruction using texture mapping hardware. 1994 symposium on volume visualization, Tysons Corner, Virginia, ACM SIGGRAPH, pp 91–98

Chang LT, Herman GT (1980) A scientific study of filter selection for a fan-beam convolution algorithm. Siam J Appl Math 39:83–105

Chidlow K, Möller T (2003) Rapid emission tomography reconstruction. Workshop on volume graphics (VG03), pp 15–26

Cormack AM (1963) Representation of a function by its line integrals, with some radiological applications. J Appl Phys 34:2722–2727

Defrise M, Clack R (1994) A cone-beam reconstruction algorithm using shift variant filtering and cone-beam backprojection. IEEE Trans Med Imaging 13:186–195

Donoho DL, Johnstone IM, Hoch JC, Stern AS (1992) Maximum entropy and the near black object. J R Statist Ser B 54:41–81

Euler E, Wirth S, Pfeifer KJ, Mutschler W, Hebecker A (2000) 3D-imaging with an isocentric mobile C-Arm. Electromedica 68:122–126

Feldkamp LA, Davis LC, Kress JW (1984) Practical cone-beam algorithm. J Opt Soc Am A 1:612–619

Geman D, Yang C (1995) Nonlinear image recovery with half-quadratic regularization. IEEE Trans Image Processing 4:932–946

Gordon R, Bender R, Herman GT (1970) Algebraic reconstruction techniques (ART) for three-dimensional electron microscopy and X-ray photography. J Theor Biol 29:471–481

Grangeat P (1991) Mathematical framework of cone-beam 3D reconstruction via the first derivative of the Radon transform. In: Herman GT, Louis AK (eds) Mathematical methods in tomography, lecture notes in mathematics. Springer, Berlin Heidelberg New York, pp 66–97

Hounsfield GN (1973) Computerized traverse axial scanning (tomography). Part I. Description of system. Br J Radiol 46:1016–1022

Hsiao IT, Rangarajan A, Gindi G (2002) A provably convergent OS-EM like reconstruction algorithm for emission tomography. Proc SPIE 4684:10–19

Hudson HM, Larkin RS (1994) Accelerated image reconstruction using ordered subsets of projection data. IEEE Trans Med Imaging 13:601–609

Kachelriess M, Schaller S, Kalender WA (2000) Advanced single slice rebinning in cone-beam spiral CT. Med Phys 27:754–772

Kalender WA, Seissler W, Klotz E, Vock P (1990) Spiral volumetric CT with single-breathhold technique, continuous transport, and continuous scanner rotation. Radiology 176:181–183

Kudo H, Saito T (1994) Derivation and implementation of a cone-beam reconstruction algorithm for non-planar orbits. IEEE Trans Med Imaging 13:196–211

Lange K, Carson R (1984) EM reconstruction algorithms for emission and transmission tomography. J Comput Tomogr 8:306–316

Mueller K (1998) Fast and accurate three-dimensional reconstruction from cone-beam projection data using algebraic methods. PhD dissertation, Ohio State University

Persson M, Bone D, Elmqvist H (2001) Total variation norm for three-dimensional iterative reconstruction in limited view angle tomography. Phys Med Biol 46:853–866

Qi J, Leahy RM (2000) Resolution and noise properties of MAP reconstruction for fully 3-D PET. IEEE Trans Med Imaging 19:493–506

Rudin LI, Osher LI, Fatemi E (1992) Nonlinear total variation-based noise removal algorithms. Physica D 60:259–268

Schaller S, Flohr T, Steffen P (1997) New, efficient Fourier-reconstruction method for approximate image reconstruction in spiral cone-beam CT at small cone angles. SPIE Med Imag Conf Proc 3032:213–224

Shepp L, Logan BF (1974) Reconstructing interior head tissue from X-ray transmissions. IEEE Trans Nucl Sci 21:228–236

Stsepankou D, Müller U, Kornmesser K, Hesser J, Männer R (2003) FPGA-accelerated volume reconstruction from X-ray. World Conference on Medical Physics and Biomedical Engineering, Sydney, Australia

Tam KC, Samarasekera S, Sauer F (1998) Exact cone-beam CT with a spiral scan. Phys Med Biol 43:1015–1024

Tikhonov AN, Leonov AS, Yagola A (1997) Nonlinear ill-posed problems. Kluwer, Dordrecht

Toft P (1996) The radon transform: theory and implementation. PhD thesis, Department of Mathematical Modelling, Technical University of Denmark

Villain N, Goussard Y, Idier J, Allain M (2003) Three-dimensional edge-preserving image enhancement for computed tomography. IEEE Trans Med Imaging 22:1275–1287

Yan XH, Leahy RM (1992) Cone-beam tomography with circular, elliptical, and spiral orbits. Phys Med Biol 37:493–506

Yu DF, Fessler JA (2002) Edge-preserving tomographic reconstruction with nonlocal regularization. IEEE Trans Med Imaging 21:159–173

3 Processing and Segmentation of 3D Images

Georgios Sakas and Andreas Pommert

CONTENTS

3.1 Introduction

For a very long time, ranging approximately from early trepanizations of heads in Neolithic ages until little more than 100 years ago, the basic principles of medical practice did not change significantly. The application of X-rays for gathering images from the body interior marked a major milestone in the history of medicine and introduced a paradigm change in the way humans understood and practiced medicine.

The revolution introduced by medical imaging is still evolving. After X-rays, several other modalities have been developed allowing us new, different, and more complete views of the body interior: tomography (CT, MR) gives a very precise anatomically clear view and allows localization in space; nuclear medicine gives images of metabolism; ultrasound and inversion recovery imaging enable non-invasive imaging; and there are many others.

All these magnificent innovations have one thing in common: they provide images as primary information, thus allowing us to literally "see things" and to capitalize from the unmatched capabilities of our visual system. On the other hand, the increasing number of images produces also a complexity bottleneck: it becomes continuously more and more difficult to handle, correlate, understand, and archive all the different views delivered by the various imaging modalities.

Computer graphics as an enabling technology is the key and the answer to this problem. With increasing power of even moderate desktop computers, the present imaging methods are able to handle the complexity and huge data volume generated by these imaging modalities.

While in the past images were typically two-dimensional – be they X-rays, CT slices or ultrasound scans – there has been a shift towards reproducing the three-dimensionality of human organs. This trend has been supported above all by the new role of surgeons as imaging users who, unlike radiologists (who have practiced "abstract 2D thinking" for years), must find their way around complicated structures and navigate within the body (Hildebrand et al. 1996).

Modern computers are used to generate 3D reconstructions of organs using 2D data. Increasing computer power, falling prices, and general availability have already established such systems as the present standard in medicine. Legislators have also recognized this fact. In the future, medical software may be used (e.g. commercially sold) in Europe only if it displays a CE mark in compliance with legal regulations (MDD, MPG). To this end, developers and manufacturers must carry out a risk analysis in accordance with EN 60601-1-4 and must validate their software. As soon as software is used with humans, this is true also for research groups, who desire to disseminate their work for clinical use, even if they do not have commercial ambitions.

The whole process leading from images to 3D views can be organized as a pipeline. An overview of the volume visualization pipeline as presented in this chapter and in Chap. 4 is shown in Fig. 3.1. After the acquisition of one or more series of tomographic images, the data usually undergo some pre-processing such as image filtering, interpolation, and image fusion, if data from several sources are to be used. From this point, one of several paths may be followed.

G. Sakas, PhD
Fraunhofer Institute for Computer Graphics (IGD),
Fraunhoferstrasse 5, 64283 Darmstadt, Germany
A. Pommert, PhD
Institut für Medizinische Informatik (IMI),
Universitätsklinikum Hamburg-Eppendorf,
Martinistrasse 52, 20246 Hamburg, Germany

The more traditional surface-extraction methods first create an intermediate surface representation of the objects to be shown. It can then be rendered with any standard computer-graphics utilities. More recently, direct volume-visualization methods have been developed which create 3D views directly from the volume data. These methods use the full image intensity information (gray levels) to render surfaces, cuts, or transparent and semi-transparent volumes. They may or may not include an explicit segmentation step for the identification and labeling of the objects to be rendered.

Extensions to the volume visualization pipeline not shown in Fig. 3.1, but covered herein, include the visualization of transformed data and intelligent visualization.

3.2
Pre-processing

The data we consider usually comes as a spatial sequence of 2D cross-sectional images. When they are put on top of each other, a contiguous image volume is obtained. The resulting data structure is an orthogonal 3D array of volume elements or voxels each representing an intensity value, equivalent to picture elements or pixels in 2D. This data structure is called the voxel model. In addition to intensity information, each voxel may also contain labels, describing its membership to various objects, and/or data from different sources (generalized voxel model; HÖHNE et al. 1990).

Many algorithms for volume visualization work on isotropic volumes where the voxel spacing is equal in all three dimensions. In practice, however, only very few data sets have this property, especially for CT. In these cases, the missing information has to be approximated in an interpolation step. A very simple method is linear interpolation of the intensities between adjacent images. Higher-order functions, such as splines, usually yield better results for fine details (MARSCHNER and LOBB 1994; MÖLLER et al. 1997).

In windowing techniques only a part of the image depth values is displayed with the available gray values. The term "window" refers to the range of CT numbers which are displayed each time (HEMMINGSSON et al. 1980; WARREN et al. 1982). This window can be moved along the whole range of depth values of the image, displaying each time different tissue types in the full range of the gray scale achieving this way better image contrast and/or focusing on material with specific characteristics (Fig. 3.2). The new brightness value of the pixel Gv is given by the formula:

$$Gv = \left(\frac{Gv_{max} - Gv_{min}}{We - Ws} \right) \cdot (Wl - Ws) + Gv_{min}$$

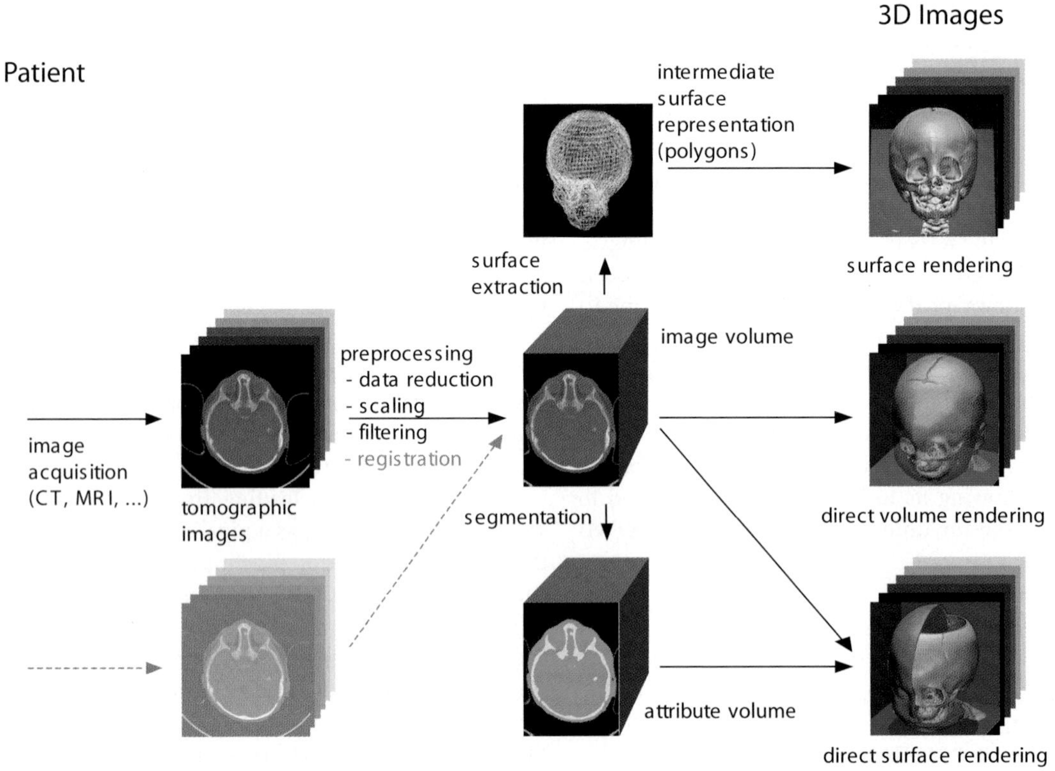

Fig. 3.1. The general organisation of processing, segmentation and visualisation steps

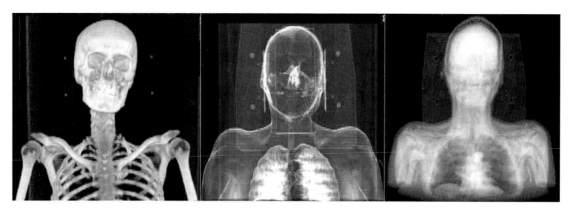

Fig. 3.2. by applying different windowing techniques different aspects of the same volume can be emphasized

where $[Gv_{max}, Gv_{min}]$, is the gray-level range, $[Ws, We]$ defines the window width, and Wl the window center. This is the simplest case of image windowing. Often, depending on the application, the window might have more complicated forms such as double window, broken window, or non-linear windows (exponential or sinusoid or the like).

The sharpening when applied to an image aims to decrease the image blurring and enhance image edges. Among the most important sharpening methods, high-emphasis masks, unsharp masking, and high-pass filtering (Fig. 3.3) should be considered.

There are two ways to apply these filters on the image: (a) in the spatial domain using the convolution process and the appropriate masks; and (b) in the frequency domain using high-pass filters.

Generally, filters implemented in the spatial domain are faster and more intuitive to implement, whereas filters in the frequency domain require prior transformation of the original, e.g., by means of the Fourier transformation. Frequency domain implementations offer benefits for large data sets (3D volumes), whereas special domain implementations are preferred for processing single images.

Image-smoothing techniques are used in image processing to reduce noise. Usually in medical imaging the noise is distributed statistically and it exists in high frequencies; therefore, it can be stated that image-smoothing filters are low-pass filters. The drawback of applying a smoothing filter is the simultaneous reduction of useful information, mainly detail features, which also exist in high frequencies (SONKA et al. 1998).

Typical filters (Fig. 3.4) here include averaging masks as well as Gaussian and median filtering. Averaging and Gaussian filtering tend to reduce the sharpness of edges, whereas median filters preserve edge sharpness. Smoothing filters are typically implemented in the spatial domain.

In MRI, another obstacle may be low-frequency intensity inhomogeneities, which can be corrected to some extent (ARNOLD et al. 2001).

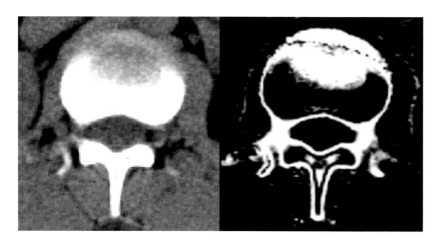

Fig. 3.3. Original CT slice (left) and contour extraction (right)

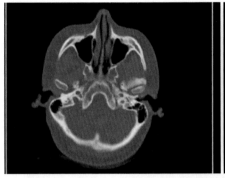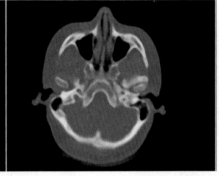

Fig. 3.4. CT image of a head (left: original image, right: after applying a median filter of size 7×7)

3.3
Segmentation

An image volume usually represents a large number of different structures obscuring each other. To display a particular one, we thus have to decide which parts of the volume we want to use or ignore. A first step is to partition the image volume into different regions, which are homogeneous with respect to some formal criteria and correspond to real (anatomical) objects. This process is called segmentation. In a subsequent interpretation step, the regions may be identified and labeled with meaningful terms such as "white matter" or "ventricle." While segmentation is easy for a human expert, it has turned out to be an extremely difficult task for the computer.

All segmentation methods can be characterized as being either binary or fuzzy, corresponding to the principles of binary and fuzzy logic, respectively (WINSTON 1992). In binary segmentation, the question of whether a voxel belongs to a certain region is always answered by either yes or no. This information is a prerequisite, e.g., for creating surface representations from volume data. As a drawback, uncertainty or cases where an object takes up only a fraction of a voxel (partial-volume effect) cannot be handled properly. Strict yes/no decisions are avoided in fuzzy segmentation, where a set of probabilities is assigned to every voxel, indicating the evidence for different materials. Fuzzy segmentation is closely related to the direct volume-rendering methods (see below).

Following is a selection of the most common segmentation methods used for volume visualization, ranging from classification and edge detection to recent approaches such as snakes, atlas registration, and interactive segmentation. In practice, these basic approaches are often combined. For further reading, the excellent survey on medical image analysis (DUNCAN and AYACHE 2000) is recommended.

3.3.1
Classification

A straightforward approach to segmentation is to classify a voxel depending on its intensity, no matter where it is located. A very simple but nevertheless important example is thresholding: a certain intensity range is specified with lower and upper threshold values. A voxel belongs to the selected class if – and only if – its intensity level is within the specified range. Thresholding is the method of choice for selecting air, bone or soft tissue in CT. In direct volume visualization, it is often performed during the rendering process itself so that no explicit segmentation step is required. Image 3.5 gives such an example.

Instead of a binary decision based on a threshold, DREBIN et al. use a fuzzy maximum likelihood classifier which estimates the percentages of the different materials represented in a voxel, according to BAYES' rule (DREBIN et al. 1988). This method requires that the gray-level distributions of different materials be different from each other and known in advance.

A similar method is the region growing algorithm and its numerous derivates (ZUCKER et al. 1976). In this case the user has to select a point within a structure, which is regarded to be "characteristic" for the structure of interest. The algorithm compares the selected point with its "neighbors." If a pre-defined similarity criterion is fulfilled, the checked neighbor is accepted as new member of the data set and becomes himself a new seed point. The points selected by this method form a set, which grows to the point where no similar neighbors can be found – then the algorithm terminates.

There are numerous variations of this principal idea, which works equally in 2D and 3D space. The principal problem of this method consists in identifying neighbors with "similar", but not "good" similarity, a case common in medical imaging. In this case the growing process stops too early. In the opposite

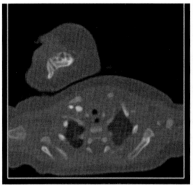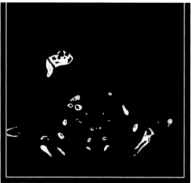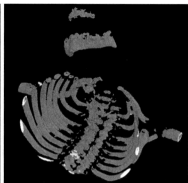

Fig. 3.5. CT thoracic data set (left: original axial image, middle: axial image after thresholding, right: direct volume rendered 3D view after thresholding)

site, a "leakage" on the boundary of the segmented object immediately creates parasitic branches or whole structures, which obviously do not belong to the structure of interest. Additional constrains can be used for reducing this effect. A common one is to require a "smoothness" or a "continuity" of the generated region, such as continuity in curvature. Image 3.6 displays such an example: starting from a central point in the middle, the algorithm grows until it finds secure points, in this case the bone boundary in the upper and lower image half. The uncertain parts to the left and right are interpolated (middle) by requiring a concave shape connecting the upper and the lower boundary segments. A snake-approach can be user here, see also paragraph 3.3.2 below. Similarly "gaps" between vertebral bodies in 3D space can be interpolated as well.

Simple classification schemes are not suitable if the structures in question have mostly overlapping or even identical gray-level distributions, such as different soft tissues from CT or MRI. Segmentation becomes easier if multi-spectral images are avail-

able, such as T1- and T2-weighted images in MRI, emphasizing fat and water, respectively. In this case, individual threshold values can be specified for every parameter. To generalize this concept, voxels in an n-parameter data set can be considered as n-dimensional vectors in an n-dimensional feature space. This feature space is partitioned into subspaces, representing different tissue classes or organs. This is called the training phase: in supervised training, the partition is derived from feature vectors, which are known to represent particular tissues (CLINE et al. 1990; POMMERT et al. 2001). In unsupervised training, the partition is generated automatically (GERIG et al. 1992). In the subsequent test phase, a voxel is classified, according to the position of its feature vector in the partitioned feature space.

With especially adapted image-acquisition procedures, classification methods have successfully been applied to considerable numbers of two- or three-parametric MRI data volumes (CLINE et al. 1990; GERIG et al. 1992). Quite frequently, however, isolated voxels or small regions are classified incorrectly such

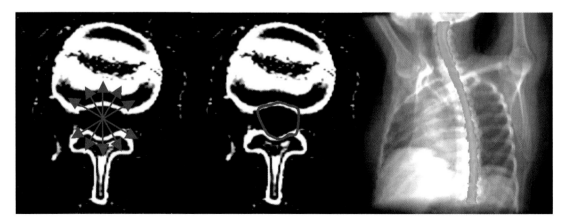

Fig. 3.6. 2D based segmentation. In case of homogeneous objects (right) and extension to 3D space is straight-forward

as subcutaneous fat in the same class as white matter. To eliminate these errors, a connected components analysis may be carried out to determine whether the voxels which have been classified as belonging to the same class are part of the same (connected) region. If not, some of the regions may be discarded.

Instead of intensity values alone, tissue textures may be considered, which are determined using local intensity distributions (SAEED et al. 1997). A survey of intensity-based classification methods is presented by CLARKE et al. (1995).

3.3.2
Edge Detection

Another classic approach to segmentation is the detection of edges, using first or second derivatives of the 3D intensity function. These edges (in 3D, they are actually surfaces; it is, however, common to refer to them as edges) are assumed to represent the borders between different tissues or organs.

There has been much debate over what operator is most suitable for this purpose. The Canny operator locates the maxima of the first derivative (CANNY 1986). While the edges found with this operator are very accurately placed, all operators using the first derivative share the drawback that the detected contours are usually not closed, i.e., they do not separate different regions properly. An alternative approach is to detect the zero crossings of the second derivative. With a 3D extension of the Marr-Hildreth operator, the complete human brain was segmented and visualized from MRI for the first time (BOMANS et al. 1987). A free parameter of the Marr-Hildreth opera-

tor has to be adjusted to find a good balance between under- and oversegmentation.

Snakes (KASS et al. 1987) are 2D image curves that are adjusted from an initial approximation to image features by a movement of the curve caused by simulated forces (Fig. 3.7). Image features produce the so-called external force. An internal tension of the curve resists against highly angled curvatures, which makes the Snakes' movement robust against noise. After a starting position is given, the snake adapts itself to an image by relaxation to the equilibrium of the external force and internal tension. To calculate the forces an external energy has to be defined. The gradient of this energy is proportional to the external force. The segmentation by Snakes is due to its 2D definition performed in a slice-by-slice manner, i.e., the resulting curves for a slice are copied into the neighboring slice and the minimization is started again. The user may control the segmentation process, by stopping the automatic tracking, if the curves run out of the contours and define a new initial curve.

3.3.3
2.5-D Boundary Tracking

The main drawback of the above-mentioned method is its typical limitation to 2D structures; thus, the snakes (Fig. 3.8) work well on a plane image but do not capitalize from the principal coherence of structures in the 3D space. The "boundary tracking" (BT) algorithm tries to encompass this difficulty. The main assumption of BT is that the shape of an organ does not change significantly between adjacent slices; thus,

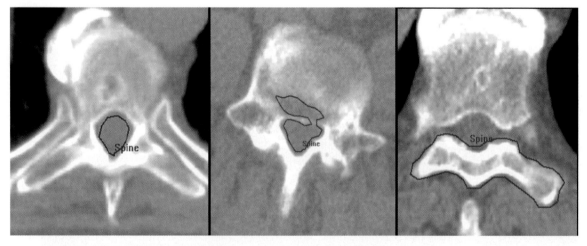

Fig. 3.7. The principle of segmentation using Snakes. The spinal canal in the left image has been correctly identified, whereas in the middle and the right examples the algorithm tracked the wrong boundary!

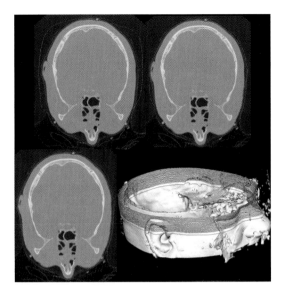

Fig. 3.8. Adaptation of a snake on one slice and propagation to the next slice

a contour found on one slice can be copied in the subsequent one as a reasonable approximation of the new shape. The snake algorithm is then re-started in order to fine-tune and adjust this initial approximation on the new slice until the whole data set has been processed (HERMAN 1991).

The quality of the result depends fundamentally on the similarity of two adjacent slices. Normally, this varies within a data set. This can be partly compensated by re-estimating the initial value for each new-segmented shape. However, this is not always possible; therefore, in regions with low similarity, the slices to be segmented by the interactive method must be selected tightly.

An alternative method is a combination of 2D and 3D approach. The segmentation approach works at one image level at a time in order to find a local contour, and once this has been done, the found structure "jumps" on the next slice in the volume. In the case of tomographic modalities, such as CT and MRI, the algorithm is applied on the original (axial) cross-sectional images. The algorithm requires an initial point to start the tracing of the edge of the object under investigation. The initial point travels to the vertical or horizontal direction until and edge of the investigated object is reached (region growing). Then the algorithm will start to exam the surrounding pixel of that edge and check whether they belong to the current edge or not. The algorithm uses a constant threshold selection. Once the shape is filled, the "mid-point" (center of gravity for convex shapes, center of largest included

ellipse for concave ones) of the shape is copied as the starting value on the next slice and the process starts again. If on the slice under investigation a shape fulfilling the constrains can not be found, the slice is temporally ignored and the same process is applied on the next slice etc. When a valid contour is found, intermediate shapes for the ignored slices are calculated by shape interpolation and the gap is filled, in other case the algorithm terminates. This process has been used in the example of figure 3.6 for 3D-segmentation of the spinal canal on the right image.

The main drawback of the BT is that it is a binary approach and hence is very sensitive to gray-value variations. If the threshold value is not selected properly, the system will fail to detect the appropriate shape. An error that usually occurs is when the user attempts to define the starting point for the algorithm: e.g. selecting a good point is possible on image 3.6 left, but impossible in the middle slice. Due to the restrictions of the BT mentioned above, in this case it is not possible to initialize the tracing process from an arbitrary slice. Due to this limitation, the user must be trained under the trial-and-error principle until the desired contour is found.

3.3.4
Geodesic Active Contours

Generally, geodesic active contours (GAC) is based on fast marching and level sets. The GAC expects two inputs, the initial level set (zero level set) and a feature image (edge image), and works in 2D and 3D data sets. Following the forces defined by the edge image, the level set iteratively approximates the area to be segmented. The result is binarized by thresholding and the binary image is overlapped in the 2D view. The processing of the original MR images is schematically described in Fig. 3.9. Usually an anisotropic diffusion filter is used to reduce noise in MR images. A sigmoid filter for contrast enhancement shows the interesting edges. After a gradient-magnitude (edge detection) filter, a second sigmoid filter is used to inverse the image.

In addition to this automated image pre-processing, a number of seed points have to be selected in the data. Approximately 10–15 seed points located within the liver are used in this example. These seed points can be placed within one or more slices. In addition, we generate a potential distance map around each image contour. With these two inputs a fast-marching algorithm is started generating all level sets, includ-

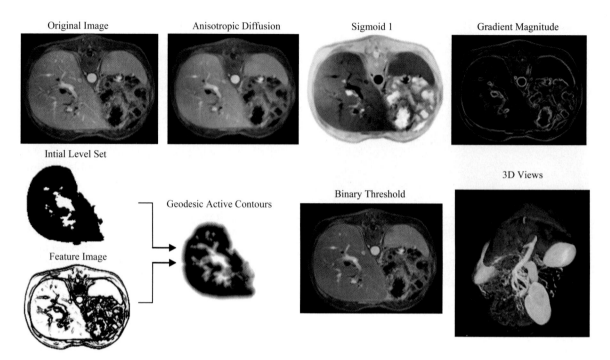

Original Image   Anisotropic Diffusion   Sigmoid 1   Gradient Magnitude

Intial Level Set

3D Views

Feature Image   Geodesic Active Contours   Binary Threshold

Fig. 3.9. Result of a liver segmentation based on MRI. Fast marching produces the initial level set and the GAC algorithm segments the contour of the liver.

ing the zero level set, which is used for the starting point called initial level set. The process is described in detail in Caselles (1997).

3.3.5
Extraction of Tubular Objects

For the segmentation of tubular objects, such as vessels, specialized segmentation algorithms have been developed (Kirbas et al. 2003). These techniques take into account the geometrical structure of the objects that have to be segmented. Besides conventional approaches, such as pattern-recognition techniques (e.g., region growing, mathematical morphology schemes), or model-based approaches (e.g., deformable models, generalized cylinders) there is the group of tracking-based algorithms. The latter ones start usually from a user-given initial point and detect the center line and the boundaries by finding edges among the voxels orthogonal to the tracking direction. Prior to the segmentation step, the data are pre-processed for lowering noise and enhancing edges within the image. The most sophisticated tracking-based algorithms generate a twofold output, the center line as well as the boundaries of the tubular object (Fig. 3.10), allowing for a further analysis of the segmented structure. That output may be gener-

ated in a single step (Verdonck et al. 1995) or in an iterative manner where first an approximate estimation of the center line is computed and afterwards corrected repeatedly by detecting the boundaries orthogonal to that line (Wesarg et al. 2004).

3.3.6
Atlas Registration

A more explicit representation of prior knowledge about object shape are anatomical atlases (see below). Segmentation is based on the registration of the image volume under consideration with a pre-labeled volume that serves as a target atlas. Once the registration parameters are estimated, the inverse transformation is used to map the anatomical labels back on to the image volume, thus achieving the segmentation.

In general, these atlases do not represent one individual – but "normal" –anatomy and its variability in terms of a probabilistic spatial distribution, obtained from numerous cases. Methods based on atlas registration were reported suitable for the automatic segmentation of various brain structures, including lesions and objects poorly defined in the image data (Arata et al. 1995; Collins et al. 1999; Kikinis et al. 1996); however, registration may not be very ac-

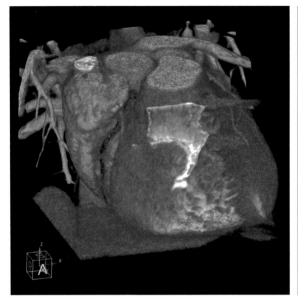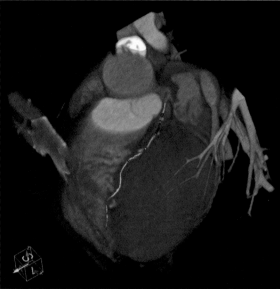

Fig. 3.10. Tracking-based segmentation of coronary arteries in cardiac CT data sets (left: vessel boundary, right: colour-coded centreline corresponding to the vessel's diameter)

curate. For improved results, atlas registration may be complemented with intensity-based classification (COLLINS et al. 1999).

3.3.7 Interactive Segmentation

Even though there are a great variety of promising approaches for automatic segmentation, "… no one algorithm can robustly segment a variety of relevant structures in medical images over a range of datasets" (DUNCAN and AYACHE 2000). In particular, the underlying model assumptions may not be flexible enough to handle various pathologies; therefore, there is currently a strong tendency to combine simple, but fast, operations carried out by the computer with the unsurpassed recognition capabilities of the human observer (OLABARRIAGA and SMEULDERS 2001).

A practical interactive segmentation system was developed by HÖHNE and HANSON (1992) and later extended to handle multiparametric data (POMMERT et al. 2001; SCHIEMANN et al. 1997). Regions are initially defined with thresholds. The user can subsequently apply connected components analysis, volume editing tools, or operators from mathematical morphology. Segmentation results are immediately visualized on orthogonal cross-sections and 3D images in such a way that they may be corrected or further refined in the next step. With this system, segmentation of gross structures is usually a matter of minutes (Fig. 3.11).

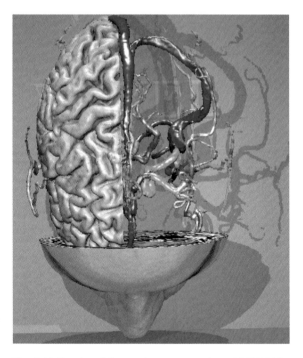

Fig. 3.11. Results of interactive segmentation of MRI (skin, brain) and Magnetic Resonance Angiography (vessels) data.

References

References for this chapter are included in the References for chapter 4.

4 3D Visualization

Georgios Sakas and Andreas Pommert

CONTENTS

4.1
Introduction

After segmentation, the choice of which rendering technique to use must be made. The more traditional surface-based methods first create an intermediate surface representation of the object to be shown. It may then be rendered with any standard computer graphics method. More recently, volume-based methods have been developed which create a 3D view directly from the volume data. These methods use the full gray-level information to render surfaces, cuts, or transparent and semi-transparent volumes. As a third way, transform-based rendering methods may be used.

G. Sakas, PhD
Fraunhofer Institute for Computer Graphics (IGD),
Fraunhoferstr 5, 64283 Darmstadt, Germany
A. Pommert, PhD
Institut für Medizinische Informatik (IMI),
Universitätsklinikum Hamburg-Eppendorf,
Martinistraße 52, 20246 Hamburg, Germany

4.2
Cut Planes

Once a surface view is available, a very simple and effective method of visualizing interior structures is cutting. When the original intensity values are mapped onto the cut plane, they can be better understood in their anatomical context (Höhne et al. 1990). A special case is selective cutting, where certain objects are left untouched (Fig. 4.1).

4.3
Surface Rendering

Surface rendering bridges the gap between volume visualization and more traditional computer graphics (Foley et al. 1995; Watt 2000). The key idea is to create intermediate surface descriptions of the relevant objects from the volume data. Only this information is then used for rendering images. If triangles are used as surface elements, this process is called triangulation.

An apparent advantage of surface extraction is the potentially very high data reduction from volume to surface representations. Resulting computing times can be further reduced if standard data structures, such as polygon meshes, are used which are supported by standard computer graphics hard- and software.

On the other hand, the extraction step eliminates most of the valuable information on the cross-sectional images. Even simple cuts are meaningless because there is no information about the interior of an object, unless the image volume is also available at rendering time. Furthermore, every change of surface definition criteria, such as adjusting a threshold, requires a recalculation of the whole data structure.

The classic method for surface extraction is the marching-cubes algorithm, developed by Lorensen and Cline (1987). It creates an iso-surface, approximating the location of a certain intensity value in the data volume. This algorithm basically considers cubes

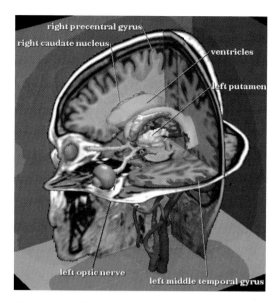

Fig. 4.1. Brain from MRI. Original intensity values are mapped onto the cut planes.

of 2×2×2 contiguous voxels. Depending on whether one or more of these voxels are inside the object (i.e., above a threshold value), a surface representation of up to four triangles is placed within the cube. The exact location of the triangles is found by linear interpolation of the intensities at the voxel vertices. The result is a highly detailed surface representation with sub-voxel resolution (Fig. 4.2).

Various modifications of the marching-cubes algorithm have been developed; these include the correction of topological inconsistencies (NATARAJAN 1994), and improved accuracy by better approximating the true isosurface in the volume data, using higher order curves (HAMANN et al. 1997) or an adaptive refinement (CIGNONI et al. 2000).

As a major practical problem, the marching-cubes algorithm typically creates hundreds of thousands of triangles when applied to clinical data. As has been shown, these numbers can be reduced considerably by a subsequent simplification of the triangle meshes (CIGNONI et al. 1998; SCHROEDER et al. 1992; WILMER et al. 1992).

4.4
Direct Volume Visualization Methods

In direct volume visualization, images are created directly from the volume data. Compared with surface-based methods, the major advantage is that all gray-level information which has originally been acquired is kept during the rendering process. As shown by HÖHNE et al. (1990) this makes it an ideal technique for interactive data exploration. Threshold values and other parameters which are not clear from the beginning can be changed interactively. Furthermore, volume-based rendering allows a combined display of different aspects such as opaque and semi-transparent surfaces, cuts, and maximum intensity projections. A current drawback of direct volume visualization is that the large amount of data which has to be handled allows only limited real-time applications on present-day computers.

4.4.1
Scanning the Volume

In direct volume visualization, we basically have the choice between two scanning strategies: pixel by pixel (image order) or voxel by voxel (volume order). These strategies correspond to the image and object order rasterization algorithms used in computer graphics (FOLEY et al. 1995).

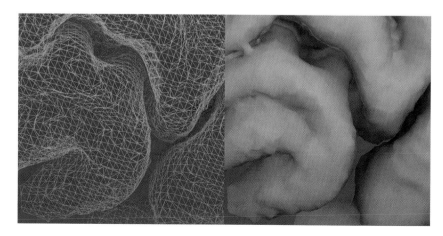

Fig. 4.2. Triangulated (*left*) and shaded (*right*) portion of the brain from MRI, created with the marching-cubes algorithm

In image order scanning, the data volume is sampled on rays along the viewing direction. This method is commonly known as ray casting. The principle is illustrated in Fig. 4.3. At the sampling points, the intensity values are interpolated from the neighboring voxels, using tri-linear interpolation or higher-order curves (MARSCHNER and LOBB 1994; MÖLLER et al. 1997). Along the ray, visibility of surfaces and objects is easily determined. The ray can stop when it meets an opaque surface. YAGEL et al. (1992) extended this approach to a full ray-tracing system, which follows the viewing rays as they are reflected on various surfaces. Multiple light reflections between specular objects can thus be handled.

Image-order scanning can be used to render both voxel and polygon data at the same time, known as hybrid rendering (LEVOY 1990). Image quality can be adjusted by choosing smaller (oversampling) or wider (undersampling) sampling intervals (POMMERT 2004). Unless stated otherwise, all 3D images shown in this chapter were rendered with a ray-casting algorithm.

As a drawback, the whole input volume must be available for random access to allow arbitrary viewing directions. Furthermore, interpolation of the intensities at the sampling points requires a high computing power. A strategy to reduce computation times is based on the observation that most of the time is spent traversing empty space, far away from the objects to be shown. If the rays are limited to scanning the data, only within a pre-defined bounding volume around these objects, scanning times are greatly reduced (ŠRÁMEK and KAUFMAN 2000; TIEDE 1999; WAN et al. 1999).

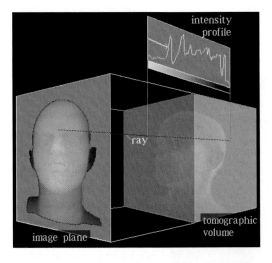

Fig. 4.3. Principle of ray casting for volume visualization. In this case, the object surface is found using an intensity threshold.

In volume-order scanning, the input volume is sampled along the lines and columns of the 3D array, projecting a chosen aspect onto the image plane in the direction of view. The volume can either be traversed in back-to-front (BTF) order from the voxel with maximal distance to the voxel with minimal distance to the image plane, or vice versa in front-to-back (FTB) order. Scanning the input data as they are stored, these techniques are reasonably fast even on computers with small main memories; however, implementation of display algorithms is usually much more straightforward using the ray-casting approach.

4.4.2
Splatting in Shear-Warp Space

The shear-warp factorization-rendering algorithm belongs to the fastest object space-rendering algorithms. The advantages of shear-warp factorization are that sheared voxels are viewed and projected only along ±X, ±Y, and ±Z axis, i.e., along the principal viewing directions in sheared object space, rather than an arbitrary viewing direction, which makes it possible to traverse and accumulate slice by slice. Neighbor voxels share the same footprint and are exactly one pixel apart.

In the original shear-warp algorithm (LACROUTE and LEVOY 1994) only slices are sheared rather than each voxel itself (see the dotted line in Fig. 4.4, left). In other words, slices are displaced (sheared) relatively to each other, but the voxels within each slice are remaining orthogonal cubes. We call this case "projective shear warp" in order to discriminate it from "splatting shear warp" introduced in CAI and SAKAS (1998).

In splatting shear warp the mathematically correct shear of the object space is calculated. The complete data set is regarded to be a continuous space sheared by the shearing transformation; thus, each voxel is sheared as well, resulting in parallelepipeds rather than cubes. Therefore, the projection area of a voxel is now in general greater than 1.0. This difference is illustrated in Fig. 4.4 (right: in projective shear-warp pixel A only accumulates the value of voxel $i+1$ (dashed line). In splatting shear warp instead, pixel A accumulates values from both voxel i and $i+1$ (solid line). The implication of this is that in the latter case a sub-voxel splatting sampling calculation according to an individual splatting table (footprint) becomes necessary.

The general footprint table is established by digitizing and scanning the 2×2 shear footprint area under different reconstruction kernel, as seen in Fig. 4.5.

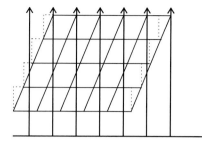

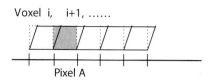

Voxel i,   i+1, ...

Pixel A

Fig. 4.4.The difference between voxel splatting and projection in shear warp

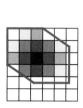

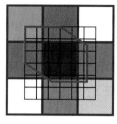

Fig. 4.5. Shear footprint and its convolution matrix

4.5
Visualization Primitives in Direct Volume Rendering

The size of the table is 2N×2N, where N×N is the digitization degree, i.e., the number of subpixels within one pixel, which is usually selected between 10 and 20 or even more, depending on the required accuracy. The weight of each subpixel is calculated by employing the integral of next equation at the mid-point of the subpixel.

$$W_i = \int_{Z_0}^{Z_1} h_v(x_0, y_0, z)\,dz$$

where (x_0, y_0) is the center of subpixel, $h_v(x, y, z)$ is the volume reconstruction kernel, and (Z_0, Z_1) is the integral range. In Cai and Sakas (1998), different digitally reconstructed radiography (DRR) rendering algorithms, ray casting, projective shear warp, and splatting shear warp, are compared with each other under different sampling methods, nearest-neighbor interpolation, and tri-linear interpolation.

Once one decides the principal traversing method (FTB or BTF) and chooses a principal algorithm (ray casting or splatting), one is able to traverse the volume visiting the voxels of interest. Now a decision has to be taken about how the value (density, material, property, etc.) represented by each voxel will be transferred to visible characteristics on the image place. The following sections summarize the methods most commonly used in medical applications.

4.5.1
Maximum and Minimum Intensity Projection

For small bright objects, such as vessels from CT or MR angiography, maximum intensity projection (MIP) is a suitable display technique (Fig. 4.6). Along each ray through the data volume, the maximum gray level is determined and projected onto the imaging plane. The advantage of this method is that neither segmentation nor shading are needed, which may fail for very small vessels. But there are also some drawbacks: as light reflection is totally ignored, maximum intensity projection does not give a realistic 3D impression. Sampling

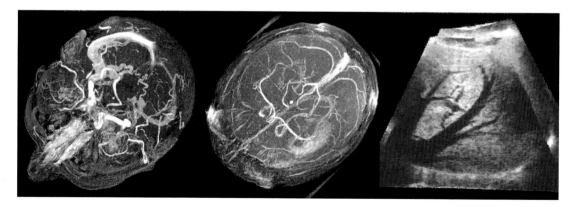

Fig. 4.6. Maximum intensity projection for brain (*left*, *middle*) and minimum intensity projection for US vessels (*right*)

of small vessels can be subject to errors, accurate and fast solutions to this problem have been proposed in (Sakas, 1995). Spatial perception can be improved by real-time rotation (Mroz et al. 2000) or by a combined presentation with other surfaces or cut planes (Höhne et al. 1990).

4.5.2
Digitally Reconstructed Radiographies

The optical model in DRR volume rendering is the so-called absorption only (Max 1995), in which particles only absorb the energy of incident light. If I_0 is the intensity of incident ray, the light intensity after penetrating distance s within the medium is

$$I(s) = I_0 \exp(-\int_0^s K_\lambda(t)dt)$$

where K_λ is the attenuation coefficient and λ is the wavelength.

In Fig. 4.7, the intensity when the ray arrives at the screen is

$$I_p = I_0 * (1 - T) + I_{background} * T$$

where $T = \exp(\Gamma(P_0 P_1))$ is the transparency and

$$\Gamma(P_0 P_1) = \int_{P_0}^{P_1} K_\lambda(t)dt$$

is the optical length.

The main computation cost in DRR volume rendering is to calculate the integration of optical length, i.e.,

$$\Gamma(s) = \sum K_\lambda(s)\Delta s, \ s \in [p_0, p_1]$$

where $\Delta s = L / N, L = |P_0 P_1|$ and N is the number of sampling points (consider even distance sampling). Thus,

$$\Gamma(s) = \frac{L}{N} \sum_{i=0}^{N} K_\lambda(i) = L(\frac{1}{N} \sum_{i=0}^{N} K_{\lambda,m}\rho)$$

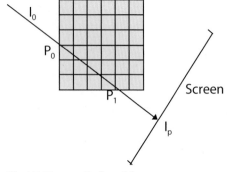

Fig. 4.7. X-ray optical model

where $K_{\lambda,m}$ is the mass attenuation coefficient and ρ is the density. See Cai and Sakas (1999) for a discussion of the transfer functions.

DRRs are extremely useful in numerus medical applications. Figure 4.8 shows examples of DRRs generated from CT data for a virtual cancer treatment simulation (Cai 1999, Cai 2000, Zamboglou 2003, Zamboglou 2004). Most of the DRR rendering algorithms are ray casting (called X-ray casting), which is image space-rendering algorithm; however, also shear-warp algorithms can be used. They have superior speed, however algorithmic advantages.

4.5.3
Direct Surface Rendering

Using one of the scanning techniques described, the visible surface of an object can be rendered directly from the volume data. This approach is called direct surface rendering (DSR). To determine the surface position, a threshold or an object membership label may be used, or both may be combined to obtain a highly accurate iso-surface (Pommert 2004; Tiede et al. 1998; Tiede 1999). Typically with CT data a threshold value is employed for extracting the location of the surface, whereas the local gradient approximates the surface normally used for shading. Note that the position of the observer may also be inside the object, thus creating a virtual endoscopy.

For realistic display of the surface, one of the illumination models developed in computer graphics may be used. These models, such as the Phong shading model, take into account both the position and type of simulated light sources, as well as the reflection properties of the surface (Foley et al. 1995; Watt 2000). A key input into these models is the local surface inclination, described by a normal vector perpendicular to the surface. Depending on the selected threshold, skin or bone or other surfaces can be visualized without having to explicitly segment them in a pre-processing step. Fig. 4.9 left and middle shows examples of direct volume rendering from CT and MRI datasets displaying the possibilities of displaying surfaces of various impression.

As shown by Höhne and Bernstein (1986), a very accurate estimate of the local surface normal vectors can be obtained from the image volume. Due to the partial-volume effect, the intensities in the 3D neighborhood of a surface voxel represent the relative proportions of different materials inside these voxels. The surface inclination is thus described by the local gray-level gradient, i.e., a 3D vector of the partial derivatives. A number of methods to calculate the gray-level gradient are

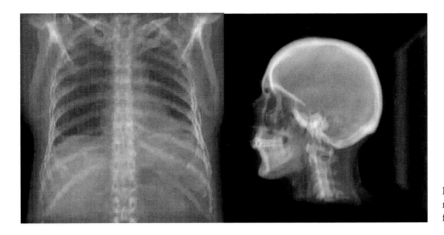

Fig. 4.8. Digitally reconstructed radiography (DRR) generated from CT data sets

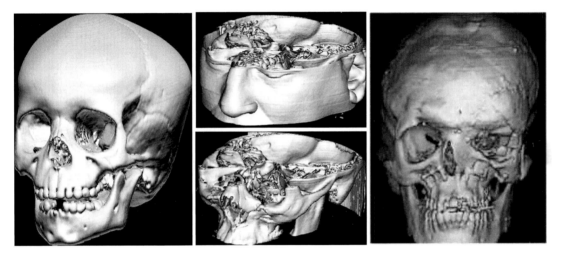

Fig. 4.9. Direct rendering shaded volume from CT and MRI data. Left and middle surface rendering, right semi-transparent rendering

presented and discussed elsewhere (MARSCHNER and LOBB 1994; TIEDE et al. 1990; TIEDE 1999).

4.5.4
Direct Semi-Transparent Volume Rendering

Direct volume rendering (DVR), or volume rendering for short, is the visualization equivalent of fuzzy segmentation (see section 1.2.2). For medical applications, these methods were first described by DREBIN et al. (1988) and LEVOY (1988). A commonly assumed underlying model is that of a colored, semi-transparent gel with suspended reflective particles. Illumination rays are partly reflected and change color while traveling through the volume.

Each voxel is assigned a color and opacity. This opacity is the product of an object-weighting function and a gradient-weighting function. The ob-ject-weighting function is usually dependent on the intensity, but it can also be the result of a more sophisticated fuzzy segmentation algorithm. The gradient-weighting function emphasizes surfaces for 3D display. All voxels are shaded, using, for example, the gray-level gradient method. The shaded values along a viewing ray are weighted and summed up.

A simplified recursive equation which models frontal illumination with a ray-casting system is given as follows:

I intensity of reflected light
p index of sampling point on ray
      (0 ... max. depth)
L fraction of incoming light (0.0 ... 1.0)
α local opacity (0.0 ... 1.0)
s local shading component

$$I(p,L) = \alpha(p)Ls(p) + (1.0-\alpha(p))I(p+1,(1.0-\alpha(p))L)$$

The total reflected intensity as displayed on a pixel of the 3D image is given as *I (0, 1.0)*. Since binary decisions are avoided in volume rendering, the resulting images are very smooth and show a lot of fine details, see e. g. the shape of the fine heart vessels in Fig. 4.10. Another advantage is that even coarsely defined objects can be rendered (TIEDE et al. 1990). On the other hand, the more or less transparent images produced with volume rendering are often hard to understand so that their value is sometimes questionable. To some extent, spatial perception can be improved by rotating the object.

Concluding, all visualization methods listed here have benefits and drawbacks and emphasize different aspects of the examined dataset as shown in Fig. 4.12. A selection of the "correct" method has to be done by the end-user on a case-by-case basis.

4.5.5
Volume Rendering Using Transfer Functions

An improvement over conventional semi-transparent rendering is the use of transfer functions for assigning optical properties such as color and opacity to the original values of the data set as shown on Fig. 4.11. If, for instance, the true colors for the organs that are included in the rendered scene are known, a very realistic rendering result can be obtained. For that, the relationship between CT number (Hounsfield unit), gray, red, green, and blue values of the tissues, and their refractive indices have to be retrieved (BISWAS and GUPTA 2002). As a result of those measurements a table that assigns CT number to gray, red, green, and blue values describing the corresponding relations for different parts of the body (brain, abdomen, thorax, etc.) can be compiled. Direct volume rendering using a color transfer function that is based on such a table reflects more or less the true colors of the tissue.

An extension of the assignment of color or opacity only to gray value that represents in fact a 1D transfer function is the usage of multi-dimensional transfer functions (KNISS et al. 2002). There, in addition to a voxel's scalar value, the first and second derivative of the image data set are taken into account. This allows for a better separation of different tissues for the purpose of direct volume rendering; however, using multi-dimensional transfer functions requires interacting in a multi-dimensional space. This task can be facilitated dramatically if those transfer functions are generated semi-automatically (KINDLMANN and DURKIN 1998). It has been shown that using multi-dimensional transfer functions for assigning opacity in direct volume rendering results in a smoother and more "correct" visualization of the image data, since complex boundaries are better separated from each other than if only a 1D transfer function is used.

4.6
Transform-Based Rendering

While both surface extraction and direct volume visualization operate in a 3D space, 3D images may be created from other data representations as well. One

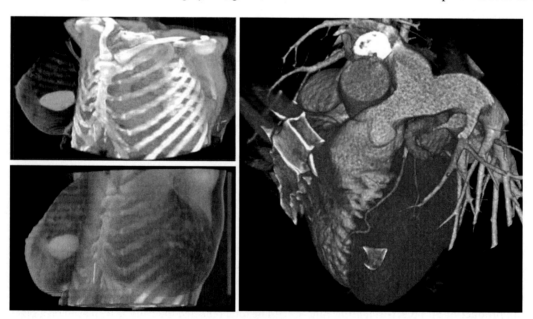

Fig. 4.10. Direct semi-transparent volume rendering technique

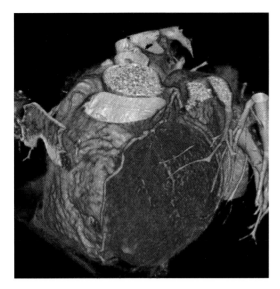

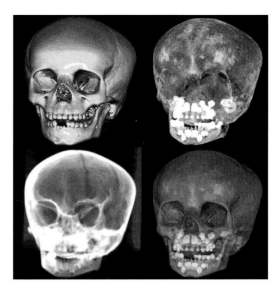

Fig. 4.11. Direct volume-rendered shaded cardiac CT data set using a color transfer function based on the measured true colors of the thoracic tissue

Fig. 4.12. Comparison of surface, maximum intensity projection, DRR, and semi-transparent rendering for the same data set. Each method emphasizes different aspects of the data set.

such method is frequency domain volume rendering, which creates 3D images in Fourier space, based on the projection-slice theorem (Totsuka and Levoy 1993). This method is very fast, but the resulting images are somewhat similar to X-ray images, lacking real depth information.

The Fourier projection slice theorem states that the inverse transformation of a slice extracted from the frequency domain representation of a volume results in a projection of the volume, in a direction perpendicular to the slice. Based on this theorem, the description of the FDR can be summarized in the following three steps (Fig. 4.13):

1. Transformation of the 3D volume from spatial domain to the frequency domain, using an FFT. Supposing that f(x, y, z) is the description of the volume in spatial domain, the resulting volume F(i, j, k) in frequency domain will be taken, with the application of the 3D fast Fourier transformation:

$$F(i, j, k) =$$

$$\sum_{x=0}^{N-1} \sum_{y=0}^{N-1} \sum_{z=0}^{N-1} f(x, y, z)\exp[-\hat{\jmath}2\pi(ix+jy+kz)/N]$$

where $\hat{\jmath} = \sqrt{-1}$ and i, j, and k vary from 0 to N-1.

2. Extraction of a 2D slice from the 3D spectrum

along a plane which includes the origin and is perpendicular to the viewing plane, resulting in an F(u, v) slice.

3. Inverse transformation of the 2D extracted spectrum to the spatial domain using a 2D IFFT:

$$f(l, m) =$$

$$\frac{1}{N} \sum_{u=0}^{N-1} \sum_{v=0}^{N-1} F(u, v)\exp[\hat{\jmath}2\pi(ul + vm))/N]$$

where $\hat{\jmath} = \sqrt{-1}$ and l, m vary from 0 to N-1.

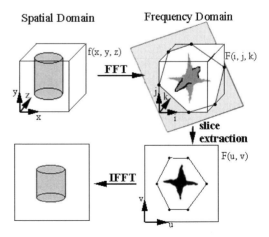

Fig. 4.13. General description of the frequency domain volume rendering method

The time-consuming 3D FFT is a data pre-processing step that is done only once and is finished before rendering; thus, the rendering time consists of the costs from the second and third steps.

A more promising approach is wavelet transforms. These methods provide a multi-scale representation of 3D objects, with the size of represented detail locally adjustable. The amount of data and rendering times may thus be reduced dramatically. Application to volume visualization is shown by HE et al. (1998).

4.7
Image Fusion

For many applications, it is desirable to combine or fuse information from different imaging modalities. For example, functional imaging techniques, such as magnetoencephalography (MEG), functional MRI (fMRI), or PET, show various physiological aspects but give little or no indication for the localization of the observed phenomena. For their interpretation, a closely matched description of the patient's morphology is required, as obtained by MRI. Considering only soft tissue anatomy, bone morphology or functional information in separate 3D data sets is not sufficient anymore in clinical practice. Pre-processing and visualizing all this complex information in a way that is easy to handle for clinicians is required to exploit the benefit of the unique clinical information of each of the modalities.

In general, image volumes obtained from different sources do not match geometrically (Fig. 4.14). Variations in patient orientation and differences in resolution and contrast of the modalities make it almost impossible for a clinician to mentally fuse all the image information accurately. It is therefore required to transform one volume with respect to the other, i.e., in a common coordinate frame. This process is known as image registration.

Image registration can be formulated as a problem of minimizing a cost function that quantifies the match between the images of the two modalities. In order to determine this function, different common features of those images can be used. MAINTZ and VIERGEVER (1998) and VAN DEN ELSEN et al. (1993) have given detailed surveys about the classification of the registration process. MAINTZ et al. describe the classification of the registration procedures based on nine different criteria:

1. The dimensionality (e.g., 3D to 3D or 2D to 3D)
2. The nature of the registration basis (e.g., intrinsic or extrinsic)
3. The nature of the transformation (e.g., rigid or curved)
4. The domain of the transformation (e.g., local or global)
5. The interaction (e.g., manual of automatic)
6. The optimization procedure (e.g., iterative closest point or simulated annealing)
7. The modalities involved (e.g., CT, MR, or PET)

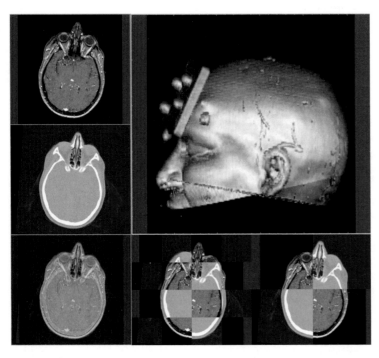

Fig. 4.14. Fusion of a CT and MR data set illustrates the differences of the patient alignment in both acquisitions.

8. The subject (e.g., inter-subject, intra-subject, or atlas)
9. The object (e.g., head or abdomen)

When considering the nature of the registration base, for example, the transformation may be defined using corresponding landmarks in both data sets. In a simple case, artificial markers attached to the patient are available which are visible on different modalities (BROMM and SCHAREIN 1996); otherwise, pairs of preferably stable matching points, such as the AC–PC line, may be used. A more robust approach is to interactively match larger features such as surfaces (SCHIEMANN et al. 1994). Figure 4.15 shows the result of the registration of a PET and an MRI data set.

Segmentation-based registration approaches can be divided into those using rigid models, such as, points, curves, or surfaces, and those using deformable models (e.g., snakes or nets). In all cases the registration accuracy of this method is limited to the accuracy of the segmentation step. Herewith any two modalities can be registered given the fact that the structures are visible within both. As an example, in some cases the target area in US images is not as visible as it should be to ensure a qualitatively high treatment of prostate cancer, whereas other modalities, such as CT, provide a better image. Unfortunately, CT cannot be used in a live-imaging procedure in the treatment room during intervention such as ultrasound imaging. To overcome these limitations, the images of both modalities can be registered in a unique data set, i.e., gathering pre-operatively a CT volume and using it in combination with the intra-operative US-guided live procedure. To realize this, the separate volumes of CT and US can be registered with respect to each other based on the geometry of the urethra or by mutual information based on their greylevels as shown in Fig. 4.16 (FIRLE et al. 2003).

In a fundamentally different approach, the results of a registration step are evaluated at every point of the combined volume, based on intensity values (STUDHOLME et al. 1996; WELLS et al. 1996). Starting from a coarse match, registration is achieved by adjusting position and orientation until the mutual information ("similarity") between both data sets is maximized. These methods are fully automatic, do not rely on a possibly erroneous definition of landmarks, and seem to be more accurate than others (WEST et al. 1997).

Mutual information, originating in the information theory, is a voxel-based similarity measure of the statistical dependency between two data sets, which has been proposed by COLLIGNON et al. (1995) and VIOLA and WELLS (1997). It evaluates the amount of infor-

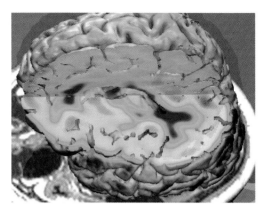

Fig. 4.15. Fusion of different imaging modalities for therapy control in a clinical study of obsessive-compulsive disorder. Magnetic resonance imaging shows that morphology is combined with a positron emission tomography (PET) scan, which shows glucose metabolism. Since the entire volume is mapped, the activity can be explored at any location of the brain.

mation that one variable contains about the other. By superimposing two data sets of the same object, but from different modalities, this method states that they are correctly aligned if the mutual information of geometrically corresponding gray values is maximal. Since no assumptions are made about the two signals, this method is not restricted to specific modalities and does not require the extraction of features in a preprocessing step. A recent survey about mutual information-based registration approaches was given by PLUIM et al. (2003; Figs. 4.17, 4.18).

The calculation of the transformation based on mutual information is a very time-consuming optimization process. Exploiting the coarse-to-fine resolution strategy (pyramidal approach) is one common possibility to speed up the registration process (PLUIM et al. 2001). Another approach, when allowing only rigid transformations, is the usage of the "3D cross model" (FIRLE et al. 2004). This partial-volume based matching assumes that the center of the volume comprises the majority of the overlapping information between both images. The data from all three directions through the reference image (MAES et al. 1997) is taken without any high sub-sampling factors or lowering the number of histogram bins (CAPEK et al. 2001). Figures 4.17 and 4.18 depict the registration result of a whole-body CT and PET data set.

4.8
3D Anatomical Atlases

Whereas in classical medicine knowledge about the human body is represented in books and atlases,

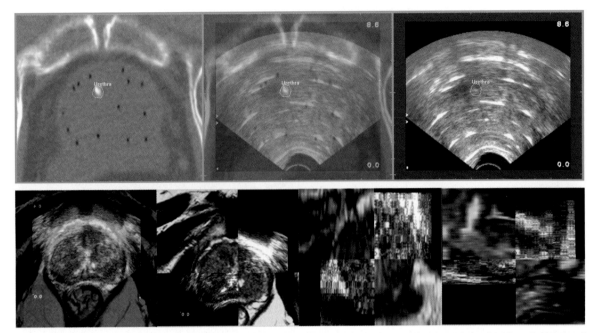

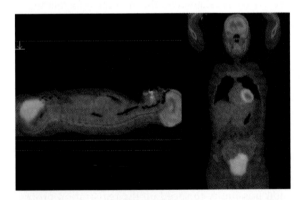

Fig. 4.16. Fusion of CT and 3D US volumes based on the urethra geometry (upper) and MRI with 3D U/S based on mutual information (lower)

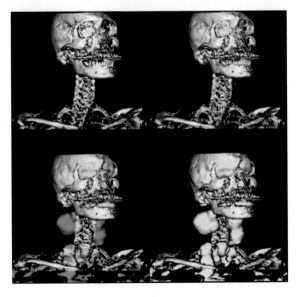

Fig. 4.17. While CT identifies the precise size, shape, and location of a mass, PET detects changes in the metabolism caused by the growth of abnormal cells in it. Fusion of the whole-body scans of both modalities

Fig. 4.18. A 3D image fusion after mutual-information-based registration of a CT and PET data set. Blending of the PET into the CT volume using image-level intermixing and different opacity settings

present-day computer science allows for new, more powerful and versatile computer-based representations of knowledge. The most straightforward example are multimedia CD-ROMs containing collections of classical pictures and text, which may be browsed arbitrarily. Although computerized, such media still follow the old paradigm of text printed on pages, accompanied by pictures.

Using methods of volume visualization, spatial knowledge about the human body may be much more efficiently represented by computerized 3D models. If such models are connected to a knowledge base of descriptive information, they can even be interrogated or disassembled by addressing names of organs (BRINKLEY et al. 1999; GOLLAND et al. 1999; HÖHNE et al. 1996; POMMERT et al. 2001).

A suitable data structure for this purpose is the intelligent volume (HÖHNE et al. 1995), which combines a detailed spatial model enabling realistic visualization with a symbolic description of human anatomy (Fig. 4.19). The spatial model is represented as a 3D volume as described above. The membership of voxels to an object is indicated by labels which are stored in attribute volumes congruent to the image volume. Different attribute volumes may be generated, e.g., for structure or function. Further attribute volumes may be added which contain, for example, the incidence of a tumor type or a time tag for blood propagation on a per-voxel basis.

The objects themselves bear attributes as well. These attributes may be divided into two groups: firstly, attributes indicating meaning such as names, pointers to text or pictorial explanations, or even features such as vulnerability or mechanical properties, which might be important (e.g. for surgical simulation); secondly, attributes defining their visual appearance, such as color, texture, and reflectivity. In addition, the model describes the interrelations of the objects with a semantic network. Examples for relations are part of or supplied by.

Once an intelligent volume is established, it can be explored by freely navigating in both the pictorial and descriptive worlds. A viewer can compose arbitrary views from the semantic description or query semantic information for any visible voxel of a picture. Apart from educational purposes, such atlases are also a powerful aid for the interpretation of clinical images (KIKINIS et al. 1996; NOWINSKI and THIRUNAVUUKARASUU 2001; SCHIEMANN et al. 1994; SCHMAHMANN et al. 1999).

Because of the high computational needs, 3D anatomical atlases are not yet suitable for present-day personal computers. SCHUBERT et al. (1999) make such a model available for interactive exploration via precomputed Intelligent QuickTime virtual-reality videos, which can be viewed on any personal computer. A 3D atlas of regional, functional, and radiological anatomy of the brain based on this technique has been published (HÖHNE et al. 2001), along with a high-resolution atlas of the inner organs, based on the Visible Human (HÖHNE et al. 2003). A screenshot is shown in Fig. 4.20.

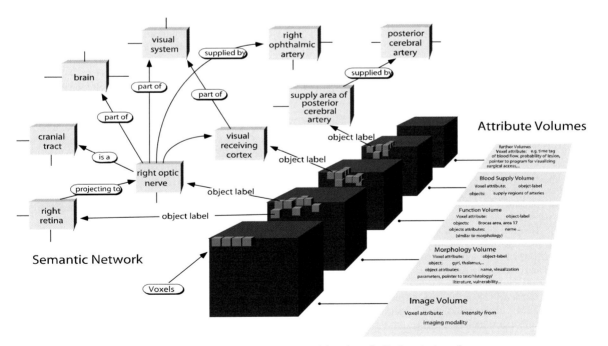

Fig. 4.19. Basic structure of the intelligent volume, integrating spatial and symbolic description of anatomy

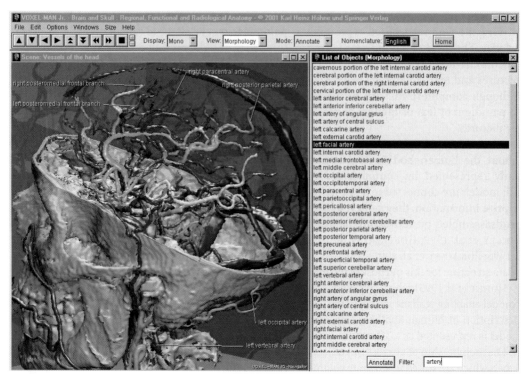

Fig. 4.20. User interface of VOXEL-MAN 3D-Navigator: brain and skull, a PC-based 3D atlas of regional, functional, and radiological anatomy. The user may navigate freely in both the pictorial (*left*) and descriptive (*right*) context.

To date, most 3D anatomical atlases are based on the data derived from one individual only. The interindividual variability of organ shape and topology in space and time is thus not yet part of the model. Methods for measuring and modeling variability are currently being developed (Mazziotta et al. 1995; Styner and Gerig 2001; Thompson et al. 2000).

Acknowledgements
We thank C. Dold, E. Firle, K.-H. Höhne, G. Karangelis, T. Schiemann, U. Tiede, and S. Wesarg for their contributions to this work.

References

Arata LK, Dhawan AP, Broderick JP, Gaskil-Shipley MF, Levy AV, Volkow ND (1995) Three-dimensional anatomical model-based segmentation of MR brain images through principal axes registration. IEEE Trans Biomed Eng 42:1069–1078

Arnold JB, Liow J-S, Schaper KA, Stern JJ, Sled JG, Shattuck DW, Worth AJ, Cohen MS, Leahy RM, Mazziotta JC, Rottenberg DA (2001) Qualitative and quantitative evaluation of six algorithms for correcting intensity nonuniformity effects. Neuro-Image 13:931–943

Biswas TK, Gupta AK (2002). Retrieval of true color of the internal organ of CT images and attempt to tissue characterization by refractive index: initial experience. Ind J Radiol Imaging 12:169–178

Bomans M, Riemer M, Tiede U, Höhne KH (1987) 3D-Segmentation von Kernspin-Tomogrammen. In: Paulus E (ed) Mustererkennung 1987. Proc of the 9th DAGM symposium. Informatik-Fachberichte, vol 149. Springer, Berlin Heidelberg New York, pp 231–235

Brinkley JF, Wong BA, Hinshaw KP, Rosse C (1999) Design of an anatomy information system. IEEE Comput Graphics Appl 19:38–48

Bromm B, Scharein E (1996) Visualisation of pain by magnetoencephalography in humans. In: Höhne KH, Kikinis R (eds) Visualization in biomedical computing. Proc VBC '96. Springer, Berlin Heidelberg New York, pp 477–481 (Lect Notes Comput Sci, vol 1131)

Cai W, Sakas G (1998) Maximum intensity projection using splatting in sheared object space. Comput Graph Forum 17:113–124

Cai W, Sakas G, (1999) Transfer functions in DRR volume rendering. In: Lemke HU et al. Computer assisted radiology and surgery (CARS). Proceedings. Amsterdam, Lausanne:Elsevier, pp 284–289 (International Congress Series 1191).

Cai W, Karangelis G, Sakas G (1999)Volume interaction techniques in the virtual simulation of radiotherapy treatment planning. In: Keldysh Institute of Applied Mathematics: Graphicon. Proceedings. Moscow, pp 231–239

Cai W, Walter S, Karangelis G, Sakas G (2000) Collaborative virtual simulation environment for radiotherapy treatment planning. In: Computer Graphics Forum 19, 3 pp. C-379 –C-390

Canny J (1986) A computational approach to edge detection. IEEE Trans Pattern Anal Machine Intell 8:679–698

Capek M, Mroz L, Wegenkittl R (2001) Robust and fast medical registration of 3D multi-modality data sets, Medicon 2001, IXth Mediterranean conference on medical and biological engineering and computing, Pula, Croatia, pp 515–518

Caselles V, Kimmel R, Sapiro G (1997). Geodesic active contours. International Journal on Computer Vision, 22(1):61–97, 1997

Cignoni P, Montani C, Scopigno R (1998) A comparison of mesh simplification algorithms. Comput Graph 22:37–54

Cignoni P, Ganovelli F, Montani C, Scopigno R (2000) Reconstruction of topologically correct and adaptive trilinear isosurfaces. Comput Graph 24:399–418

Clarke LP, Velthuizen RP, Camacho MA, Heine JJ, Vaidyanathan M, Hall LO, Thatcher RW, Silbiger ML (1995) MRI segmentation: methods and applications. Magn Reson Imaging 13:343–368

Cline HE, Lorensen WE, Kikinis R, Jolesz F (1990) Three-dimensional segmentation of MR images of the head using probability and connectivity. J Comput Assist Tomogr 14:1037–1045

Collignon A, Maes F, Delaere D, Vandermeulen D, Suetens P, Marchal G, Viola P, Wells W (1995) Automated multimodality medical image registration using information theory. Fourteenth International Conference on Information Processing in Medical Imaging. Kluwer, Boston, pp 263–274 (Computational Imaging and Vision, vol 3)

Collins DL, Zijdenbos AP, Barré WFC, Evans AC (1999) ANIMAL+INSECT: improved cortical structure segmentation. In: Kuba A, Samal M, Todd-Pokropek A (eds) Information processing in medical imaging. Proc IPMI 1999. Springer, Berlin Heidelberg New York, pp 210–223 (Lect Notes Comput Sci, vol 1613)

Drebin RA, Carpenter L, Hanrahan P (1988) Volume rendering. Comput Graphics 22:65–74

Duncan JS, Ayache N (2000) Medical image analysis: progress over two decades and the challenges ahead. IEEE Trans Pattern Anal Machine Intell 22:85–105

Firle E, Wesarg S, Karangelis G, Dold C (2003) Validation of 3D ultrasound–CT registration of prostate images. Medical Imaging 2003. Proc SPIE 5032, Bellingham, pp 354–362

Firle E, Wesarg S, Dold C (2004) Fast CT/PET registration based on partial volume matching. Computer assisted radiology and surgery. Proc CARS 2004. Elsevier, Amsterdam, pp 31–36

Foley JD, van Dam A, Feiner SK, Hughes JF (1995) Computer graphics: principles and practice, 2nd edn. Addison-Wesley, Reading, Massachusetts

Gerig G, Martin J, Kikinis R, Kübler O, Shenton M, Jolesz FA (1992) Unsupervised tissue type segmentation of 3D dual-echo MR head data. Image Vision Comput 10:349–360

Golland P, Kikinis R, Halle M, Umans C, Grimson WEL, Shenton ME, Richolt JA (1999) Anatomy browser: a novel approach to visualization and integration of medical information. Comput Aided Surg 4:129–143

Hamann B, Trotts I, Farin G (1997) On approximating contours of the piecewise trilinear interpolant using triangular rational-quadratic Bézier patches. IEEE Trans Visual Comput Graph 3:215–227

He T (1998) Wavelet-assisted volume ray casting. Pac Symp Biocomput 1998, pp 153–164

Hemmingsson A, Jung B (1980) Modification of grey scale in computer tomographic images. Acta Radiol Diagn (Stockh) 21:253–255

Herman GT (1991) The tracking of boundaries in multidimensional medical images. Comput Med Imaging Graph 15:257–264

Hildebrand A, Sakas G (1996) Innovative 3D-methods in medicine. Korea Society of Medical and Biomedical Engineering. Advanced Medical Image Processing Proceedings 1996, Seoul, Korea

Höhne KH, Bernstein R (1986) Shading 3D-images from CT using gray level gradients. IEEE Trans Medical Imaging MI-5:45–47

Höhne KH, Hanson WA (1992) Interactive 3D-segmentation of MRI and CT volumes using morphological operations. J Comput Assist Tomogr 16:285–294

Höhne KH, Bomans M, Pommert A, Riemer M, Schiers C, Tiede U, Wiebecke G (1990) 3D-visualization of tomographic volume data using the generalized voxel-model. Visual Comput 6:28–36

Höhne KH, Pflesser B, Pommert A, Riemer M, Schiemann T, Schubert R, Tiede U (1995) A new representation of knowledge concerning human anatomy and function. Nat Med 1:506–511

Höhne KH, Pflesser B, Pommert A, Riemer M, Schiemann T, Schubert R, Tiede U (1996) A virtual body model for surgical education and rehearsal. IEEE Comput 29:25–31

Höhne KH, Petersik A, Pflesser B, Pommert A, Priesmeyer K, Riemer M, Schiemann T, Schubert R, Tiede U, Urban M, Frederking H, Lowndes M, Morris J (2001) VOXEL-MAN 3D navigator: brain and skull. Regional, functional and radiological anatomy. Springer Electronic Media, Heidelberg (2 CD-ROMs, ISBN 3-540-14910-4)

Höhne KH, Pflesser B, Pommert A, Priesmeyer K, Riemer M, Schiemann T, Schubert R, Tiede U, Frederking H, Gehrmann S, Noster S, Schumacher U (2003) VOXEL-MAN 3D navigator: inner organs. Regional, systemic and radiological anatomy. Springer Electronic Media, Heidelberg (DVD-ROM, ISBN 3-540-40069-9)

Kass M, Witkin A, Terzopoulos D (1987) Snakes: active contour models. Proc 1st ICCV, June 1987, London, pp 259–268

Kikinis R, Shenton ME, Iosifescu DV, McCarley RW, Saiviroonporn P, Hokama HH, Robatino A, Metcalf D, Wible CG, Portas CM, Donnino RM, Jolesz FA (1996) A digital brain atlas for surgical planning, model driven segmentation, and teaching. IEEE Trans Visual Comput Graphics 2:232–241

Kindlmann G, Durkin JW (1998) Semi-automatic generation of transfer functions for direct volume rendering, volume visualization. IEEE symposium on 19–20 October 1998, pp 79–86, 170

Kirbas C, Quek FKH (2003) Vessel extraction techniques and algorithms: a survey, bioinformatics and bioengineering 2003. Proc 3rd IEEE Symposium, 10–12 March 2003, pp 238–245

Kniss J, Kindlmann G, Hansen C (2002) Multidimensional transfer functions for interactive volume rendering, visualization and computer graphics. IEEE Trans 8:270–285

Lacroute P, Levoy M (1994) Fast volume rendering using a shear-warp factorization of the viewing transformation. Proc SIGGRAPH 1994, Orlando, Florida, July 1994, pp 451–458

Levoy M (1988) Display of surfaces from volume data. IEEE Comput Graph Appl 8:29–37

Levoy M (1990) A hybrid ray tracer for rendering polygon and volume data. IEEE Comput Graph Appl 10:33–40

Lorensen WE, Cline HE (1987) Marching cubes: a high resolution 3D surface construction algorithm. Comput Graph 21:163–169

Maes F, Collignon A, Vandermeulen D, Marchal G, Suetens P (1997) Multimodality image registration by maximization of mutual information. IEEE Trans Med Imaging 16:187–198

Maintz JBA, Viergever M (1998) A survey of medical image registration. Med Image Anal 2:1–36

Marschner SR, Lobb RJ (1994) An evaluation of reconstruction filters for volume rendering. In: Bergeron RD, Kaufman AE (eds) Proc IEEE visualization 1994. IEEE Computer Society Press, Los Alamitos, Calif., pp 100–107

Max N (1995) Optical models for direct volume rendering, IEEE Trans Visual & Comput Graph, 1(2):99–108

Mazziotta JC, Toga AW, Evans AC, Fox P, Lancaster J (1995) A probabilistic atlas of the human brain: theory and rationale for its development. NeuroImage 2:89–101

Möller T, Machiraju R, Mueller K, Yagel R (1997) Evaluation and design of filters using a Taylor series expansion. IEEE Trans Visual Comput Graph 3:184–199

Mroz L, Hauser H, Gröller E (2000) Interactive high-quality maximum intensity projection. Comput Graphics Forum 19:341–350

Natarajan BK (1994) On generating topologically consistent iso-surfaces from uniform samples. Visual Comput 11:52–62

Nowinski WL, Thirunavuukarasuu A (2001) Atlas-assisted localization analysis of functional images. Med Image Anal 5:207–220

Olabarriaga SD, Smeulders AWM (2001) Interaction in the segmentation of medical images: a survey. Med Image Anal 5:127–142

Pluim J, Maintz J, Viergever M (2001) Mutual information matching in multiresolution contexts. Image Vision Comput 19:45–52

Pluim J, Maintz J, Viergever M (2003) Mutual information based registration of medical images: a survey. IEEE Trans Med Imaging 22:986–1004

Pommert A (2004) Simulationsstudien zur Untersuchung der Bildqualität für die 3D-Visualisierung tomografischer Volumendaten. Books on Demand, Norderstedt 2004 (zugleich Dissertation, Fachbereich Informatik, Universität Hamburg)

Pommert A, Höhne KH, Pflesser B, Richter E, Riemer M, Schiemann T, Schubert R, Schumacher U, Tiede U (2001) Creating a high-resolution spatial/symbolic model of the inner organs based on the visible human. Med Image Anal 5:221–228

Saeed N, Hajnal JV, Oatridge A (1997) Automated brain segmentation from single slice, multislice, or whole-volume MR scans using prior knowledge. J Comput Assist Tomogr 21:192–201

Sakas G, Grimm M, Savopoulos A (1995) Optimized maximum intensity projection (MIP). In: Hanrahan P et al. 6th Eurographics workshop on rendering. Proceedings. Eurographics, pp. 81-93

Schiemann T, Höhne KH, Koch C, Pommert A, Riemer M, Schubert R, Tiede U (1994) Interpretation of tomographic images using automatic atlas lookup. In: Robb RA (ed) Visualization in biomedical computing 1994. Proc SPIE 2359, Rochester, Minnesota, pp 457–465

Schiemann T, Tiede U, Höhne KH (1997) Segmentation of the visible human for high quality volume based visualization. Med Image Anal 1:263–271

Schmahmann JD, Doyon J, McDonald D, Holmes C, Lavoie K, Hurwitz AS, Kabani N, Toga A, Evans A, Petrides M (1999) Three-dimensional MRI atlas of the human cerebellum in proportional stereotaxic space. NeuroImage 10:233–260

Schroeder WJ, Zarge JA, Lorensen WE (1992) Decimation of triangle meshes. Comput Graph 26:65–70

Schubert R, Pflesser B, Pommert A et al.(1999) Interactive volume visualization using "intelligent movies". In: Westwood JD, Hoffman HM, Robb RA, Stredney D (eds) Medicine meets virtual reality. Proc MMVR 1999. IOS Press, Amsterdam, pp 321–327 (Health Technology and Informatics, vol 62)

Sonka M, Hlavac V, Boyle R (1998) Image processing, analysis, and machine vision, 2nd edn. PWS Publishing, Boston, Mass

Šrámek M, Kaufman A (2000) Fast ray-tracing of rectilinear volume data using distance transforms. IEEE Trans Visual Comput Graph 6:236–251

Studholme C, Hill DLG, Hawkes DJ (1996) Automated 3-D registration of MR and CT images of the head. Med Image Anal 1:163–175

Styner M, Gerig G (2001) Medial models incorporating object variability for 3D shape analysis. In: Insana MF, Leahy RM (eds) Information processing in medical imaging. Proc IPMI 2001. Springer, Berlin Heidelberg New York, pp 502–516 (Lect Notes Comput Sci, vol 2082)

Thompson PM, Woods RP, Mega MS, Toga AW (2000) Mathematical/computational challenges in creating deformable and probabilistic atlases of the human brain. Hum Brain Mapping 9:81–92

Tiede U (1999) Realistische 3D-Visualisierung multiattributierter und multiparametrischer Volumendaten. PhD thesis, Fachbereich Informatik, Universität Hamburg

Tiede U, Höhne KH, Bomans M, Pommert A, Riemer M, Wiebecke G (1990) Investigation of medical 3D-rendering algorithms. IEEE Comput Graph Appl 10:41–53

Tiede U, Schiemann T, Höhne KH (1998) High quality rendering of attributed volume data. In: Ebert D, Hagen H, Rushmeier H (eds) Proc IEEE Visualization 1998. IEEE Computer Society Press, Los Alamitos, Calif., pp 255–262

Totsuka T, Levoy M (1993) Frequency domain volume rendering. Comput Graph 27:271–278

Van den Elsen P, Pol E, Viergever M (1993) Medical image matching: a review with classification. IEEE Eng Med Biol 12:26–39

Verdonck B, Bloch I, Maître H, Vandermeulen D, Suetens P, Marchal G (1995) Blood vessel segmentation and visualization in 3D MR and spiral CT angiography HU. In: Lemke (ed) Proc CAR 1995. Springer, Berlin Heidelberg New York, pp 177–182

Viola P, Wells W (1997) Alignment by maximization of mutual information. Int J Comput Vis 24:137–154

Wan M, Kaufman A, Bryson S (1999) High performance presence-accelerated ray casting. Proc IEEE Visualization 1999, San Francisco, Calif., pp 379–386

Warren RC, Pandya YV (1982) Effect of window width and viewing distance in CT display. Br J Radiol 55:72–74

Watt A (2000) 3D computer graphics, 3rd edn. Addison-Wesley, Reading, Massachusetts

Wells WM III, Viola P, Atsumi H, Nakajima S, Kikinis R (1996) Multi-modal volume registration by maximization of mutual information. Med Image Anal 1:35–51

Wesarg S, Firle EA (2004) Segmentation of vessels: the corkscrew algorithm. Proc of SPIE medical imaging symposium 2004, San Diego (USA), vol 5370, pp 1609–1620

West J, Fitzpatrick JM, Wang MY, Dawant BM, et al. (1997) Comparison and evaluation of retrospective intermodality brain image registration techniques. J Comput Assist Tomogr 21:554–566

Wilmer F, Tiede U, Höhne KH (1992) Reduktion der Oberflächenbeschreibung triangulierter Oberflächen durch Anpassung an die Objektform. In: Fuchs S, Hoffmann R (eds) Mustererkennung 1992. Proc 14th DAGM symposium. Springer, Berlin Heidelberg New York, pp 430–436

Winston PH (1992) Artificial intelligence, 3rd edn. Addison-Wesley, Reading, Massachusetts

Yagel R, Cohen D, Kaufman A (1992) Discrete ray tracing. IEEE Comput Graph Appl 12:19–28

Zamboglou N, Karangelis G, Nomikos I, Zimeras S, Helfmann T, Uricchio R, Martin T, Röddiger S, Kolotas C, Baltas D, Sakas G (2003) EXOMIO virtual simulation: oropharynx, prostate and breast cancers. In: Mould, RF (Ed) Progress in CT-3D simulation. Bochum: Medical Innovative Technology, 2003, pp 1–18

Zamboglou N, Karangelis G, Nomikos I, Zimeras S, Kolotas C, Baltas D, Sakas G (2004) Virtual CT-3D simulation using Exomio: With special reference to prostate cancer. In: Nowotwory Journal of Oncology 54, 6 pp 547–554

Zucker S (1976) Region growing: childhood and adolescence. Comput Graph Image Proc 5:382–399

5 Image Registration and Data Fusion for Radiotherapy Treatment Planning

MARC L. KESSLER and MICHAEL ROBERSON

CONTENTS

5.1 Introduction

Treatment planning for conformal radiotherapy requires accurate delineation of tumor volumes and surrounding healthy tissue. This is especially true in inverse planning where trade-offs in the dose-volume relationships of different tissues are usually the driving forces in the search for an optimal plan. In these situations, slight differences in the shape and overlap regions between a target volume and critical structures can result in different optimized plans. While X-ray computed tomography (CT) remains the primary imaging modality for structure delineation, beam placement and generation of digitally reconstructed radiographs, data from other modalities, such as magnetic resonance imaging and spectroscopy (MRI/MRSI) and positron/single photon emission tomography (PET/SPECT), are becoming increasingly prevalent for tumor and normal tissue delineation (Fig. 5.1).

M. L. KESSLER, PhD, Professor
M. ROBERSON, BS
Department of Radiation Oncology
The University of Michigan Medical School
1500 East Medical Center Drive
Ann Arbor, MI 48109-0010
USA

There are many reasons to include image data from other modalities into the treatment-planning process. In many sites, MRI provides superior soft tissue contrast relative to CT and can be used to enhance or suppress different tissues such as fat and conditions such as edema. Magnetic resonance imaging also permits imaging along arbitrary planes which can improve visualization and segmentation of different anatomic structures. More recently, MRI has been used to acquire localized information about relative metabolite concentrations, fluid mobility, and tissue microstructure. With a variety of tracer compounds available, PET and SPECT can provide unique information about different cellular and physiologic processes to help assess normal and diseased tissues.

In addition to static treatment planning, volumetric image data is playing a larger role in treatment delivery and adaptive radiotherapy. Modern treatment machines can now be equipped with imaging devices that allow acquisition of volumetric CT data at the time of treatment. These "treatment-delivery CT" studies can be used to adapt a treatment plan based on the patient's anatomy at the time of treatment and allow more accurate tracking of delivered dose. Furthermore, data from magnetic resonance and nuclear medicine acquired during the course of therapy may also help assess the efficacy of therapy and indicate prescription changes (CHENEVERT et al. 2000; MARDOR et al. 2003; BARTHEL et al. 2003; BRUN et al. 2002; ALLAL et al. 2004).

To fully realize the benefits of the information available from different imaging studies, the data they provide must be mapped to a single coordinate system, typically that of the treatment planning CT. This process is called image registration. Once they are all linked to a common coordinate system, data can be transferred between studies and integrated to help construct a more complete and accurate representation of the patient. This process is called data fusion. This chapter describes the mechanics of image registration and data fusion processes.

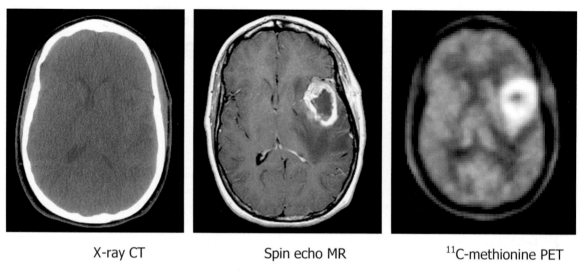

X-ray CT   Spin echo MR ^{11}C-methionine PET

Fig. 5.1. Examples of different multimodality imaging data available for treatment planning

5.2
Image Registration

Image registration is the process of determining the geometric transformation that maps the coordinates between identical points in different imaging studies. With this mapping, information can be transferred between the studies or fused in various ways (Fig. 5.2).

For the discussion that follows, we describe the mechanics of automated image registration using two data sets which are labeled Study *A* and Study *B*. Study *A* is the base or reference data set and is held fixed and Study *B* the homologous data set that is manipulated to be brought into geometric alignment with Study *A*. Study *B'* is the transformed data from Study *B*.

Numerous techniques exist for image registration. The choice of technique depends on the imaging modalities involved, the anatomic site, and the level of control over the imaging conditions. A detailed review is given by MAINTZ and VIERGEVER (1998). The general approach in each of the methods is to devise a registration metric that measures the degree of mismatch (or similarity) between one or more features in two data sets and to use standard numerical optimization methods to determine the parameters of a

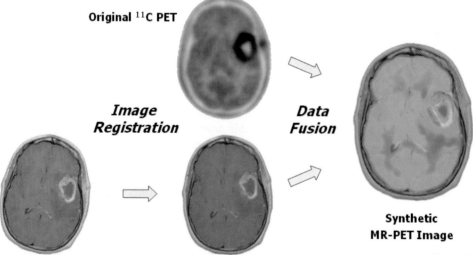

Original ^{11}C PET

Image Registration

Data Fusion

Synthetic
MR-PET Image

Original MR   Registered MR

Fig. 5.2. The image registration and data fusion processes

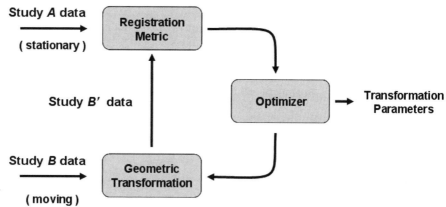

Fig. 5.3. The image registration process

geometric transformation that minimize (maximize) the metric. Differences between various techniques include the metric used, the features used to measure the mismatch, the particular form of the geometric transformation used, and the optimization method used for computing the required parameters.

The image registration process illustrated in Fig. 5.3 can be either automated or manual. Automated image registration is analogous to inverse treatment planning, with the registration metric replacing the "plan cost function" and the parameters of the transformation replacing the plan defining parameters such as beamlet intensities or weights. In both cases the parameters are iterated using an optimizer until an "optimal" set of parameters is found. Manual image registration is more like forward treatment planning. The "optimizer" is a human with a suite of interactive tools that let the user perform various image transformations and visualize the results in real time. While manual registration tools typically only support rotation and translation, they can also be used to initialize more complex automated registrations.

5.2.1
Geometric Transformations

The fundamental task of image registration is to find the geometric transformation, T, which maps the coordinates of a point in Study A to the coordinates of the corresponding point in Study B. In general, this transformation can be written as

$$\mathbf{x}_B = T(\mathbf{x}_A, \{\beta\}),$$

where $\mathbf{x}_A$ is the coordinate of the point in Study A, $\mathbf{x}_B$ is the coordinate of the same anatomic point in Study

B, and $\{\beta\}$ is the set of parameters of the transformation. The output of the image registration process is the parameters $\{\beta\}$ for a particular pair of imaging studies (Fig. 5.3). The number of parameters required to determine the transformation depends on the form of T, which in turn depends on the clinical site, clinical application, and the modalities involved.

In the ideal case, where the patient is positioned in an identical orientation in the different imaging studies and the scale and center of the coordinate systems coincide, T is simply an identity transformation I and $\mathbf{x}_B = \mathbf{x}_A$ for all points in the two imaging studies. This situation most closely exists for combined imaging modality devices such as PET–CT machines, especially if physiologic motion is controlled or absent. Unfortunately, it is far more common for the orientation of the patient to change between imaging studies, making more sophisticated transformations necessary.

For situations where the anatomy of interest can be assumed to move as a rigid body, the set of parameters consists of three rotation angles $(\theta_x, \theta_y, \theta_z)$ and three translations (t_x, t_y, t_z). The rigid body transformation is then written as

$$\mathbf{x}_B = T_{rigid}(\mathbf{x}_A, \{\beta\}) = A\,\mathbf{x}_A + \mathbf{b},$$

where A is a 3×3 rotation matrix and $\mathbf{b}$ is a 3×1 translation vector. This transformation is simply the familiar one dimensional function "$y = m\cdot x + b$", except in three dimensions.

In matrix notation this can be written as

$$\begin{bmatrix} x_B \\ y_B \\ z_B \end{bmatrix} = \begin{bmatrix} & & \\ & A & \\ & & \end{bmatrix} \begin{bmatrix} x_A \\ y_A \\ z_A \end{bmatrix} + \begin{bmatrix} \\ b \\ \end{bmatrix}.$$

If the scales of the two data sets are not identical, it is necessary to also include scale factors (s_x, s_y, s_z) into the matrix **A**. This is usually a device calibration issue rather than an image registration problem, but if these factors exist and are not compensated for by a preprocessing step, they must be determined during the registration process.

The rigid and scaling transformations are special cases of the more general *affine* transformation. A *full* affine transformation includes parameters for rotation, translation, scale, shear, and plane reflection. One attribute of affine transformations is that all points lying on a line initially still lie on a line after transformation and "parallel lines stay parallel."

Another notation used to specify rigid or affine transformations is to combine the 3×3 matrix **A** and the 3×1 translation vector **b** into a single 4×4 matrix, i.e.,

$$\begin{bmatrix} x_B \\ y_B \\ z_B \\ 1 \end{bmatrix} = \begin{bmatrix} & & & \vdots & \\ & A & & \vdots & b \\ & & & \vdots & \\ \hline 0 & 0 & 0 & \vdots & 1 \end{bmatrix} + \begin{bmatrix} x_A \\ y_A \\ z_A \\ 1 \end{bmatrix}.$$

The DICOM imaging standard uses this representation for affine transformations to specify the spatial relationship between two imaging studies (National Electrical Manufacturers Association 2004).

The assumption of global rigid movement of anatomy is often violated, especially for sites other than the head and large image volumes that extend to the body surface. Differences in patient setup (arms up versus arms down), organ filling, and uncontrolled physiologic motion confound the use of a single affine transform to register two imaging studies. In some cases where *local* rigid motion can be assumed, it may be possible to use a rigid or affine transforma-tion to register sub-volumes of two imaging studies. For example, the prostate itself may be considered rigid, but it can move relative to the pelvis depending on the filling of the rectum and bladder. By considering only a limited field-of-view that includes just the region of the prostate, it is often possible to use an affine transformation to accurately register the prostate anatomy in two studies. One or more sub-volumes can be defined by simple geometric cropping or derived from anatomic surfaces (Fig. 5.4).

Even with a limited field-of-view approach, there are many sites in which affine registration techniques are not powerful enough to achieve acceptable alignment of anatomy. In these sites, an organ's size and shape may change as a result of normal organ behavior or the motion of surrounding anatomy. For example, the lungs change in both size and shape during the breathing cycle, and the shape of the liver can be affected by the filling of the stomach. When registering data sets that exhibit this kind of motion, a deformable model must be used to represent the transformation between studies.

One class of deformation model is called a spline, which is a curve that interpolates points in space based on a set of control points. A set of parameters associated with each control point defines the exact shape of this interpolation. The number and location of control points determine the extent of deformation that a spline can express. Two types of splines commonly used in biologic and medical imaging applications are thin-plate splines and B-splines (Bookstein 1989; Unser 1999).

Thin-plate spline transformations model the deformations of an infinite thin plate. The parameters associated with this transformation consist of a displacement at each control point. Interpolation

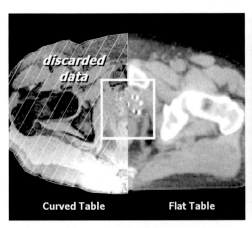 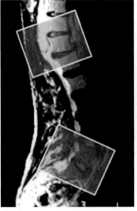 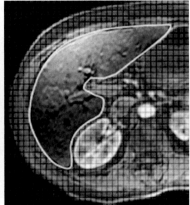

simple geometric cropping   piecewise cropping   anatomic-based cropping

Fig. 5.4. Different types of cropping for limited field-of-view registration

between control points is done by minimizing the "bending energy" of this thin plate while leaving the control point displacements intact. This concept is not limited to flat "plates" and can be extended to three dimensions. The deformation at an arbitrary point depends on its distance from each control point, so a change in any control point affects the deformation of all points in the image volume (except the other control points). Thin-plate splines are therefore considered a global deformation model. Because of this property, they perform well with relatively few control points but do suffer from increased computation time when many (>50–100) control points are used.

B-spline transformations use control points called knots, arranged in a grid. A piecewise polynomial function is used to interpolate the transformation between these knots. Any degree polynomial can be used, but in medical image registration cubic B-splines are typical. A B-spline transformation is expressed as a weighted sum

$$x_B = x_A + \sum w_i\, B(x_A - k_i),$$

where each k_i is the location of knot i, each w_i is a weight parameter associated with knot i, and $B(x)$ is a basis function. Figure 5.5 illustrates this weighted sum in a one-dimensional cubic B-spline example. Note that the basis function has a limited extent, so each knot only affects a limited region of the overall deformation. In this way, B-splines are considered a local deformable model. This property of locality allows B-spline models to use very fine grids of thousands of knots with only a modest increase in computation time. Each knot adds more control over the transformation (more degrees of freedom), so using

many knots greatly enhances the ability of B-splines to model complex deformations. Unfortunately, increasing the number of parameters to optimize during the registration process can increase the difficulty of finding the optimal solution.

Other deformable models that are possible include freeform deformations (used with physical or optical flow models) and finite element methods (THIRION 1998; BHARATHA et al. 2001).

5.2.2
Registration Metrics

The goal of the image registration process is to determine the parameters of geometric transformation that optimally align two imaging studies. To achieve this goal, a registration metric is devised which quantifies the degree to which the pair of imaging studies are aligned (or mis-aligned). Using standard optimization techniques the transformation parameters are manipulated until this metric is maximized (or minimized) (Fig. 5.6). Most registration metrics in use presently can be classified as either geometry based or intensity based. Geometry-based metrics make use of features extracted from the image data such as anatomic or artificial landmarks and organ boundaries, whereas intensity-based metrics use the image data directly.

Geometry-Based Metrics

The most common geometry-based registration metrics involve the use of point matching or surface matching. For point matching, the coordinates of pairs of corresponding points from Study A and Study B

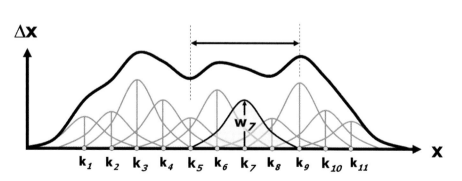
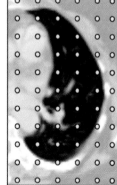

Fig. 5.5. B-spline deformation model. *Left:* 1D example of the cubic B-spline deformation model. The displacement Δx as a function of x is determined by the weighted sum of basis functions. The *double arrow* shows the region of the overall deformation affected by the weight factor w_7. The 3D deformations are constructed using 1D deformations for each dimension. *Right:* B-spline knot locations relative to image data for lung registration using deformation

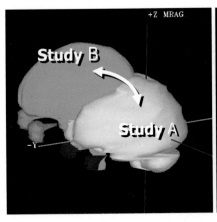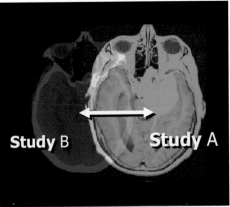

Fig. 5.6. Examples of image registration using geometric data (*left*) and image data (*right*). Geometry-based registration aligns points or surfaces while intensity-based registration aligns image intensity values.

are used to define the registration metric. These points can be anatomic landmarks or implanted or externally placed fiducial markers. The registration metric is defined as the sum of the squared distances between corresponding points:

$$R = \sum (p_A - p_{B'})^2 / N,$$

p_A is the coordinate of the a point in Study A, $p_{B'}$ is the coordinate of the transformed point from Study B and N is the number of pairs of points.

To compute the rotations and translations for a rigid transformation, a minimum of three pairs of points are required. For affine transformations, a minimum of four pairs of non-coplanar points are required. Using more pairs of points reduces the bias that errors in the delineation of any one pair of points has on the estimated transformation parameters; however, accurately identifying more than the minimum number of corresponding points can be difficult as different modalities often produce different tissue contrasts (a major reason why multiple modalities are used in the first place) and placing or implanting larger numbers of markers is not always possible or desirable.

Alternatively, surface matching does not require a one-to-one correspondence of specific points but instead tries to maximize the overlap between corresponding surfaces extracted from two imaging studies, such as the brain or skull surface or pelvic bones. These structures can be easily extracted using automated techniques and minor hand editing. The surfaces from Study A are represented as a binary volume or as an explicit polygon surface and the surfaces from Study B are represented as a set of points sampled from the surface (Fig. 5.7). The metric, which represents the degree of mismatch between the two datasets, can be computed as the sum or average of the squared distances of closest approach from the points from Study B to the surfaces from Study A. It is written as

$$R = \sum \text{dist}(p_{B'}, S_A)^2 / N,$$

where $\text{dist}(p_{B'}, S_A)$ computes the (minimum) distance between point $p_{B'}$ and the surfaces S_A.

As with defining pairs of points, it may be inherently difficult or time-consuming to accurately delineate corresponding surfaces in both imaging studies. Furthermore, since the extracted geometric features are surrogates for the entire image volume, any anatomic or machine-based distortions in the image data away from these features are not taken into account during the registration process.

Intensity-Based Metrics

To overcome some of the limitations of using explicit geometric features to register image data, another class of registration metric has been developed which uses the numerical gray-scale information directly to measure how well two studies are registered. These metrics are also referred to as similarity measures since they determine how similar the distributions of corresponding voxel values from Study A and a transformed version of Study B are. Several mathematical formulations are used to measure this similarity. The more common similarity measures in clinical use include: sum of squared differences; cross correlation; and mutual information.

The sum of squared differences (SSD) metric is computed as the average squared intensity difference between Study A and Study B', i.e.,

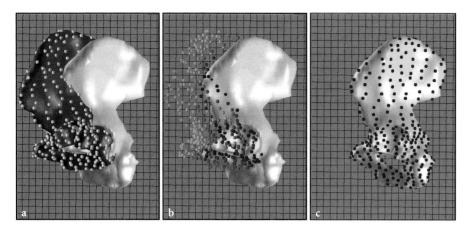

Fig. 5.7. a Extracted surface from Study *A* and extracted surface and surface points from Study *B*. **b** Points colorized based on computed distance of closest approach. **c** Study *B* points registered to Study *A* surface

$$SSD = \sum (I_A - I_{B'})^2 / N.$$

This metric is simple to compute and is effective for registering two imaging studies which have essentially identical intensities for corresponding anatomy, such as serial or 4D CT data.

When this condition is not met but there is still a linear relationship between the intensities of Study *A* and Study *B*, the cross correlation (CC) metric may be used. Rather than minimizing the intensity difference, cross correlation registration maximizes the intensity product:

$$CC = \sum (I_A * I_{B'}) / N.$$

A normalized version of the cross correlation metric exists and is called the correlation coefficient metric (KIM and FESSLER 2004).

For data from different modalities where the pixel intensities of corresponding anatomy are typically (and inherently) different, registration metrics based on simple differences or products of intensities are not effective. In these cases, sophisticated metrics based on intensity statistics are more appropriate. When using these metrics, there is no dependence on the absolute intensity values. One such metric that has proved very effective for registering image data from different modalities is called mutual information (MI). As the name implies, this metric is based on the information content of the two imaging studies and is computed directly from the intensity distributions of the studies. Since this metric is widely used in clinical image registration systems, it is described in detail here (see also WELLS et al. 1996).

According to information theory, the information content *H* of a "signal" is measured by the expectation (of the log) of the probability distribution function (PDF) of the signal values (ROMAN 1997). For image data, the signal values are the gray-scale intensities and the PDF is the normalized histogram of these intensities. The information content in the image data is

$$H(I_A) = - E [\log_2 p(I_A)] = -\sum p(I_A) \log_2 p(I_A),$$

where $p(I_A)$ is the probability distribution function of the intensities I_A of Study *A* (Fig. 5.8).

The joint or combined information content of two imaging studies has the same form and represents the information content of the two studies fused together. This is computed as

$$H(I_A, I_{B'}) = -\sum \sum p(I_A, I_{B'}) \log_2 p(I_A, I_{B'})$$

where $p(I_A, I_{B'})$ is the 2D joint probability distribution function of the intensities I_A of Study *A* and $I_{B'}$ of Study *B'* (Fig. 5.9). This PDF is constructed from the pairs of gray-scale values at each common point in Study *A* and Study *B'*.

The joint or total information content for the two imaging studies is always less than or equal to the sum of the individual information contents:

$$H(I_A, I_{B'}) \leq H(I_A) + H(I_{B'}).$$

If there is no redundant information in the pair of imaging studies (e.g., they are completely independent), the joint information of the pair is simply the sum of the information in Study *A* and Study *B'*:

$$H(I_A, I_{B'}) = H(I_A) + H(I_{B'}).$$

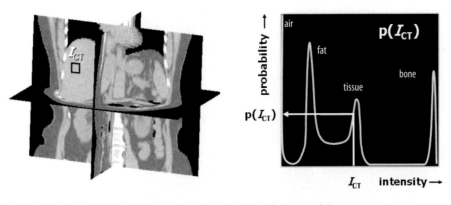

Fig. 5.8. *Left:* 3D image volume; *Right:* probability density function of the image intensities

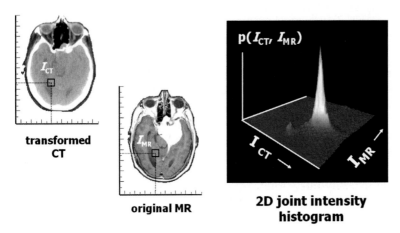

Fig. 5.9. Two-dimensional joint-intensity histogram constructed from an MR scan (Study *A*) and a transformed (reformatted) CT (Study *B'*)

If there is some redundant information, then the joint information content will be less than the sum of the information in the two studies:

$$H(I_A, I_{B'}) < H(I_A) + H(I_{B'}).$$

The amount of shared or mutual information is just the difference between the sum of the individual information contents and the joint information content,

$$MI(I_A, I_{B'}) = H(I_A) + H(I_{B'}) - H(I_A, I_{B'}).$$

Solving for MI from the above equations,

$$MI(I_A, I_{B'}) = \sum \sum p(I_A, I_{B'}) \log_2 [p(I_A, I_{B'}) / p(I_{A'}) p(I_{B'})].$$

The mutual information between two imaging studies can be thought of as the information in Study *B'* that is also present in Study *A*. Accordingly,

one way to describe mutual information is as the amount of information in Study *B'* that can be determined (or predicted) from Study *A*. To completely predict Study *B'* from Study *A*, each intensity value in Study *A* must correspond to exactly one intensity value in Study *B'*. When this is the case the joint intensity histogram has the same distribution as the histogram of Study *A*, and H($I_A, I_{B'}$) equals H(I_A). The MI is therefore equal to H($I_{B'}$), and Study *B'* at this point can be thought of as a "recolored" version of Study *A*.

A major advantage of mutual information is that it is robust to missing or incomplete information. For example, a tumor might show up clearly on an MR study but be indistinct on a corresponding CT study. Over the tumor volume the mutual information is low, but no prohibitive penalties are incurred. In the surrounding healthy tissue the mutual information can be high, and this becomes the dominant factor in the registration.

5.3
Data Fusion

The motivation for registering imaging studies is to be able to map information derived from one study to another or to directly combine or fuse the imaging data from the studies to create displays that contain relevant features from each modality. For example, a tumor volume may be more clearly visualized using a specific MR image sequence or coronal image plane rather than the axial treatment-planning CT. If the geometric transformation between the MR study and the treatment-planning CT study is known, the clinician is able to outline the tumor using images from the MR study and map these outlines to the images of the CT study. This process is called structure mapping.

Figure 5.10 illustrates the structure mapping process. The contours for a target volume are defined using the images from an MR imaging study which has been registered to the treatment-planning CT. Next, a surface representation of the target volume is constructed by tessellating or "tiling" the 2D outlines. Using the computed transformation, the vertices of the surface are mapped from the coordinate system of the MR study to the coordinate system of the CT study. Finally, the transformed surface is inserted along the image planes of the CT study. The result

is a set of outlines of the MR-defined structure that can be displayed over the CT images. These derived outlines can be used in the same manner as other outlines drawn directly on the CT images.

Another approach to combining information from different imaging studies is to directly map the image intensity data from one study to another so that at each voxel there are two (or more) intensity values rather than one. Various relevant displays are possible using this multi-study data. For example, functional information from a PET imaging study can be merged or fused with the anatomic information from an MR imaging study and displayed as a colorwash overlay. This type of image synthesis is referred to as image fusion.

The goal of this approach is to create a version of Study B (Study B') with images that match the size, location, and orientation of those in Study A. The voxel values for Study B' are determined by transforming the coordinates of each voxel in Study B using the appropriate transformation and interpolating between the surrounding voxels. The result is a set of images from the two studies with the same *effective* scan geometry (Fig. 5.11). These corresponding images can then be combined or fused in various ways to help elucidate the relationship between the data from the two studies (Fig. 5.12).

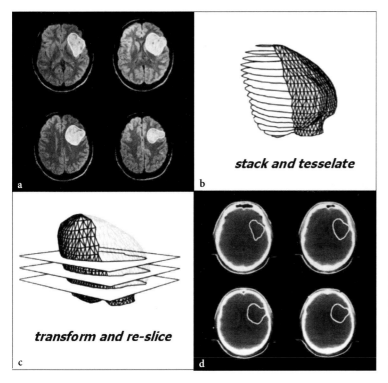

Fig. 5.10. Structure mapping. **a)** Tumor volume outlined on MR. **b)** Outlines stacked and tessellated to create a surface representation. **c)** The MR-based surface is mapped to the CT coordinate system and re-sliced along the image planes of the CT study. **d)** The derived contours are displayed over the CT images.

A variety of techniques exist to present fused data, including the use of overlays, pseudo-coloring, and modified gray scales. For example, the hard bone features of a CT imaging study can be combined with the soft tissue features of an MR imaging study by adding the bone extracted from the CT to the MR data set. Another method is to display anatomic planes in a side-by-side fashion (Fig. 5.12). Such a presentation allows structures to be defined using both images simultaneously.

In addition to mapping and fusing image intensities, 3D dose distributions computed in the coordinate system of one imaging study can be mapped to another. For example, doses computed using the treatment planning CT can be reformatted and displayed over an MR study acquired after the start of therapy. With this data, regions of radiologic abnormality post-treatment can be readily compared with the planned doses for the regions. With the introduction of volumetric imaging on the treatment units,

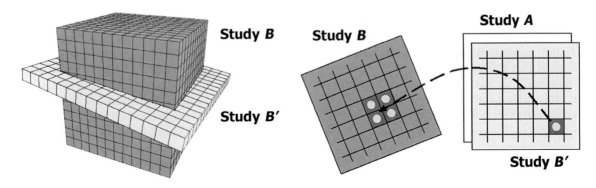

Fig. 5.11. Study *B* is reformatted to match the image planes of Study *A* to produce Study *B'*. Because the center of a pixel in one study will not usually map to the exact center of another, interpolation of surrounding pixel values is required.

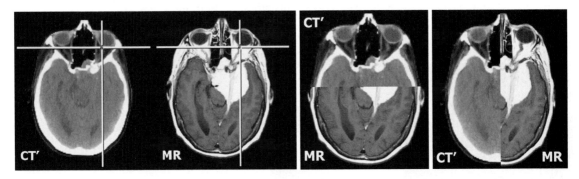

Fig. 5.12. Different approaches to display data which has been registered and reformatted

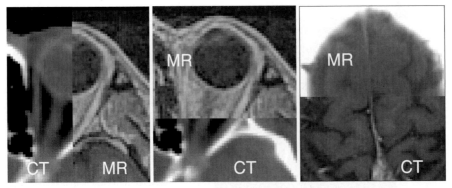

Fig. 5.13. Image–image visual validation using split-screen displays of native MR and reformatted CT study

treatment-delivery CT studies can now be acquired to more accurately determine the actual doses delivered. By acquiring these studies over the course of therapy and registering them to a common reference frame, doses for the representative treatments can be reformatted and accumulated to provide a more likely estimate of the delivered dose. This type of data can be used as input into the adaptive radiotherapy decision process.

5.4
Validation

It is important to validate the results of a registration before making clinical decisions using fused images or mapped structures. To do this, most image registration systems provide numerical and visual verification tools. A common numerical evaluation technique is to define a set of landmarks for corresponding anatomic points on Study *A* and Study *B* and compute the distance between the actual location of the points defined in Study *A* and the resulting transformed locations of the points from Study *B'*. This calculation is similar to the "point matching" metric, but as discussed previously it may be difficult to accurately and sufficiently define the appropriate corresponding points, especially when registering multimodality data. Also, if deformations are involved, the evaluation is not valid for regions away from the defined points.

Regardless of the output of any numerical technique used, which may only be a single number, it is important for the clinician to appreciate how well in three dimensions the information they define in one study is mapped to another. There are many visualization techniques possible to help qualitatively evaluate the results of a registration; most of these are based on the data-fusion techniques already described. For example, paging through the images of a split-screen display and moving the horizontal or vertical divider across regions where edges of structures from both studies are visible can help uncover even small areas of mis-registration. Another interesting visual technique involves switching back and forth between corresponding images from the different studies at about once per second and focusing on particular regions of the anatomy to observe how well they are aligned.

In addition to comparing how well the images from Study *A* and Study *B'* correspond at the periphery of anatomic tissues and organs, outlines from one study can be displayed over the images of the other. Figure 5.14 shows a brain surface which was automatically segmented from the treatment-planning CT study and mapped to the MR study. The agreement between the CT-based outlines at the different levels and planes of the MR study demonstrate the accuracy of the registration.

In practice, the accuracy of the registration process depends on a number of factors. For multimodality registration of PET/CT/MR data in the brain, registration accuracy on the order of a voxel size of the imaging studies can be achieved. Outside the head many factors confound single-voxel level accuracy, such as machine-induced geometric and intensity distortions as well as dramatic changes in anatomy and tissue loss or gain. Nevertheless, accuracy at the level of a few voxels is certainly possible in many situations.

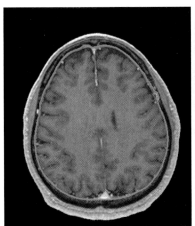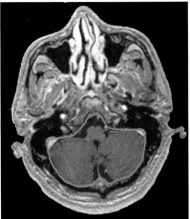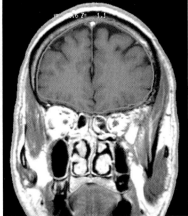

Fig. 5.14. Image-geometry visual validation structure overlay of CT-defined brain outlines (*green*) over MR images

5.5
Conclusion

Accurate delineation of tumor volumes and critical structures is a vital component of treatment planning. The use of a single imaging study to perform the delineation is not always adequate and multiple studies may need to be combined. In order to use data from multiple studies, the spatial alignment of the studies must be determined. Image registration, the process of finding a coordinate transformation between two studies, can recover rigid, affine, and deformed transformations. Automatic registration requires a similarity measure or registration metric. The metric may be specialized for particular types of registration (e.g., single modality) or may be generally applicable. Manual registration can use a metric as well, or it can be based on interactive visual inspection. Deformations are generally not manually registered. Once the coordinate transformation between the imaging studies has been found, various structure mapping and data fusion techniques can be used to integrate the data. Before the resulting data is used in the clinic, a validation process should take place. This might include numerical measurements such as comparisions of landmark positions, but should always include a visual inspection across the entire data set.

The techniques described in this chapter are tools to help use the information from different imaging studies in a common geometric framework. These techniques apply to both time-series single modality and multimodality imaging studies. Most modern radiotherapy treatment planning systems support the use of functional as well as multimodality anatomic imaging using one or more of the techniques presented. These tools, however, cannot replace clinical judgment. Different imaging modalities image the same tissues differently, and although tools may help us better understand and differentiate between tumor and non-tumor, they cannot yet make the ultimate decision of what to treat and what not to treat. These decisions still lie with the clinician, although they now have more sophisticated tools to help make these choices.

References

Allal AS, Slosman DO, Kebdani T, Allaoua M, Lehmann W, Dulguerov P (2004) Prediction of outcome in head-and-neck cancer patients using the standardized uptake value of 2-[18F]fluoro-2-deoxy-D-glucose. Int J Radiat Oncol Biol Phys 59:1295–1300

Barthel H, Cleij MC, Collingridge DR, Hutchinson OC, Osman S, He Q, Luthra SK, Brady F, Price PM, Aboagye EO (2003) 3'-deoxy-3'-[18F]fluorothymidine as a new marker for monitoring tumor response to antiproliferative therapy in vivo with positron emission tomography. Cancer Res 63:3791–3798

Bharatha A, Hirose M, Hata N, Warfield SK, Ferrant M, Zou KH, Suarez-Santana E, Ruiz-Alzola J, D'Amico A, Cormack RA, Kikinis R, Jolesz FA, Tempany CM (2001) Evaluation of three-dimensional finite element-based deformable registration of pre- and intraoperative prostate imaging. Med Phys 28:2551–2260

Bookstein FL (1989) Principal warps: thin-plate splines and the decomposition of deformations. IEEE Trans Pattern Anal Mach Intell Vol 11. 6:567–585

Brun E, Kjellen E, Tennvall J, Ohlsson T, Sandell A, Perfekt R, Perfekt R, Wennerberg J, Strand SE (2002) FDG PET studies during treatment: prediction of therapy outcome in head and neck squamous cell carcinoma. Head Neck 24:127–135

Chenevert TL, Stegman LD, Taylor JM, Robertson PL, Greenberg HS, Rehemtulla A, Ross BD (2000) Diffusion magnetic resonance imaging: an early surrogate marker of therapeutic efficacy in brain tumors. J Natl Cancer Inst 92:2029–2036

Kim J, Fessler JA (2004) Intensity-based image registration using robust correlation coefficients. IEEE Trans Med Imaging 23:1430–1444

Maintz JB, Viergever MA (1998) A survey of medical image registration. Med Image Anal 2:1–36

Mardor Y, Pfeffer R, Spiegelmann R, Roth Y, Maier SE, Nissim O, Berger R, Glicksman A, Baram J, Orenstein A, Cohen JS, Tichler T (2003) Early detection of response to radiation therapy in patients with brain malignancies using conventional and high b-value diffusion-weighted magnetic resonance imaging. J Clin Oncol 21:1094–1100

National Electrical Manufacturers Association (2004) DICOM, part 3, PS3.3 – service class specifications. Rosslyn, Virgina

Roman S (1997) Introduction to coding and information theory. Undergraduate texts in mathematics. Springer, Berlin Heidelberg New York

Thirion JP (1998) Image matching as a diffusion process: an analogy with Maxwell's demons. Med Image Anal 2:243–260

Unser M (1999) Splines: a perfect fit for signal and image processing. IEEE Sign Process Mag Vol 16. 6:22–38

Wells WM 3rd, Viola P, Atsumi H, Nakajima S, Kikinis R (1996) Multi-modal volume registration by maximization of mutual information. Med Image Anal Vol 1. 1:35–51

6 Data Formats, Networking, Archiving and Telemedicine

Karsten Eilertsen and Dag Rune Olsen

CONTENTS

dollar range impeded a widespread introduction of PACS into clinical practice. During the 1990s all aspects of the technology matured and the development of "filmless" digital networked enterprises with a PACS (Dreyer et al. 2002) solution as the key component took off: primarily within the field of radiology, but the technology soon gained foothold in other medical disciplines as well. As such, digital image networking is not merely a technological issue, but can contribute to improved health care as imaging modalities become readily available across traditional departmental barriers. This development has had, and will have in many years to come, a dramatic impact on the working practice of medical imaging.

In this chapter we address various aspects of digital image networking with a special focus on radiotherapy.

6.1 Introduction

Over the past two decades, a tremendous growth in digital image acquisition systems, display workstations, archiving systems and hospital/radiology information systems has taken place. The need for networked picture archiving and communications systems (PACS) is evident.

The conception of such a system dates back to the early 1970s (Lemke 1991); however, for many years the lack of basic technology to provide required network infrastructure (e.g. network bandwidth, data communication standards, workflow management) as well as initial capital outlays in the multimillion

K. Eilertsen, PhD
Department of Medical Physics, The Norwegian Radium Hospital, Montebello, 0310 Oslo, Norway
D. R. Olsen, PhD
Institute of Cancer Research, The Norwegian Radium Hospital, Montebello, 0310 Oslo, Norway

6.2 Data Formats

There are many different data types in use in the hospital environment ranging from comprehensive cine sequences, 3D image sets, voice recordings, to textual reports, prescriptions and procedures. They all play important roles in the field of electronic health where the electronic patient record is one of the cornerstones. Restricting the scope to that of imaging, four types of information are generally present in such data sets: image data (which may be unmodified or compressed); patient identification and demographics; and technical information about the imaging equipment in use as well as the exam, series and slice/image. The formats used for storing these images may depend on the needs of equipment-specific reviewing applications, e.g. to facilitate rapid reload of the images into dedicated viewing consoles. There are three basic families of formats in use: the fixed format (the layout is identical in each file); the block format (the header contains pointers to information); and the tag- or record-based format (each item contains its own length). Extracting image data from such files is usually easy, even if a proprietary formatting is used, but to decipher every detail may require detailed insight into the format specification. Examples of standard file formats in widespread use are Tagged Image File Format (TIFF), Graphic Interchange Format (GIF), JPEG file interchange format (JIFF), MPEG (movies), WAV and MIDI (voices). Recently, the Portable Network Graphic (PNG) has gained popularity, especially for Internet web applications. The stream of DICOM messages stored in a file may also be considered an image format and has become a common way of keeping medical images. The reading of such files may require detailed knowledge about the streaming syntax and the underlying communication protocols that were used (see below).

Extensive use of different image formats restricts the ability of cross-platform data sharing in a networked environment, as dedicated file readers and viewers are needed. Important image characteristics, such as resolution, gray and/or color scale interpretation, contrast and brightness, may deteriorate or even be lost due to inherent limitations of the format in use. The creation of sufficiently comprehensive image formats and the subsequent network transport from the modality to (any) application in a standardized fashion is therefore a major endeavor. In the next sections we explore one solution to this challenge.

6.3
Networking: Basic Concepts

The International Organization for Standardization has defined the Open System Interconnection (OSI) reference model to be used as an architectural framework for network communication. The OSI model describes how data in one application is transported trough a network medium to another application. The model concept consists of seven different layers, each layer specifying a particular network function (Table 6.1). The functions of the different layers are fairly self-contained, and the actual implementation of these functions (often called protocols) makes possible the communication or transport of data between the layers.

The design of a PACS network within a particular hospital environment would constitute a typical Local Area Network (LAN) where an Ethernet topology (or Fast Ethernet, Gigabit Ethernet) with TCP/IP is utilized to facilitate networked communica-

Table 6.1. The seven OSI model layers and an OSI model realization with Ethernet and TCP/IP

Layer	OSI model	Ethernet with TCP/IP
7	Application	Telnet, ftp, SMTP
6	Presentation	Data formats (e.g. JPEG, MPEG, ASCII)
5	Session	Session Control Protocol (SCP), DECNet
4	Transport	Transmission Control Protocol (TCP)
3	Network	Internet Protocol (IP)
2	Data link	Ethernet Network Interface Card (NIC)
1	Physical	Twisted pair CAT 5 cabling, FiberChannel

tion. Furthermore, several hospitals within the same health care organization can be connected into Wide-Area Networks (WAN).

Having defined the basic framework for network operation, the challenge is to exploit this topology to facilitate smooth connectivity between the multivendor modalities.

6.3.1
Network Connectivity: The DICOM Standard

An apparently seamless exchange of data between different computer applications (or modalities; Fig. 6.1) in the hospital network has traditionally been restricted to vendor-specific equipment that applies proprietary standards. In a modern multimodality hospital environment with a multitude of digital systems from many different manufacturers, proprietary solutions offer little flexibility. They are costly and cumbersome to operate as custom interfaces must be developed and maintained, possibly error prone and safety critical since data consistency can be jeopardized as data must be reformatted to suit a given equipment specification. An additional maintenance level may be required to assure data quality when, for instance, equipment is upgraded or replaced. The use of such solutions may therefore represent a serious obstacle to the progress and introduction of state-of-the art network technology.

During the 1980s the need for simplification and standardization became apparent in order to ensure and maintain vital connectivity and interoperability of all pieces of equipment. The medical equipment industry, represented by the National Electrical Manufacturers Association (NEMA) and the medical community, represented by the American College of Radiology (ACR), joined forces to develop the *Digital Imaging and Communications in Medicine* standard (DICOM). The intention was to create an

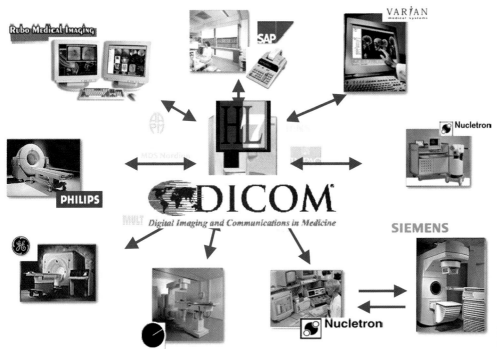

Fig. 6.1. A typical oncology LAN

industry standard to which all vendors of medical equipment could conform. This would establish a win–win situation for all involved parties: if DICOM support is built into a medical imaging device, it can be directly connected to another DICOM-compatible device, eliminating the need for a custom interface – DICOM defines the interface. Even though the first versions of the standard, ACR-NEMA 1.0 (1985) and ACR-NEMA 2.0 (1989), never became very popular among vendors, the later DICOM v3.0 (DICOM) is by present-day standards ubiquitous. One important reason for this development is that the concepts of the OSI model (e.g. the use of standard network protocols and topology) were utilized when drafting the basic network functionality of DICOM. On one hand, this makes possible the use commercial off-the-shelf hardware and software, and on the other, this ensures that DICOM remains an open standard that encourages both users and vendors to get involved in its development.

DICOM was first developed to address connectivity and inter-operability problems in radiology, but presently there are parts of the DICOM standard which define service classes for many other modalities. During the RSNA conference in 1994, a meeting was held at which a clear need was expressed for standardization of the way radiotherapy data (such as external beam and brachytherapy treatment plans, doses and images) are transferred from one piece of equipment to another. The importance of such a standard was clear. As a result of the RSNA meeting, an ad-hoc Working Group, later to become Working Group 7 (Radiotherapy Objects) was formed under the auspices of NEMA. Participating members of this group include many manufacturers of radiotherapy equipment, some academics and also members involved with the IEC. The DICOM v3.0 standard is large and consists of 16 different parts, each part addressing a particular functional side of DICOM. The standard defines fundamental network interactions such as:

- **Network Image Transfer:** Provides the capability for two devices to communicate by sending objects, querying remote devices and retrieving these objects. Network transfer is currently the most common connectivity feature supported by DICOM products.
- **Open Media Interchange:** Provides the capability to manually exchange objects and related information (such as reports or filming information). DICOM standardizes a common file format, a medical directory and a physical media. Examples include the exchange of images for a publication and mailing a patient imaging study for remote consultation.
- **Integration within the Health Care Environment:** Hospital workflow and integration with other hospital information systems have been addressed

with the addition services such as Modality Worklist, Modality Performed Procedure Step, and Structured Reporting. This allows for scheduling of an acquisition and notification of completion.

To facilitate the desired network functionality, DICOM defines a number of network services (Service Classes). These services are described in brief below (Table 6.2).

Table 6.2 DICOM service classes and their functional description

DICOM service class	Task
Storage	Object transfer/archiving
Media storage	Object storage on media
Query/retrieve	Object search and retrieval
Print management	Print service
Patient, study, results management	Create, modify
Worklist management	Worklist/RIS connection
Verification	Test of DICOM connection

6.3.2
The DICOM Radiotherapy Model

In 1997 four radiotherapy-specific DICOM objects and their data model were ratified. In 1999 three additional objects were added to the DICOM standard, along with CD-R support for the storage of all radiotherapy objects (Fig. 6.2). The seven DICOM RT objects are as follows:
- **RT Structure Set,** containing information related to patient anatomy, for example structures, mark-

ers and isocenters. These entities are typically identified on devices such as CT scanners, physical or virtual simulation workstations or treatment planning systems.
- **RT Plan,** containing geometric and dosimetric data specifying a course of external beam and/or brachytherapy treatment, e.g. beam angles, collimator openings, beam modifiers, and brachytherapy channel and source specifications. The RT Plan entity may be created by a simulation workstation and subsequently enriched by a treatment-planning system before being transferred to a record-and-verify system or treatment device. An instance of the RT Plan object usually references an RT Structure Set instance to define a coordinate system and set of patient structures.
- **RT Image,** specifying radiotherapy images that have been obtained on a conical imaging geometry, such as those found on conventional simulators and (electronic) portal imaging devices. It can also be used for calculated images using the same geometry, such as digitally reconstructed radiographs (DRRs).
- **RT Dose,** containing dose data generated by a treatment-planning system in one or more of several formats: three-dimensional dose data; isodose curves; DVHs; or dose points.
- **RT Beams Treatment Record, RT Brachy Treatment Record** and **RT Treatment Summary Record,** containing data obtained from actual radiotherapy treatments. These objects are the historical record of treatment and are linked with the other "planning" objects to form a complete picture of the treatment.

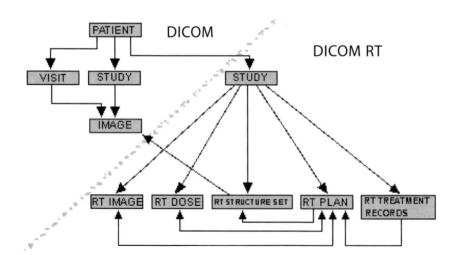

Fig. 6.2. The DICOM RT data model

Working Group 7 is constantly involved in the maintenance of the existing radiotherapy objects and is examining potential uses of newer DICOM extensions in the radiotherapy context. Presently, the DICOM RT information objects provide a means of standardized transfer of most of the information that circulates in the radiotherapy department; however, there are at present few manufacturers of radiotherapy equipment that fully support the DICOM RT standard. In particular, the full exploitation of the DICOM RT data model has only limited support.

6.3.3
A Radiotherapy Example

The equipment scenario shown in Fig. 6.3 is used to illustrate how DICOM objects are produced and furthermore utilized during patient treatment. A sequence of possible steps is listed below along with their associated specified DICOM objects:

1. The patient is scanned on a CT scanner, producing a DICOM CT image study. Other DICOM imaging modalities, such as MR, could also be involved.
2. A virtual simulation application queries the scanner using DICOM, retrieves the images and performs a virtual simulation. An RT Structure Set object is produced, containing identified structures such as the tumor and critical organs. An associated RT Plan is also created, containing beam-geometry information. Digitally reconstructed radiographs (DRRs) may also be created as RT Image objects.

3. A treatment-planning system then reads the CT images, RT Structure Set and RT Plan. It adds beam modifiers, modifies the beam geometries where necessary, and also calculates dosimetric data for the plan. A new RT Plan object is created, and RT Image DRRs may also be produced.
4. A record-and-verify system then obtains the completed RT Plan object and uses the data contained within it to initialize a treatment. Alternatively, the treatment machine itself could make use of the object directly. An EPID can create RT image verification images and compare acquired images with DRRs created by the above steps.
5. Periodically during the course of treatment, the treatment machine or record and verify system creates Treatment Record objects, generally one for each treatment session.
6. At the end of the treatment, the entire DICOM set of DICOM objects is pushed to a dedicated DICOM Archive.

The above sequence illustrates just one scenario. In reality there is a wide variety of different utilizations possible, and the DICOM RT objects have been designed with this flexibility in mind.

6.3.4
The DICOM Conformance Statement

The standard specifies that the manufacturer of any device claiming DICOM conformance shall provide a *DICOM Conformance Statement* that describes the

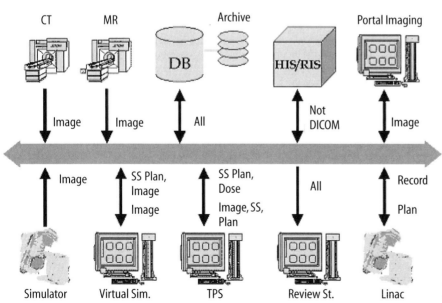

Fig. 6.3. Scenario displays the different DICOM modalities that may be involved in patient treatment

DICOM capabilities of the device (cf. part 2 of the DICOM standard). Many manufacturers make their conformance statements available on the Internet. Potential connectivity between two pieces of equipment can therefore be evaluated in advance by reading Conformance statements which provide a foundation to determine connectivity and assess the potential inter-operability of two products, and in some cases identify potential problems without ever having physically connected them.

It is not sufficent for a vendor to simply claim conformance to DICOM. The statement "This product is DICOM" has even less meaning in the radiotherapy domain, in which inter-operability is a very complex issue. For RT applications, it is usually not possible to determine inter-operability a priory – this must be established through extensive testing. Still, radiotherapy professionals should insist upon a conformance statement for any device that claims to be DICOM conformant. Even so, DICOM conformance is voluntary and there is no authority that approves or may enforce conformance; however, by conforming to DICOM, one can develop safe, reliable computer applications with a high degree of built-in connectivity.

6.3.5
DICOM Problems

Even a standard such as DICOM does not completely eliminate connectivity issues, and the inherent flexibility of the standard is a common source of confusion and frustration. The DICOM standard, and the RT parts in particular, contains numerous so-called type-2 and type-3 attributes. Type-2 attributes must be present in a DICOM message for the message to be valid, but the attribute value may be sent empty if unknown, i.e. it is left with the application vendor to fill in the value. Type-3 attributes are optional, i.e. they may or may not be present in a message. Some of these attributes can be crucial for the functionality of other applications. One example would be the tabletop positions (type 3) in the RT plan. These attributes are very useful to ensure a correct setup of the patient during external-beam radiotherapy. The attributes should be provided by the planning system for use by the record and verify system when setting up the patient at treatment. Presently, there are very few planning systems that provide these values. Another common problem is the different ways vendors organize 3D image sets (CT, MR, etc.) into series. Some applications, such as treatment-planning

systems and virtual simulators, rely on all images of a given type (e.g. all axial images) to exist in the same series in order to create a 3D reconstructed volumetric data representation. Especially older CT/MR scanners tend to split such data sets into several series or even put localizer and axial images in the same series.

6.3.6
How Are the Manufacturers Doing Today?

DICOM is now a mature standard. After a lot of hard work understanding, developing and testing product inter-operability in the radiotherapy context, a large number of manufacturers now have products available that support one or more of the radiotherapy DICOM objects. Vendors who have such products available, or have demonstrated them as works in progress, include Elekta, General Electric Medical Systems, IMPAC, Merge Technologies, Multidata, NOMOS, Nucletron, Picker International, ROCS, Siemens Medical Systems, SSGI, CMS and Varian Medical Systems.

6.4
Archiving

Historically the term "archive" refers to an institution or facility that undertakes the task of preserving records for longer periods of time, sometimes indefinitely. The core tasks of such an archive are typically to provide means for access control, ensure long-term media stability and readability, and to preserve record authenticity, in addition to disaster safe storage. The records in such archives have traditionally been data on analogue media such as paper and photographic film that can be visually inspected. This description also applies to medical archives that typically contain patient records and X-ray films.

The growth of information that exists in digital format within the medical environment poses a tremendous challenge to the traditional way of archiving. The main concern is probably the media on which the data is being stored. Media storage technology is evolving rapidly and there is an inherent risk that such media is outdated in a matter of few years (10–20 years) mainly because the hardware and software components required to access the storage media are no longer manufactured and supported.

A fundamental prerequisite for digital archives is therefore that they easily can accommodate new storage technology as well as to easily scale with growing demands in capacity. In other words, the archive should have built-in technology that, for instance, facilitates automatic data migration from one media/storage technology to another. The cost of archiving digital information should thus include not only the running costs of technical maintenance and support, but also the cost of keeping up with the continuous changes in technology.

The role of the archives is also changing. Contrary to the historical "archive", an important feature of a digital archive, is potential ease and speed of access. In a PACS implementation the archive serves as a common location for receiving all images as well as the source for distribution of images. To achieve this, the notions of on-line cache or storage (meaning fast access, low capacity), and long-term or secondary storage (meaning slow access but huge capacity), are often used. The present trend is that that these two concepts tend to merge as the storage media become cheaper.

Another important aspect in the digital archive is reliability and robustness. A common measure of reliability is a system's "uptime", and it is generally accepted that this should be greater than 99% per year, i.e. less than 3.6 days/year downtime. Still, the archive is in essence the hub in the PACS network and, if unavailable, production is jeopardized. It is therefore important to invest in hardware and software with built-in redundancy that can mitigate the effect of technology failures. In addition, special procedures and plans should be developed to handle catastrophic events such as fires, earthquakes, etc., for instance, by establishing remote vaults for off-site data storage.

6.4.1
Media Storage Technology

Storage media has for years enjoyed continuous progress in increased capacity and reduction in prices. This applies to all of the most popular storage media. It makes no sense to quote absolute figures as these will soon become dated; however, a crude estimate on a relative scale would roughly be (2004) 1000, 10, 5 and 1, for solid-state memory, magnetic disk, magnetic tape storage (DLT) and optical storage (DVD), respectively, considering the media price only.

The choice of storage technology is usually a trade-off between storage price/capacity and access time. The faster access times required, the more expensive the storage. For archiving purposes there has been a tradition to use magnetic disk for the on-line storage and optical/magnetic tape for secondary storage; however, advances in both storage and LAN technology along with lower prices has demonstrated a trend towards extensive use of large-scale magnetic-disk arrays for both on-line as well as long-term storage. The concepts of Network Attached Storage (NAS) and Storage Area Networks (SAN) have furthermore promoted this trend.

6.4.2
Data Formats for Archiving

When using DICOM the problem of different or proprietary data formats is reduced to having access to a DICOM object/message viewer. If one cannot use the application that generated the object initially to review the object, there exist several applications – either as freeware or for a small fee – that can be downloaded from the internet to accomplish this task. Some of these viewer applications include built-in DICOM storage providers that can be used to receive the DICOM message. In the field of radiotherapy proper viewers are still scarce probably due to the inherent complexity of the RT DICOM objects and the amount on non-image data.

The demands on the data format used for storage in an archive are different. The amount of data to be stored is potentially enormous and many archives apply compression to allow for more data to be stored. From a user's point of view the archive can be considered a black box that talks DICOM. As long as the archive gives back what once was stored, the user is satisfied and the internal storage format of the archive is as such irrelevant; however, if compression is used, the quality of the returned image may be reduced with respect to the original and this may not be acceptable. The DICOM committee has deemed lossless JPEG and lossy JPEG acceptable techniques for compressing medical images, and several archive vendors have implemented strategies for the use of these techniques. A common option is to make the compression technique in use dependent on the age of the object, i.e. newer objects are only made subject to lossless JPEG (two to three times reduction of most images), whereas older objects are compressed using lossy JPEG (10- to 100-fold reduction in image size). This implies that the archive continuously migrates data from lossless to lossy JPEG. Another strategy is to use different compression techniques for different image types. A noisy 512×512 radiotherapy portal image may suffer considerable loss in quality if lossy JPEG at a ratio of 10:1 is used, whereas a 4000×8000 MB chest X-ray may be compressed satisfactorily at a ratio of 60:1. Only careful testing can reveal what is the appropriate compression level.

Evidently, it is important that the user defines this strategy in accordance with health care regulations.

6.5
Workflow Management: From Connectivity to Integration

The computer network of the hospital should provide the technical infrastructure required for rapid transmission and exchange of data between the different modalities used for diagnosis, planning and treatment. The use of DICOM will eventually provide excellent connectivity between the interacting parties. But DICOM has as of yet limited support for workflow management and data integration. Custom-made solutions may be required, and DICOM was not intended to solve this task.

In order to understand what DICOM can and cannot provide, it is important to distinguish between DICOM *connectivity* and *application interoperability*. This is especially true in the domain of radiotherapy where the working process is very dynamic. DICOM connectivity refers to the DICOM message exchange standard responsible for establishing connections and exchanging properly structured messages so that an information object sent from one node will be completely received by the receiving node. In other words, the successful transfer of information: the successful "plug and exchange" between two pieces of equipment.

Beyond connectivity lies application interoperability: the ability to process and manipulate information objects. DICOM radiotherapy objects play a crucial role in enabling such interoperability, but sometimes "plug and play" at this level requires more than the standardized definition and coding of information provided by DICOM. Specification and testing of the clinical application capabilities and data flow needs to be performed by the health care facility to ensure effective integration of the various DICOM applications. For example, transfer of IMRT (intensity-modulated) data from an IMRT-capable treatment-planning system requires a record-and-verify or treatment system capable of managing such dynamic treatments. DICOM requires implementers to explicitly specify these application-specific informa-

tion needs in a DICOM Conformance Statement that will provide the basis for achieving such application interoperability.

The service classes "Modality Worklist Management (MWL)", "Modality Performed Procedure Step (MPPS)" and "Storage Commitment (SC)" were all defined to facilitate the communication between information systems (RIS/HIS), PACS and the modalities. In principle, these services are designed to work independently but may also be set up to work together. The MWL enables scheduling information to be conveyed at the modalities and supplies the DICOM objects with HIS/RIS data such as patient demographics; the latter is very useful in avoiding typing errors at the modalities. The MPPS is used to update a schedule when a scheduled procedure step commences, as well as notifying the PACS when a scheduled procedure has been completed. In addition, details describing the performed procedure can be included such as a list of images acquired, accession numbers, radiation dose, etc. The SC facilitates automated or simplified deletion of the images on the modalities as the PACS confirms their safe storage.

It is customary to implement what is called a "broker" to deal with the intricate communication between the HIS/RIS, PACS and the modalities, i.e. to fully exploit the possibilities provided by MWL, MPPS and SC. The broker is often a proprietary third-party software that has been designed to provide a dedicated solution to a user-specified workflow.

6.6
Telemedicine Applications in Radiation Therapy

The advances in modern radiotherapy throughout the past decade have, to a large extent, relied on technological development, in general, and in imaging and computer technology in particular. The role of CT images in treatment planning of radiation therapy is evident, and imaging modalities, such as MR and PET, represent functional or physiological, and biological or even molecular, information that will become essential in modern radiotherapy treatment planning. Moreover, imaging tools have been developed and implemented in treatment verification and for adaptive treatment strategies. Digital representation of the image information, development of image information standards, such as DICOM, establishment of networks and protocols and improved connectivity between modalities, are all crucial elements for the utilization of image information within a clinical radiotherapy environment; however, the development of information standards, protocols and network connectivity do not merely allow digital information flow within a single department, it also advocates communication between and among different institutions, and the formation of inter-institutional networks has thus become feasible.

Telemedicine has been widely adopted and provides increased access to medical expertise in a variety of medical applications. The most common areas for telemedicine have traditionally been radiology, pathology and dermatology – all examples of medical specialties where medical images play a key role. Other areas of medical practice, where digital medical information is available, e.g. emergency medicine and cardiology, have also adopted the concept of telemedicine. Telemedicine is most commonly defined as long-distance communication between medical centers. Telemedicine is, however, evolving conceptually, and The Mayo Clinic has introduced the term "telehealthcare" to include all aspects of communication between medical centers for patient care and limits the term "telemedicine" to communication between centers for purposes of individual patient care.

Although both radiology and radiation therapy are medical disciplines that utilize medical images to a large extent, telemedicine has until recently rarely been adopted in radiotherapy as compared with radiology; thus, the experience is limited and only a few reports worldwide document clinical applications (HASHIMOTO et al. 2001a, b; SMITH et al. 1998; EICH et al. 2004); however, it is expected that applications of telemedicine in radiation therapy will become equally important in improving the quality and standardization of radiotherapy procedures. Telemedicine may especially play a key role in distributed radiotherapy services, in rural areas and possibly in developing countries, but also in the provision of high-end radiation therapy, such as proton treatment, and in treatment of rare cancers; thus, telemedicine will be an appropriate tool in maintaining high-quality, decentralized radiotherapy services, and in preventing professional isolation (REITH et al. 2003). Moreover, telemedicine may facilitate collaboration between highly specialized centers of excellence with respect to rare conditions. The role of telemedicine in radiation therapy is to provide a tool for apparent seamless dialogue between clinical experts in the following:

- Treatment planning and simulation of individual patients
- Treatment verification of individual patients
- Follow-up and clinical trial management

Treatment planning and simulation of individual patients is perhaps the most evident role of telemedicine in radiation therapy. There are at least two steps in this process where remote consultation may be of clinical importance: (a) delineation of the target volume based on 3D medical imaging, e.g. CT, MR and PET; and (b) the discussion of beam setup and evaluation of the plan options. Delineation of the target volume is perhaps the most critical part of the treatment planning and with respect to clinical outcome, moreover, an inter- as well as intra-observer variability in target-volume delineation has been well documented. Lastly, imaging modality, settings and parameters are known to influence target-volume delineation. In rare cancers or at smaller, satellite radiotherapy clinics the required expertise may not be available for the optimal use of medical imaging in radiation therapy planning. Telemedicine in such situations may be the appropriate tool for consulting remote expertise. The ideal scenario is a real-time, on-line telemedicine service, where the target volume can be jointly delineated by the two physicians. Either a complete set of data must to be available at both centers or on-line transferred, e.g. as video signal, for simultaneous display. In addition, the drawing device must be operable from both centers. A few systems have been developed dedicated to remote treatment planning and virtual simulation (NTASIS et al. 2003; HUH et al. 2000; STITT et al. 1998; EICH et al. 2004). A less attractive, but still useful, alternative is transfer of data from one center to the other for target delineation by an expert team. This mode of operation is less technology demanding, and requires merely that exported data sets from the one institution be successfully imported by the other; however, the dialogue between the professionals is not facilitated by this procedure. Identical modes of operations are relevant with respect to beam setup and dose computation. Final plan evaluation, or selection of the preferred plan, often take place within a larger group of professionals including not only the oncologist but also the medical physicist, the dosimetrist/RTT and sometimes the radiologist and surgeon, in addition to the oncologist. If more centers are involved, telemedicine may be used to include all the professionals at the different institutions in a clinical discussion of the final treatment plan (Fig. 6.4).

Virtual simulation has become more common and frequently used in a number of radiotherapy clinics, and to a certain degree has replaced conventional simulation; however, conventional simulation is still

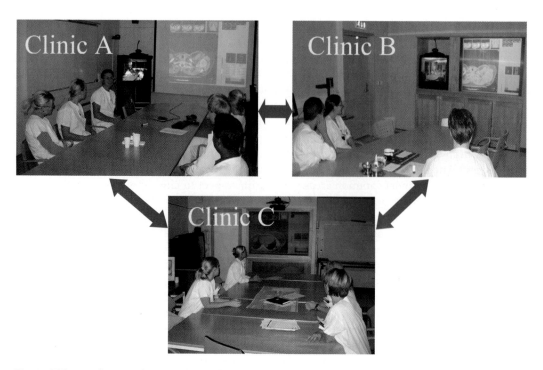

Fig. 6.4. Video-conferencing between three different radiotherapy institutions, demonstrating weekly clinical conference where a treatment plans are discussed, and electronic portal images, DRR and simulator images are reviewed

widely employed for simpler treatments where full 3D treatment planning and dose computation may be superfluous. At Hokkaido University School of Medicine, the THERAPIST system, has been applied for a number of years with success for remote simulation of emergency radiotherapy of spinal cord compression (HASHIMOTO et al. 2001a).

Treatment verification of individual patient involves most often acquisition of electronic portal images (EPI) and comparison with either digital simulator images of DRR. This task may very well be conducted by individuals on dedicated workstations, but compliance between intended and actual treatment is also often discussed during clinical conferences, again involving a larger group of the staff, and may represent an important arena for quality management of individual patient treatment within the framework of decentralized radiotherapy services, involving more regional or satellite units. Teleconferencing facilitates such an activity and satisfies the education aspects of quality management (Fig. 6.5). Hashimoto and co-workers have shown that remote consulting involving both DRRs and EPIs is both feasible and of value to clinical practice (HASHIMOTO et al. 2001b).

Follow-up and clinical trial management by telemedicine applications is a further development of the telemedicine concept in clinical radiation oncology. Telemedicine will allow multicenter participation in clinical trials that require strict adherence to protocols of complex treatment planning and verification. An example of this is the initiation of the German teleradiotherapeutic network for lymphoma trial (EICH et al. 2004) and the US dose escalation trial for early-stage prostate cancer (PURDY et al. 1996). All treatment plans, including dose-volume statistics, and treatment verifications data, both for dummy runs and actual patient treatment, are submitted to study coordination centers, such as RTOG, for protocol compliance verification. Also, efficient and consistent data collection pave the way for elaborate analysis on larger patient population materials than are commonly available; however, participation and data collection from a multitude of centers are most demanding with respect to data formats and network connectivity.

Classification of Telemedicine Functionality in Radiation Therapy

Level-1 (Table 6.3) telemedicine in radiotherapy has been defined as teleconferencing and the display of radiotherapy related information, which facilitates discussions of target volumes and organs-at-risk de-

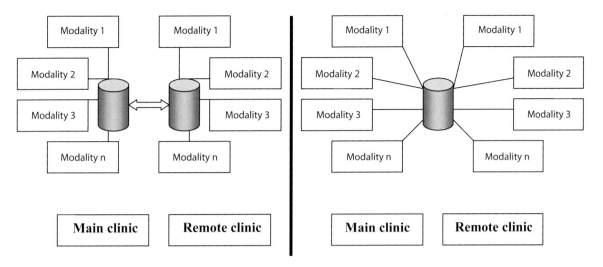

Fig. 6.5. Two different networking and data-storage strategies applicable to level-2 telemedicine functionality, allowing inter-institutional exchange of data

lineation, treatment techniques and beam arrangement, dose distributions and image-based treatment verification (OLSEN et al. 2000). Remote conventional simulation of single portals is another example of level-1 functionality. Remote, on-line operations are not supported. Level-1 functionality may be based on ISDN communication and video-signal technology, and is thus a low-cost service. The disadvantages are mostly related to its functional limitations.

Level-2 (Table 6.3) telemedicine features data transfer between institutions, and limited remote image handling. Remote treatment planning, non-real time, is an example of a level-2 operation that requires transfer of data between the participating institutions. Different networking and data storage strategies may be implemented. At some institutions all data are stored in a central DICOM database that communicates with all the modalities, including those at the remote clinic. Others have chosen to establish DICOM databases at each clinic, which are replicated at certain intervals. Irrespectively of networking and storage strategy data transfer, compliant with level-2 operations, higher-speed communication than that

provided by ISDN is often required at this level of operation. Finally, it is pointed out that level-2 applications may raise medico-legal issues with respect to responsibility for treatment planning of the patient.

Level-3 (Table 6.3) telemedicine featuring remote, real-time operations and joint delineation of target volumes is an example of a level-3 functionality. The direct interaction and discussion that is feasible at this level may be of particular importance when a radiologist's review is required for target-volume delineation, or when a discussion is desirable for educational purposes. Level-3 application faces the same disadvantages as level-2 services with respect to costs and medico-legal issues.

References

Dreyer KJ, Metha A, Thrall JH (2002) PACS: a guide to the digital revolution. Springer, Berlin Heidelberg New York

DICOM (1998) Digital imaging and communications in medicine, version 3.0. NEMA, Rosslyn, Virginia 2000, URL: http://medical.nema.org

Eich HT, Muller RP, Schneeweiss A, Hansemann K, Semrau R, Willich N, Rube C, Sehlen S, Hinkelbein M, Diehl V (2004) Initiation of a teleradiotherapeutic network for patients in German lymphoma studies. Int J Radiat Oncol Biol Phys 58:805–808

Hashimoto S, Shirato H, Kaneko K, Ooshio W, Nishioka T, Miyasaka K (2001a) Clinical efficacy of telemedicine in emergency radiotherapy for malignant spinal cord compression. J Digit Imaging 14:124–130

Table 6.3. Functions featured by a telemedicine system in radiation oncology. (Adapted from OLSEN et al. 2000)

	Tele conference	Image display	Data exchange	Real-time operations
Level 1	+	+	–	–
Level 2	+	+	+	–
Level 3	+	+	+	+

Hashimoto S, Shirato H, Nishioka T, Kagei K, Shimizu S, Fujita K, Ogasawara H, Watanabe Y, Miyasaka K (2001b) Remote verification in radiotherapy using digitally reconstructed radiography (DRR) and portal images: a pilot study. Int J Radiat Oncol Biol Phys 50:579–585

Huh SJ, Shirato H, Hashimoto S, Shimizu S, Kim DY, Ahn YC, Choi D, Miyasaka K, Mizuno J (2000) An integrated service digital network (ISDN)-based international telecommunication between Samsung Medical Center and Hokkaido University using telecommunication helped radiotherapy planning and information system (THERAPIS). Radiother Oncol 56:121–123

Lemke HU (1991) The Berlin Communication System (BERKOM). In: Huang HK (ed) Picture Archiving and Communication Systems (PACS). Medicine, series F. Springer, Berlin Heidelberg New York (Computer and Systems Science, vol 74)

Ntasis E, Maniatis TA, Nikita KS (2003) Secure environment for real-time tele-collaboration on virtual simulation of radiation treatment planning. Technol Health Care 11:41–52

Olsen DR, Bruland S, Davis BJ (2000) Telemedicine in radiotherapy treatment planning: requirements and applications. Radiother Oncol 54:255–259

Purdy JA, Harms WB, Michalski J, Cox JD (1996) Multi-institutional clinical trials: 3-D conformal radiotherapy quality assurance. Guidelines in an NCI/RTOG study evaluating dose escalation in prostate cancer radiotherapy. Front Radiat Ther Oncol 29:255–263

Reith A, Olsen DR, Bruland O, Bernder A, Risberg B (2003) Information technology in action: the example of Norway. Stud Health Technol Inform 96:186–189

Smith CL, Chu WK, Enke C (1998) A review of digital image networking technologies for radiation oncology treatment planning. Med Dosim 23:271–277

Stitt JA (1998) A system of tele-oncology at the University of Wisconsin Hospital and Clinics and regional oncology affiliate institutions. World Med J 97:38–42

3D Imaging for Radiotherapy

7 Clinical X-Ray Computed Tomography

Marc Kachelriess

CONTENTS

7.1 Introduction

Since its introduction in 1972 by Godfrey N. Hounsfield, the importance of CT for the medical community has increased dramatically. Major technical improvements have taken place in the meantime (Fig. 7.1). Now whole-body scans with isotropic submillimeter resolution are acquired routinely during a single breath-hold.

The first step toward true 3D data acquisition was the introduction of spiral CT in 1989 by W.A. Kalender. This scan mode is based on a continuous rotation of the gantry while simultaneously translating the patient along the axis of rotation. The resulting scan trajectory is a spiral and, by symmetry, means truly 3D data acquisition. The z-interpolation step allows selection of the longitudinal position (z-position) of the reconstructed images arbitrarily and retrospectively. The continuous axial sampling is required for high-quality 3D displays and has led to a renaissance of CT (KALENDER 2001). Multislice spiral CT (MSCT), which allows simultaneous scanning of

M slices, further improved the scanner's volume coverage, z-resolution, and scan speed. For example, typical chest exams are carried out with collimations of 1×5 mm in 36 s with single-slice, 4×1 mm in 30 s with 4-slice, and 16×0.75 mm in 10 s with 16-slice scanners, and in the near future 64×0.6-mm scan modes will be used (Fig. 7.2).

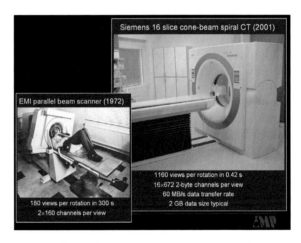

Fig. 7.1. Subsecond true 3D cone-beam scanner with submillimeter resolution

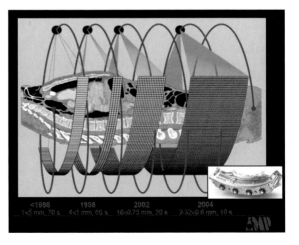

Fig. 7.2. Generations of fan-beam CT scanners: from single-slice to multislice to true cone-beam CT with up to 64 slices

M. KACHELRIESS, PhD
Institute of Medical Physics, Henkestrasse 91, 91052 Erlangen, Germany

Recently, patient dose has become an increasingly important issue: increasing the information density of a scan, e.g., the spatial resolution, requires increase of the photon density (absorbed photons per volume element) by at least the same factor if the signal-to-noise ratio (SNR) is to be maintained. Currently, less than 5% of all X-ray exams are performed with CT. The cumulative dose of CT, however, is of the order of 40% of all X-ray exams. Dose is especially critical in pediatric CT (BRENNER et al. 2001). Practical hardware- and software-based solutions to reduce patient dose are strongly required.

The CT developers have made tremendous efforts in developing new techniques to achieve the high image quality as we now know it. Especially the improvement of the three key components – X-ray production, X-ray detection, and image reconstruction – has been most challenging.

This chapter gives an overview of clinical CT. Basic reconstruction and image display principles are reviewed. The principle of spiral CT is introduced and single-slice, multislice, and true cone-beam CT scanner generations are addressed. Furthermore, this chapter describes why doses went up and how dose values can be reduced. A section about tube and detector technology, and a section about cardiac CT, conclude this chapter.

7.2
Basics of X-Ray CT

Clinical X-ray CT is the measurement of an object's X-ray absorption along straight lines. As long as a desired object point is probed by X-rays under an angular interval of 180°, image reconstruction of that point is possible; therefore, clinical CT scanners have an X-ray focal spot that rotates continuously around the patient. On the opposing side of the X-ray tube a cylindrical detector, which consists of about 10^3 channels per slice, is mounted (Figs. 7.2, 7.3). The number of slices that are simultaneously acquired is denoted as M. In the longitudinal direction (perpendicular to the plane of rotation) the size of the detectors determines the thickness S of the slices that are acquired. During a full rotation 10^3 readouts of the detector are performed. Altogether about 10^6 intensity measurements are taken per slice and rotation. The negative logarithm p of each intensity measurement I corresponds to the line integral along line L of the object's linear attenuation coefficient distribution $\mu(x, y, z)$:

$$p(L) = -\ln\frac{I(L)}{I_0} = \int_L dL\, \mu(x, y, z)$$

Here, I_0 is the primary X-ray intensity and is needed for proper normalization.

The CT image $f(x, y, z)$ is intended to be a close approximation to the true distribution $\mu(x, y, z)$. The process of computing the CT image from the set of measured projection values $p(L)$ is called image reconstruction and is one of the key components of a CT scanner. For single-slice CT scanners ($M=1$) that perform measurements within one slice at a time only (at a fixed z-position) image reconstruction is simple. It consists of a convolution of the projection data with the reconstruction kernel followed by a backprojection into image domain:

$$f(x, y) = \int_0^\pi d\vartheta\, p(\vartheta, \xi) * k(\xi)\Big|_{\xi = x\cos\vartheta + y\sin\vartheta}.$$

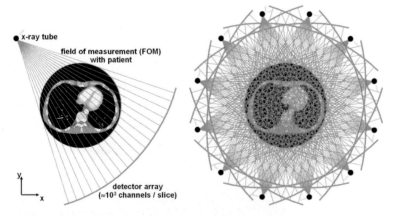

Fig. 7.3. Clinical CT is the measurement of X-ray photon attenuation along straight lines. The in-plane scan geometry is the fan-beam geometry with one point-like source and many detector elements. During a rotation of the gantry many line integrals are measured, enough to perform image reconstruction

The algorithm is called the filtered backprojection (FBP) and is implemented in all clinical CT scanners. Several reconstruction kernels $k(\xi)$ are available to allow modifying image sharpness (spatial resolution) and image-noise characteristics (Fig. 7.4).

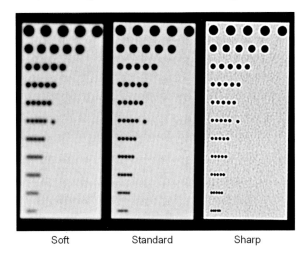

| Soft | Standard | Sharp |

Fig. 7.4. The effect of the reconstruction kernel on spatial resolution and image noise. The object scanned and reconstructed is a high-contrast-resolution phantom

The reconstructed image $f(x, y, z)$ is expressed in CT values. They are defined as a linear function of the attenuation values. The linear relation is based on the demand that air (zero attenuation) has a CT value of –1000 HU (Hounsfield units) and water (attenuation μ_{water}) has a value of 0 HU; thus, the CT value is given as a function of μ as follows:

$$CT = \frac{\mu - \mu_{water}}{\mu_{water}} 1000\,\mathrm{HU}.$$

The CT values, also known as Hounsfield values, were introduced by G.N. Hounsfield to replace the handling with the inconvenient μ values by an integer-valued quantity. Since the CT value is directly related to the attenuation values, which are proportional to the density of the material, we can interpret the CT value of a pixel or voxel as being the density of the object at the respective location. Typical clinical CT values are given in Fig. 7.5; they range from –1000 to 3000 HU, except for very dense materials such as dental fillings or metal implants. It is noted that CT, in contrast to other imaging modalities, such as magnetic resonance imaging or ultrasound, is highly quantitative regarding the accuracy of the reconstructed values. The reconstructed attenuation map can therefore serve for quantitative diagnosis such as bone-densitometry measurements and is optimal for treatment planning in radiation therapy.

The CT images are usually displayed as gray-scale images. A mapping of the CT values to gray values must be performed when displaying the data. To optimize contrast a pure linear mapping is avoided; a truncated linear response curve is used instead. In clinical CT the display window is parameterized by the two parameters, center C and width W. Values below $C-W/2$ are displayed black, values above $C+W/2$ are displayed white, and values within the window are linearly mapped to the gray values ranging from black to white (see Fig. 7.6). For example, the window (0, 500) means that it is centered at 0 HU and has a width of 500 HU; thus, values in the range from –250 to 250 HU are mapped to the gray values, values below –250 HU are displayed black, and values above 250 HU are displayed white.

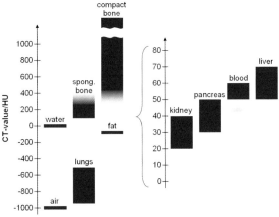

Fig. 7.5. Ranges of CT values of the most important organs

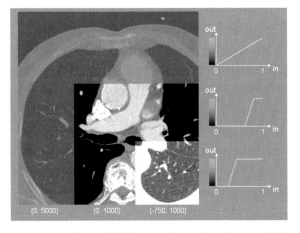

Fig. 7.6. One slice with different window settings (in Hounsfield units). The window setting is given in the format (center, width). Modifying the window values directly corresponds to modifying the truncated linear response curves (*right*)

7.3
Conventional CT

The conventional scan mode consists of a rotation about the stationary object. Reconstruction yields a single image corresponding to one z-position (longitudinal patient position). In order to acquire a complete volume several acquisitions must be performed with a short table movement in-between. This scan mode is called conventional CT or step-and-shoot CT. We also refer to the combination of several circle scans as a sequence scan or a sequential scan.

During the interscan delay the table is translated by a short distance. The value of this table increment is typically determined by the collimated slices. For a single-slice scanner of slice thickness S the table is transported by just the same distance. This allows to sample a complete volume with equidistant images of distance S.

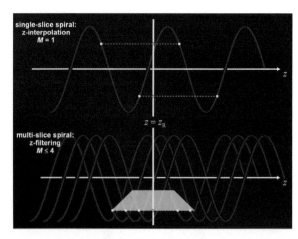

Fig. 7.7. The spiral z-interpolation principle is a linear interpolation of line integrals measured before and after the desired reconstruction plane $z=z_R$. It can be extended to scanners with up to M=4 slices without impairing image quality. Nevertheless, some manufacturers use these approaches for more than four slices

7.4
Spiral CT

In the late 1980s, just when continuously rotating scanners became available, a new scan mode was introduced by W.A. KALENDER (2001; KALENDER et al. 1989, 1990). In spiral CT data acquisition is performed continuously while the patient moves at constant speed through the gantry. Viewed from the patient the scan trajectory is a spiral (c.f. Fig. 7.2). Although not obvious – motion is regarded as a source of artifacts – spiral scans turned out to yield better image quality than conventional scans. This, however, requires addition of the z-interpolation as an additional image reconstruction step. Given a desired reconstruction plane z_R the z-interpolation uses projection data acquired at positions adjacent to that plane to synthesize virtual scan data corresponding to a circular scan at $z=z_R$. Typically, but not necessarily, linear interpolation is used and a similar principle holds for multislice scanners with $M=4$ slices (c.f. Fig. 7.7). As soon as the virtual circle scan has been synthesized, the final image is obtained using FBP.

There are two key features of the spiral scan. Firstly, the continuous data acquisition yields continuous data even in the case of patient movement or breathing. There is no interscan delay as in the step-and-shoot case where anatomical details could be lost or imaged twice. No discontinuities between adjacent images will result in spiral CT.

Secondly, the possibility of retrospectively selecting the reconstruction positions z_R allows reconstruction of images at finer intervals than dictated by the collimated slice thickness S. It is recommended to reconstruct images at an interval that is equal to or lower than half of the collimated slice thickness (Nyquist criterion). Only then will 3D displays be of high quality and show no step artifacts. Only then can multiplanar reformations (MPRs) yield image quality equivalent to the primary, transaxial images. It is this feature of improved sampling in the z-direction that led to the new application of CT angiography (CTA) in the 1990s.

A new scan parameter has been introduced in spiral CT. The table increment d is the distance the table travels during one rotation of the gantry. In conventional CT this value has typically been equal to the collimated thickness (adjacent but nonoverlapping images). In spiral CT we can choose d smaller or greater than $M \cdot S$. The so-called pitch value is used to relativize the definition of the table increment (IEC 1999):

$$p = \frac{d}{W_{tot}}.$$

W_{tot} is the total collimation and usually equal to the product of the number of slices times nominal slice thickness, i.e., $W_{tot}=M \cdot S$. Small pitch values yield overlapping data and the redundancy can be used to decrease image noise (more quanta contribute to one z-position) or to reduce artifacts. Large pitch values increase the scan speed and complete anatomical ranges can be covered very fast. Typical values are between 0.3 and 1.5. For single-slice scanners pitch values up to 2 are allowed.

7.5
Cone-Beam CT

Recently, scanners that simultaneously acquire more than one slice have become available. The run for more slices started in 1998 with the introduction of four-slice scanners.[1] The advantages of scanning more than one slice are obvious. For a given range scan time can be reduced by a factor of M. Alternatively, spatial resolution can be increased by the same factor. In fact, the typical scan modes were a combination of both options and are illustrated in Fig. 7.2. Curiously, many sites switched from the 1×5 mm collimation to the 4×1 mm collimation and improvements in spatial resolution often seemed more important than improvements in scan time.

The most challenging component of cone-beam CT is the image reconstruction algorithm. In contrast to single-slice or four-slice scanners, where each acquired slice can be regarded as being (approximately) perpendicular to the axis of rotation (c.f. Fig. 7.7), the cone-beam nature in scanners with many slices must not be neglected. Two object points located at the same z- but at different x- or y-positions will contribute to different slices (Fig. 7.8); hence, image reconstruction must take the cone-angle explicitly into account, now.

Two dedicated cone-beam algorithms are implemented in the product software of the present clinical 16- to 64-slice CT scanners: adapted versions of the Feldkamp algorithm and modifications of the advanced single-slice rebinning (ASSR) algorithm.

The Feldkamp algorithm is an extension of the 2D reconstruction to three dimensions (FELDKAMP et al. 1984). It also consists of the convolution of the acquired data followed by a backprojection. In contrast to the 2D case, the backprojection is 3D for the Feldkamp algorithm and during image reconstruction a complete volume must be kept in memory (Fig. 7.9).

---

[1] Before 1998 some scanners, such as the EMI scanner, the Siemens SIRETOM, the Imatron C-100, and the Elscint Twin scanner, were able to acquire two slices simultaneously. The detector and image reconstruction technology used, however, did not allow to further increase the number of slices towards true cone-beam scans.

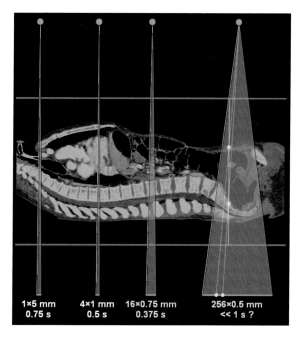

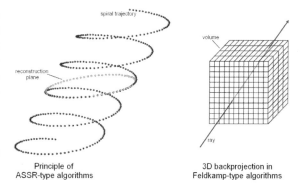

1×5 mm	4×1 mm	16×0.75 mm	256×0.5 mm
0.75 s	0.5 s	0.375 s	<< 1 s ?

Fig. 7.8. The longitudinal cut illustrates the cone-beam problem. For scanners with many slices (M>4) it is prohibitive, with respect to image quality, to reconstruct the slices separately. The demand for cone-beam reconstruction algorithms has led to new approaches in CT image reconstruction

Fig. 7.9. Basic principles of advanced single-slice rebinning (*ASSR*)-type and Feldkamp-type reconstruction

The ASSR uses the fact that 180° segments of circles can be accurately fitted to the spiral trajectory as long as tilting of these circles with respect to the x–y plane is allowed (KACHELRIESS et al. 2000a). The ASSR resorts the acquired cone-beam data to correspond to virtual 2D scans along these tilted circles. Then, a standard 2D image reconstruction is performed on these reconstruction planes to obtain tilted images. The ASSR uses many adjacent and overlapping reconstruction planes to sufficiently cover the whole volume. As soon as enough tilted images are available, the data are resampled in spatial domain to obtain a standard Cartesian volume that can be stored and viewed.

The advantage of the ASSR approach over the Feldkamp-type reconstruction is its high computational efficiency. Backprojection, which is the most time-consuming component during image reconstruction, is done in two dimensions with ASSR-type algorithms, whereas a 3D backprojection is required for Feldkamp-type algorithms. Furthermore, ASSR allows use of available 2D reconstruction hardware and thereby minimizes development costs; however, the ASSR approach is limited to approximately M=64 slices (KACHELRIESS et al. 2000) and data for scanners with far more slices should be reconstructed using a modified Feldkamp algorithm such as the extended parallel backprojection (EPBP; KACHELRIESS et al. 2004). Only then can cone-beam artifacts be avoided.

7.6
Scan and Reconstruction Parameters

Before performing a CT scan on a clinical scanner, several parameters must be defined by the operator.

The X-ray spectrum is determined by the tube voltage. Values between 80 and 140 kV are typical. Smaller values are ideal for smaller cross-sections, and high tube voltages are used for thicker patients or when metal implants are present.

The number of X-ray quanta is determined by the product of tube current and scan time $I \cdot t$ which is also called the mAs product. Since image quality depends on how many quanta contribute to one slice (and not to the complete volume) one defines the effective mAs product as the independent parameter; thus, the user chooses $(I \cdot t)_{eff}$ as the scan parameter. It is defined as the mAs product per rotation divided by the pitch value:

$$(I \cdot t)_{eff} = (I \cdot t)_{360}/p.$$

For sequence scans it is equal to the product of tube current and rotation time: in this case the table increment d equals the total collimation $M \cdot S$ and therefore $p=1$. Typical effective mAs values range from 10 to 500 mAs. Low values yield high image noise, and vice versa. It is important to emphasize that the selection of the effective mAs value yields an image quality independent of the pitch value, i.e., independent of the degree of scan overlap. To give evidence we have conducted a simple experiment which is shown in Fig. 7.10.

Another important scan parameter is the rotation time. Low values of t_{rot} are desired to reduce motion artifacts and to decrease the total scan time. High values are needed to increase the number of quanta that contribute to one slice (e.g., for obese patients). Presently, the rotation time ranges from 0.33 to 2 s. Note that for fast rotations the centrifugal forces acting on the rotating components can be up to 15g or more.

Resolution in the longitudinal direction is determined by the slice-thickness values. The colli-

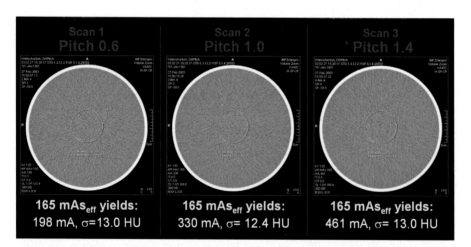

Fig. 7.10. Multislice scanners compensate for the effect of scan overlap by adjusting the actual (physical) tube current. The scans shown were performed with 165 mAs$_{eff}$ a rotation time of 0.5 s, and three different pitch values. The actual tube current chosen by the scanner ranges from 198 to 461 mA such that image noise and thus dose and image quality remain constant. Thus, for scanners that adapt the tube current we can state: same noise, same image quality, and same dose with MSCT regardless of the pitch value

mated slice thickness S determines the thickness of one cross section as it is measured during the scan. During image reconstruction the so-called effective slice thickness S_{eff} can be used to reconstruct images thicker than S. Thicker images show reduced image noise. Typical values for the collimated slice thickness range from 0.5 to 2.5 mm on modern 16-slice CT scanners. The effective slice thickness typically ranges up to 10 mm.

For spiral scans the table increment d can further be set or, alternatively, the spiral pitch value p. Pitch values range from about 0.3 to about 1.5. Values lower than 1 mean overlapping data acquisition where each z-position is covered by more than a 360° rotation. Low pitch values are used for thick patients to accumulate dose and, more important, for cardiac CT. High pitch values are used to increase scan speed.

7.7
Image Quality and Radiation Dose

By image quality in CT one often means the subjective impression of the image. This image-quality criterion depends on many parameters; among those is the image content and the diagnostic task to perform. For example, an image showing a patient's liver may be rated as insufficient if the pixel noise exceeds a certain value; however, an image of the lung region acquired and reconstructed with the same scan parameters is likely to be rated as adequate. This subjective image quality further strongly depends on the observer and on the observer's experience.

We now regard CT image quality from the physicist's point of view. Image quality is a function of dose and can be quantified by giving two parameters: (a) the pixel noise σ that is measured as the standard deviation of the pixel values within a homogeneous image region; and (b) the spatial resolution. Due to the circular symmetry of CT scans, one usually distinguishes between the in-plane resolution Δr and the axial- or z-resolution S_{eff}. For example, an image noise of $\sigma = 30$ HU, an in-plane resolution of $\Delta r = 0.7$ mm, and a z-resolution of $S_{eff} = 0.8$ mm are typical values for present standard exams. (Note that the resolution values are not identical to the voxel size of the image data, which must be significantly smaller than the desired resolution elements.)

Dose is usually quantified by specifying the organ dose or the effective dose (values in milliSievert). For our purposes it is sufficient to regard the effective mAs value $(I \cdot t)_{eff}$ as being the characteristic dose pa-

rameter since this parameter is proportional to the physical dose values.

Quantitatively, the proportionality

$$\sigma^2 \propto \frac{1}{(I \cdot t)_{eff} \cdot S_{eff} \cdot \Delta r^3} \qquad (1)$$

holds. The constant of proportionality depends in a complicated way on the patient size and density, on the tube voltage, and on the prefiltration.

Prior to the scan, the user specifies the three parameters effective mAs value, effective slice thickness, and in-plane resolution (via the name of the reconstruction kernel) at the scanner console. The user is thereby aiming for his preferred image quality. The image noise × is the only dependent parameter since it cannot be controlled directly.

Equation (1) has important consequences: increasing the spatial resolution requires dramatic increase in the effective mAs value and thereby the dose, given that the same object is scanned and that image noise is held constant. For example, doubling spatial resolution (i.e., reducing S_{eff} and Δr to 50%) requires increasing the patient dose by a factor of 16 if the pixel noise remains at its original level. Since such dose increases are hardly acceptable, scan protocols must carefully balance between spatial resolution and image noise; therefore, many applications are designed to either aim at good low-contrast resolution (low image noise) or to produce high spatial resolution; the first is the case in tumor detection tasks where lower spatial resolution is acceptable. In contrast, lung exams and, especially, exams of the inner ear, require scanning with the highest spatial resolution available.

The image noise obtained is strongly object dependent, and therefore Eq. (1) cannot be used to estimate image quality as a function of object size. (The constant of proportionality changes as a function of the object.) To demonstrate the dependence on the object size we performed a simple experiment. A thorax phantom (www.qrm.de) equipped with extension rings of different thicknesses was used to mimic a standard and an obese patient. The diameter of the obese patient is 5 cm larger than the normal patient's diameter. The obese patient was scanned with 200 mAs at 80, 120, and 140 kV, respectively. The normal patient's mAs value was adjusted to yield the same image quality (image noise) as for the obese patient. The results are given in Fig. 7.11.

Obviously, a significant mAs reduction of more than 60% is indicated, although the patient sizes differ by only 15%. In our case the effective half-value layer (HVL$_{eff}$) – the thickness of a layer that reduces

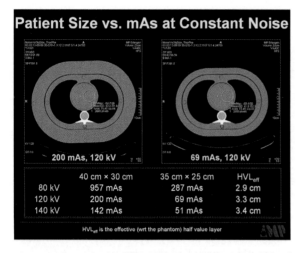

Fig. 7.11. The optimal milliampere value and thus the minimal patient dose can be achieved only if the patient shape is taken into account. To do so it is important to have an idea of the half-value layer of typical cross sections. In the thorax region the half-value layer is of the order of 3–5 cm

the X-ray intensity by a factor of two – is of the order of 3 cm. In general, an increase/decrease of patient size of the order of 3–5 cm requires to increase/decrease the mAs value by a factor of 2.

7.8
Ways to Reduce Dose

The demand for higher spatial resolution, higher image quality, and for increasingly more examinations inevitably yields to higher patient dose. To compensate for this trend several techniques to reduce dose or, equivalently, to improve dose efficiency have become available.

Automatic exposure control (AEC) is an automatic method that controls the tube current according to the patient cross section. This automatic adaptation decreases the tube current for rays of low attenuation (typically the anteroposterior direction) and increases the tube current for rays of high attenuation (lateral projections; GIES et al. 1999; KALENDER et al. 1999a). It further compensates for the global changes in the patient cross section. For example, a lower mean tube current value is used in the head and neck regions than in the shoulder and thorax regions. The method is illustrated in Fig. 7.12 (KACHELRIESS et al. 2001d; LEIDECKER et al. 2004). Significant dose reduction values are achievable with AEC.

Furthermore, so-called adaptive filtering approaches that seek to reduce image noise either by

directly manipulating the reconstructed images or by performing dedicated raw-data preprocessing prior to image reconstruction have become available (EKLUNDH and ROSENFELD 1981; KESELBRENER et al. 1992; LAURO et al. 1990; HSIEH 1994, 1998). A highly effective method is multidimensional adaptive filtering (MAF) that is especially suited for multislice and cone-beam CT scanners (Fig. 7.13). It allows reduction of image noise in cross sections of high eccentricity such as the pelvis, thorax, or shoulder without affecting spatial resolution (KACHELRIESS et al. 2001c; KACHELRIESS and KALENDER 2000). Thereby, the visibility of low-contrast objects can be greatly improved (BAUM et al. 2000).

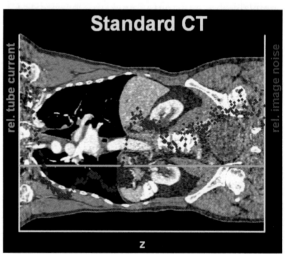
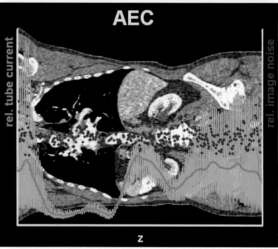

Fig. 7.12. Tube current modulation and control techniques optimize the tube current as a function of anatomy. Whereas a constant tube current (*green curve*) yields varying image noise (*red dots*), in the standard case the automatic exposure control can achieve a desired image noise profile with reduced radiation. Dose reduction factors of the order of 30% are typical. *AEC* automatic exposure control

The largest dose reduction potential, however, lies in scan protocol optimization. Especially the parameters scan length and effective mAs value lend themselves to optimization. Often the scan length exceeds the anatomical region of interest by 5 or even 10 cm. For short organs (e.g., the heart has a longitudinal extension of about 15 cm) this can mean a waste of dose. Whereas this parameter can be minimized easily, the optimization of the tube current is very difficult since it depends on many parameters, as we have seen. Only highly experienced operators can minimize the mAs value and still obtain images of sufficient diagnostic values. To experience the effect of mAs optimization one can use the dose tutor software (www.vamp-gmbh.de). It uses acquired raw data and adds virtual noise to each projection value to convert the data to a lower mAs value. This tool helps to iteratively seek for the optimal scan parameters and to learn more about the relationship between patient shape and image noise (Fig. 7.14).

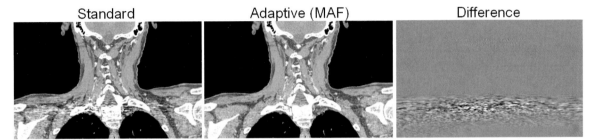

Fig. 7.13. Multidimensional adaptive filtering significantly reduces image noise and correlated noise structure (*streaks*). The subtraction image shows that there is no loss in resolution

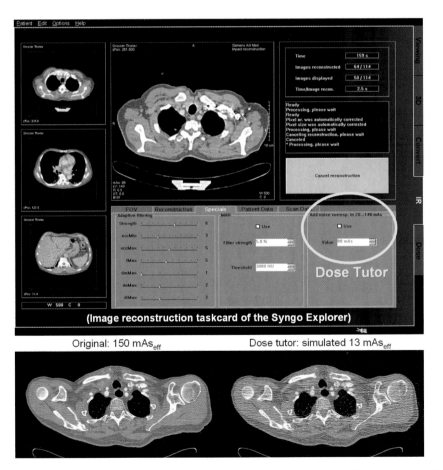

Fig. 7.14. Dose tutoring also means experiencing the dependence of image quality as a function of milliampere value. Software tools, such as the dose tutor shown here, allow manipulation of the milliampere value retrospectively by adding noise to the raw data

Individual dose assessment will play a significant role in the future. For example, the European Union Directive (EUROPEAN COMMUNITIES 1997) demands the specification of the individual patient dose for every CT examination. To directly assess scanner- and organ-specific dose values, dose calculators have become available. Based on Monte Carlo dose calculations (SCHMIDT and KALENDER 2002) they allow computation of organ dose and effective dose values as a function of the scanner type and scan protocol (Fig. 7.15). The effective dose is regarded as the best indicator of stochastic risk such as induction of malignancy. It expresses the summed dose value over all organs weighted by their radiation sensitivities. The weighting factors are recommended by the International Commission on Radiation Protection (ICRP) (ICRP 1990).

Putting all the dose-reduction efforts, including dose training and dose information, together it is likely that the increase in patient dose will be lower than predicted by Eq. (1), although spatial resolution and image quality will improve.

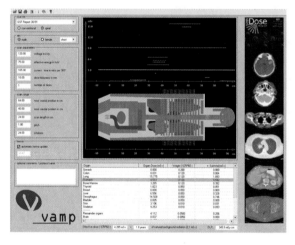

Fig. 7.15. Dose calculators compute organ-dose values and the weighted effective dose as a function of the anatomical region scanned and the scan parameters. (From www.vamp-gmbh.de; KALENDER et al. 1999b)

7.9
Technology

The decisive technological steps towards multislice and cone-beam scanning are image reconstruction as well as tube and detector technology. Image reconstruction techniques are demanding not only because the cone-beam problem poses a mathematically completely different situation than 2D scanning, but also because reconstruction speed must keep pace with the increased scan speed. Even with modern computing, power algorithms must be carefully designed and implemented. Two reconstructed slices per second (frames per second) are the minimum acceptable reconstruction speed on modern scanners. Since a complete anatomic volume of, for example, 50 cm length and 0.6 mm resolution can be covered in less than 15 s of the order of several hundred to 1000 images must be reconstructed. Manufacturers use highly parallelized hardware to face this task.

The drastic increase in scan speed demands the same increase in tube power since the same number of photons must be provided in a shorter time. To ensure high patient throughput cooling delays must be reduced in the same order. The drastic increase in rotation speed – rotation times of 0.4 s per rotation are typical presently and will certainly go down in the near future – is demanding with respect to the centrifugal forces that affect the tube housing and bearings. A typical tube is located 60 cm off the rotation axis. With 0.4 s per rotation, the resulting forces are of the order of 15g.

A typical X-ray tube consists of a vacuum-filled tube envelope. Inside the vacuum is the cathode and a rotating anode with one bearing (Fig. 7.16). This conventional concept has the disadvantage that tube cooling is not very efficient and that the one-sided bearing cannot tolerate high forces. Further on, it is difficult to lubricate bearings in vacuum. To improve these conventional tubes vendors try to maximize the heat capacity (expressed in mega heat-units MHU) to protract cooling delays as long as possible. Recently, a new CT tube has become available where the cathode, the anode, and the envelope together rotate in the cooling medium (Fig. 7.16). Due to the direct contact to the cooling oil, there is no need for storing the heat and there will be no cooling delays.

X-ray photon detection in modern CT scanners is performed using a scintillator layer that converts the X-rays into visible light.[2] Photodiodes convert the light to an electrical current that is amplified and digitized. As soon as the number of detector rows exceeds 2, the arrangement of the electronic components that read out, amplify, and digitize the signal becomes nontrivial and increasingly difficult with increasingly more detector rows. Another challenging, and often limiting, aspect of multislice detector technology is the data rates that must be transferred from

---

[2] The previous Xenon gas-chamber technology does not lend itself to dense packing and is not used in multislice scanners.

conventional tube
(rotating anode)

high performance tube
(rotating cathode, anode + envelope)

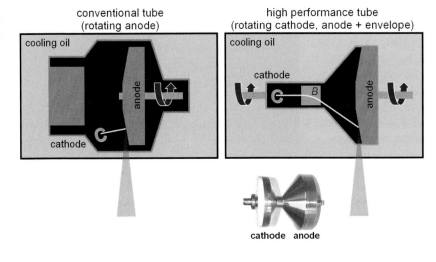

cathode   anode

Fig. 7.16. Conventional tubes have stationary envelope and stationary cathode. Recently, high-performance tubes have become available that have a rotating envelope and therefore allow for direct cooling of the anode

the gantry to the acquisition computer. If we assume a rotation time of 0.4 s and let the scanner simultaneously acquire 16 slices each of 800 two-byte channels per projection, with about 1000 projections per rotation a data transfer rate of 64 MB/s is required.

Due to these challenges, the present detectors typically have more detector rows than can be read out simultaneously. Electronic combination (binning) of neighboring rows is used to generate thicker collimated slices and to make use of all detector rows available. (Note that 16-slice CT scanners turned out to either have 24 or 32 detector rows, but none of them allow to simultaneously acquire more than 16 slices.) The 32-row scanners are of matrix-array type and combine every other row to generate thicker slices. The 24-row scanners are of adaptive array type and use thicker rows for the outermost slices. The adaptive array technology has a higher X-ray sensitivity and requires less patient dose than the matrix array detectors. Illustrations of these concepts are given in Figs. 7.17 and 7.18.

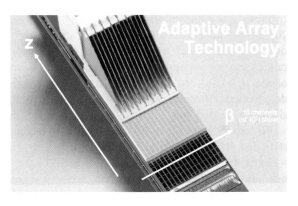

Fig. 7.17. Photo of a 64-slice adaptive array detector. (Data courtesy of Siemens Medical Solutions, Forchheim, Germany)

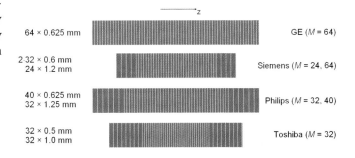

64 × 0.625 mm	GE (M = 64)
2·32 × 0.6 mm 24 × 1.2 mm	Siemens (M = 24, 64)
40 × 0.625 mm 32 × 1.25 mm	Philips (M = 32, 40)
32 × 0.5 mm 32 × 1.0 mm	Toshiba (M = 32)

7.10
Cardiac CT

One of the most prominent special applications at present is retrospectively gated CT of the heart. Improvements in CT technology, such as the introduction of spiral CT, subsecond rotation times, and multislice data acquisition, have stimulated cardiac CT imaging within the past decade. Cardiac spiral CT started with the introduction of dedicated phase-correlated reconstruction algorithms for single-slice spiral CT in 1997 (KACHELRIESS and KALENDER 1997,

Fig. 7.18. Detector concepts announced for 2004. Three of four manufacturers will provide adaptive array detectors to maximize dose efficiency for thicker slices. GE's announcement of its 64-row detector remained unclear with respect to the maximum number of simultaneous acquired slices. It may be restricted to reading out M=32 slices and all 64 rows would only be needed after pair-wise binning. The Siemens detector provides only 32 high resolution detector rows but generates 64 high-resolution slices due to a flying focal spot that samples each row twice. Thereby it is the only detector that fulfils the Nyquist sampling criterion in the z-direction

Kachelriess M, Kalender WA (1998) Electrocardiogram-correlated image reconstruction from subsecond spiral CT scans of the heart. Med Phys 25:2417–2431

Kachelriess M, Kalender WA (2000) Computertomograph mit reduzierter Dosisbelastung bzw. reduziertem Bildpunktrauschen. Deutsches Patent- und Markenamt. Germany. Patent specification DE 198 53 143

Kachelriess M, Kalender WA (2002) Extended parallel backprojection for cardiac cone-beam CT for up to 128 slices. Radiology 225:310

Kachelriess M, Kalender WA, Karakaya S, Achenbach S, Nossen J, Moshage W, et al. (1998) Imaging of the heart by ECG-oriented reconstruction from subsecond spiral CT scans. In: Glazer G, Krestin G (eds) Advances in CT IV. Springer, Berlin Heidelberg New York, pp 137–143

Kachelriess M, Schaller S, Kalender WA (2000a) Advanced single-slice rebinning in cone-beam spiral CT. Med Phys 27:754–772

Kachelriess M, Ulzheimer S, Kalender WA (2000b) ECG-correlated imaging of the heart with subsecond multi-slice spiral CT. IEEE Trans Med Imaging (Spec Issue) 19:888–901

Kachelriess M, Fuchs T, Lapp R, Sennst D-A, Schaller S, Kalender WA (2001a) Image to volume weighting generalized ASSR for arbitrary pitch 3D and phase-correlated 4D spiral cone-beam CT reconstruction. Proc 2001 International Meeting on Fully 3D Image Reconstruction, Pacific Grove, Calif., pp 179–182

Kachelriess M, Watzke O, Kalender WA (2001b) Generalized multi-dimensional adaptive filtering for conventional and spiral single-slice, multi-slice, and cone-beam CT. Med Phys 28:475–490

Kachelriess M, Sennst D-A, Kalender WA (2001c) 4D phase-correlated spiral cardiac reconstruction using image to volume weighting generalized ASSR for a 16-slice cone-beam CT. Radiology 221:457

Kachelriess M, Leidecker C, Kalender WA (2001d) Image quality-oriented automatic exposure control (iqAEC) for spiral CT. Radiology 221:366

Kachelriess M, Sennst D-A, Maxlmoser W, Kalender WA (2002) Kymogram detection and kymogram-correlated image reconstruction from sub-second spiral computed tomography scans of the heart. Med Phys 29 : 1489–1503

Kachelriess M, Knaup M, Kalender WA (2004) Extended parallel backprojection for standard 3D and phase-selective 4D axial and spiral cone-beam CT with arbitrary pitch and dose usage. Med Phys 31:1623–1641

Kalender WA (2001) Computed tomography. Wiley, New York

Kalender WA, Seissler W, Vock P (1989) Single-breath-hold spiral volumetric CT by continuous patient translation and scanner rotation. Radiology 173:414

Kalender WA, Seissler W, Klotz E, Vock P (1990) Spiral volumetric CT with single-breath-hold technique, continuous transport, and continuous scanner rotation. Radiology 176:181–183

Kalender WA, Wolf H, Suess C (1999a) Dose reduction in CT by anatomically adapted tube current modulation II. Phantom measurements. Med Phys 26:2248–2253

Kalender WA, Schmidt B, Zankl M, Schmidt M (1999b) A PC program for estimating organ dose and effective dose values in computed tomography. Eur Radiol 9:555–562

Keselbrener L, Shimoni Y, Akselrod S (1992) Nonlinear filters applied on computerized axial tomography. Theory and phantom images. Med Phys 19:1057–1064

Lauro KL, Heuscher DJ, Kesavan H (1990) Bandwidth filtering of CT scans of the spine. Radiology 177:307

Leidecker C, Kachelriess M, Kalender WA (2004) Comparison of an attenuation-based automatic exposure control (AEC) to alternative methods utilizing localizer radiographs. Eur Radiol 14 (Suppl 2):247

Schmidt B, Kalender WA (2002) A fast voxel-based Monte Carlo method for scanner- and patient-specific dose calculations in computed tomography. Phys Med XVIII:43–53

Sourbelle K, Kachelriess M, Kalender WA (2002) Feldkamp-type reconstruction algorithm for spiral cone-beam (CB) computed tomography (CT). Radiology 225:451

Taguchi K, Anno H (2000) High temporal resolution for multislice helical computed tomography. Med Phys 27:861–872

8   4D Imaging and Treatment Planning

Eike Rietzel and George T.Y. Chen

CONTENTS

8.1
Introduction

Respiratory motion can introduce significant errors in radiotherapy imaging, treatment planning, and treatment delivery (BALTER et al. 1996; CHEN et al. 2004; SHIMIZU et al. 2000). Currently, respiratory motion is incorporated into the radiotherapy process by target expansion. According to ICRU 52 and ICRU 60 (ICRU 50 1993; ICRU 62 1999), the gross target volume (GTV) is expanded to a clinical target volume (CTV) to include possible microscopic spread of malignant cells. Different sources of error in the radiotherapy process are accounted for by expansion of the

E. RIETZEL, MD
Siemens Medical Solutions, Particle Therapy, Henkestrasse 127, 91052 Erlangen, Germany
G. T. Y. CHEN, MD
Department of Radiation Oncology, Massachusetts General Hospital, 30 Fruit Street, Boston, MA 02114, USA

CTV to the planning target volume (PTV). The PTV should account for daily patient positioning errors and interfractional as well as intrafractional target motion. In this chapter, we focus on intrafractional motion induced by respiration. Interfractional organ motion is covered in another chapter on adaptive radiotherapy.

Significant imaging errors can result from the CT scanning of moving objects. In general, CT data acquisition is serial in nature, or slice by slice. Even helical scanning protocols reconstruct images at one couch position at a specific time, utilizing projection data acquired during continuous table movement. Asynchronous interplay between the advancing imaging plane and internal organ motion can result in severe geometric distortions of the imaged object. Ross et al. (1990) reported imaging of respiratory motion in 1990. In that era, image acquisition was performed one slice at a time. With the present technology (2004), 4D computed tomography (4D CT) can provide insight into organ motion during respiration, with volumetric anatomic data sets at time intervals of 0.5 s or less. Typically 10–20 different respiratory states are imaged over a respiratory cycle, with full volumetric information at each motion state. Various 4D CT techniques have been described in the literature (FORD et al. 2003; LOW et al. 2003; VEDAM et al. 2003b). We do not quantitatively compare these techniques but instead describe the protocol implemented at the Massachusetts General Hospital in collaboration with General Electric and Varian (PAN et al. 2004; RIETZEL et al. 2005b). The protocol was evaluated and refined in an IRB approved study between September 2002 and September 2003. This protocol has been in routine clinical use for liver and lung cancer patients since fall 2003.

Based on the volumetric information due to respiratory motion, several different options in treatment planning arise. 4D treatment planning – explicitly accounting for internal target motion during treatment planning – can be implemented at various levels of complexity (RIETZEL et al. 2005a). The simplest approach is to generate a composite target volume that encompasses the CTV throughout organ motion during the respiratory cycle.

More advanced 4D treatment planning requires deformable image registration techniques that relate different respiratory states of a patient to each other on a voxel-by-voxel basis. It is essential to track individual anatomical voxels throughout the respiratory cycle in order to accurately score dose delivered to each anatomical voxel. Such registrations often include two components: (a) an initial affine registration to capture gross motion including translation, rotation, scaling, and shearing; and (b) local deformations. While composite target volumes are in routine clinical use, non-rigid registration techniques have been implemented on a research basis. In order to calculate 4D dose distributions, we have extended the functionality of the CMS treatment planning software, FOCUS, to recalculate dose distributions at different respiratory states of breathing (RIETZEL et al. 2005a).

8.2
Computed Tomography

8.2.1
Motion Artifacts in Computed Tomography

In general, CT data acquisition is serial. This is obvious in axial scanning modes where projection data are acquired and reconstructed slice by slice, with couch movement in between slices while X-rays are off. Even in helical scanning, data at adjacent couch positions are in principle acquired serially during continuous couch movement while the X-rays are continuously on. Collection of projection data one slice after another in combination with motion of the scanned object usually leads to significant interplay effects. Scanning of organs during respiratory motion with modern multi-slice scanners has a high potential of introducing significant artifacts. Fast helical scanners can image the thorax in a region near a medium-sized lung tumor in a few seconds (2–3 s), and the entire lung in ~20 s. Assuming a respiratory period of 5 s, the tumor can move through nearly one half of its respiratory motion cycle during the time required to scan this region. Depending on the relative motion of the advancing scan plane and the tumor, very different artifacts can be imaged. If tumor motion is parallel to scan-plane advancement, the tumor will be imaged as an elongated object. If tumor motion is antiparallel to scan-plan advancement, the tumor will appear shortened. Figure 8.1 shows surface renderings of artifacts obtained when axially CT scanning a sinusoidally moving spherical object. All scan and motion parameters were kept constant. Only the relative phase between data acquisition and object motion was varied by starting the scans at varying object motion phases. The left image shows the bottom of the sphere twice, the top once, and the central part not at all. In the next scan (middle) object motion was in part parallel to the scan plane motion, and the upper part of the sphere was captured twice in adjacent regions. The complexity of possible artifacts is increased due to the sinusoidal motion. The sphere's velocity changes depending on its actual amplitude/position. Figure 8.1, right, shows the top and the bottom of the sphere fused to each other in the center of the trajectory. Additionally, other portions of the sphere were imaged separately, below its bottom.

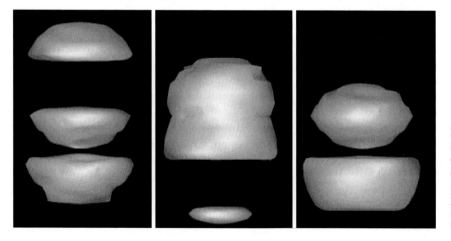

Fig. 8.1. Axial CT scans of a sinusoidally moving spherical object. Different motion artifacts are imaged depending on the relative motion phase between advancing scan plane and object motion

8.2.2
Breath-Hold and Gated CT Scans

Respiratory motion artifacts can be eliminated by scanning patients during breath hold (BALTER et al. 1998). This is possible with modern multi-slice CT scanners that can capture the full extent of the lung in less than 20 s. To image the extremes of motion, CT scans could be acquired at tidal end inhalation and end exhalation during breath hold; however, our experience shows that anatomy during breath-hold scans does not necessarily represent actual end inhale/exhale during normal respiration. Furthermore, no information on the tumor trajectory is obtained during breath hold.

Another alternative to reduce motion artifacts during scanning is to acquire respiratory-gated CT scans (RITCHIE et al. 1994; WAGMAN et al. 2003). In this approach, data acquisition at each couch position is triggered by a specific phase of the patient's respiration. One possibility for triggering is described later. The drawback of gated CT scans is total scan time. The time required to record a respiratory-gated CT scan is determined by the patient's respiratory period. For coverage of 25 cm with a 4-slice CT scanner at a slice thickness of 2.5 mm, the scan duration per volume is of the order of 125 s for a respiratory period of 5 s. After acquisition begins at each couch position, the time for CT tube rotation during data acquisition is required. Then the couch advances to the next scan position. The total time for acquisition and couch movement is usually shorter than a respiratory period; therefore, at the next couch position data acquisition can start again at the specified respiratory phase. The time of a full respiratory cycle per couch position is needed to acquire a gated CT scan. If several CT volumes, each representing a different respiratory patient state, are to be imaged, several independent scans must be acquired, each requiring 125 s. To image, for example, ten respiratory states with gated CT acquisition would require more than 20 min not including couch movement and scan setup between the scans.

8.2.3
4D Computed Tomography

8.2.3.1
Data Acquisition

The 4D CT technique images multiple respiratory states within one data acquisition. The basic principle involves temporally oversampling data acquisition at each couch position. At Massachusetts General Hospital we evaluated and clinically implemented a 4D CT scanning protocol developed by General Electric. At each couch position, data are acquired in axial cine mode. In this approach the CT tube rotates continuously for the duration of the respiratory cycle and acquires projection data through all respiratory states. The rotation is performed as rapidly as possible (depending on the CT scanner), and therefore several complete revolutions during the respiratory cycle are achieved (typically five to ten). The projection data then contain information about a full respiratory cycle. A temporal reconstruction window within these data is selected. This window has to encompass at least 180° plus fan angle in order to reconstruct an image. In our implementation currently, full 360° rotations are used. Within these oversampled data at each couch position, different reconstruction windows are set at different acquisition times – corresponding to various respiratory phases – to reconstruct multiple anatomical states. Usually images are evenly distributed over the cine acquisition time. The reconstructed images then represent different respiratory states of the patient that are evenly distributed over a respiratory cycle. After image projection data at a given couch position have been acquired, the X-rays are turned off and the couch is advanced to the next position to begin data acquisition again. This process is repeated until full coverage of the area to be scanned has been obtained. The 4D CT data acquisition is fully automated to obtain images at all couch positions.

Typically 10–20 images per slice are reconstructed representing 10–20 different respiratory phase states. This results in a total number of images between 1000 and 2000 per 4D CT study. The time required for 4D CT data acquisition with the parameters listed above is on the order of 150 s. As newer scanners now have 8–16 slices, the acquisition time for 4D continues to decrease almost linearly with numbers of rings of detectors. In contrast with the acquisition of ten gated CT scans, 4D CT scanning is acceptable for use in clinical routine.

8.2.3.2
External Sorting Signal: Abdominal Surface Motion

After 4D CT data acquisition and image reconstruction, 10–20 images at each slice index are contained within the set of 1000–2000 images. In order to sort these images into specific temporally coherent volumes, additional information is required. For this reason the patient's abdominal surface motion is monitored during 4D CT data acquisition with the Varian RPM system (Ramsey et al. 2000; Vedam et al. 2003a). The basis for this approach lies in the fact that the rise and fall of the abdominal surface is a surrogate for respiratory motion.

The RPM system is shown in Fig. 8.2. The main components are a marker block with two infrared reflecting dots, a CCD camera with an infrared filter, and an infrared light source. The marker block is placed on the patient's abdomen and illuminated by an infrared light source. The video camera records the block motion during respiration and sends the video stream to a dedicated PC. The RPM software tracks the motion of the infrared reflecting dots in real time. From the video stream, block-motion amplitudes are extracted, and relative respiratory phases are calculated based on these amplitudes. Abdominal motion and 4D CT data acquisition are accurately correlated temporally via signals sent from the CT scanner to the RPM PC during data acquisition.

8.2.3.3
Retrospective Image Sorting

The General Electric Advantage4D software is used to retrospectively sort the images into multiple temporally coherent volumes. The software loads the 4D CT images as well as the respiratory trace recorded by the RPM system. Based on the data acquisition time stamps in the image Dicom headers and the correlation signal in the RPM trace, a specific respiratory phase can be assigned to each image. Figure 8.3

shows a display of the Advantage4D software. In the lower panel of the user interface, the respiratory trace is displayed. In the upper portion of the screen, all reconstructed images are shown. The user selects a specific respiratory phase, which is shown as a red line in the respiratory trace display. The software then automatically identifies images at the respiratory phase closest to the desired respiratory phase for each slice location. The phase tolerance window within which images can be selected is marked in brown around the red line. Then the complete volumetric data for the specified respiratory phase is exported as a Dicom CT volume.

It is necessary to have a phase tolerance since data are acquired independently per couch position; therefore, at each couch position the respiratory phases at which images are reconstructed vary slightly. For example, reconstruction of ten images per slice could lead to a reconstructed image every 10% of the respiratory cycle; however, images at the first couch position could be reconstructed at relative phases of 2%, 12%, etc., and at the next couch position at 8%, 18%, etc. This example shows that temporal consistency depends on the number of reconstructed images per slice. The more images are reconstructed, the smaller the required phase tolerance for resorting will be. The other parameter that influences temporal coherence within resorted volumes is the cine duration at each couch position. If, for example, the cine duration equals twice the respiratory period and ten images per slice are reconstructed, the resulting sampling rate is only 20% because data of two cycles were acquired. To achieve good results we therefore monitor

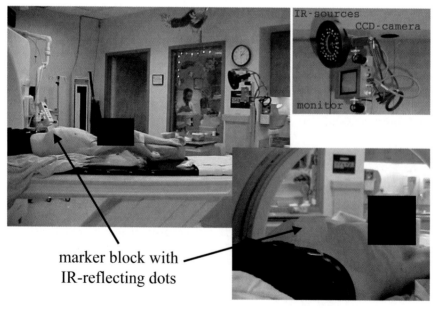

marker block with
IR-reflecting dots

Fig. 8.2. The Varian RPM system consisting of a CCD camera, an infrared light source, and marker block with infrared reflecting dots. The motion of the patient's abdominal surface is recorded on a dedicated PC running RPM software

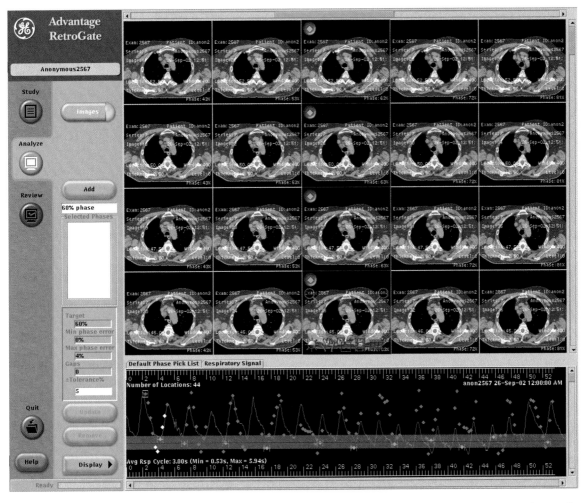

Fig. 8.3. General Electric Advantage4D software for retrospective sorting of 4D CT data into temporally coherent volumes. The *main panel* shows all images reconstructed. The ones selected for the current respiratory phase are marked with a *light blue square*. The *lower panel* shows the patient's respiratory trace during 4D CT data acquisition. The currently selected respiratory phase is indicated by a *red line*. Phase tolerances required for re-sorting a complete volume are shown as a *brown band*. Different respiratory phases can be selected in the control panel on the left

the patient's respiratory cycle for several minutes before 4D CT data acquisition starts.

During the 4D CT retrospective sorting process, data acquired at different respiratory cycles are assembled into one volume, since one respiratory cycle per couch position is required. Figure 8.4 shows respiratory traces of different patients. The upper row shows the complete respiratory trace during a 4D CT data acquisition, end inhalation at the top, and end exhalation at the bottom. Note the rapidly changing variations at peak inhalation from breath to breath and the slow drifting baseline at peak exhalation. Currently it is not clear how well an external motion surrogate, such as the marker block, represents internal anatomic motion. Abdominal motion is measured at one point only so differences between chest and abdominal breathing

cannot be easily detected. Using relative respiratory phases for data re-sorting allows assembling full volumes for different phases. Using the respiratory amplitudes instead would result in many volumes with missing slices. On the other hand, it could very well be that respiratory phases represent internal motion better than amplitudes observed at a single point only. This is an area of current active research.

The mid and lower graphs in Fig. 8.4 show enlarged regions of the respiratory trace for two other patients. Whereas the patient in the middle breathed with a reproducible wave form, the patient in the lower graph held his/her breath several times during data acquisition. The resulting re-sorted volumes therefore showed poor temporal coherence. The CT images from these examples are presented in Fig. 8.8.

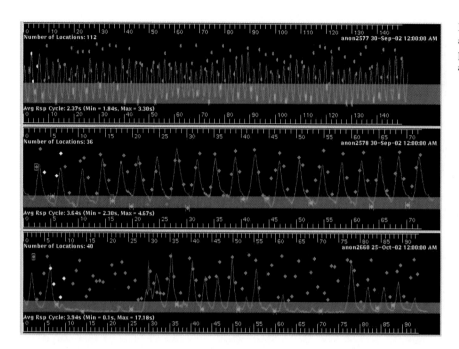

Fig. 8.4. Respiratory traces acquired for three different patients during 4D CT data acquisition

8.2.3.4
Visualization

Figure 8.5 shows a comparison of a standard helical CT scan acquired under light breathing (left) and one of the respiratory phases obtained from 4D CT scanning (right). Respiratory artifacts are obvious in the helical scan data, especially at the lung–diaphragm interface. The lung tumor shows severe artifacts as well. In comparison with the 4D CT data on the right, the severity of motion artifacts can be seen. The 4D CT presents a more realistic shape of the tumor, confirmed with breath hold CT scans (data not shown).

To visualize internal tumor/organ motion, dedicated software is required. Initially we coded our own 4D browser. Vendors now provide software to perform 4D review. Multiple respiratory phase CT volumes can be loaded. The 4D review allows browsing data in 3D as well as time/phase. The CT data can be visualized in axial, sagittal, and coronal views manually selecting the respiratory phase or playing temporal movie loops. This functionality is useful to assess internal motion by visual inspection or using a measurement tool.

Figure 8.6 shows sagittal views of different CT volumes for a patient with a hepatocellular tumor. The sagittal image displayed at top left was acquired in standard helical scanning mode under light breathing. Note that respiratory artifacts are visualized as a jagged abdominal skin surface of the patient and as artifacts at boundaries of internal organs, e.g., the

posterior edge of the liver. The five other images show different respiratory phases of the patient as acquired with 4D CT. A full respiratory cycle is displayed, starting at the top middle with end inhalation. A red rectangle (fixed to image coordinates) is added to guide the eye. The patient had gold fiducials implanted for image-guided therapy. One of these markers is shown in the images visualizing craniocaudal liver motion. In comparison with the standard helical CT scan, respiration-induced artifacts are largely absent. Note that during helical CT data acquisition the marker was imaged twice at different slice positions.

Similar images are shown in sagittal view for a lung tumor in Fig. 8.7. Here the bounding box containing the tumor throughout the respiratory cycle is plotted in magenta. Individual contours were manually drawn on each respiratory phase as well as the standard helical scan acquired under light breathing. Contours were drawn on axial slices. Note the heavily distorted shape of the tumor in the helical data. These represent a mixture of different respiratory phases per slice. In the 4D CT data contouring problems for lung tumors can be observed. Due to manual contouring variation, densities that were included in the contours vary from respiratory phase to respiratory phase. This illustrates that quantitative analysis of 4D CT data/target volumes based on manually drawn contours is problematic. On the other hand, manual contouring is the current gold standard and therefore new techniques are evaluated against it. Another interesting aspect of Fig. 8.7 is the quasi-static bones, e.g., ribs. Although

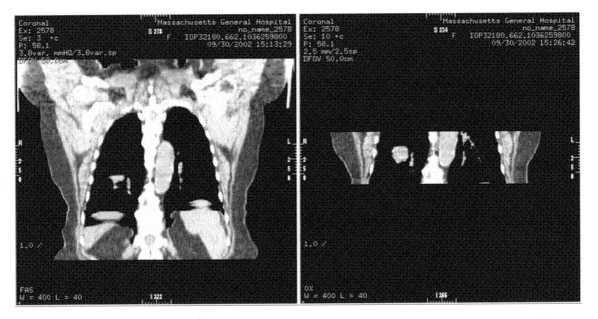

Fig. 8.5. Standard helical CT scan acquired under light breathing (*left*) in comparison with one respiratory phase of a 4D CT scan (*right*). For the 4D CT scan, scan length was reduced. Typical motion artifacts seen in the standard helical scan are absent in the 4D CT volume

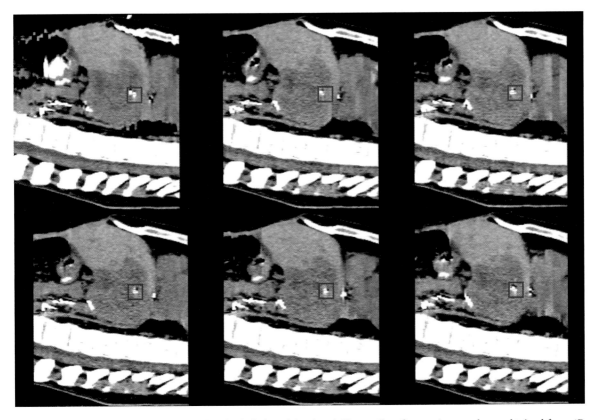

Fig. 8.6. Standard helical CT scan acquired under light breathing (*top left*) as well as five respiratory phases obtained from 4D CT scanning. The 4D CT phases are displayed for a full respiratory cycle. The patient had radio-opaque markers implanted in the liver for image guidance. A *box* was plotted to visualize respiratory motion. Note that the displayed marker is imaged in two adjacent slices in the helical scan

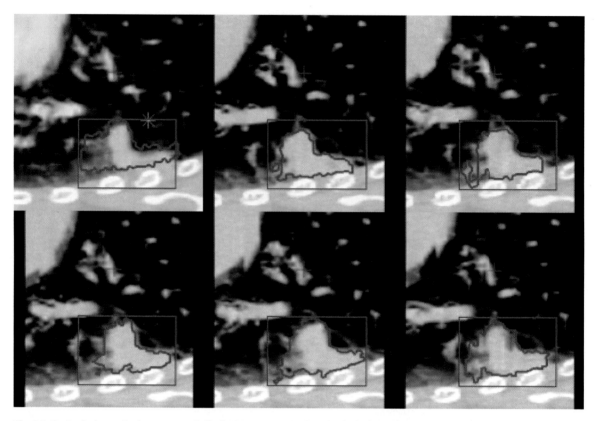

Fig. 8.7. Sagittal views of a lung tumor: helical CT scan acquired under light breathing (*top left*) and phases of a respiratory cycle acquired with 4D CT. Contours were drawn manually. The *box* is plotted to guide the eye

the tumor moves by approximately 1 cm, the ribcage remains virtually static. The tumor slides along the pleural interface along with the lung tissue.

It should be emphasized that to obtain 4D CT volumes with the described approach, data acquisitions during several respiratory cycles are synthesized to obtain full volumetric information; therefore, data quality depends strongly on the reproducibility of amplitude and periodicity of the respiratory cycle during acquisition. Recall that at each couch position (4 slices for our scanner) projection data for a full respiratory cycle need to be collected. Figure 8.4, bottom panel, shows the very irregular respiratory trace of a patient. The corresponding 4D CT data are shown in Fig. 8.8. Due to several breath-hold periods, volumetric image data cannot completely be obtained for all respiratory phases. If data are sorted regardless, selecting the phases closest to the desired target phases in order to obtain full volumetric information, severe motion artifacts occur. Note the jagged patient skin surfaces in Fig. 8.8. Only volumes close to end exhalation (middle) appear artifact free. Since the patient held his/her breath at exhalation, these volumes are not impacted by irregular respiration.

8.3
4D Treatment Planning

The explicit inclusion of temporal effects in radiotherapy treatment planning is referred to as 4D treatment planning. In general, there are no significant differences between 3D and 4D treatment planning. Simply an additional degree of freedom has to be considered; however, 4D radiotherapy is still in its early days and requires much more research before full implementation in clinical routine is possible. In general, interfractional as well as intrafractional motion components should be included in 4D radiotherapy, although herein only intrafractional respiration-induced organ motion is considered. Inclusion of interfractional organ motion, usually referred to as adaptive radiotherapy, is addressed in another chapter.

Intrafractional motion can be included in treatment planning at different levels. The simplest level is to generate composite target volumes encompassing the target throughout the respiratory cycle. More sophisticated approaches include multiple-dose calculations and deformable registration. In the fu-

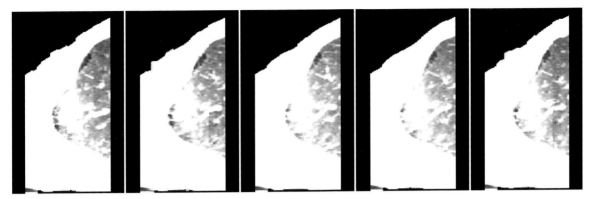

Fig. 8.8. Sagittal views of different respiratory phases acquired with 4D CT. Note the residual artifacts due to irregular respiration during 4D data acquisition

ture, motion information may be included directly in treatment plan optimization and treatment delivery.

8.3.1
Target Delineation

Target delineation is a tedious task in 4D treatment planning. In principle, the target should be contoured on each individual respiratory phase, increasing the workload typically by a factor of ~10. Such contours have been presented in Fig. 8.7 for a lung tumor. Figure 8.9 shows additional patient studies (contours not shown). Figure 8.9a and d displays data acquired during light breathing with helical CT scanning. Figure 8.9b and e shows data close to end inhalation and end exhalation (Fig. 8.9c and f). Contours were drawn manually. The outlined gross target volumes (GTV) were 5.5 cc and 149.3 cc on the helical scan data. On average, the 4D CT GTVs were 2.8 cc and 158 cc. Respiratory artifacts are obvious in the helical CT data for both patients. In comparison with the 4D CT volumes, different effects were observed. For the patient shown in Fig. 8.9a–c motion artifacts increased the GTV volume by a factor of ~2 in comparison with the 4D CT volumes. The tumor volume was imaged moving in parallel to the scan plane for the full range of motion. In contrast, the target volume for the patient in Fig. 8.9d–f was slightly reduced due to imaging artifacts. These geometric distortion effects are magnified when planning target volumes (PTV) are designed with margin expansions.

In standard lung tumor planning at Massachusetts General Hospital, uniform margins from GTV to PTV of 20 mm were used. These margins in principle included microscopic clinical spread (clinical target volume, CTV), and inter- and intrafractional motion.

After 4D CT imaging was implemented in the clinical routine, video loops are used to determine the tumor position extremes in 3D. Then targets at the two extreme respiratory phases are contoured. These GTVs per phase are fused to obtain a composite GTV. This composite GTV is overlaid on the corresponding video loops that include all respiratory phases. This quality-assurance check is used to ensure full coverage of the target throughout the respiratory cycle. In case there is undercoverage, additional target volumes are contoured to generate the composite GTV.

For our initial patient studies, we compared reduced margins from GTV to PTV from 20 mm to 17.5 mm when performing 4D CT and using composite target volumes. This represents a very conservative margin reduction; however, margin expansions of composite GTVs by 17.5 mm in comparison with 20 mm of GTVs obtained from helical CT scanning showed approximately the same PTV volumes for almost all of our initial 10 patient studies. In order to keep complication ratios at approximately the same level as before, while decreasing the chances for geometrical misses, we chose to use these expansions of 17.5 mm from GTV to PTV. While the resulting PTV volumes remain approximately the same, PTV centroids are often shifted and PTV aperture shapes differ. For the patients shown in Fig. 8.9 the resulting 4D PTVs were 82.2 cc (Fig. 8.9a–c) and 721.2 cc (Fig. 8.9d–f) in comparison with PTVs obtained from helical CT scans with margins of 20 mm which were 137.4 cc and 797.7 cc, respectively. For both patients the PTV size was reduced. This is especially apparent for the patient with the small lung tumor, where the reduction in PTV volume is significant (~60%). These example data illustrate that target design based on standard helical CT scans is problematic. Similar studies have recently been published by Allen et al. (2004).

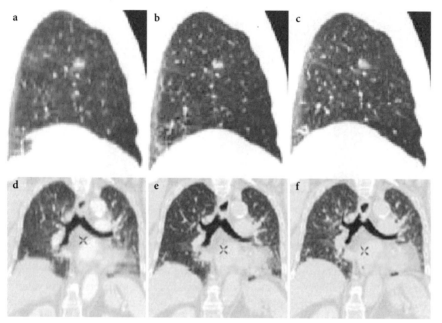

Fig. 8.9. Sagittal (**a–c**) and coronal (**d–f**) views of two lung tumor patients. **a, d** Standard helical CT scans acquired under light breathing. **b, e** End inspiration. **c, f** End expiration of 4D CT data

8.3.2
Fast Segmentation of Lung Tumors

Manual contouring of several respiratory phases for research is feasible. In clinical routine, however, other segmentation techniques need to be developed. One possibility was described above, that is contouring selected extreme phases and fusing two targets. For lung tumors, the large density difference between tumor and surrounding lung can be exploited to propose an efficient segmentation approach to multiple CT data sets from 4D CT data. The calculation of a maximum intensity volume (MIV) as shown in Fig. 8.10 can be used for contouring a composite target volume on one CT data set. A maximum intensity volume is calculated based on all respiratory phases available. For each individual voxel throughout the time varying image volume, the maximum voxel density at each point from all respiratory phase volumes is determined. This procedure identifies all high-density voxels in the resulting volume and therefore shows the full extent of the tumor throughout the respiratory cycle. Figure 8.10, left, shows an axial slice of a maximum intensity volume. Figure 8.10, right, shows the corresponding end-inhalation and end-exhalation respiratory states of the 4D CT volumes color coded. The tumor at inspiration is shown in green, at expiration in red. Wherever densities match the overlay image appears in yellow. The example shows the potential of MIVs, and the lung tumor, including respiratory motion, can be manually contoured

with a single contour rather than ten (one for each phase); however, to date, the MIV concept has not been fully tested. Currently it should only be used for initial contouring in combination with validation as overlay on respiratory phase videos. When a tumor is adjacent to other high-density objects, such as the mediastinum, MIV-based contouring of the composite GTV might be inappropriate, since portions of the moving mediastinum can occlude the tumor in some respiratory phases.

8.3.3
Digitally Reconstructed Fluoroscopy

Traditionally organ/tumor motion is often assessed by fluoroscopy on the conventional simulator. For many soft tissue tumors, direct visualization of tumors is not possible with projection radiography/ fluoroscopy. In such cases, motion of surrounding organs or implanted fiducial markers can be used as surrogates for tumor motion (Balter et al. 2001). Utilizing 4D CT, marker motion can directly be related to tumor motion in 3D. It is possible to estimate lung tumor motion in some cases with fluoroscopy; however, the precise extent and shape of the lung tumor often cannot be visualized.

Calculation of digitally reconstructed radiographs (DRR) at each respiratory phase based on 4D CT volumes can help in determining the exact position or extent of the tumor in 2D projections. Figure 8.11

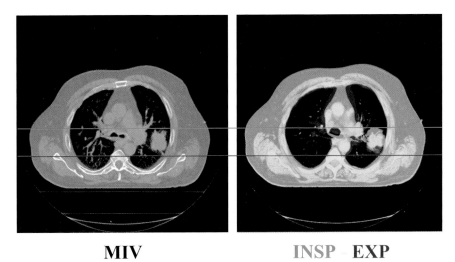

MIV   **INSP   EXP**

Fig. 8.10. Maximum intensity volume (*MIV*) of a lung tumor (*left*) and corresponding end-inhalation (*green*) and end-expiration (*red*) breath-hold CT scans as color overlay (*right*)

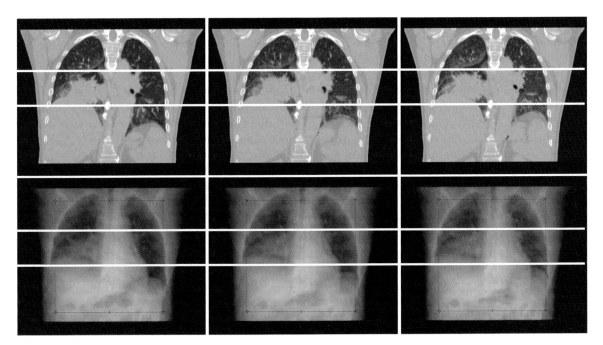

Fig. 8.11. Coronal views of 4D CT data for three respiratory phases of a lung tumor patient (*top*) and corresponding digitally reconstructed fluoroscopy images (*bottom*)

shows three respiratory phases of a lung cancer patient. The upper row displays coronal views through the center of the GTV, the lower row shows images of the corresponding DRRs. Without 4D CT data, the DRRs are difficult to interpret and the tumor position and extent are ambiguous. With additional information from the 4D data, displayed side by side, tumor outlines are more easily detectable. Furthermore, it is useful to project GTV contours delineated at each respiratory phase on the corresponding DRRs to even better describe the tumor position in the DRRs.

An animation of several DRRs can be considered digitally reconstructed fluoroscopy (DRF). The DRF is useful for beam gating or tracking based on fluoroscopic data acquired directly during treatment delivery. The fluoroscopy stream can then directly be compared with the DRF that corresponds to respiratory states of the patient known from 4D CT. Another possibility to use DRFs is in validating breathing models of an individual patient. Rather than acquiring multiple 4D CTs, several respiratory cycles could be recorded with fluoroscopy and compared with the

DRF. With recently available treatment technology, specifically on-board imaging on linear accelerators, this may even be performed on a daily basis prior to treatment delivery.

8.3.4
Dose Calculations

4D CT data provide the ability to study the impact of respiratory motion on dose distributions. We have implemented 4D dose calculations in our commercial treatment planning system FOCUS (CMS). Since current treatment planning systems do not offer 4D dose calculations, the multiple dose calculations are set up in the underlying Unix file system using a set of shell scripts and C-code. Based on a specific set of treatment plan parameters (e.g., beam angles, weights, MUs per field, etc.), dose calculations can be performed at each respiratory phase.

The current implementation is on a research basis only. A treatment plan is optimized for one CT volume state. Then this treatment plan is copied to the study sets of all respiratory phases. We have implemented multiple dose calculations for 3D conformal radiotherapy, IMRT, and proton therapy. For proton therapy the dose calculations can be submitted to a

Linux cluster in order to perform all dose calculations in parallel. For photons the individual dose calculations are initiated through the standard graphical user interface of the treatment planning system. For IMRT, interplay effects between organ motion and dynamic beam delivery have not yet been implemented. The current implementation for 4D IMRT dose calculations is equivalent to IMRT as delivered through a compensator. Interplay effects cannot yet be implemented easily since precise temporal modeling of organ motion and MLC motion during irradiation are required.

Figure 8.12 shows coronal views at different respiratory phases of a 4D dose distribution. A treatment plan was optimized based on a standard helical CT scan for 3D conformal radiotherapy (data not shown). Then dose distributions were re-calculated for several respiratory phases obtained from 4D CT. Parameters of the treatment plan were not altered.

Changes in anatomy, especially the lung tumor position, as well as slight distortions of the displayed isodose lines, are visible; however, the shape of the isodose lines and their relative position to patient anatomy remain nearly identical, predominantly in medium and low-dose regions. Typically, changes in dose deposition are overestimated when examining isodose lines. Even small variations in dose deposi-

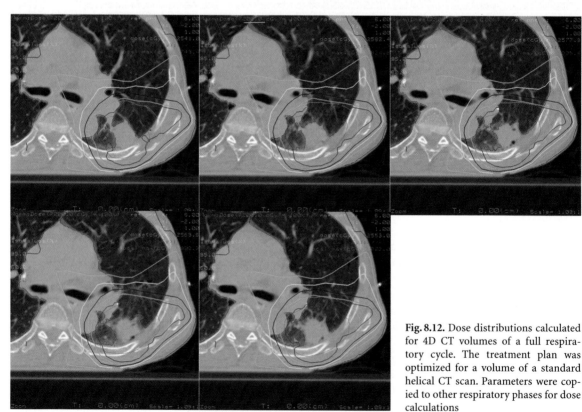

Fig. 8.12. Dose distributions calculated for 4D CT volumes of a full respiratory cycle. The treatment plan was optimized for a volume of a standard helical CT scan. Parameters were copied to other respiratory phases for dose calculations

tion, e.g., from 100.1 to 99.9%, will change the resulting contours. Expected changes in dose deposition due to density changes caused by respiration have been reported to be small, in the 5% range (ENGELSMAN et al. 2001). We have observed variations in dose deposition of the order of 3–4%. It is noted that the underlying dose calculation algorithm plays a significant role in such analysis. We used the CMS convolution algorithm for our studies.

Once multiple dose calculations have been performed, the resulting dose volume histograms (DVH) can be inspected for target coverage. This analysis assumes that target volumes have been determined for all respiratory phases of interest. Figure 8.13 shows DVHs for the patient and dose distribution shown in Fig. 8.12. The prescribed dose to the target was 72 Gy, and the expansions from GTV to CTV and CTV to PTV were 10 mm each. Recall that the initial treatment plan was optimized for a helical CT scan acquired under light breathing. The DVHs for all respiratory-phase CT volumes show some decrease in dose; however, for this patient the clinical target volume (CTV) is covered with the prescribed dose throughout the respiratory cycle, although variations can be observed. This shows that (a) the respiratory artifacts in the helical CT data were not too severe; and (b) the expansion of the target volume to include possible errors were sufficient to compensate for intrafractional target motion.

Neglecting delineation and systematic imaging errors, at least intrafractional motion as well as setup errors should be included in the PTV design. Fig. 8.14 shows DVHs for the same patient and treatment plan but now assuming a setup error of 5 mm. The volumes used to compile the DVHs were generated by uniform expansion of the individual CTVs. Although it is not possible that the patient is systematically mispositioned in all directions, this process is in accordance to current standard and ICRU recommendations. Setup errors are in general incorporated in margin design by uniform expansions.

If assuming a setup error of 5 mm the target volume is covered only by 95% of the prescribed dose for one of the respiratory phases as shown in Fig. 8.14. If the patient had been treated based on this treatment plan, partial underdosing of the CTV would have been expected.

The dose analysis example presented shows a possible next step in the implementation of full 4D treatment planning. Composite target volumes can be designed based on multiple respiratory phases. Then the target volume is expanded to incorporate the remaining sources of error. Dose distributions are calculated for multiple phases. Either dose distributions can be inspected directly or DVHs can be analyzed for multiple respiratory phases.

Display of multiple DVHs is possible with commercial treatment planning systems. Dynamic display of dose distributions, on the other hand, is not. We have coded our own 4D dose browser that can display anatomy and contours as well. Figure 8.15 shows an overview of the graphical user interface as well as the target region zoomed for different respiratory phases. The browser allows the user to display cuts through the patient in the three major axes, selecting the cutting plane via a slider bar. There is a slider bar for selecting the specific respiratory phase statically. The image data can also be displayed dynamically in

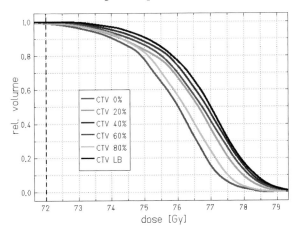

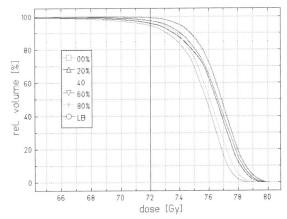

Fig. 8.13. Dose volume histograms (DVH) for different respiratory phases (compare Fig. 8.12). Respiratory phases are given as relative numbers, 0% respiratory phase corresponding to end inhalation. *LB* is the DVH for the standard helical CT scan

Fig. 8.14. Dose volume histograms assuming a possible setup error of 5 mm by uniform target expansion according to ICRU recommendations. Respiratory phases are given as relative numbers, 0% respiratory phase corresponding to end inhalation. *LB* is the DVH for the standard helical CT scan

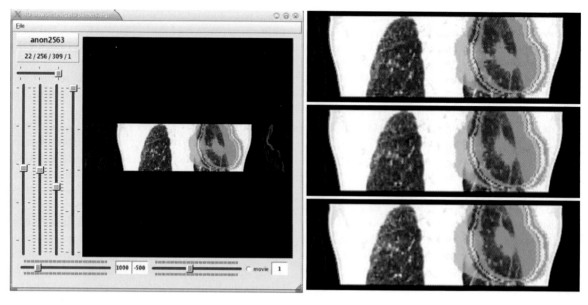

Fig. 8.15. Four-dimensional browsing software. Displayed are anatomy as well as dose distribution (*left*). Three different respiratory phases are shown (*right*)

a video loop. Dose distributions are overlaid in color-wash mode. The browser is useful for fast inspection of 4D dose distributions.

8.3.5
Non-Rigid Registration

To add dose distributions of different respiratory cycles non-rigid registration is needed. Each individual anatomical voxel position must be tracked throughout the respiratory cycle in order to accurately assess deposited dose. The tracking of voxel positions can be accomplished by registration of CT volumes which includes the modeling of local deformations.

Non-rigid registrations of liver have recently been reported in the literature by BROCK et al. (2003) using thin-plate splines. We have chosen to use the freeware tool vtkCISG developed at King's College in London (HARTKENS et al. 2002). The toolkit registers organ volumes at different (respiratory/deformed) states to each other. Full affine transformations including translations, rotations, scaling, and shearing as well as local deformations are supported. Local deformations are modeled with B-splines (RUECKERT et al. 1999). A control point grid is superimposed to the patient anatomy. Each of the knots of the grid is moved during registration. The underlying anatomy follows the motion of the translated knots. Registration in general is an optimization problem. The parameter to be optimized is the similarity between the two volumes to be registered. Within the optimization loop the control points are translated and the resulting deformations to the anatomy are calculated. Then the similarity between the target volume and the deformed volume is computed. A variety of similarity measures is offered in vtkCISG. We have chosen to use sum of squared differences (SSD) as similarity measure for most registrations. To obtain the similarity between two volumes voxels at corresponding positions are subtracted. Then the sum over all differences squared is used to assess overall similarity.

Figure 8.16 shows the results obtained for registration of end inhalation to end exhalation for a patient with a hepatocellular tumor. The top row displays the patient anatomy at these respiratory states. To compare the CT data visually, color overlays are used. In the bottom row end exhalation is color coded in yellow, end inhalation in blue. Wherever densities between the different respiratory states match the resulting color is on a gray scale. Transformation parameters were optimized for a full-affine registration followed by modeling local deformations as described above. The arrows indicate the transformations from end inhalation to end exhalation for different regions of the anatomy.

Figure 8.16 shows the good performance of the registration process; high-density structures are in good agreement. The different lengths and directions of the displacement vectors illustrate that modeling of local deformations is essential to obtain precise registration results.

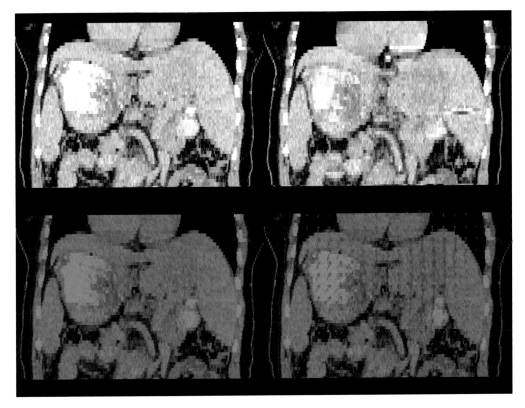

Fig. 8.16. Non-rigid registration of internal patient anatomy. Top: end exhalation (*left*) and end inhalation (*right*); bottom: color overlays before (left) and after registration (*right*). Transformations are indicated by red arrows

One important step in obtaining good registration results has not yet been addressed: adequate pre-processing of the data. The internal anatomy of the patient was segmented prior to registration. This is necessary because the ribcage, for example, hardly moves during respiration, whereas the liver right next to it moves significantly. Due to intrinsic smoothness constraints B-splines cannot model such abrupt anatomical changes. Independently moving anatomy has to be separated from each other prior to registration; otherwise, different parts compete against each other during registration. This would then lead to insufficient registration results as well as breaking of bones, for example.

8.3.6
Total Dose Distributions in the Presence of Respiratory Motion

Once dose calculations per respiratory phase have been performed and non-rigid registration parameters between different respiratory phases have been optimized, total dose distributions, including respiratory motion, can be calculated. Dose distributions are deformed according to the parameters obtained from non-rigid CT–CT registration to ensure tracking of the dose within each voxel throughout the respiratory cycle. For the patient presented in Fig. 8.15, Fig. 8.17 shows the original dose distribution for end inhalation as well as the dose distribution for end exhalation mapped to end inhalation. In this frame of reference patient anatomy is static, whereas dose distributions move according to the underlying respiratory motion. This is most obvious when comparing the relative position of the CTV (purple) to the isodose lines. In addition, the shape of individual isodose lines changes, most obvious left of the CTV.

After non-rigid transformation to one reference respiratory phase dose distributions can be added on a voxel-by-voxel basis to obtain a total dose distribution. Accurate tracking of the dose deposited per individual voxel throughout the respiratory cycle is guaranteed by the transformations. The DVHs for the dose distributions in Fig. 8.17 are plotted in Fig. 8.18 for individual respiratory phases as well as for the total dose distribution including respiratory motion.

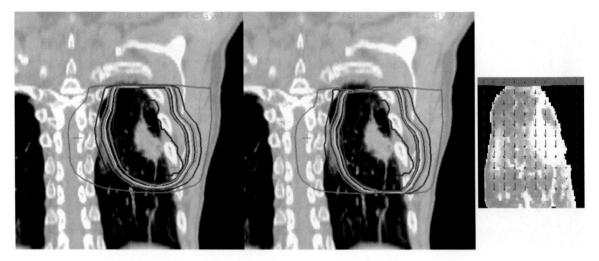

Fig. 8.17. Deformation of a dose distribution according to non-rigid CT–CT registration parameters to a reference respiratory phase. Corresponding transformations are indicated by *red arrows* (*right*)

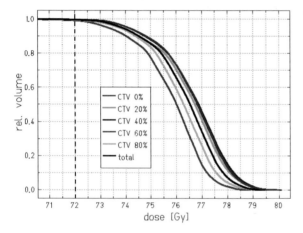

Fig. 8.18. Dose volume histograms for five respiratory phases and the total dose distribution including respiratory motion. Respiratory phases are given as relative numbers, 0% respiratory phase corresponding to end inhalation

8.4
Conclusion

The presence of intrafractional target motion is a major challenge in radiotherapy. This chapter has surveyed the technical issues to utilize 4D imaging data in treatment planning of respiration induced organ motion. 4D imaging can be applied to markedly reduce the artifacts that can occur in scanning moving objects. The imaging technique used at Massachusetts General Hospital captures anatomy at multiple respiratory states, increasing the image data by a factor of ~10. The benefit of 4D imaging is

that it provides a more accurate estimate of the true target shape and its trajectory as a function of time. This information can help improve the precision of external beam irradiation.

The initial method we implemented in utilizing 4D CT involved defining composite volumes that combined target volumes at the extremes of motion observed. Our observations from this implementation indicate that 4D target volumes differ from those derived by conventional helical scanning. What can change is that the shapes of volumes of interest and their centroids change, and they are more accurate from 4D CT. Utilization of 4D CT image data in treatment planning has also been implemented and provides information on how organ motion can perturb dose distributions. A key technology in the calculation of dose to moving organs is deformable image registration. Other technical challenges that are needed include visualization and segmentation tools that are designed to deal with dynamic image data.

There are a number of caveats to be kept in mind when considering dose to moving targets. Our focus has been solely on the effects of respiratory motion. We do not consider cardiac motion, or interfractional motion, due to either setup errors or variations in physiologic state, e.g., stomach-size variations that could influence positions of adjacent organs. The data acquired in 4D CT is synthesized from multiple breaths during an acquisition time of a few minutes. Reproducibility of this pattern during each treatment fraction is implicitly assumed when analyzing the resulting 4D dose distributions. Possible variations may be monitored by examining the respiratory trace on a daily basis – either the abdominal surface or even better the target volume directly, e.g., using fluoroscopy.

In the future, techniques to deliver 4D treatment will be explored, developed, and refined to take full advantage of the new knowledge provided by 4D CT. Image-guided therapy in the treatment room could well include 4D cone-beam CT, if appropriate. The global view of irradiation of moving targets involves knowledge and control of the entire process from imaging for planning to daily irradiation. This knowledge will lead to methods to mitigate the dose-perturbing effects of motion, and possibly lead to safe decrease of geometric margins and increased therapeutic gain.

References

Allen AM, Siracuse KM, Hayman JA et al. (2004) Evaluation of the influence of breathing on the movement and modeling of lung tumors. Int J Radiat Oncol Biol Phys 58:1251–1257

Balter JM, Ten Haken RK, Lawrence TS et al. (1996) Uncertainties in CT-based radiation therapy planning associated with patient breathing. Int J Radiat Oncol Biol Phys 36:167–174

Balter JM, Lam KL, McGinn CJ et al. (1998) Improvement of CT-based treatment planning model of abdominal targets using static exhale imaging. Int J Radiat Oncol Biol Phys 41:939–943

Balter JM, Dawson LA, Kazanjian S et al. (2001) Determination of ventilatory liver movement via radiographic evaluation of diaphragm position. Int J Radiat Oncol Biol Phys 51:267–270

Brock KK, McShan DL, Ten Haken RK et al. (2003) Automated generation of a four-dimensional model of the liver using warping and mutual information. Med Phys 30:1128–1133

Chen GT, Kung JH, Beaudette KP (2004) Artifacts in computed tomography scanning of moving objects. Semin Radiat Oncol 14:19–26

Engelsman M, Damen EM, de Jaeger K et al. (2001) The effect of breathing and set-up errors on the cumulative dose to a lung tumor. Radiother Oncol 60:95–105

Ford EC, Mageras GS, Yorke E et al. (2003) Respiration-correlated spiral CT: a method of measuring respiratory-induced anatomic motion for radiation treatment planning. Med Phys 30:88–97

Hartkens T, Rueckert D, Schnabel JA et al. (2002) VTK CISG Registration Toolkit. An open source software package for affine and non-rigid registration of single- and multimodal 3D images. In: Meiler M, Saupe D, Kruggel F, Handels H, Lehmann T (eds) Bildverarbeitung für die Medizin Algorithmen - Systeme - Anwendungen. Workshop proceedings, Leipzig March 2002, series: Informatik aktuell. Springer, Berlin Heidelberg New York

ICRU50 (1993) Prescribing, recording and reporting photon beam therapy. ICRU, Bethesda, Maryland

ICRU62 (1999) Prescribing, recording and reporting photon beam therapy (supplement to ICRU Rep 50). ICRU, Bethesda, Maryland

Low DA, Nystrom M, Kalinin E et al. (2003) A method for the reconstruction of four-dimensional synchronized CT scans acquired during free breathing. Med Phys 30:1254–1263

Pan T, Lee TY, Rietzel E et al. (2004) 4D-CT imaging of a volume influenced by respiratory motion on multi-slice CT. Med Phys 31:333–340

Ramsey CR, Scaperoth D, Arwood D et al. (2000) Clinical experience with a commercial respiratory gating system. Int J Radiat Oncol Biol Phys 48:164–165

Rietzel E, Chen GTY, Choi NC et al. (2005a) 4D image based treatment planning: target volume segmentation and dose calculations in the presence of respiratory motion. Int J Radiat Oncol Biol Phys 61:1535–1550

Rietzel E, Pan T, Chen GTY (2005b) 4D computed tomography: image formation and clinical protocol. Med Phys 32:874–889

Ritchie CJ, Hsieh J, Gard MF et al. (1994) Predictive respiratory gating: a new method to reduce motion artifacts on CT scans. Radiology 190:847–852

Ross CS, Hussey DH, Pennington EC et al. (1990) Analysis of movement of intrathoracic neoplasms using ultrafast computerized tomography. Int J Radiat Oncol Biol Phys 18:671–677

Rueckert D, Sonoda LI, Hayes C et al. (1999) Non-rigid registration using free-form deformations: application to breast MR images. IEEE Trans Med Imaging 18:712–721

Shimizu S, Shirato H, Kagei K et al. (2000) Impact of respiratory movement on the computed tomographic images of small lung tumors in three-dimensional (3D) radiotherapy. Int J Radiat Oncol Biol Phys 46:1127–1133

Vedam SS, Kini VR, Keall PJ et al. (2003a) Quantifying the predictability of diaphragm motion during respiration with a noninvasive external marker. Med Phys 30:505–513

Vedam SS, Keall PJ, Kini VR et al. (2003b) Acquiring a four-dimensional computed tomography data set using an external respiratory signal. Phys Med Biol 48:45–62

Wagman R, Yorke E, Giraud P et al. (2003) Reproducibility of organ position with respiratory gating for liver tumors: use in dose-escalation. Int J Radiat Oncol Biol Phys 55:659–668

9 Magnetic Resonance Imaging for Radiotherapy Planning

LOTHAR R. SCHAD

CONTENTS

9.1 Correction of Spatial Distortion in Magnetic Resonance Imaging for Stereotactic Operation/Treatment Planning of the Brain

9.1.1 Aim

To measure and correct the spatial distortion in magnetic resonance imaging. Depending on the individual magnetic resonance (MR) system, inhomogeneities and nonlinearities induced by eddy currents during the pulse sequence can distort the images and produce spurious displacements of the stereotactic coordinates in both the x–y plane and the z axis. If necessary, these errors in position can be assessed by means of two phantoms placed within the stereotactic guidance system – a 2D phantom display-ing "pincushion" distortion in the image, and a 3D phantom displaying displacement, warp and tilt of the image plane itself. The pincushion distortion can be "corrected" by calculations based on modelling the distortion as a fourth-order 2D polynomial.

9.1.2 Introduction

Accurate planning of stereotactic neurosurgery (SCHLEGEL et al. 1982), interstitial radiosurgery or radiotherapy (SCHLEGEL et al. 1984; BORTFELD et al. 1994) requires precise spatial information. By virtue of good soft tissue contrast and multiplanar tomographic format, MR is a logical choice providing for display of the pertinent anatomy and pathology. Accurate spatial representation demands uniform main magnetic fields and linear orthogonal field gradients. Inhomogeneity of the main magnetic field and nonlinearity of the gradients produce image distortion (O'DONNELL and EDELSTEIN 1985). A significant source of these geometric artefacts are eddy currents produced during the imaging sequence. Correction of these distortions is usually unnecessary for clinical diagnosis but is important for stereotaxy and in planning radioisotopic or radiation therapy.

This article chapter deals with the assessment of and correction for geometric distortions of MR images using phantom measurements.

9.1.3 2D Phantom Measurement

The first phantom, the 2D phantom, is used to measure the geometrical distortions within the imaging plane (Fig. 9.1a). It consists of a water-filled cylinder 17 cm in radius and 10 cm in depth, containing a rectangular grid of plastic rods spaced 2 cm apart and oriented in the z direction (i.e. perpendicular to the imaging plane). Since the exact positions (x,y) of these rods are known a priori, their positions (u,v) in the 2D phantom reflect the geometric distortion in the imag-

L. R. SCHAD, PhD
Abteilung Medizinische Physik in der Radiologie, Deutsches Krebsforschungszentrum, Postfach 101949, 69009 Heidelberg, Germany

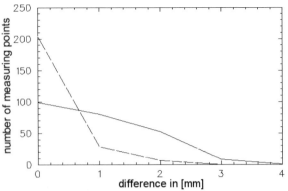

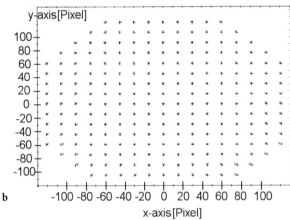

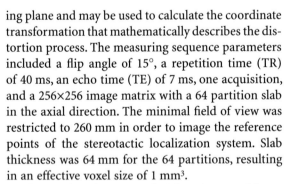

Fig. 9.1. a The two-dimensional (2D) phantom for measuring the geometrical distortions within the imaging plane. It consists of a water-filled cylinder 17 cm in radius and 10 cm in depth, containing a rectangular grid of plastic rods spaced 2 cm apart and oriented in the z direction (i.e., perpendicular to the imaging plane). b Typical example of a 2D phantom measurement using a velocity-compensated fast imaging with steady precession sequence. The a priori known regular grid of the plastic rods (+ symbols: calculated points) is deformed to a pincushion-like pattern (x symbols: measured points) from which the 2D-distortion polynomial can be derived. Note that the origin of the coordinate system of both the calculated and measured points lies in the centre of the image, the area free of distortion. c Distributions of lengths of distortion vector (magnitude of positional errors) of the 2D phantom pins in the uncorrected (*solid line*) and in corrected image (*broken line*) calculated with N=4 (expansion order of the 2D polynomial). Magnitude of positional errors in uncorrected image is about 2–3 mm at outer range of phantom (i.e., at the position of the reference points of the stereotactic guidance system). These errors are reduced to about 1 mm in corrected image which corresponds to the pixel resolution of the image. (From SCHAD 1995)

ing plane and may be used to calculate the coordinate transformation that mathematically describes the distortion process. The measuring sequence parameters included a flip angle of 15°, a repetition time (TR) of 40 ms, an echo time (TE) of 7 ms, one acquisition, and a 256×256 image matrix with a 64 partition slab in the axial direction. The minimal field of view was restricted to 260 mm in order to image the reference points of the stereotactic localization system. Slab thickness was 64 mm for the 64 partitions, resulting in an effective voxel size of 1 mm³.

The x–y plane distortion of axial MR images of the 2D phantom can be "corrected" (reducing displacements to the size of a pixel) by calculations based on modelling the distortion as a fourth-order 2D polynomial:

$$u_k = \sum_{i,j=0}^{N} U_{ij} \cdot x_{ki} \cdot y_{kj}, \qquad (1)$$

$$v_k = \sum_{i,j=0}^{N} V_{ij} \cdot x_{ki} \cdot y_{kj} \text{ with } k = 1 \dots M, \qquad (2)$$

where (x,y) are the true pin positions of the 2D phantom (Fig. 9.1b, "+"symbols); (u,v) are the distorted positions measured on the image (Fig. 9.1b, "x" symbols), N (=4) is the order of the polynomial, and M (=249) is the total number of pin positions of the 2D phantom. Note that the origin of the u,v coordinate system corresponds to the origin of the x,y coordinate system and lies in the centre of the image, the area free of distortion.

Equations 9.1 and 9.2 applied to all M pin positions of the 2D phantom form a system of simultaneous linear equations from which the coefficients U_{ij} and V_{ij} can be calculated and the distortion corrected. Thereby, the selection of N=4 is a compromise between computational burden and reduction in distortion. A more quantitative, direct measurement of distortion is provided by the distortion vector which measures the discrepancy between the true $(x,y)_k$ position of a pin in the 2D phantom and its apparent position in the image $(u,v)_k$: $d_k = |(x,y)_k - (u,v)_k|$ (i.e. the vector distance between the true and apparent positions). Reduction in distortion is demonstrated in Fig. 9.1c, a comparison of the distribution of d_k in the uncorrected and corrected images. The distribution of d_k in the corrected 2D phantom image shows that the positional errors are reduced from about 2–3 mm (Fig. 9.1c, solid line) to about 1 mm across the entirety of the 2D phantom (Fig. 9.1c, broken line). No further attempts were made to reduce this residual error since this approximates the dimensions of the image pixels (256×256 matrix).

9.1.4
3D Phantom Measurement

After correcting for distortion in the imaging plane, the 3D position of the imaging plane was assessed with the 3D phantom (Fig. 9.2a). This is necessary

Fig. 9.2. a The three-dimensional (3D) phantom mounted in the stereotactic guidance system for measuring 3D position of MR imaging plane. This is comprised of a regular grid of water-filled rectangular boreholes, with oblique water-filled boreholes in between. This produces a pattern of reference points surrounded by measurement points in axial image from which the 3D position of the imaging plane can be reconstructed. **b** A 3D-phantom measurement. The distance d between reference point (rectangular water-filled borehole) and measurement point (oblique water-filled borehole) is a direct measure of the z coordinate of the imaging plane at the reference point, since the angle between oblique water-filled borehole is $\alpha = 2 \arctan (0.5)$. Systematic discrepancy in the distance measurements of $d1 > d2 > d3 > d4 > d5 > d6$ would be detected if the imaging plane were tilted. **c** Axial image of the 3D phantom. The 3D position of the imaging plane can be reconstructed from the measured points. **d** Typical example of the 3D position of the imaging plane with properly adjusted shims. Deviations in z direction are reduced to about 1 mm which approximates again the dimensions of the image pixels (1 mm slice thickness). Mathematical correction of these discrepancies in the z direction was not pursued since properly adjusted shims avoid deformation or tilting of the imaging plane. **e** A tilting of the 3D phantom of about 3° from transversal to sagittal and coronal direction can be clearly detected. (From SCHAD 1995)

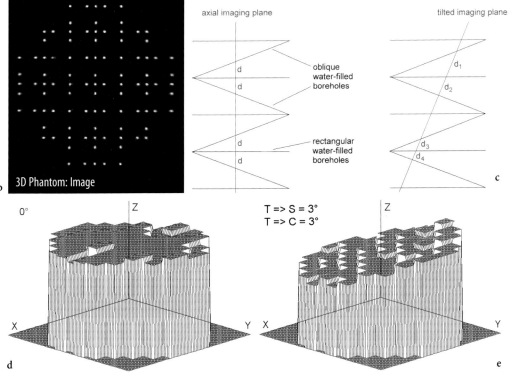

a

b 3D Phantom: Image

axial imaging plane

oblique water-filled boreholes

rectangular water-filled boreholes

tilted imaging plane

c

0°

T => S = 3°
T => C = 3°

d

e

and comparing these with those obtained using the same stereotactic system with CT (Schad et al. 1987a,b, 1992, 1995). This proves that after correction, the locations in CT and MR correspond to the same anatomic focus. After correction, the accuracy of the geometric information is limited only by the pixel resolution of the image (=1 mm). For example, such a correction provides for the more accurate transfer of anatomical/pathological informations and target point coordinates to the isocentre of the therapy machine so essential for the stereotactic radiation technique, or in case of stereotactic-guided operation planning.

9.4
Functional Structures at Risk

Functional MRI with its blood oxygen level dependent (BOLD) contrast (Ogawa et al. 1990) can contribute to a further improvement of stereotactic planning by identifying functional structures at risk. In cases of brain lesions in the motor or visual cortices, for example, these functional structures can then be spared out (Fig. 9.6c). The effect of EPI correction is most important at locations of functional areas close to tissue/air interfaces. After correction (Fig. 9.6b, c) both functional areas coincide well with the underly-

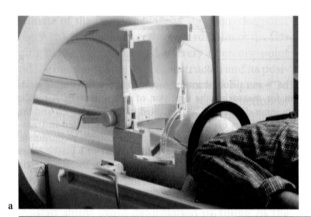

a

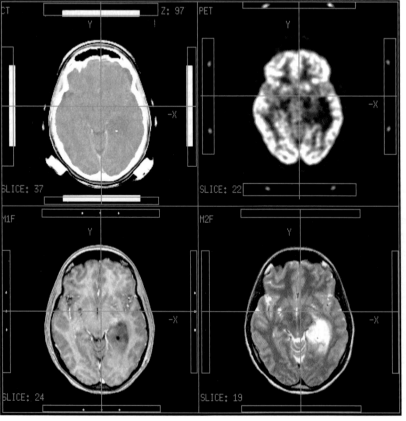

b

Fig. 9.5. a Patient positioning by mask fixation of the head in the stereotactic guidance system in the head coil of the MR scanner. b Definition of the target volume on a set of CT, PET ([18]FDG) and MRI (T1-weighted and T2-weighted) images in a patient with astrocytoma, grade II. The target volume can only be defined by the therapist interactively with a digitizer. Thereby the computer calculates automatic the stereotactic z coordinate (=slice position) and the stereotactic coordinate system in all axial slices with the help of reference points seen as landmarks in the images. After correction of geometrical distortions in MRI data, an accurate transfer of features, such as target volume, calculated dose distribution, or organs at risk can be done from one set of imaging data to another. Complicating factors, such as pixel size, slice position or slice thickness differences are taken into account by the stereotactic coordinate system. (From Schad 2001)

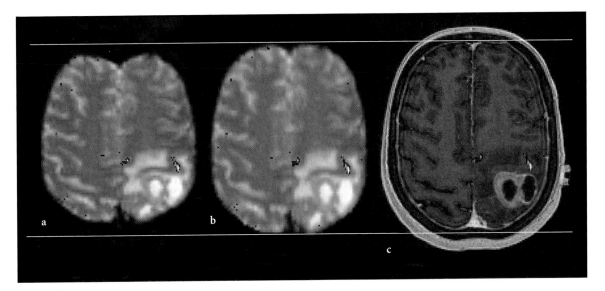

Fig. 9.6a–c. Echo-planar imaging correction in a patient with a glioblastoma. **a** Finger tapping fMRI original EPI image with errors of about 10 mm in phase-encoding direction. **b** Corresponding corrected image. **c** Overlaid corrected EPI-fMRI on corresponding post-contrast T1-weighted 3D fast low-angle-shot image. (From SCHAD 2001)

ing morphological grey-matter structure measured by spin-echo imaging, which is not the case without EPI correction (Fig. 9.6a).

9.5
Target Volume Definition: Tissue Oxygenation

The BOLD contrast can contribute to a further improvement of stereotactic planning while areas of high oxygenation and vascularity can be identified, which subdivides our target volume into an area of higher oxygen supply and tumour activity (Fig. 9.7). This area can then be defined as high-dose region and dose painted by an intensity-modulated therapy concept.

On the other hand, oxygen partial pressure is a critical factor for the malignancy and response of tumours to radiation therapy; therefore, a good diagnostic localization of oxygen supply in tumours is of great interest in oncology. In this context new methods of physiological MRI can help to determine important parameters tightly connected to tissue oxygenation (HORSMAN 1998). Currently, there are several promising approaches under investigation to enable the measurement of oxygenation parameters of tissues.

All of these methods rely on the direct or indirect dependency of relaxation rates on oxygen content within the probe. One example for a direct dependency is the proportionality of the T1 relaxation rate

of Fluor (^{19}F) on oxygen partial pressure, which can in principle be used to map oxygen content (ABOAGYE et al. 1997). Since special fluorinated contrast agents and dedicated scanner hardware are needed for this spectroscopic imaging method, it remains somewhat exotic and is not clinically available at present.

The most promising NMR methods for oxygen mapping are the BOLD methods (OGAWA et al. 1990). These are based on the fact that the magnetic properties of blood depend on its oxygenation state. This causes variations in the transverse relaxation times (i.e. T2 or T2*) of ^{1}H nuclei which can be detected by dedicated MR-sequence techniques optimized for maximum susceptibility sensitivity. While T2*-weighted sequences can be used to qualitatively detect the localization of areas in different oxygenation states (GRIFFITHS et al. 1997), multi-echo techniques can achieve quantitative information about oxygen extraction in tissues (VAN ZIJL et al. 1998).

One interesting approach to map oxygen supply is the detection of localized signal responses in a series of T2*-weighted images when blood oxygen content is varied by inhalation of different oxygen-rich gases (HOWE et al. 1999). From such an image series parameter maps of relative signal enhancement due to the oxygen inhalation can be calculated. As an example, a resulting parameter map of the signal response in an astrocytoma (WHO III) during inhalation of carbogen gas (95%O_2, 5%CO_2) is shown in Fig. 9.7. The parameter map is overlaid on a T2-weighted image. The red areas depict significant signal enhancements

Problems appear in patients with AVMs close to tissue/air interfaces. In this case, the malformation cannot be delineated clearly on MR venography and the size of the lesion is underestimated due to susceptibility artefacts caused by the presence of adjacent bony structures. Another severe problem appears in patients after subacute bleeding. Thereby contrast-enhanced subtraction and TOF MRA are able to depict residual pathological vessels. The MR venography showed a large area of signal loss due to signal dephasing caused by haemosiderin. The size and shape of the AVM and the origin of the pathological vessels were difficult to assess.

9.6.2
Brain Tumours

In cases of patients with brain tumours, MR venography shows a complex and variable venous pattern.

In case of patients with glioblastoma conventional spin-echo imaging shows a typical pattern of contrast enhancement on T1-weighted imaging, surrounded by a large oedema seen on T2-weighted images. On MR venography a large area of signal loss is present in the necrotic part of the tumour with a different pattern of small veins in the part of the contrast enhancement of the lesion (Fig. 9.10).

9.7
Target Volume Definition: Tissue Perfusion

Perfusion measurements using MRI (e.g. contrast-enhanced T1-weighted MRI, arterial spin labelling) are the methods of choice for detection and delineation of cerebral metastases during diagnostic work-up and treatment planning, allowing for accurate definition of the treatment target volume. After

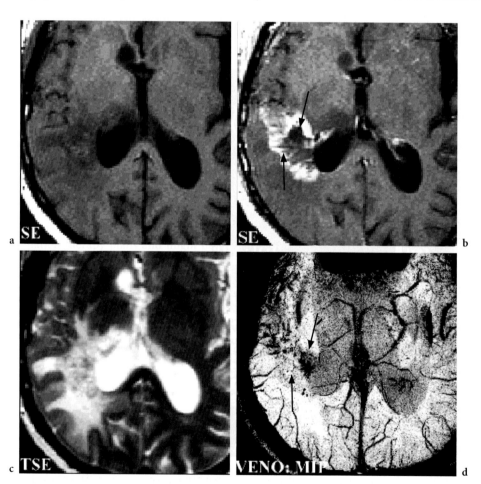

Fig. 9.10a–d. High-resolution BOLD venography (**d**) in comparison with conventional T1-weighted (**a, b**) and T2-weighted (**c**) imaging in patients with brain tumours. Patient with glioblastoma shows an area of signal loss in BOLD venography in the necrotic part (**d**) surrounded by a complex venous pattern in the contrast enhancing part of the tumour (**b**). (From Schad 2005)

therapy, MRI is used for monitoring the response to treatment and for evaluating radiation-induced side effects. Ideally, MRI would be able to distinguish between local tumour recurrence, tumour necrosis, and radiation-related deficiencies of the blood–brain barrier and would be able to predict treatment outcome early. Pulsed arterial spin-labelling techniques (DETRE et al. 1992) have the ability to measure perfusion in an image slice of interest without additional application of contrast agent. Intrinsic water molecules in arterial blood are used as a freely diffusible contrast agent carrying the labelled magnetization flow.

Arterial spin labeling (ASL) techniques, such as flow-sensitive alternating inversion recovery (FAIR; KIM 1995), provide a tool to assess relative perfusion (Fig. 9.11). These techniques acquire inversion recovery images with a nonselective inversion pulse which flips the magnetization of water in arterial blood. The image is acquired after the inflow time TI to allow the labelled blood to reach the image slice. To eliminate static tissue signal, these images are subtracted from inversion recovery images with a slice-selective inversion pulse placed over readout slice and acquired after the inflow time TI. The difference image in Fig. 9.11 containing the signal of arterial blood delivered to the image voxel within the inflow time TI is proportional to relative perfusion. Using arterial spin labelling techniques, absolute perfusion can be assessed using sophisticated MRI sequences (e.g. Q2TIPS, ITS-FAIR) in combination with a general kinetic model (BUXTON et al. 1998; LUH et al. 1999) based on a one-compartment model. A clinical example of a patient with glioblastoma is shown in Fig. 9.12. Perfusion-weighted imaging using the Q2TIPS method shows a clear correlation to contrast-enhanced post-Gd-DTPA T1-weighted imaging in areas of high tumour vascularity (WEBER et al. 2004).

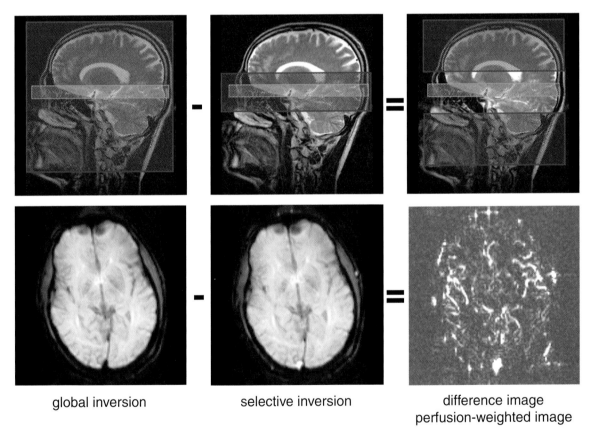

global inversion   selective inversion   difference image
                                                         perfusion-weighted image

Fig. 9.11. Principle of flow-sensitive alternating inversion recovery arterial spin-labeling technique: inversion slab (*red*), image slice (*green*). The difference between global and selective inversion results in a perfusion-weighted image. (Courtesy of A. KROLL)

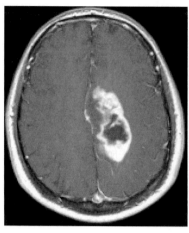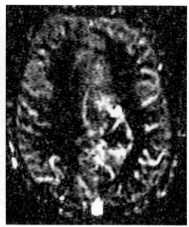

Fig. 9.12. Comparison of T1-weighted contrast-enhanced post Gd-DTPA (*left*) and perfusion-weighted imaging using the Q2TIPS method (*right*) in a patient with glioblastoma. Perfusion-weighted imaging shows a clear correlation to post Gd-DTPA T1-weighted imaging in areas of high tumour vascularity. (From WEBER et al. 2004)

9.8
Discussion

Our preliminary results show that BOLD imaging is a very attractive tool in the high-precision, high-dose radiotherapy concept of brain lesions. Functional MRI can help to define functional structures at risk, which have to be spared out. Thereby MRI quality assurance is essential in the stereotactic therapy concept, and geometric image distortions, due to the measuring sequence or to magnetic inhomogeneities, have to be corrected.

In 17 patients with angiographically proven cerebral AVMs, MR venography was able to detect all AVMs, whereas TOF MRA failed in 3 patients. In the delineation of venous drainage patterns MR venography was superior to TOF MRA; however, the method failed in the detection of about half of the feeding arteries. Due to susceptibility artefacts at air/tissue boundaries and interference with paramagnetic hemosiderin, MR venography was limited with respect to the delineation of the exact nidus sizes and shapes in 10 patients with AVMs located close to the skull base or having suffered from previous bleeding. Although the visualization of draining veins is an important prerequisite in the surgical and radiosurgical treatment planning of cerebral AVMs, MR venography was able to detect and delineate the exact venous drainage pattern in 20 of 25 draining veins (80%). Because delineation of the venous structures is of great importance for the treatment planning, MR venography may be a valuable diagnostic modality in this process.

The presentation of the venous drainage pattern also offers the possibility of an exact AVM grading and may help in the selection of the appropriate therapeutic approach. After therapy, the loss of ve-

nous structures may be used as an indicator of haemodynamic changes within the malformation which has to be evaluated in further studies. The HRBV has a limited value in the diagnostic work-up of those patients, where haemosiderin or air/tissue boundaries cause severe susceptibility artefacts, but it may be, on the other hand, of special importance in the detection and assessment of small AVMs, which are difficult to diagnose with other MR methods.

In patients with brain tumours high-resolution BOLD venography shows a complex and variable venous pattern in different parts of the lesion (oedema, contrast-enhancing area, central and/or necrotic part of the lesion). An interesting aspect of MR venography is the potential to probe tumour angiogenesis, which could be considered by our intensity-modulated treatment planning concept. In this aspect, the translation of a BOLD signal pattern measured in the tumour to a dose distribution is still an unanswered question and should be addressed in further clinical studies.

References

Aboagye EO, Maxwell RJ, Kelson AB, Tracy M, Lewis AD, Graham MA, Horsman MR, Griffiths JR, Workman P (1997) Preclinical evaluation of the fluorinated 2-nitroimidazole N-(2-hydroxy-3,3,3-trifluoropropyl)-2-(2-nitro-1-imidazolyl) acetamide (SR-4554) as a probe for the measurement of tumor hypoxia. Cancer Res 57:3314–3318

Baudelet C, Gallez B (2002) How does blood oxygen level-dependent (BOLD) contrast correlate with oxygen partial pressure (pO2) inside tumors? Magn Reson Med 48:980–986

Bortfeld T, Boyer AL, Schlegel W, Kahler DL, Waldron TJ (1994) Realization and verification of three-dimensional conformal radiotherapy with modulated fields. J Radiat Oncol Biol Phys 30:899–908

Buxton RB, Frank LR, Wong EC, Siewert B, Warach S, Edelman RR (1998) A general kinetic model for quantitative perfusion imaging with arterial spin labeling. Magn Res Med 40:383–396

Detre JA, Leigh JS, Williams DS, Koretsky AP (1992) Perfusion imaging. Magn Reson Med 23:37–45

Edelman RR, Siewert B, Adamis M, Gaa J, Laub G, Wielopolski P (1994) Signal targeting with alternating radiofrequency (STAR) sequences: application to MR angiography. Magn Reson Med 31:233–238

Essig M, Engenhart R, Knopp MV, Bock M, Scharf J, Debus J, Wenz F, Hawighorst H, Schad LR, van Kaick G (1996) Cerebral arteriovenous malformations: improved nidus demarcation by means of dynamic tagging MR-angiography. Magn Reson Imaging 14:227–233

Essig M, Reichenbach JR, Schad LR, Schoenberg SO, Debus J, Kaiser WA (1999) High-resolution MR venography of cerebral arteriovenous malformations. Magn Reson Imaging 17:1417–1425

Griffiths JR, Taylor NJ, Howe FA, Saunders MI, Robinson SP, Hoskin PJ, Powell ME, Thoumine M, Caine LA, Baddeley H (1997) The response of human tumors to carbogen breathing, monitored by gradient-recalled echo magnetic resonance imaging. Int J Radiat Oncol Biol Phys 39:697–701

Horsman MR (1998) Measurement of tumor oxygenation. Int J Radiat Oncol Biol Phys 42:701–704

Howe FA, Robinson SP, Rodrigues LM, Griffiths JR (1999) Flow and oxygenation dependent (FLOOD) contrast MR imaging to monitor the response of rat tumors to carbogen breathing. Magn Reson Imaging 17:1307–1318

Jezzard P, Balaban RS (1995) Correction of geometric distortion in echo planar imaging from B0 field variations. Magn Reson Med 34:65–73

Kim S-G (1995) Quantification of relative cerebral blood flow change by flow-sensitive alternating inversion recovery (FAIR) technique: application to functional mapping. Magn Reson Med 34:293–301

Luh WM, Wong EC, Bandettini PA, Hyde JS (1999) QUIPSS II with thin-slice TI_1 periodic saturation: a method for improving accuracy of quantitative perfusion imaging using pulsed arterial spin labeling. Magn Reson Med 41:1246–1254

Mansfield P, Pykett IL (1978) Biological and medical imaging by NMR. J Magn Reson 29:355–373

O'Donnell M, Edelstein WA (1985) NMR imaging in the presence of magnetic field inhomogeneities and gradient field nonlinearities. Med Phys 12:20–26

Ogawa S, Lee TM, Nayak AS, Glynn P (1990) Oxygenation-sensitive contrast in magnetic resonance image of rodent brain at high magnetic fields. Magn Reson Med 14:68–78

Reichenbach JR, Venkatesan R, Schillinger DJ, Kido DK, Haacke EM (1997) Small vessels in the human brain: MR venography with deoxyhemoglobin as an intrinsic contrast agent. Radiology 204:272–277

Schad LR (1995) Correction of spatial distortion in magnetic resonance imaging for stereotactic operation/treatment planning in the brain. In: Hartmann GH (ed) Quality assurance program on stereotactic radiosurgery. Springer, Berlin Heidelberg New York, pp 80–89

Schad LR (2001) Improved target volume characterization in stereotactic treatment planning of brain lesions by using high-resolution BOLD MR-venography. NMR Biomed 14:478–483

Schad LR, Lott S, Schmitt F, Sturm V, Lorenz WJ (1987a) Correction of spatial distortion in MR imaging: a prerequisite for accurate stereotaxy. J Comput Assist Tomogr 11:499–505

Schad LR, Boesecke R, Schlegel W, Hartmann G, Sturm V, Strauss L, Lorenz WJ (1987b) Three dimensional image correlation of CT, MR, and PET studies in radiotherapy treatment planning of brain tumours. J Comput Assist Tomogr 11:948–954

Schad LR, Ehricke HH, Wowra B, Layer G, Engenhart R, Kauczor HU, Zabel HJ, Brix G, Lorenz WJ (1992) Correction of spatial distortion in magnetic resonance angiography for radiosurgical planning of cerebral arteriovenous malformations. Magn Reson Imaging 10:609–621

Schad LR, Bock M, Baudendistel K, Essig M, Debus J, Knopp MV, Engenhart R, Lorenz WJ (1996) Improved target volume definition in radiosurgery of arteriovenous malformations by stereotactic correlation of MRA, MRI blood bolus tagging, and functional MRI. Eur Radiol 6:38–45

Schlegel W, Scharfenberg H, Doll J, Pastyr O, Sturm V, Netzeband G, Lorenz WJ (1982) CT-images as the basis of operation planning in stereotactical neurosurgery. Proceedings of First International Symposium on Medical Imaging and Image Interpretation. IEEE Computer Society, Silver Spring, pp 172–177

Schlegel W, Scharfenberg H, Doll J, Hartmann G, Sturm V, Lorenz WJ (1984) Three dimensional dose planning using tomographic data. Proc 8th International Conference on the Use of Computers in Radiation Therapy. IEEE Computer Society, Silver Spring, pp 191–196

Van Zijl PCM, Eleff SM, Ulatowski JA, Oja JME, Ulug AM, Traystman RJ, Kauppinen RA (1998) Quantitative assessment of blood flow, blood volume and blood oxygenation effects in functional magnetic resonance imaging. Nature Med 4:159–167

Weber MA, Thilmann C, Lichy MP, Günther M, Delorme S, Zuna I, Bongers A, Schad LR, Debus J, Kauczor HU, Essig M, Schlemmer HP (2004) Assessment of irradiated brain metastases by means of arterial spin-labeling and dynamic susceptibility-weighted contrast-enhanced perfusion MRI: initial results. Invest Radiol 39:277–287

10 Potential of Magnetic Resonance Spectroscopy for Radiotherapy Planning

Andrea Pirzkall

CONTENTS

10.1 Introduction

Metabolic imaging has been gaining increased interest and popularity among the radiation oncology community. The ongoing development of such powerful focal treatment delivery techniques such as intensity modulated radiation therapy (IMRT), radiosurgery, and HDR brachytherapy has generated renewed interest in dose escalation, altered fractionation schemes, and integrated boost techniques as ways to increase radiation effectiveness while reducing side effects. These focused radiation therapy (RT) techniques offer great potential in redefining the value of RT in certain disease sites, however, they also demand a more precise picture of the tumor, its extent and heterogeneity, in order to direct more or less dose to the appropriate areas while sparing as much normal tissue as possible (disease-targeted RT). This

A. Pirzkall, MD
Assistant Adjunct Professor, Departments of Radiation Oncology, Radiology and Neurological Surgery, University of California, San Francisco, 505 Parnassus Avenue, L-75, San Francisco, CA 94143, USA

"dose painting," a phrase appropriately coined by Clifton Ling, only makes sense if one knows the "lay of the land." Both CT and MRI have been providing the underlying sketch (black-and-white picture) used for target definition and treatment planning for several decades. They describe the anatomic picture of the tumor and surrounding structures well but they fail to provide information (color) about the functional status, the degree of tumor activity and cellular composition, and the presence of infiltrative disease or distant spread.

Functional or metabolic imaging is beginning to be used to fill this gap and provide the needed color. Techniques for acquiring such data include imaging that exploits glucose or amino acids labeled with radioactive isotopes [positron emission tomography (PET), single photon emission computed tomography (SPECT), and magnetic resonance (MR)-based imaging techniques (perfusion- and diffusion-weighted MRI, magnetic resonance spectroscopy imaging (MRSI)].

This chapter examines multi-voxel proton (^{1}H) MRSI, the most advanced of the MR spectroscopy applications. Three-dimensional multivoxel MRSI has improved the discrimination of cancer from surrounding healthy tissue and from necrosis by adding metabolic data to the morphologic information provided by MRI. Two major disease sites have been studied at University of California, San Francisco (UCSF) with respect to the potential and actual incorporation of MRS imaging into the treatment planning process for RT and are the subject of this chapter: prostate cancer and brain gliomas. Brain gliomas have traditionally proved difficult to image using standard means due to their literally "non-visible" infiltrative behavior and heterogeneous composition. The extent and location of prostate cancer are similarly difficult to determine without the aid of MRSI because, although MRI has good sensitivity, it has relatively low specificity. In these two disease sites, the combination of MRI and MRSI provide the structural and metabolic information required to support an improved assessment of aggressiveness, location, extent and spatial distribution.

10.2
MR Spectroscopy Imaging

10.2.1
Background

Magnetic resonance imaging (MRI) is a noninvasive imaging technique that makes use of the fact that certain atomic nuclei, such as ^1H, ^{31}P, ^{19}F, and ^{13}C, have inherent spin properties that allow them to acquire discrete amounts of energy in the presence of a static magnetic field. The application of electromagnetic fields (nonionizing radiofrequency radiation) at right angles to a static magnetic field causes these nuclei to jump to higher energy levels. After removal of the electromagnetic fields, the nuclei subsequently drop back to their original spin states by emitting electromagnetic radiation at a rate that can be characterized by their T1 (spin-lattice) and T2 (spin–spin) relaxation times. A receiver coil detects the emitted radiation and records the time domain of the MR signal that, once processed using a fourier transform, reveals the spectrum of intensities and frequencies of the nuclei from different chemical species within the excited volume. The location of peaks in the spectrum defines the chemicals within the sample. The peak intensity reflects their concentration. Conventional MRI uses the properties of the protons from water to obtain information about their spatial distribution in different tissues. Specialized radiofrequency pulses and magnetic field gradients are used to label the water signal as a function of space and, after appropriate post-processing, provide an anatomic image of the changes in proton density and relaxation properties.

The MR spectroscopic data is typically acquired by suppressing the large signal from water and allowing the properties of other compounds to be recorded and analyzed. Water suppressed ^1H spectroscopy techniques are commercially available for obtaining spectra from selected regions within the brain and prostate and can be combined with additional localization techniques to produce either a single spectrum from a region of interest (single-voxel MRS) or a multidimensional array of spectra from the region of interest [3D multivoxel MRS, MRSI, chemical-shift imaging (CSI)]. The peaks in individual spectra reflect the relative concentrations of cellular chemicals within that spatial location. The peak heights and/or areas under the curve relate to the concentration of the respective metabolites. Differences in these concentrations can be used to distinguish "normal/healthy" tissue from neoplastic or necrotic tissue. As

an efficient method for obtaining arrays of spatially localized spectra at spatial resolutions of 0.2–1 cc, 3D multivoxel MRSI is of greater potential value than single-voxel MRS in target delineation and monitoring response to therapy and allows the generation of maps of the spatial distribution of cellular metabolites. This is an ideal representation for integrating the information into RT treatment planning.

A significant advantage of ^1H-MRSI over other metabolic imaging techniques is that the data can be obtained as part of a conventional MRI and can be directly overlaid upon each other. This enhances the display of metabolic data and allows it to be correlated with the anatomy as revealed by MRI. This allows areas of anatomic abnormality to be directly correlated with the corresponding area of metabolic abnormality.

10.2.2
MRSI Technique

10.2.2.1
Brain

Details of the MRI and MRSI examinations that were developed at UCSF have been published previously (NELSON 2001; NELSON et al. 2002). Briefly, MR data are acquired on a 1.5-T scanner (General Electric Medical Systems, Milwaukee, Wis.) using a quadrature head coil. The MR images include axial T1-weighted [pre- and post-gadolinium contrast-enhanced 3D spoiled gradient-echo (SPGR), 1.5-mm slice thickness] and axial T2-weighted sequences [fluid-attenuated inversion recovery (FLAIR), 3-mm slice thickness]. At the conclusion of the MRI sequences, a 3D multivoxel MRSI sequence with a 1-cc nominal resolution is performed (TR/TE=1000/144 ms; 12×12×8 or 16×8×8 phase-encoding matrix) applying PRESS (Point REsolved Spatial Selection) volume localization and very selective saturation (VSS) bands for outer-voxel suppression. In order to allow for the acquisition of control spectra, the PRESS-box volume is positioned so as to extend beyond the suspected disease and to include normal brain from the contralateral hemisphere. The PRESS-box also is positioned to avoid areas of subcutaneous lipid and varying magnetic susceptibility that might compromise the quality of the spectra. In some cases this means that it is impossible to cover the entire region of suspected disease. All data are aligned to the MR images obtained immediately prior to the MRSI data (usually the post-contrast T1-weighted image).

The raw spectral data are reconstructed using software developed at the UCSF Magnetic Resonance Science Center. The peak parameters (height, width, area) for Cho, Cr, NAA and, with the recent implementation of lactate edited sequences, Lac and Lip are estimated on a voxel-by-voxel basis within the excited region. The brain MRSI protocol has an average acquisition time of 17 min and is performed as an add-on to the conventional MRI exam. This means that the total examination time is approximately 1 h, including patient positioning.

10.2.2.2
Prostate

Details of the MRI/MRSI technique have been published previously (Hricak et al. 1994; Kurhanewicz et al. 2000). Briefly, patients are scanned on a 1.5-T scanner (GE Medical Systems, Milwaukee, Wis.) in the supine position using the body coil for excitation and four pelvic phased-array coils (GE Medical Systems, Milwaukee, Wis.) in combination with a balloon-covered expandable endorectal coil (Medrad, Pittsburgh, Pa.) for signal reception. Axial spin-echo T1-weighted images are obtained from the aortic bifurcation to the symphysis pubis, using the following parameters: TR/TE=700/8 ms; slice thickness=5 mm; interslice gap=1 mm; field of view=24 cm; matrix 256×192; frequency direction transverse; and 1 excitation. Thin-section high spatial resolution axial and coronal T2-weighted fast spin-echo images of the prostate and seminal vesicles are obtained using the following parameters: TR/effective; TE 6000/96 ms; eCho-train length=16; slice thickness=3 mm; interslice gap=0 mm; field of view=14 cm; matrix 256×192; frequency direction anteroposterior (to prevent obscuration of the prostate by endorectal coil motion artifact); and 3 excitations.

After review of the axial T2 weighted images, a spectroscopic imaging (MRSI) volume is selected to maximize coverage of the prostate, while minimizing the inclusion of periprostatic fat and rectal air. This is achieved using a water- and lipid-suppressed double-spin-echo PRESS sequence, in which both 180° pulses are replaced with dualband phase compensating spectral/spatial pulses designed for robust spectroscopic volume selection and water and lipid suppression (Schricker et al. 2001). The ability to include all of the prostate in the spectroscopic volume while minimizing spectral contamination from outside the gland has recently been improved by applying very selective suppression (VSS) pulses to better define the sides of the PRESS volume and to conform the

selected volume to fit the shape of the prostate (Tran et al. 2000). Subsequent to volume selection, 16×8×8 phase encoding is applied to produce spectral arrays having spectroscopic voxels with a nominal spatial resolution of 0.3 cm^3, TR/TE=1000/130 ms.

The raw spectral imaging data is reconstructed using in-house developed software tools. This postprocessing involves automatic phasing, frequency alignment, and baseline correction, thereby revealing the "hidden" information on the chemical composition of tissues of interest within each spatial element (voxel). Prostate MRSI protocols have an acquisition time of 17 min, giving a total examination time of 1 h, including patient positioning and coil placement. Display and interpretation of the data is achieved using custom design software developed using IDL language (Research Systems, Boulder, Colo.)

10.2.3
Application of MRSI

10.2.3.1
Brain

General

Brain gliomas represent the most common brain tumor in adults. The outcome for high-grade gliomas (HGG) remains dismal with a mean survival of 9–12 months for grade IV and 20–36 months for grade III, despite multimodality treatment approaches (Bleehen and Stenning 1991). The tumors are poorly demarcated from surrounding tissues, having a nearly undecipherable zone of infiltration that has been documented by histopathologic correlation studies (Kelly et al. 1987; Burger et al. 1988).

Efforts to increase local control and ultimately survival have failed in newly diagnosed GBM (Hochberg and Pruitt 1980; Liang et al. 1991) except for a relatively small number of recent studies. Sneed et al. (1996) could demonstrate a dose-response relationship for interstitial brachytherapy boost irradiation employing temporary 125-I sources after resection and fractionated RT. They determined that a higher minimum brachytherapy tumor dose was strongly associated with better local control and ultimately survival. Adverse effects and life-threatening necrosis, however, were observed after exceeding an optimal tumor dose of 47 Gy (corresponding to about 65 Gy to 95% of the tumor volume). Fitzek et al. (1999) reported not only a change in recurrence patterns but also significantly improved survival after hyperfractionated mixed photon/proton RT in patients with GBM after surgical resection of various extent. In this case, however, the need for repeat surgical removal due to necrosis was quite considerable. On the other hand, escalating dose in the conventional fashion of 2 Gy increments per day beyond 70 Gy does not improve outcome according to the latest published experience from the University of Michigan (CHAN et al. 2002). Patients in the dose escalation arms 80 and 90 Gy did not demonstrate superior outcome compared with the 70 Gy arm, but they did experience increased incidence of side effects.

These recent results encourage the exploration of different fractionation schemes coupled with a refined target definition. IMRT offers the ability to deliver simultaneously different doses to different regions. For instance, while the CTV could be treated to the conventional 2.0 Gy (or 1.8 Gy), the GTV simultaneously could receive an increased dose of 2.5–3.0 Gy, or even higher (PIRZKALL 2005). Such a dose regimen would significantly increase the biologically effective dose to defined tumor areas but would not result in an increase in dose to surrounding organs compared

with conventional 3D stereotactic conformal RT. This should not increase the rate of complications, but there would be an increased inhomogeneity at the border of the GTV and CTV (THILMANN et al. 2001).

Changing the target may be necessary in order to garner the greatest benefit from differential dose delivery techniques such as IMRT or RS. In this case, it is critical that the regions identified for special attention be defined accurately; namely, that the areas of active tumor (suitable for high dose) must be identified separately (as gross tumor volume, GTV) from areas suspicious for tumor extension that are appropriate to receive a lower dose (classified as clinical target volume, CTV). MRSI may be able to provide the information required to make such target definitions.

MRSI Analysis

At long echo times (TE=144 ms), MRSI provides information about tumor activity based on the levels of cellular metabolites such as choline (Cho), creatine (Cr), N-acetylaspartate (NAA), and lactate/lipid (Fig. 10.1).

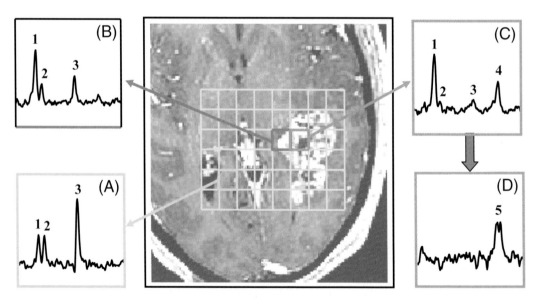

Fig. 10.1a–d. Patient with left temporo-occipital glioblastoma multiforme (GBM). Axial T1-weighted post-contrast MRI shows typical contrast enhancement (CE) with central necrosis. Three-dimensional magnetic resonance spectroscopy imaging (MRSI) performed within area of excitation [Point REsolved Spatial Selection (PRESS) box, represented by *gridlines*] reveals metabolic signature of examined brain tissue. Four examples of spectral patterns are given detecting the following metabolites: *1* choline (Cho); *2* creatine (Cr); *3* N-acetylaspartate (NAA); *4* lipid (Lip); and *5* lactate (Lac). Normal brain tissue is marked by high peak of NAA and low peak of Cho. The resulting Cho-to-NAA ratio is therefore low (about 1:2). Cho and Cr are exhibiting similar peak heights. Tumor spectrum is characterized by an increase of Cho and a decrease in NAA as compared with the normal tissue voxel (**a**). The Cho-to-NAA ratio is high (about 1:0.5). Note that the spectrum is derived from a single voxel that contains only partially CE. **c, d** Mix of tumor and necrosis revealed by lactate-edited sequences which are postprocessed to separate Lip and Lac that overlap due to resonating at the same frequency. **c** Summed spectrum shows peaks of high Cho and extremely diminished NAA (Cho-to-NAA ratio is about 1:0.3), and in addition, the presence of Lip. Note that the spectrum is derived from a single voxel that contains partially CE and macroscopic necrosis. **d** Difference spectrum allows quantification of Lac as a marker of hypoxia. The Cho-to-NAA index (CNI) values for the above respective voxels are: **a** –0.6; **b** 3.6; **c, d** 3.4

Choline is a membrane component that is increased in tumors; NAA is a neuronal metabolite, not present in other CNS cells, that is decreased in tumors; creatine is necessary for cellular bioenergetic processes and osmotic balance and might be a good marker for cell density; lactate is an end product of anaerobic metabolism; and lipid peaks are generally correlated with necrosis and represent cellular breakdown (NELSON et al. 2002). Typical spectra for normal brain tissue, tumor, and necrosis are displayed in Fig. 10.1.

Work in our department has shown that combinations of changes in these metabolites, rather than changes in the levels of individual metabolites themselves, may be of even greater value in determining the presence and extent of CNS disease. At UCSF, metabolic indices, such as CNI (Cho to NAA index), CCrI (Cho to Cr index), and CrNI (Cr to NAA index), are calculated within each voxel for each study using an automated regression analysis process developed in our laboratory (MCKNIGHT et al. 2001). This technique allows us to describe the number of standard deviations of difference between the ratio of two metabolites within one voxel and the mean ratio in control voxels from that same exam (i.e. from the contralateral side. For instance, a CNI of 2 or greater (95% confidence limit) has been shown to correspond to tumor (MCKNIGHT et al. 2002). Work is ongoing to determine the value and meaning of other ratios.

Image Registration

In order to be useful for comparative data analysis and for treatment planning, the MRSI data have to be correlated with the CT treatment planning data set. We register the MR images from the MRSI examination to the CT. A combination of internal markers as well as volume- and surface-matching techniques have been implemented to achieve this task (GRAVES et al. 2001; NELSON et al. 2002). The MRI data are then reformatted and the MRSI data reconstructed to correspond to the plane and orientation of the CT images. Values of metabolic indices are then calculated, interpolated to match the resolution of the MRI, translated into contour lines corresponding to metabolite and ratio values, and superimposed on the anatomic images (Fig. 10.2). Combined MRI/MRSI data are transferred to a clinical PACS workstation using a DICOM protocol from where they can be directly uploaded onto the treatment planning workstation.

Similarly, CNI contours are superimposed onto treatment planning MR images for gamma knife (GK) RS. Ideally, those contours are generated at the day of treatment based on the MRI/S data acquired

that same morning. If that was not possible due to scheduling issues, then the CNI contours are generated based on a MRSI exam acquired a few days prior to the actual treatment. In this case, prior alignment of the MRSI and the treatment planning MRI has to be performed as described above.

Tumor Analysis

We performed several studies using these metabolic indices to compare the spatial extent and heterogeneity of metabolic (MRSI) and anatomic (MRI) information in patients with newly diagnosed (PRIZKALL et al. 2001, 2002) and surgically resected (PIRZKALL et al. 2004) gliomas in order to explore the value that MRSI might have for defining the target for radiation therapy in brain gliomas. In particular, significant differences have been found between anatomic and metabolic determinants of volume and spatial extent of the neoplastic lesion for patients with newly diagnosed GBM (PIRZKALL et al. 2001). Similar but more pronounced findings were also observed for grade-III gliomas. Taken together, these findings suggest that MRSI-derived volumes are likely to be more reliable in defining the location and volume of microscopic and actively growing disease when compared with conventional MRI.

It was also shown that MRSI is more sensitive in identifying the presence of residual disease compared with MRI alone for patients studied after surgical resection of their HGG (PIRZKALL et al. 2004). These differences likely have an enormous impact on the management of gliomas in general, especially for malignant lesions. Any type of local or targeted, nonsystemic therapy that relies solely on anatomic information may suffer from insufficient information about the extent and composition of tumor, resulting in potential over and/or under treatment of involved regions.

Preliminary evaluation of MRSI follow-up exams that were performed post-RT has shown a predictive value for MRSI with respect to focal recurrence (PIRZKALL et al. 2004). For 10 patients without contrast-enhancing residual disease following surgical resection we were able to establish a spatial correspondence between areas of new CE, developed during follow-up, and areas of CNI abnormality, as assessed after surgery but prior to RT. We found a very strong inverse correlation between the volume of the CNI abnormality and the time to onset of new contrast enhancement – the greater the volume of CNI, the shorter the time to recurrence. The MRSI has also proved to be of value in predicting overall

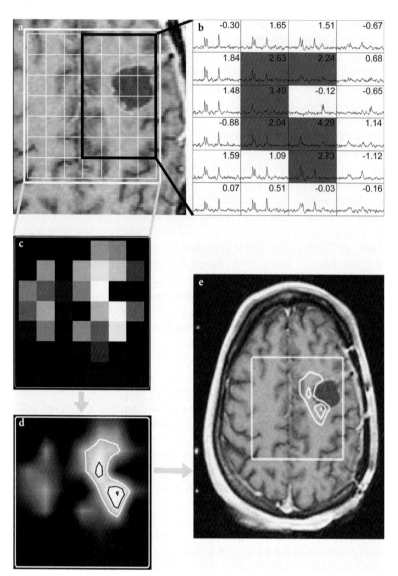

Fig. 10.2a–e. Patient with recurrent, initially low-grade glioma; status post-resection and fractionated RT with 59.4 Gy. A subsequent boost with gamma knife (GK) radiosurgery was planned based upon MRI/MRSI. **a)** T1-weighted axial MRI with superimposed MRSI area of excitation. **b)** Enlarged spectra and actual Cho-to-NAA index (CNI) for a subset of voxels in immediate vicinity of the resection cavity. *Shaded voxels* highlight those with a CNI of ≥ 2. **c)** Gray-scale CNI image. The brighter the voxels, the higher the respective CNI. **d)** High-resolution CNI image resulting from sampling the low-resolution CNI image (**c**) to match the resolution of the MR image. Superimposed are CNI contours of 2 (*bright line*), 3 (*dark middle contour*), and 4 (*dark inner contour*) as a result of interpolation. **e)** The CNI contours of 2, 3, and 4 superimposed onto the respective MRI slice in preparation for treatment planning

survival in patients with GBM; the larger the volume of the CNI abnormality the shorter the survival (OH et al. 2004). Similarly, a low value of normalized apparent diffusion coefficient (ADC) within the T2 hyperintensity corresponded to shorter survival in the same patient cohort (OH et al. 2004). These findings go along with a prior study that showed a significant inverse correlation of ADC and cell density (GUPTA et al. 2000). Other MRSI measures associated with poorer outcome are higher ratios of Cho-to-Cr, Cho-to-NAA, LL and a lower ratio of Cr-to-NAA (LI et al. 2004).

Additional metabolic indices have been evaluated by LI et al. (2002): Cho-to-NAA index (CNI), Cho-to-Cr index (CCrI), Cr-to-NAA index (CrNI), and lactate/lipid index (LLI). These studies suggest that tumor burden, as measured with either the volume of the metabolic abnormalities or the maximum magnitude of the metabolic indices, correlates with the degree of malignancy. The spatial heterogeneity within the tumor, and the finding that metabolic disease activity appears to extend beyond MRI changes, may be responsible for the continuing failure of current treatment approaches.

Impact on Treatment Planning

The greatest clinical value of a targeted therapeutic intervention for gliomas will be achieved only if "all" of the active disease is addressed and if normal tissue is spared as much as possible. Accurate target definition thus becomes of prime importance. The variability in tumor cell distribution makes it difficult to define an effective yet safe margin for purposes of RT treatment planning. A uniform margin can be expected to cover either too much noninfiltrated brain or to leave out small areas of tumor infiltration from the treatment volume; the latter is undesirable because it will increase the likelihood of local recurrence; the former is equally unacceptable because it is clear that radiation-induced CNS toxicity is related not only to dose but to volume as well (MARKS et al. 1981). Since focal RT appears to decrease neuropsychologic sequelae when compared with large-volume RT (HOCHBERG and PRUITT 1980; MAIRE et al. 1987), and since the goal of any revised treatment protocol is to prolong survival, a margin that is tailored to tumor extension, rather than one that is uniform, could be presumed to improve quality of survival as well as local control.

10.2.3.1.1
High-Grade Glioma

In the treatment of high-grade gliomas with RT, the standard definition of the target volume is the contrast-enhancing area, as determined based on a contrast-enhanced T1-weighted MRI or a CT scan, plus a margin of 1–4 cm to account for "invisible" tumor infiltration (BLEEHEN and STENNING 1991; LIANG et al. 1991; GARDEN et al. 1991; WALLNER et al. 1989). The RTOG guidelines recommend to treat the T2 hyperintensity plus a 2-cm margin to 46 Gy and deliver the remaining 14 Gy to a total dose of 60 Gy after performing a cone-down to the contrast enhancement plus a 2.5-cm margin. Dose escalation protocols deliver an additional dose to the CE area itself. A lower dose may be delivered to the T2-weighted region of hyperintensity plus a variable margin (FITZEK et al. 1999; NAKAGAWA et al. 1998).

We have conducted a series of studies to determine the impact the use of MRSI might have on treatment plans for gliomas. In a preliminary study of 12 newly diagnosed grade-IV gliomas, the regions with CNI 2, 3, and 4 were compared with the MRI enhancing volume (PIRZKALL et al. 2001). Adding the volume where the CNI index is greater than 2, 3, or 4 and extends beyond the volume of CE to the volume of CE would increase the size of the dose escalation target by 150, 60, and 50%, respectively. The median distances that the CNI regions extended outside the enhancing volume were 1–2 cm. Conversely, the volume where the CNI was greater than or equal to 2 was on average only about 50% of the volume of T2 hyperintensity. While it did extend beyond the T2 hyperintensity in

some cases, this would only have extended the T2 lesion by about 10% if it had been included for targeting purposes.

The analysis of metabolic lesions for 22 grade-III gliomas show even more dramatic results than for the grade-IV gliomas (PIRZKALL et al. 2001). Adding the volume of CNI extending outside the volume of CE would increase the size of the target by 500, 300, and 150% for CNI abnormalities of 2, 3, and 4, respectively. The median distances that the CNI regions extended outside the enhancing volume were 2–3 cm. The volume where the CNI was 2 or greater was on average about 70% of the T2 lesion, and while it did extend beyond the T2 lesion, it would have only extended the target by about 15%.

10.2.3.1.2
Low-Grade Glioma

The definition of target volumes for low-grade gliomas is equally difficult. Margins are defined usually for the purpose of encompassing suspected microscopic spread of the tumor (CTV). The size of the margin required to define the CTV remains problematic since these tumors are poorly demarcated with respect to normal surrounding tissue and are infiltrative in nature. A differentiation between normal brain, edema, and tumor infiltration based on MRI is therefore quite difficult. In a study similar to the one reported for high-grade gliomas, analysis of the CNI contours for 20 grade-II gliomas showed that the metabolic abnormality was usually within the T2 lesion. When it extended outside T2 (in 45% of patients), the extension was relatively small and usually directed along white matter tracks (PIRZKALL et al. 2002). In the two grade-II patients who had gadolinium enhancement, the maximum CNI corresponded with the small volume of patchy enhancement. If the treatment volume was modified from the common target volume of T2 hyperintensity plus a 2–3 cm margin (MORRIS et al. 2001), such that it included only the T2 lesion plus the region of CNI greater than 2, there would have been a substantial reduction in the target volume and hence, presumably, in the radiation damage to normal brain tissue.

Although it is not clear how planning margins might be adjusted for low-grade gliomas based on MRSI findings, it is clear that MRSI would provide a different estimate of gross, clinical, and boost target volumes depending on how those are defined for a given patient. We believe that our findings should be taken as a justification to consider a reduction in treatment planning margins for patients with low-grade glioma, paying careful attention to preferential avenues of spread of disease. It is also interesting to note that even low-grade gliomas demonstrate regions of varying metabolic activity within their volume. The possible differentiation of more or less radiosensitive/resistant areas, as defined by metabolic measures, may play an important role in the future diagnosis and treatment of low-grade gliomas.

In addition to being of value in determining the overall volume appropriate for treatment, the metabolic heterogeneity defined by MRSI might be of particular interest when deciding on "differential" therapy to delineated regions within brain gliomas. MRSI might offer valuable information on the relative radiosensitivity or radioresistance of certain tissue types within a neoplastic lesion and thus on desired dose. For instance, the presence of lactate can be considered an indicator of anaerobic metabolism, suggestive of poor blood supply and therefore the presence of hypoxic cells within the tumor. Targeting these areas with focal higher doses may be appropriate. In contrast, regions with increased Cr relative to NAA reflect high cellularity and active bioenergetic processes (phosphorylation of ADP); therefore, regions of tumor with a CrNI ≥ 2 may indicate radiosensitive areas that might be adequately treated with a lower dose.

10.2.3.2
Radiosurgery for Recurrent Gliomas

The idea of combining IMRT and MRSI data to deliver extremely sophisticated treatment plans with differential doses delivered across the target volume is still quite "theoretical" for the treatment of brain gliomas; however, its value in radiosurgery (RS) for recurrent malignant gliomas has a stronger bias. Although the prognosis for recurrent gliomas is poor in general, focal re-treatment has been shown to prolong survival. Radiosurgery, although seemingly counterintuitive for an infiltrating disease such as recurrent gliomas, offers an effective and noninvasive treatment option, with median survival times after radiosurgery for grade-IV gliomas in the order of 8–13 months (CHAMBERLAIN et al. 1994; HALL et al. 1995; LARSON et al. 1996). Recurrence patterns, however, are predominantly local and marginal in up to 80% (HALL et al. 1995; SHRIEVE et al. 1995).

GRAVES et al. (2000) studied the prognostic value of MRSI in gamma knife (GK) RS of recurrent malignant gliomas. Thirty-six patients with recurrent gliomas were retrospectively divided into two groups depending on whether the metabolic lesion was con-

fined to the radiosurgical target or whether it extended well outside the target. Among the patients with recurrent GBM, the most significant finding was a poorer median survival of 36 weeks in patients with metabolic extension vs 96 weeks for patients with no metabolic abnormality beyond the radiosurgical target at the time of treatment. In addition, a significant difference was found between the two groups in the volume of CE as assessed throughout the follow-up period, with an increase in CE in the patient group with metabolic abnormality beyond the region of treatment. In some cases the metabolic lesion outside the treatment area would have been small enough to be included within the GK target, but in others the lesion would have been too large to be treated adequately. A more recent study by CHAN et al. (2004) employed the Cho-to-NAA Index (CNI) to examine the degree of overlap between the radiosurgically treated volume and the volume where the CNI was greater than or equal to 2. Based on degree of overlap, patients were divided into two risk groups and it was found that the survival for the low-risk group was significantly longer than for the high-risk group.

This analysis prompted us to look at the potential impact that MRSI would have on patient selection and treatment plans for RS of recurrent gliomas (PIRZKALL et al. 2001). In 18 patients with recurrent glioma (4 anaplastic astrocytomas and 14 GBM) treated with GK at UCSF, the radiosurgical target volume (contrast enhancement plus 1–2 mm) was compared with the CNI volumes derived from the prior to GK acquired MRSI as described above. The CNI abnormality was found to extend beyond the CE to some degree in all patients. This extension was significant (>6 cc) in 11 of 18 patients (61%). In 4 of these 11 patients, the metabolic abnormality was completely outside the CE. If adding the CNI abnormality to the volume of contrast enhancement, the standard target volume would increase by 9–277% (0.5–27 cc). Such target definition would require a change in the dose prescription according to the GK dose-volume relationship in 12 of 18 patients (67%). In fact, 4 patients (22%) would have had a treatment volume exceeding 20 cc and would therefore not be considered eligible for RS in most institutions.

As an extension to this study, we are currently examining the follow-up studies in patients who exhibited metabolic activity beyond the volume of contrast enhancement that remained outside the radiosurgical target volume in order to assess recurrence patterns. These data are now available (CHUANG et al. 2004).

We are also developing an institutional protocol that will investigate the incorporation of CNI con-

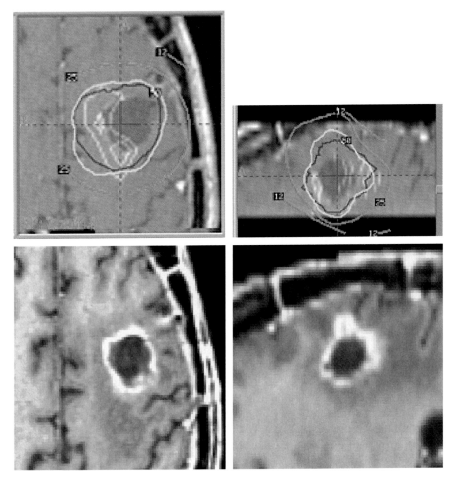

Fig. 10.3. Same case as in Fig. 10.2. *Upper row:* GK treatment plan for the respective case in axial and sagittal plane. The *inner dark line* shows the radiosurgical target that encompasses the resection cavity and tumor-suggestive areas according to the CNI. Isodose lines (IDL) of 50 (prescription IDL), 25, and 12.5%, respectively, are shown as *brighter lines* surrounding the target (from inside to outside). *Bottom row:* Follow-up MRI at 4 months post-GK shows contrast enhancement particularly in the medio-posterior and superior aspect worrisome of tumor recurrence. Continued increase in contrast enhancement over the subsequent 2 months prompted a re-resection. Histopathology revealed gliosis and necrotic tissue compatible with radiation effect

tours into the selection process and the target definition for RS (see Fig. 10.3), where appropriate (e.g. noneloquent brain regions), and will implement a phase-I study that will assess toxicity of such approach.

10.2.3.3
Treatment Protocol for Newly Diagnosed GBM

Combining metabolic imaging guidance by means of MRSI (and proton-weighted MR, diffusion-weighted MR) with the powerful capability of IMRT to increase dose selectively while simultaneously prescribing a conventional-like dose to areas at lower risk has been the goal of our group for several years. We have tested the feasibility of incorporating MRSI data into the IMRT treatment planning process and established the necessary image data analysis and transfer (GRAVES et al. 2001; NELSON et al. 2002). We previously discussed the possibility of defining the GTV or boost volume according to the contrast enhancement enlarged by the CNI abnormality and the CTV according to the T2 hyperintensity plus the CNI, a proposal that seemed reasonable based on the assumption that the CNI correlates with cell density. This latter fact has been confirmed for grade-III gliomas (MCKNIGHT et al. 2002) but is currently still under investigation for GBM samples. Recent studies have suggested that the use of CNI abnormalities to enlarge the definition of the GTV may not be the optimal approach. These studies showed that the addition of CNI abnormality to the volume of contrast enhancement would increase its average volume by 60% (CNI ≥ 3) and 50% (CNI ≥ 4), if the contrast enhancement alone would have been defined (PIRZKALL et al. 2001). In addition, preliminary patterns of recurrence analysis in presumed gross totally resected GBM after conventional RT with 60 Gy indicate that the first onset of new contrast enhancement during follow-up was contained within the overall extent of pre-RT metabolic abnormality (CNI 2) in 7 of 8 patients (PIRZKALL et al. 2004), but even the volumes of CNI greater than or equal to 3 or 4 were still much larger than the volume of new contrast enhancement at first onset. This suggests that the volume of CNI 3 and 4 would be too generous for many patients. Correlation to other metabolic indices, such as CrNI, CCrI, and LL, did not reveal a uniform pattern either. Further evaluation of recurrence patterns in this patient group is currently ongoing based on a larger patient population with the addition of new pulse sequences so that lactate and lipid resonances can be separated for improved interpretation.

Another obstacle that still needs to be addressed is extending the coverage of the tumor and surrounding areas while maintaining sufficient coverage of normal-appearing brain tissue. This is part of a technical development effort that will be undertaken shortly. Once these limitations are addressed, a clinical phase-I study that uses metabolic imaging-enhanced target definition and dose prescription by means of IMRT will be initiated for newly diagnosed patients with GBM.

10.2.3.4
Assessing Response to RT

The interpretation of MRI changes following RT continues to pose a problem for all clinicians who are trying to decide whether a tumor has recurred and needs additional treatment. A continuous increase in contrast enhancement of more than 25% has been considered by most protocols as a benchmark for assessing tumor recurrence. In more recent studies, additional MR parameters are being looked at, including MRSI, as well as perfusion- and diffusion-weighted MRI.

Several groups are currently investigating the value of MRSI for assessing response to RT. Patients found to respond to therapy have been reported as showing a reduction in Cho, whereas patients who progressed exhibit a retention or increase in Cho (WALD et al. 1997; TAYLOR et al. 1996; SIJENS et al. 1995). Another indicator of therapy response is the occurrence or increase in the Lac/Lip peak. Those changes have to be interpreted in light of changes in normal tissue as well. VIRTA (2000) et al. describe transient reductions in NAA and relative increases in Cho in normal-appearing white matter following RT. Late effects of radiation were studied by ESTEVE et al. (1998) who reports on reductions in Cho, Cr, and NAA in regions of T2 hyperintensity and reductions in Cho in normal-appearing white matter. LEE et al. (2004) tried to quantify the dose dependent effect in normal tissue post RT. The Cho-to-NAA ratio for instance, exhibited its steepest dose-dependent increase to 117–125% of its original value by 2 months post-

RT and underwent a dose-dependent recovery later on but did not return to its original value within the 6 month follow-up. A similar trend in the Cho-to-Cr ratio and separate metabolite analyses suggested that Cho underwent the most significant changes. These normal tissue changes have to be taken into account when trying to assess tumor response to therapy.

10.2.4
Prostate

General

Modern RT delivers doses of more far more than 70 Gy in patients with T1–T3 prostate in response to findings of (a) increased biopsy-proven local recurrences after CRT of 70 Gy and less (CROOK et al. 1995), and (b) decreased probability of relapse with increasing doses (POLLACK and ZAGARS 1997). Three-dimensional CRT and IMRT have been employed as a means to escalate dose.

ZELEFSKY et al. (2001) propose to treat localized prostate cancer with 81 Gy claiming superior control and decreased toxicity if treated with IMRT based on preliminary data. Even though it remains to be seen whether the shapes of the dose-response curve and complication-probability curves justify doses above 75.6 Gy, being able to direct/intensify dose at the area of highest risk appears beneficial.

On T2-weighted MRI, regions of cancer within the prostate demonstrate lower signal intensity relative to healthy peripheral zone tissue as a result of loss of normal ductal morphology and associated long T2 water signal. Efforts to identify the location of tumor within the prostate have included digital rectal examination, prostate-specific antigen (PSA), PSA density, systematic biopsy, transrectal ultrasound (TRUS), and MRI, but none of those have provided satisfying results for disease targeted therapies.

Recent studies in pre-prostatectomy patients have indicated that the metabolic information provided by MRSI combined with the anatomical information provided by MRI can significantly improve the assessment of cancer location and extent of disease within the prostate, the magnitude of extracapsular spread, and the degree of cancer aggressiveness. The MRI alone has good sensitivity (78%) but low specificity (55%) in identifying the tumor location within the prostate because of a large number of false positives attributable to noncancerous changes such as post-biopsy hemorrhage, prostatitis, and therapeutic effects (HRICAK et al. 1994). The addition of MRSI has been shown to improve the detection of cancer within a sextant of the prostate, with a sensitivity of 95% and a specificity of 91% (SCHEIDLER et al. 1999). The addition of a positive sextant biopsy to MRI/MRSI increases specificity and sensitivity for cancer localization even further to 98 and 94%, respectively (WEFER et al. 2000). Additionally, pre- and post-therapy studies have demonstrated the potential of MRI/MRSI to provide a direct measure of the presence and spatial extent of prostate cancer after therapy, a measure of the time course of response, and information concerning the mechanism of therapeutic response.

MRSI analysis

Healthy prostatic metabolism is identified based on the presence of high levels of citrate and lower, approximately equal, levels of choline, creatine, and polyamines (KURHANEWICZ et al. 2002). Similarly, areas of benign prostatic hyperplasia (BPH) and/or atrophy can be distinguished by their own unique spectral pattern. The prostatic gland exhibits distinct

metabolic differences between central gland and peripheral zone. Currently, MRSI is limited in identifying cancer within the central gland since there is significant overlap between the metabolic pattern associated with predominately stromal BPH and prostate cancer (ZAKIAN et al. 2003). Approximately 68% of prostate cancers, however, reside in the peripheral zone (STAMEY et al. 1988) where there is a mean threefold reduction in prostate citrate levels and a twofold elevation of prostate choline levels relative to surrounding healthy tissue (KURHANEWICZ et al. 1996). As a result, it is noted that spectroscopy data refer mostly to findings within the peripheral zone.

Prostate cancer can be discriminated metabolically from the healthy peripheral zone based on significant decreases in citrate, zinc, and polyamines and an increase in choline (Fig. 10.4). The decrease in citrate is due both to changes in cellular function (COSTELLO and FRANKLIN 1991a,b) and changes in the organization of the tissue that loses its characteristic ductal morphology (KAHN et al. 1989; SCHIEBLER et al. 1989). Its decrease is closely linked with a dramatic

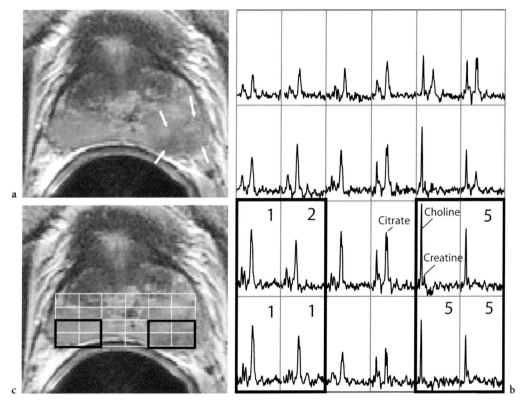

Fig. 10.4. a A representative reception-profile-corrected T2-weighted fast-spin-echo (FSE) axial image demonstrating a tumor in the left midgland to apex. **b** Superimposed PRESS-selected volume encompassing the prostate with the corresponding axial 0.3 cm³ proton spectral array. **c** Corresponding individual voxel with spectral pattern and their overall spectroscopic grading along the peripheral zone. *Marked voxels* suggest "definitely healthy" (*1*) and "probably healthy" (*2*) prostate metabolism on the right side, but "definitely cancer" (*5*) on the left side in spatial agreement with the anatomic abnormality

reduction of zinc levels, an important factor in the transformation of prostate epithelial cells to citrate oxidizing cells and the development and progression of prostate cancer. The elevation in Cho is associated with changes in cell membrane synthesis and degradation that occur with the evolution of human cancers (ABOAGYE and BHUJWALLA 1999). The polyamines spermin, spermidine, and putrescine are also abundant in healthy prostatic tissues and reduced in cancer. Polyamines have been associated with cellular differentiation and proliferation (HESTON 1991).

In order to quantify spectral metabolites in the peripheral zone a standardized five-point scale has been developed (JUNG et al. 2004). A score from 1 to 5 is assigned to each spectroscopic voxel based on changes in choline, citrate, and polyamines (Fig. 10.4). A score of 1 is considered to be definitely benign, 2 likely benign, 3 equivocal, 4 likely abnormal, and 5 definitely abnormal. Regions of three or more adjacent abnormal spectroscopic voxels with scores of 4 and/or 5 that show a correspondence with anatomic abnormality (decreased signal intensity on T2-weighted images) are read as cancer and are termed "dominant intraprostatic lesion" (DIL).

Tumor Analysis

Knowledge of possible cancer spread outside the prostate is critical for choosing the appropriate therapy. Endorectal MRI has an excellent negative predictive value for detecting seminal vesicle invasion (D'AMICO et al. 1998) but is less accurate in the assessment of extracapsular extension. This assessment can be significantly improved by combining MRI findings that are predictive of cancer spread with an estimate of the spatial extent of metabolic abnormality as assessed by MRSI (YU et al. 1999); MRSI-derived tumor volume per lobe was significantly higher in patients with extracapsular extension than in patients without. Preliminary MRI/MRSI biopsy correlation studies also have shown that increased levels of choline and decreased levels of citrate correlate with the degree of pathology as measured by the Gleason score (VIGNERON et al. 1998). Targeted postsurgical prostate tissue samples studied spectroscopically ex vivo have confirmed such a correlation (SWANSON et al. 2003).

Impact on Treatment Planning

The described capability of combined MRI/MRSI to localize intraglandular cancer, evaluate extracapsular extent of the disease, and grade cancer aggressiveness makes it of value not only for patient selection but also for disease-targeted treatments such as cryosurgery, high-intensity ultrasonography (HIFU), and especially focal RT. One motivation for incorporating MRI/MRSI into the treatment planning process for RT is that the noncontrast enhanced CT overestimates the volume of the prostate as compared with MRI, is inadequate in identifying the complex anatomy of the prostate (ROACH et al. 1996), and sheds no light on identifying those areas that might benefit from increased dose. The integration of MRI/MRSI information into the treatment plan can be used to optimize dose planning and reduce the dose delivered to surrounding organs at risk (rectum, bladder, neurovascular bundles, etc.) and hence decrease the incidence of damage to normal tissues. Perhaps more importantly, the MRI/MRSI/biopsy-identified DILs can be targeted with increased dose with the hope of increased tumor control and, at the same time, a potential decrease in overall treatment time if increased daily dose fractions could be applied safely.

Prostate cancer is usually a multifocal disease and, therefore, most patients have multiple DILs. Delivering a high dose uniformly to the entire prostate increases the risk of complications to surrounding normal tissues. Selective dose intensification techniques to the spectroscopically identified DIL(s) seems appropriate, thereby targeting the area of cancer with increased dose. Such a treatment rationale has been pursued at UCSF to date. Approaches have included: (a) static field IMRT (SF-IMRT); (b) forward planned segmental multileaf collimation (SMLC) IMRT (PICKETT et al. 1999); (c) inverse planned SMLC or sequential tomotherapy employing the MIMiC collimator (XIA et al. 2001); (d) inverse planned radiosurgical boost IMRT (PICKETT et al. 2000); (e) brachytherapy via radioactive iodine-125 seed implantation (PICKETT et al. 2004) or high-dose rate application (POULIOT et a. 2004).

In order to incorporate the spectral information into the treatment planning process for RT, MRI/MRSI data have to be co-registered to the treatment planning CT; however, one confounding factor in merging these data sets is the distortion of the prostate by the inflatable rectal coil. As expected, a comparison of the anatomy derived from CT with and without the endorectal coil has demonstrated substantial displacement differences when the endorectal coil was used (VIGNEAULT et al. 1997). The acquisition of images with and without the endorectal coil within the same exam improves the translation of the MRSI data to the nondistorted images for manual merging with the planning CT or MRI scans (POULIOT et al. 2004).

A first attempt to utilize the spectroscopic information for treatment planning was made by PICKETT et al. (1999) who demonstrated the feasibility of using MRSI to dose escalate DIL(s) with external beam RT. The CT and MRI/MRSI data were aligned manually based on bony anatomy. The prostate PTV was defined based on the largest margins of the prostate created by the endorectal probe and based on its extent on CT. With the probe in place, the prostate usually was pushed anterior and superior to the position of the prostate in the CT position; therefore, the PTV was defined according to the posterior and inferior position of the CT and the anterior and superior position of the MRI/MRSI thereby encompassing the entire prostate. The DIL was defined based on its extent on MRSI enlarged by a 2-mm margin.

Organ deformation due to the endorectal coil, however, is very difficult to account for (HIROSE et al. 2002). The UCSF has been using gold markers placed at the base, midgland, and apex of the prostate prior to MRI/MRIS and CT to help address this problem. If the CT data are acquired with the endorectal coil inserted in the rectum, direct and reproducible fusion with the MRI/MRSI based upon gold seeds and bony anatomy is possible (PICKETT et al. 2000). The immobilization of the prostate achieved by the use of the coil also allows a reduction of treatment planning margins used to account for organ movement. Based on such a reduction, the delivery of a few fraction radiosurgical-type IMRT treatment, with the goal being to intensify the dose to the DIL(s) and possibly shorten the overall treatment time, has been explored and has been shown to be feasible (Fig. 10.5; PICKETT

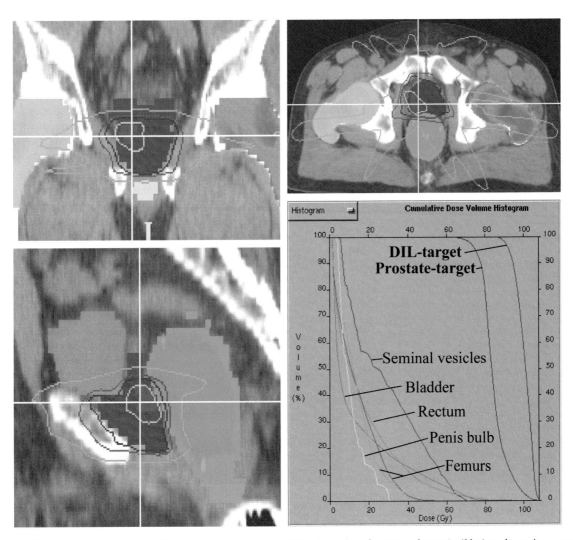

Fig. 10.5 Intensity-modulated radiation therapy prescribing 92 Gy (*green*) to the DIL and 73.8 Gy (*blue*) to the entire prostate while sparing surrounding normal structures (*red*: 60 Gy; *turquoise*: 25 Gy)

et al. 2000); however, such an approach will require a coil-like balloon to be placed in the rectum for each treatment session. Such an approach has been shown to be feasible according to experience at Baylor College of Medicine (TEH et al. 2002).

Similarly, treatment planning for high-dose rate brachytherapy is being explored currently based on MRI/MRSI data acquired with an endorectal coil. On such grounds, POULIOT et al. (2004) could demonstrate the feasibility of DIL dose escalation for prostate HDR Brachytherapy. An in-house developed inverse planning optimization algorithm is used to increase the dose delivered to the DIL to the 120–150% range by varying dwell time and dwell position of the 192-Ir source within the placed catheter while preserving prostate coverage and keeping the dose delivered to the organs at risk at the same level as compared with an inverse planned dose distribution without DIL boost (Fig. 10.6). Image registration of treatment planning CT (with catheters in place) and MRI/MRSI was carried out manually by matching corresponding anatomical structures in the transversal plane and interpolation to the nearest corresponding CT slice due to thinner MRI vs CT slices. Having proven the feasibility of safely escalating dose to DIL(s) via optimized HDR treatment planning, future efforts are directed to perform MRI/MRSI scans with catheters already implanted. The catheters would hereby provide additional landmarks that would facilitate the MRI to MRSI image registration. Moreover, the MRI/MRSI scans could be used directly for dose planning.

Researchers at MSKCC are pursuing similar avenues. ZAIDER et al. (2000) explored the combination of an optimization planning algorithm, MRSI, and tumor control probability (TCP), and found this approach to be safe enough to escalate dose and possibly improve outcome of patients treated with permanent implanted seeds.

In summary, efforts to incorporate the seemingly valuable information provided by MRI/MRSI into the treatment plan continue. A wealth of data is confirming the power of combined MRI/MRSI to detect the spatial location of prostate cancer and to direct dose escalation to such regions. In order to achieve the best possible data and high resolution necessary one has to rely on the use of the endorectal coil which introduces issues that need their own solutions. In addition, one has to account for inter-treatment (due to different filling of rectum and bladder) and intra-treatment (due to breathing) organ movement. But that is another topic in and of itself.

Assessing Response to RT

Two recent studies by PICKETT et al. have evaluated the value of MRI in assessing treatment response after external beam and permanent seed implant RT for prostate cancer (PICKETT et al. 2003, 2004). After external beam RT (EBRT) delivering 72–75.6 Gy to the entire prostate, they could show that 78% of 55 studied patients achieved their primary end point, a "negative" MRSI defined as the absence of tumor sus-

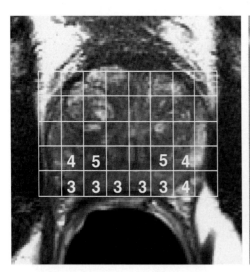
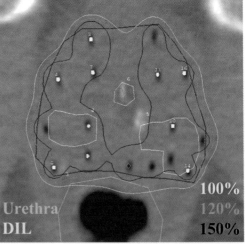

Fig. 10.6 *Left:* T2-weighted reception-profile-corrected FSE axial image and superimposed PRESS-excited volume through the midgland of a 67-year-old patient. Spectral grading suggests bilateral disease with values of 4 ("probably cancer") and 5 ("definitely cancer") and "uncertain" (3) in midline of the peripheral zone. Voxels with values of 4 and 5 were defined as dominant intraprostatic lesion (DIL) and targeted with an HDR boost. *Right:* CT treatment planning with catheters in place and resulting isodose lines after inverse planning show the desired 150% to encompass both DILs and spare urethra and rectum

picious spectra, at various time points >26 months post-RT (PICKETT et al. 2003). The time course of response is hereby of interest: the percent of voxels indicating metabolic atrophy increased from 76% at 18 months to 93% at 55 months post-RT. Mean overall time to resolution of disease was 40 months. The MRSI revealed persistent cancerous metabolism in 22% of patients at the end of evaluation and are referred to biopsy if still persistent at 36 months post-EBRT.

After permanent prostate implant (PPI), on the other hand, MRSI exams suggested complete metabolic atrophy in 46% of 65 examined patients at 6 months post-PPI and, after continued follow-up, in all patients by 48 months after PPI (PICKETT et al. 2004). Complete metabolic atrophy was seen in only 31% of the patients treated with EBRT. This dramatic difference in treatment response and its time course is likely attributable to the significantly higher prescribed dose (144 Gy PPI vs 70–75.6 Gy EBRT). In addition, both studies found that MRSI seemed to indicate therapeutic effectiveness much earlier than the PSA level reached its nadir. It was concluded that the MRSI-derived time to metabolic atrophy appears to be a very useful adjunct to PSA for assessing local control following PPI.

In a study of 21 patients with biochemical failure after external-beam radiation therapy for prostate cancer in whom subsequent biopsy confirmed locally recurrent prostate cancer, COAKLEY et al. (2004) found that the presence of three or more MRSI voxels with isolated elevated choline to creatine (ratio greater than 1.5) showed a sensitivity and specificity of 87 and 72%, respectively. This preliminary study suggests that MRSI may have a role in detecting local cancer recurrence post-radiation.

10.3
Current Limitations and Possible Solutions

Developing MRSI as a tool for routine use for brain tumors with respect to radiation therapy planning and follow-up requires improvements in coverage of the lesion and in the signal-to-noise ratio of the spectral data. The spectroscopy studies performed at our institution have been restricted to a rectangular area of excitation (PRESS box) that is of limited volume. Since gliomas are relatively large and irregular in shape, this means that in many cases we have not covered the entire lesion. The volume limitation is a function of the signal-to-noise ratio and acquisition time for a given spatial resolution. Our current studies have limited the MRSI acquisition time to 17 min so that the entire exam can be performed in conjunction with a conventional MRI exam.

Optimization of the methods for lipid suppression are critical for achieving adequate coverage of the anatomic lesion and surrounding normal tissue without compromising spectral quality. Very selective saturation bands (VSS) have been used to sharpen the edges of the PRESS volume and to suppress regions of subcutaneous lipid and bone on the edges of the selected volume. Future studies will examine alternative combinations of VSS pulses, as well as balancing the trade-off between increased imaging time and data quality. This will also be improved by utilizing a higher-strength magnet or improved radiofrequency coils.

There is also a need for further development concerning the calculation of metabolic indices; these are currently generated from the data that have a nominal spatial resolution of 1 cc and results in potentially large partial-volume effects. Higher spatial resolution can be achieved using surface or phased-array radiofrequency coils. Resolutions as small as 0.2 cc have been achieved for brain tumors and other focal lesions of the cortex at 1.5 T. Magnets as strong as 3–7 T will improve this capability even further. An additional benefit of higher field strengths is the improved spectral dispersion and increased T2 relaxation times which will allow estimation of additional metabolites that are currently not being assessed.

One of the problems in the prostate is organ deformation due to the use of inflated rectal coils. A newly designed, noninflatable, rigid endorectal coil for MRI/MRSI may reduce deformity of the prostate and the magnitude of the discrepancies between the prostate position on MRI/MRSI vs CT, making the data much more "transportable" for use in treatment planning.

Finally, it is not always clear how to interpret the metabolic data itself or which level of a metabolic index should be used as a guide for defining a separate target or boost volume that should receive a higher or perhaps lower dose. Because of partial-volume effects, a region of given metabolic index could have a few cells with very high metabolic activity and many cells with little activity and/or necrosis, or it could contain all cells with average activity. These two scenarios might suggest different therapeutic solutions. For instance, it seems intuitive that regions with greater metabolic activity (however that is defined) should be targeted for an integrated boost; however, it could also be argued that the regions with

Biehl K, Kong F, Bradley J, Dehdashti F, Mutic S, Siegel B (2004) FDG-PET definition of gross target volume for radiotherapy on NSCLC: is the use of 40% threshold appropriate? Cancer J 10 (Suppl 1):36

Bradley J, Thorstad WL, Mutic S, Miller TR, Dehdashti F, Siegel BA, Bosch W, Bertrand RJ (2004) Impact of FDG-PET on radiation therapy volume delineation in non-small-cell lung cancer. Int J Radiat Oncol Biol Phys 59:78–86

Chao KSC, Bosch WR, Mutic S, Lewis JS, Dehdashti F, Mintun MA, Dempsey JF, Perez CA, Purdy JA, Welch MJ (2001) A novel approach to overcome hypoxic tumor resistance Cu-ATSM-guided intensity-modulated radiation therapy. Int J Radiat Oncol Biol Phys 49:1171–1182

Cook GJ, Fogelman I, Maisey MN (1996) Normal physiological and benign pathological variations of 18-fluoro-2-deoxy-glucose positron-emission tomography scanning: potential for error in interpretation. Semin Nucl Med 26:308–314

Defrise M, Kinahan PE, Michel C (2003) Image reconstruction algorithms in PET. In: Valk PE, Bailey DL, Townsend DW, Maisey MN (eds) Positron emission tomography: basic science and clinical practice. Springer, Berlin Heidelberg New York, pp 91–114

Dizendorf EV, Baumert BG, Schulthess GK von, Lutolf UM, Steinert HC (2003) Impact of whole-body 18F-FDG PET on staging and managing patients for radiation therapy. J Nucl Med 44:24–29

Erdi YE, Mawlawi O, Larson SM, Imbriaco M, Yeung H, Finn R, Humm JL (1997) Segmentation of lung lesion volume by adaptive positron emission tomography image thresholding. Cancer 80:2505–2509

Eshappan J, Mutic S, Malyapa RS, Grigsby PW, Zoberi I, Dehdashti F, Miller TR, Bosch WR, Low DA (2004) Treatment planning guidelines regarding the use of CT/PET-guided IMRT for cervical carcinoma with positive para-aortic lymph nodes. Int J Radiat Oncol Biol Phys 58:1289–1297

Ezzel GA, Galvin JM, Low D, Palta JR, Rosen I, Sharpe MB, Xia P, Xiao Y, Xing L, Yu CX (2003) Guidance document on delivery, treatment planning, and clinical implementation of IMRT: report of the IMRT Subcommittee of the AAPM Radiation Therapy Committee. Med Phys 30:2089–2115

Gallagher BM, Fowler JS, Gutterson NI, MacGregor RR, Wan CN, Wolf AP (1978) Metabolic trapping as a principle of radiopharmaceutical design: some factors responsible for the biodistribution of [18F] 2-deoxy-2-fluoro-D-glucose. J Nucl Med 10:1154–1161

Garcia-Ramirez JL, Mutic S, Dempsey JF, Low DA, Purdy JA (2002) Performance evaluation of an 85-cm bore X-ray computed tomography scanner designed for radiation oncology and comparison with current diagnostic CT scanners. Int J Radiat Oncol Biol Phys 52:1123–1131

Goitein M, Abrams M (1983) Multi-dimensional treatment planning I. Delineation of anatomy. Int J Radiat Oncol Biol Phys 9:777–787

Goitein M, Abrams M, Rowell D, Pollari H, Wiles J (1983) Multi-dimensional treatment planning II. Beam's eye-view, back projection, and projection through CT sections. Int J Radiat Oncol Biol Phys 9:789–797

Grigsby PW, Perez CA, Chao KS, Herzog T, Mutch DG, Rader J (2001a) Radiation therapy for carcinoma of cervix with biopsy-proven positive para-aortic lymph nodes. Int J Radiat Oncol Biol Phys 49:733–738

Grigsby PW, Williamson JF, Chao KSC, Perez CA (2001b) Cervical tumor control evaluated with ICRU 38 reference vol-

umes an integrated reference air kerma. Radiother Oncol 58:10–23

Grigsby PW, Siegel BA, Dehdashti F, Mutch DG (2003) Post-therapy surveillance monitoring of cervical cancerr by FDG-PET. Int J Radiat Oncol Biol Phys 55:907–913

Gross MW, Weber WA, Feldman HJ, Bartenstein P, Schwaiger M, Molls M (1998) The value of F-18-fluorodeoxyglucose PET for the 3-D radiation treatment planning for malignant gliomas. Int J Radiat Oncol Biol Phys 41:989–995

Hicks RJ, MacManus MP (2003) 18F-FDG PET in candidates for radiation therapy: Is it important and how do we validate its impact? J Nucl Med 44:30–32

Jerusalem G, Hustinx R, Beguin Y, Fillet G (2003) PET scan imaging in oncology. Eur J Cancer 39:1525–1534

Klingenbeck Regn K, Schaller S, Flohr T, Ohnesorge B, Kopp AF, Baum U (1999) Subsecond multi-slice computed tomography: basics and applications. Eur J Radiol 31:110–124

Lavely WC, Scarfone C, Cevikalp H, Rui L, Byrne DW, Cmelak AJ, Dawant B, Price RR, Hallahan DE, Fitzpatrick JM (2004) Phantom validation of coregistration of PET and CT for image-guided radiotherapy. Med Phys 31:1083–1092

Levivier M, Wikier D, Goldman S, David P, Matens T, Massager N, Gerosa M, Devriendt D, Desmedt F, Simon S, van Houtte P, Brotchi J (2000) Integration of the metabolic data of positron emission tomography in the dosimetry planning of radiosurgery with the gamma knife: early experience with brain tumors. J Neurosurg 93:233–238

Ling CC, Humm J, Larson S, Amols H, Fuks Z, Leibel S, Koutcher JA (2000) Towards multidimensional radiotherapy (MD-CRT): biological imaging and biological conformality. Int J Radiat Oncol Biol Phys 47:551–560

Low DA, Nystrom M, Kalinin E, Parikh P, Dempsey JF, Bradley JD, Mutic S, Wahab SH, Islam T, Christensen G, Politte DG, Whiting BR (2003) A method for the reconstruction of 4-dimensional synchronized CT scans acquired during free breathing. Med Phys 30:1254–1263

MacManus MP, Hicks RJ, Ball DL, Kalff V, Matthews JP, Salminen E, Khaw P, Wirth A, Rischin D, McKenzie A (2001) F-18 Fluorodeoxyglucose positron emission tomography staging in radical radiotherapy candidates with nonsmall cell lung carcinoma. Cancer 92:886–895

Mah K, Caldwell CB, Ung YC, Danjoux CE, Balogh JM, Ganguli SN, Ehrlich LE, Tirona R (2002) The impact of 18FDG-PET on target and critical organs in CT-based treatment planning of patients with poorly defined non-small cell lung carcinoma: a prospective study. Int J Radiat Oncol Biol Phys 52:339–350

Malyapa RS, Mutic S, Low DA, Zoberi I, Bosch WR, Miller TR, Grigsby PW (2002) Physiologic FDG-PET three dimensional brachytherapy treatment planning for cervical cancer. Int J Radiat Oncol Biol Phys 54:1140–1146

Meikle SR, Badawi RD (2003) Quantitative techniques in PET. In: Valk PE, Bailey DL, Townsend DW, Maisey MN (eds) Positron emission tomography: basic science and clinical practice. Springer, Berlin Heidelberg New York, pp 115–146

Miller TR, Grigsby PW (2002) Measurement of tumor volume by PET to evaluate prognosis in patients with advance cervical cancer treated by radiation therapy. Int J Radiat Oncol Biol Phys 53:353–359

Mutic S, Dempsey JF, Bosch WR, Low DA, Drzymala RE, Chao KSC, Goddu SM, Cutler PD, Purdy JA (2001) Multimodality image registration quality assurance for conformal three-dimensional treatment planning. Int J Radiat Oncol Biol Phys 51:255–260

Mutic S, Grigsby PW, Low DA, Dempsey JF, Harms WR Sr, Laforest R, Bosch WR, Miller TR (2002) PET guided three-dimensional treatment planning of intracavitary gynecologic implants. Int J Radiat Oncol Biol Phys 52:1104–1110

Mutic S, Palta JR, Butker E, Das IJ, Huq MS, Loo LD, Salter BJ, McCollough CH, van Dyk J (2003a), Quality assurance for CT simulators and the CT simulation process: report of the AAPM Radiation Therapy Committee Task Group no. 66. Med Phys 30:2762–2792

Mutic S, Malyapa RS, Grigsby PW, Dehdashti F, Miller TR, Zoberi I, Bosch WR, Esthappan J, Low DA (2003b) PET-guided IMRT treatment for cervical carcinoma with positive para-aortic lymph nodes: a dose escalation treatment planning study. Int J Radiat Oncol Biol Phys 55:28–35

National Electrical Manufacturers Association (NEMA) (1998) DICOM PS 3 (set). Digital Imaging Communictaions in Medicine (DICOM)

Nuutinen J, Sonninen P, Lehikoinen P, Sutinen E, Valavaara R, Eronen E, Norrgard S, Kulmala J, Teras M, Minn H (2000) Radiotherapy treatment planning and long-term follow-up with [(11)C]methionine PET in patients with low-grade astrocytoma, Int J Radiat Oncol Biol Phys 48:43–52

Osman MM, Cohade C, Nakamoto Y, Marshall LT, Leal JP, Wahl RL (2003) Clinically significant inaccurate localization of lesions with PET/CT: frequency in 300 patients. J Nuc Med 44:240–243

Perez CA, Purdy JA, Harms W, Gerber R, Grahm MV, Matthews JW, Bosch W, Drzymala R, Emami B, Fox S (1995) Three-dimensional treatment planning and conformal radiation therapy: preliminary evaluation. Radiother Oncol 36:32–43

Pieterman, RM, van Putten JW, Meuzelaar JJ, Mooyaart EL, Vaalburg W, Koeter GH, Fidler V, Pruim J, Groen HJ (2000) Preoperative staging of non-small-cell lung cancer with positron-emission tomography. N Engl J Med 343:254–261

Poncelet AJ, Lonneux M, Coche E, Weynand B, Noirhomme P (2001) PET-FDG enhances but does not replace preoperative surgical staging in non-small cell lung carcinoma. Eur J Cardiothorac Surg 20:468–474

Purdy JA (1997) Advances in three-dimensional treatment planning and conformal dose delivery. Semin Oncol 24:655–671

Sherouse GW, Novins K, Chaney EL (1990) Computation of digitally reconstructed radiographs for use in radiotherapy treatment design. Int J Radiat Oncol Biol Phys 18:651–658

Steinert HC, Hauser M, Allemann F, Engel H, Berthold T, Schulthess GK von, Weder W (1997) Non-small cell lung cancer: nodal staging with FDG PET versus CT with correlative lymph node mapping and sampling. Radiology 202:441–446

Strauss LG (2004) Fluorine-18 deoxygluocose and false-positive results: a major problem in the diagnostics of oncological patients. Eur J Nucl Med 23:1409–1415

Townsend DW, Carney JP, Yap JT, Hall NC (2004) PET/CT today and tomorrow. J Nucl Med 45 (Suppl 1):4S–14S

Valk PE, Bailey DL, Townsend DW, Maisey MN (eds) (2002) Positron emission tomography: basic science and clinical practice. Springer, Berlin Heidelberg New York

Weissleder R, Mahmood U (2001) Molecular imaging. Radiology 219:316–333

Young H, Baum R, Cremerius U, Herholz K, Hoekstra O, Lammertsma AA, Pruim J, Price P (1999) Measurement of clinical and subclinical tumour response using [18F]-fluorodeoxyglucose and positron emission tomography: review and 1999 EORTC recommendations. European Organization for Research and Treatment of Cancer (EORTC) PET Study Group. Eur J Cancer 13:1773–1782

12 Patient Positioning in Radiotherapy Using Optical-Guided 3D Ultrasound Techniques

Wolfgang A. Tomé, Sanford L. Meeks, Nigel P. Orton, Lionel G. Bouchet, and Mark A. Ritter

CONTENTS

12.1 Introduction

Over the past decade, virtual simulation has become the standard of care for planning the majority of external beam radiotherapy treatments. There are many obvious technical advantages of virtual simulation over conventional radiotherapy simulation. Target identification is more accurate, because the target may be identified directly on CT and/or co-identified on MRI, functional imaging scans, and PET

W.A. Tomé, PhD, Associate Professor
N.P. Orton, PhD, Medical Physicist
M.A. Ritter, MD, PhD, Associate Professor
Department of Human Oncology, University of Wisconsin Medical School, CSC K4/B100, 600 Highland Avenue, Madison, WI 53792, USA
S.L. Meeks, PhD
Associate Professor, Department of Radiation Oncology, Room W189Z-GH, University of Iowa, 200 Hawkins Drive, Iowa City, IA 52242, USA
L.G. Bouchet, PhD
Vice President, Technology Development, ZMed, Inc., Unit B-1, 200 Butterfield Drive, Ashland, MA 01721, USA

scans. Furthermore, the spatial relationship of the tumor to organs at risk is more easily appreciated in virtual simulation. This allows the treatment planner to tailor the beam orientations, field sizes, and field shapes to conformally avoid these organs at risk as far as possible. Further gains in conformal avoidance of organs at risk can usually be achieved when inverse planning is added to the treatment-planning arsenal. Virtual simulation has provided the framework for 3D conformal therapy and intensity modulated radiotherapy.

While virtual simulation provides a significant improvement over conventional radiotherapy, the actual delivery of such virtually designed treatment plans has been limited by the accuracy of the fiducial systems traditionally chosen in radiotherapy. Optically guided radiotherapy systems have the potential for improving the precision of treatment delivery by providing more robust fiducial systems. The ability of optically guided systems to accurately position internal targets with respect to the linear accelerator isocenter and to provide real-time patient tracking enables one at least theoretically to significantly reduce the amount of normal tissue included in the total irradiated volume by decreasing treatment field margins. Implementation and correct use of such systems present new challenges for the clinical physicist, and it is important that one thoroughly understands the strengths and weaknesses of such systems.

Furthermore, high-precision radiotherapy outside the cranium is challenging, because the target position can shift relative to bony anatomy between the time of initial image acquisition for virtual simulation and the time of the actual delivery of treatment. Our group has recently developed and clinically implemented an optically guided 3D ultrasound system for high-precision radiation delivery. The purpose of this chapter is to describe (a) optical ultrasound-guided radiotherapy, (b) the underlying mathematics that drive optical guidance, (c) the quality assurance techniques for such systems, and (d) the clinical use of optically guided 3D ultrasound patient positioning systems.

12.2
Optical Tracking

Tracking is the process of measuring the location of instruments, anatomical structures, and/or landmarks in 3D space and in relationship to each other. Various technologies have been tested for determining an object's location, including mechanical, magnetic, acoustic, inertial, and optical position sensors. Most of these technologies have been tested for medical use in either image-guided surgery or image-guided tracking in radiation therapy.

Optical tracking systems use infrared light to determine a target's position. The target may either be active or passive. The most common active targets are infrared light emitting diodes (IRLED). Passive targets are generally spheres or disks coated with a highly reflective material that reflects infrared light from an external source. Various detectors can be used to determine the positions of an optical target; however, charged couple device (CCD) cameras are used most often. The CCD cameras are simply a collection of light sensitive cells, or pixels, arranged in either a one- or two-dimensional array. When light strikes a CCD cell, electron production is proportional to the intensity of the light incident on the cell; thus, a 2D CCD array provides a 2D digital "image" of the target, with brighter pixels in the array corresponding to a higher light intensity and darker pixels corresponding to lower light intensity. This digital image can then be analyzed to determine the pixel with the highest light intensity. Each camera in a 2D CCD array determines a ray in 3D. If two 2D CCDs are used in the camera of an optical tracking system, the intersection of the two 2D rays emanating form the CCDs determines a line in 3D space, whereas if three 2D CCDs are used, the intersection of all three 2D rays determines a point in 3D space.

12.3
Optically Guided 3D Ultrasound

In extracranial radiotherapy, soft tissue targets can move relative to bony anatomy between the time of image acquisition for treatment planning and the time of the actual treatment delivery; therefore, real-time imaging of internal anatomy is required in extracranial radiotherapy if one would like to accurately localize a target at the time of treatment delivery. We have developed and tested an optical-guided system for 3D ultrasound guidance (BOUCHET et al. 2001; BOUCHET et al. 2002; MEEKS et al. 2003; RYKEN et al. 2001; TOMÉ

et al. 2002) that is commercially available under the trade name SonArray (ZMed, Inc., Ashland, Mass.). Ultrasound was chosen because it is an inexpensive, yet flexible and high-resolution imaging modality that can easily be adapted for use in a radiation therapy treatment room. Systems have been developed that rely on 2D ultrasound probes attached to mechanical tracking systems, and these have proved effective for improving the precision of patient localization for prostate radiotherapy (TROCCAZ et al. 1995; LATTANZI et al. 1999; LATTANZI et al. 2000). The interpretation of 2D ultrasound images, however, is difficult and can be highly dependent on the skill and expertise of the operator in manipulating the transducer and mentally transforming the 2D images into a 3D tissue structure. In principle, the use of 3D ultrasound can help overcome this limitation, but the relatively high cost of true 3D ultrasound devices has prohibited their use in radiotherapy for target localization on a routine basis; however, the development of our optical-guided 3D ultrasound target localization system has decreased the cost of such a system to a level that is comparable to commercially available 2D ultrasound target localization systems, and has therefore led to a more widespread use of reconstructed 3D Ultrasound for target localization in radiotherapy. The 3D ultrasound data sets are generated through optically tracking the acquisition of free-hand 2D ultrasound images. The operator holds the ultrasound probe and manipulates it over the anatomical region of interest. The raw 2D ultrasound images are transferred to a workstation using a standard video link. During acquisition, the position and angulation of each 2D ultrasound image is determined form the position and angulation of the 2D ultrasound probe by optically tracking a rigidly attached active fiducial array that has four infrared light-emitting diodes (IRLEDs; see Fig. 12.1). The position of each ultrasound pixel can therefore be determined, and an ultrasound volume can be reconstructed by coupling the position information obtained through optical tracking with the raw ultrasound data (see Figs. 12.2, 12.3).

Therefore, in addition to building the 3D image volume, optical guidance is also used to determine the absolute position of the ultrasound volume in the treatment room coordinate system. Because the relative positions of the ultrasound volume and the ultrasound probe are fixed, the knowledge of the probe position in the treatment room coordinate system at the time of image acquisition is sufficient to determine the position of the image volume relative to the linear accelerator isocenter. The relative position of the image and probe is determined during a calibration

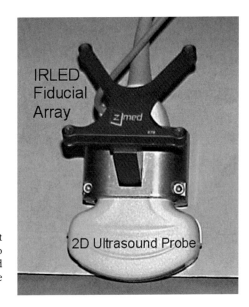

Fig. 12.1. The ultrasound probe is optically tracked using an array that consists of four infrared light-emitting diodes, which is rigidly attached to the probe. Using optical tracking a 3D ultrasound volume can be obtained and referenced to the optical tracking system origin, which is typically the linear accelerator isocenter

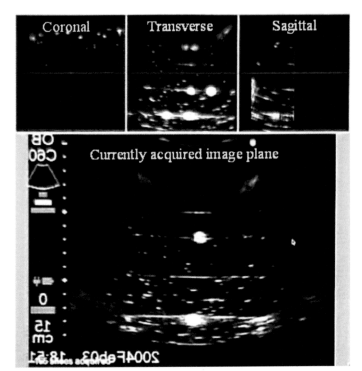

Fig. 12.2. Freehand ultrasound acquisition is used to acquire ultrasound data of an echoic spherical target at any orientation by either sliding or arcing the probe across the region of interest. One acquires arbitrary 2D ultrasound images until enough images have been acquired to fill a 3D matrix covering the volume of interest. The sagittal view (*upper right hand corner*) through the volume of interest shows the operator when a sufficient number of 2D ultrasound images have been acquired

procedure (BOUCHET et al. 2001). This calibration procedure employs an optically tracked ultrasound phantom that contains 39 echoic wires at 13 depths in an anechoic medium. Because the phantom is optically tracked, the room position of each wire is known very accurately. The ultrasound image coordinate of each wire is then determined by collecting multiple images of the phantom. From this the ultrasound image to treatment room coordinate system transformation can be easily determined. The ultrasound image to

ultrasound probe coordinate system transformation is then obtained by multiplying from the left the ultrasound image to treatment room coordinate system transformation with the inverse of ultrasound probe to treatment room coordinate system transformation. Once one has established the ultrasound image to ultrasound probe coordinate system transformation one is free to move the ultrasound probe anywhere in the treatment room coordinate system established by the optical tracking system.

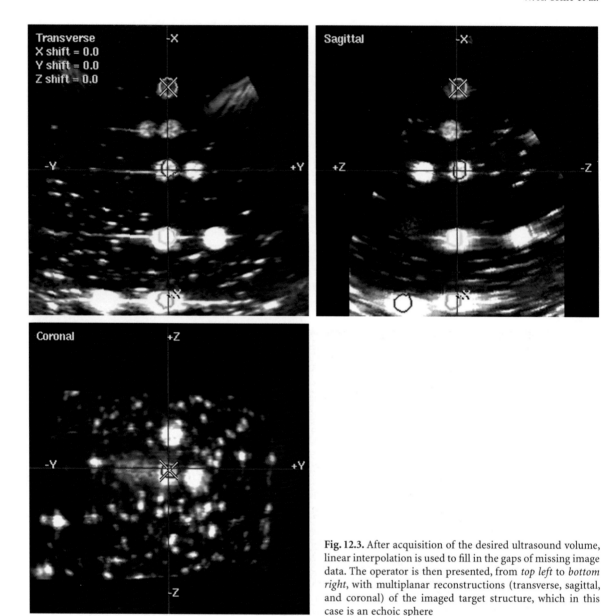

Fig. 12.3. After acquisition of the desired ultrasound volume, linear interpolation is used to fill in the gaps of missing image data. The operator is then presented, from *top left* to *bottom right*, with multiplanar reconstructions (transverse, sagittal, and coronal) of the imaged target structure, which in this case is an echoic sphere

12.4
Mathematics of Optical Tracking and Ultrasound Calibration

12.4.1
Mathematics of Optical Tracking

In optical guidance for radiation therapy, we determine the image coordinates of the passive markers and the desired isocenter location from the virtual simulation CT scan, and we determine the room coordinates of the passive markers relative to the machine isocenter from optical tracking. The mathematics required for optical guidance simply entails the determination of the relationship between these two sets of points. Let us denote the desired marker coordinates from the CT images and the optically measured room coordinates of the markers by the following two 4×1 column matrices $\bar{p}_I = \begin{pmatrix} x_I & y_I & z_I & 1 \end{pmatrix}^T$ and $\bar{p}_R = \begin{pmatrix} x_R & y_R & z_R & 1 \end{pmatrix}^T$, respectively. Since it is unlikely that $\bar{p}_I$ is equal to $\bar{p}_R$, one must mathematically determine the relationship between these two matrices, and hence, determine the rotational and translational misalignment of the patient at the time of treatment relative to the time of virtual simulation. The geometric transformation that relates $\bar{p}_I$ to $\bar{p}_R$ is given by a 4×4 matrix $E_{I \to R}$, which is an element of the Euclidian group in three dimensions E(3):

$$\bar{p}_R = E_{I \to R} \bar{p}_I. \tag{12.1}$$

It is well known that E(3) is the semidirect product of the group of rotations SO(3) and the group of linear translations T; therefore, we can write Eq. 12.1 also as:

$$\bar{p}_R = \hat{R} \bar{p}_I + \hat{T}, \tag{12.2}$$

where $\hat{R}$ is a 3×3 rotation matrix, $\hat{T}$ is a 3×1 translation matrix, and $\bar{p}_I$ and $\bar{p}_R$ are now considered as vectors in 3D space.

Only three non-colinear points with known positions in both room and image coordinates are needed to solve this equation; however, using more points reduces the statistical noise and increases the accuracy in determining the transformation matrices (YANG et al. 1999). We use a minimum of four fiducial markers in all of our fiducial arrays. Assuming that N points are available, $\hat{R}$ and $\hat{T}$ are the solution of the following least-square fit equation:

$$\varepsilon^2 = \sum_{k=1}^{N} \left\| \bar{p}_{R,k} - \left(\hat{R} \bar{p}_{I,k} + \hat{T} \right) \right\|^2. \tag{12.3}$$

It is useful to change Eq. (12.3) by referring all points to the centroids $\bar{p}_{R,C}$ and $\bar{p}_{I,C}$ of the sets of points in room and image space using the centroid coincidence theorem (YANG et al. 1999). The centroids are determined using the geometric averages of the points in both coordinate systems:

$$\bar{p}_{R,C} = \frac{1}{N} \sum_{k=1}^{N} \bar{p}_{R,k}, \text{ and} \tag{12.4}$$

$$\bar{p}_{I,C} = \frac{1}{N} \sum_{k=1}^{N} \bar{p}_{I,k}. \tag{12.5}$$

Equation 12.3 can therefore be written in terms of $\bar{p}'_{R,k} = \bar{p}_{R,k} - \bar{p}_{R,C}$ and $\bar{p}'_{I,k} = \bar{p}_{I,k} - \bar{p}_{I,C}$:

$$\varepsilon^2 = \sum_{k=1}^{N} \left\| \left(\bar{p}'_{R,k} - \hat{R} \bar{p}'_{I,k} \right) + \left(\bar{p}_{R,C} - R \bar{p}_{I,C} - \hat{T} \right) \right\|^2, \text{ or} \tag{12.6}$$

$$\varepsilon^2 = \sum_{k=1}^{N} \left\| \left(\bar{p}'_{R,k} - \hat{R} \bar{p}'_{I,k} \right) \right\|^2 - 2 \left(\bar{p}_{R,C} - R \bar{p}_{I,C} - \hat{T} \right).$$

$$\sum_{k=1}^{N} \left(\bar{p}'_{R,k} - R \bar{p}'_{I,k} \right) + N \left\| \bar{p}_{R,C} - \hat{R} \bar{p}_{I,C} - T \right\|^2. \tag{12.7}$$

Since

$$\sum_{k=1}^{N} \left(\bar{p}'_{R,k} \right) = \sum_{k=1}^{N} \left(\bar{p}_{R,k} \right) - N \left(\bar{p}_{R,C} \right) = 0, \text{ and} \tag{12.8}$$

$$\sum_{k=1}^{N} \left(R \bar{p}'_{I,k} \right) = R \sum_{k=1}^{N} \left(\bar{p}_{I,k} \right) - N \left(R \bar{p}_{I,C} \right) = 0 \tag{12.9}$$

From Eqs. (12.8) and (12.9) we find that the middle term in Eq. 12.7 is equal to zero, and therefore Eq. 12.7 simply becomes:

$$\varepsilon^2 = \sum_{k=1}^{N} \left\| \left(\bar{p}'_{R,k} - \hat{R} \bar{p}'_{I,k} \right) \right\|^2 + N \left\| \bar{p}_{R,C} - R \bar{p}_{I,C} - \hat{T} \right\|^2. \tag{12.10}$$

As one can see from Eq. (12.10) the translation minimizing the above equation is simply the difference vector of the image centroid and the rotated room centroid:

$$\hat{T} = \bar{p}_{I,C} - \hat{R} \bar{p}_{R,C}. \tag{12.11}$$

Consequently, to determine the rotation, we have to minimize:

$$\varepsilon_{\hat{R}}^2 = \sum_{k=1}^{N} \left\| \left(\bar{p}'_{R,k} - \hat{R} \bar{p}'_{I,k} \right) \right\|^2. \tag{12.12}$$

A multitude of algorithms can be used to determine the best-fit rotation $\hat{R}$ from Eq. (12.12). Since this is a minimization problem, iterative optimization algorithms can be used to find the best-fit rotation. Several different optimization algorithms have been used to solve the patient orientation problem in stereotactic radiotherapy, including simulated annealing and various downhill algorithms such as the downhill simplex and the Hooke and Jeeves pattern search algorithm (YANG et al. 1999; SHOUP and MISTREE 1987). The solution space for the stereotactic radiotherapy application has been shown to be relatively flat, and downhill methods have proven fast and effective for stereotactic radiotherapy (YANG et al. 1999).

12.4.2
Mathematics of Optical-Guided 3D Ultrasound Calibration

The geometric transformation required for ultrasound calibration is ultrasound image space to treatment room space, and the end result is still minimization of Eq. (12.12). While downhill optimization algorithms are sufficient for solution of the absolute orientation as applied to stereotactic radiotherapy, additional noise from ultrasound imaging decreases the reliability of these simple algorithms for determining the ultrasound calibration matrix (BOUCHET et al. 2001). Below are brief discussions of two closed-form solutions algorithms (the singular value decomposition algorithm and Horn's algorithm using quaternions) to the absolute orientation problem.

12.4.2.1
Singular Value Decomposition

The minimum of Eq. 12.12 is achieved when for all $k \in \{1, ..., N\}$, we have $\vec{p}_{R,k} = \hat{R}\vec{p}_{I,k}$. This can be rewritten in terms of a simple matrix equation:

$$\begin{bmatrix} X_{R,1} & X_{R,2} & .... & X_{R,N} \\ Y_{R,1} & Y_{R,2} & .... & Y_{R,N} \\ Z_{R,1} & Z_{R,2} & .... & Z_{R,N} \end{bmatrix} = \hat{R} \begin{bmatrix} X_{I,1} & X_{I,2} & .... & X_{I,N} \\ Y_{I,1} & Y_{I,2} & .... & Y_{I,N} \\ Z_{I,1} & Z_{I,2} & .... & Z_{I,N} \end{bmatrix}, \quad (12.13)$$

where (X, Y, Z) represent the Cartesian coordinates of the points p. Equation (12.13) represents an overdetermined linear set of equations that can be solved by using the singular value decomposition theorem. This theorem yields the decomposition of the resulting N×3 matrix into a product matrices of the form UWV^T where U,V are orthogonal matrices and W is a diagonal matrix whose diagonal elements are either positive or zero. After decomposition, the solution of Eq. (12.13) can be determined by inverting the corresponding orthogonal matrices. This closed-form solution gives the rotation that minimizes the least-square problem stated in Eq. (12.12).

12.4.2.2
Horn's Algorithm

Another algorithm that can be used to solve Eq. (12.12) is the closed-form solution presented by Horn using quaternion theory (HORN 1987; BOUCHET et al. 2001). Quaternions are an extension of complex numbers. Instead of just i, one has three different numbers that are all square roots of –1 labeled i, j, and k, that fulfill the following relationship:

$$i*j = \varepsilon_{ijk}k,$$

where ε_{ijk}, is the totally antisymmetric tensor. The conjugate and the magnitude of a quaternion are formed in much the same way as the complex conjugate and magnitude:

$$\dot{q} = q_0 + iq_x + jq_y + kq_z,$$

$$\dot{q}' = q_0 - iq_x - jq_y - kq_z$$

$$\|\dot{q}\| = \dot{q} * \dot{q}' = \sqrt{q_0^2 + q_x^2 + q_y^2 + q_z^2} \quad (12.14)$$

In what follows we only between will and deal with unit quaternions, i.e., quaternions for which $\|\dot{q}\| = 1$. For unit quaternions the inverse of a quaternion is equal to its conjugate. Since unit quaternions form a representation of the special unitary group SU(2) one can express any 3D rotation R in terms of unit quaternions using the following isomorphism:

$$\vec{r}_1 = R\vec{r}_0 \rightarrow \dot{r}_1 = \dot{q} * \dot{r}_0 * \dot{q}', \quad (12.15)$$

Expanding Eq. (12.12) and using the fact that rotations are isomorphisms, i.e., leave the norm unchanged, one finds:

$$\varepsilon_R^2 = \sum_{k=1}^N \|\vec{p}_{R,k}'\|^2 + \sum_{k=1}^N \|\vec{p}_{I,k}'\|^2 - 2\sum_{k=1}^N \vec{p}_{R,k}' \cdot \hat{R}\vec{p}_{I,k}'. \quad (12.16)$$

Therefore, minimizing Eq. (12.16) corresponds to maximizing the last term in this expression since the first two terms of this expression are positively definite:

$$\xi = \sum_{k=1}^N \vec{p}_{R,k}' \cdot \hat{R}\vec{p}_{I,k}' = \sum_{k=1}^N \hat{R}\vec{p}_{I,k}' \cdot \vec{p}_{R,k}', \quad (12.17)$$

which can be written in terms of quaternions using the isomorphism (Eq. (12.15)) as follows:

$$\xi = \sum_{k=1}^N \dot{q} * \dot{p}_{I,k} * \dot{q}' * \dot{p}_{R,k}. \quad (12.18)$$

Using the fact that there exists a matrix representation of quaternion multiplication, Eq. (12.18) can be rewritten as (cf. HORN (1987):

$$\xi = qNq^T, \quad (12.19)$$

where $N = \sum_{k=1}^N N_k$ is a 4×4 matrix formed from the coordinates of the points $\vec{p}_{I,k}$ and $\vec{p}_{R,k}$ and the unit quaternion q is represented as a row vector of the form $q = (q_x, q_y, q_z, q_o)$. HORN (1987) has shown that the unit quaternion maximizing Eq. (12.19) is the eigenvector corresponding to the largest positive eigenvalue of the matrix N. Since N is a 4×4 matrix, this corresponds to finding the roots of the fourth-degree characteristic polynomial of N. This unit quaternion corresponds to the closed-form solution of the rotation minimizing the least-square problem stated in Eq. (12.12).

12.5
Commissioning and Quality Assurance of 3D Ultrasound-Guided Systems

To ensure accurate localization, all possible errors in the imaging, patient localization, and treatment delivery processes must be systematically analyzed (BOUCHET et al. 2002; TOMÉ et al. 2002). Outlined below are test procedures necessary to meet the quality assurance challenges presented by an optically guided 3D ultrasound system for real-time patient localization. While all tests were performed using the SonArray system, the general philosophy and procedures are applicable to all systems utilizing this technology. Determination of absolute localization accuracy requires the user to establish a consistent stereotactic, or 3D, coordinate system in both the treatment planning system and the treatment vault. While we chose to establish this coordinate system through optical guidance, it can also be done using mechanical means, as in conventional stereotactic procedures, or other telemetry technologies that are commercially available. Regardless of the methodology utilized, it is imperative that acceptance tests be performed prior to clinical use of the system to ensure that the image-guided system allows for safe, controlled, and efficient delivery of both conventional and intensity-modulated radiotherapy.

For our testing, we used a specially designed ultrasound phantom (Fig. 12.4). This phantom consists of 12 echoic spheres embedded in a tissue-equivalent non-echoic medium (with speed of sound 1470 m s^{-1}), with a passive infrared fiducial array attached to it. The spheres are arranged at five different nominal depths: one at 30 mm; two at 50 mm; and the remaining nine arranged in groups of three at nominal depths of 70, 100, and 130 mm. A CT scan (0.49×0.49×2.0-mm resolution) of the phantom was acquired. Using the fiducial array for optical tracking, the same stereotactic coordinate system was established in the treatment planning system and in the treatment room, relative to which the positions of each sphere are known within imaging uncertainty. Each of the spheres were localized in the Pinnacle treatment planning system and its coordinates were transferred to the SonArray system as intended treatment isocenters. In order to reproduce the exact position of the ultrasound phantom at the time of the CT, a 2D couch mount has been employed in the treatment room. This couch mount has two orthogonal rotational axes (spin, tilt). Together with the four degrees of freedom of the treatment couch,

Fig. 12.4. A specifically designed ultrasound phantom that has a number of echoic spheres embedded at different depths in a non-echoic medium. This phantom can be optically tracked to determine the coordinates of the spheres in the treatment room, which enables it to be used to test an ultrasound target localization system.

three orthogonal translations (anteroposterior, lateral, vertical), and one rotational degree of freedom couch, this allows reproduction of the position of the phantom to within a predefined error tolerance. Each target sphere was positioned at the treatment machine isocenter using the following method: first the isocenter corresponding to the target sphere chosen for ultrasound localization was selected on the control computer. The phantom was then moved using optical tracking and the couch controls until the RMS error between actual and desired position of the target sphere was less than 0.2 mm. Once a target sphere had been positioned at the treatment machine isocenter using optical tracking, the ultrasound probe was fixed on top of the phantom. A 3D ultrasound volume of each sphere was acquired using optical tracking as described above. The 3D-ultrasound-based position of the target sphere was determined by finding the center of the sphere in the axial, sagittal, and coronal planes using a circle tool placed on each of the three orthogonal ultrasound views. The target localization accuracy of the 3D-ultrasound optically guided system was thus determined by comparing the experimentally determined position of each sphere to its predicted position from the treatment planning system.

Our data show that the localization error does not depend on the target depth or the ultrasound focal depth used. Table 12.1 demonstrates representative values for such a test of optically guided 3D-ultrasound target localization for a 15-cm depth ultrasound probe format, which is the typical depth one employs in ultrasound localizations of the prostate. Similar to optical guidance testing described previously, optically guided 3D-ultrasound localization should be able to localize a well-defined internal target to within the inherent imaging uncertainty; however, localization errors in each of the spatial dimensions may exceed the predicted localization error due to finite image pixel size.

Other experiments can be performed using anthropomorphic phantoms. As an example we describe below an experiment using a specially designed prostate phantom. This prostate phantom consists of three layers. The first layer contains a model of the bladder, the second a model of the prostate and urethra, and the third a model for the rectum in the form of a long cylinder. The bladder, prostate, urethra, rectum, and the background material are made of Zerdine with different echogenicity closely mimicking sound absorption and speed properties characteristic for these anatomical structures. A CT scan (0.49×0.49×2.0-mm resolution) of the phantom with a passive infrared fiducial array attached was acquired. Using optical tracking as described previously, the positions of each organ can be determined in both CT image space and treatment room space within imaging uncertainty. In the CT data set four distinct regions of interest (ROIs) – bladder, prostate, urethra, and rectum – were segmented. A treatment isocenter was chosen in the treatment planning system, and the planning CT, isocenter, and segmented ROIs were transferred to the control computer of the 3D-ultrasound target localization system. The phantom was then set up on the treatment couch as shown in Fig. 12.5.

Again, optical tracking can be used to align the planned treatment isocenter with the treatment machine isocenter within 0.2 mm RMS error. The fiducial array attached to the phantom is then covered and a separate fiducial array is attached to the treatment couch, and the position of this second array is recorded using the optical positioning sensor system. In this way a fixed "bony anatomy" is established about which introduced internal organ motion can be simulated as follows: first the fiducial array attached to treatment couch is covered and the fiducial array attached to the phantom is uncovered. Then the phantom is shifted from its starting posi-

Table 12.1. Accuracy of optically guided 3D ultrasound as a function of the depth of the target for a focal depth of 15 cm. Results are given in terms of the mean distance in the antero-posterior, lateral, and axial directions and the 95% confidence interval around that mean value

Depth (mm)	Anteroposterior distance (mm)	Lateral distance (mm)	Axial distance (mm)
30	0.8±0.6	0.5±0.3	1.0±1.7
50	0.1±0.7	0.6±0.8	1.3±1.6
70	-0.2±0.6	1.0±0.9	0.1±1.5
100	0.2±0.8	1.2±0.9	0.3±1.8
130	0.3±0.6	1.1±1.0	0.2±1.5

tion ±5 mm in the lateral and/or superior/inferior directions using a 2D translation table. These shifts are measured to within ±0.2 mm RMS error using the optical positioning sensor system. Once the shifts have been made, the fiducial array attached to the phantom is again covered and the fiducial array attached to the treatment couch is uncovered. Since the treatment couch position remains fixed, the "bony anatomy" is not changed, even though the phantom has been physically moved; hence, through the alternate use of two fiducial arrays one is able to introduce accurate apparent organ motion while maintaining a fixed "bony anatomy." This apparent organ motion can then be detected using optically guided 3D ultrasound imaging by employing the optically guided 3D ultrasound target localization sys-

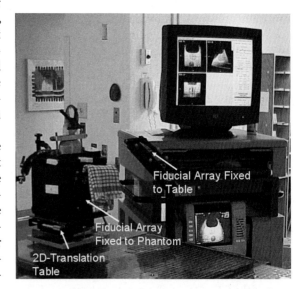

Fig. 12.5. Experimental setup of the prostate phantom in the treatment room for determining unknown induced motion in a target structure using an ultrasound target localization system

tem in clinical mode. This experiment gives a lower limit of the localization accuracy of the optically guided 3D ultrasound target localization system in daily clinical use.

The measurements for each translation of the phantom were repeated non-consecutively five times. Once an organ shift had been measured using optic guided 3D-ultrasound localization, the phantom was translated to a new position and a new 3D-ultrasound localization was performed. This procedure was followed until the required number of ultrasound localizations of each phantom shift had been obtained. Repeated performance of this experiment shows that one is able to localize an internal structure to within the inherent imaging uncertainty (cf. Table 5 of TOMÉ et al. 2002). Again, localization errors in each spatial dimension may exceed the predicted localization error due to finite image pixel size.

In the preceding paragraphs we have described two methods for quantitative testing of the accuracy of an optically guided 3D ultrasound target localization system. While many qualitative tests exist for testing these systems, we believe that it is extremely important for the physicist to perform commissioning tests of the system that quantify the errors locally. For this reason we have summarized these tests in Table 12.2 together with their suggested frequency. As mentioned in the introductory paragraph of this section, this requires the user to establish some local standard coordinate system in the treatment room. We have chosen to establish this through optical tracking, but such a standard can also be established using alternative mechanisms.

12.6
Clinical Applications

We have used daily transabdominal 3D ultrasound localization for conformal and intensity-modulated radiotherapy of the prostate employing a rectal balloon (PATEL et al. 2003). whereas CHINNAIYAN et al. (2003) have employed daily transabdominal 3D ultrasound localization in the treatment of prostate cancer in the post-operative adjuvant or salvage setting to improve the reproducibility of coverage of the intended volumes and to enhance conformal avoidance of adjacent normal structures. They studied 16 consecutive patients who received external beam radiotherapy and underwent daily localization using an optically guided 3D-ultrasound target localization system. Six of these patients were treated in a post-prostatectomy setting, either adjuvantly or for salvage, while the remaining 10 with intact prostates were treated definitively. During 3D ultrasound localization the bladder base was used as the primary localization structure for the post-prostatectomy patients (see Fig. 12.6), whereas for patients treated definitively the prostate was the primary reference structure (see Fig. 12.7). They found that the ultrasound-based displacements were not statistically different between the two groups of patients, and therefore concluded that daily transabdominal 3D-ultrasound localization is a clinically feasible method of correcting for internal organ motion in the post-prostatectomy setting.

In addition, this system has been used for patient localization for patients undergoing extracranial radiosurgery for a variety of abdominal, paraspinal,

Table 12.2. Possible quality assurance tests and their suggested frequency

Initially (perform tests in sequence)	Daily before first treatment	Annually and with each software update
Verify camera calibration	Verify camera calibration	Verify camera calibration
Verify ultrasound calibration	Verify ultrasound calibration	Verify ultrasound calibration
Known target test (verify entire process on sphere phantom)		Known target test (verify entire process on sphere phantom)
Acquire new CT image set		Acquire new CT image set
Load into TPS and create contours and isocenter		Load into TPS and create contours and isocenter
Send to US system		Send to US system
Verify isocenter and contours		Verify isocenter and contours
Localize spheres		Localize spheres
Unknown target test (verify entire process on prostate phantom as outlined above)		Unknown target test (verify entire process on prostate phantom as outlined above)

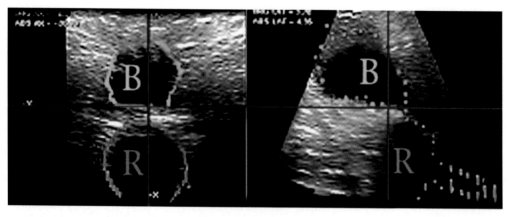

Fig. 12.6. Target localization using optical-guided 3D ultrasound in the post-prostatectomy setting. As shown in the ultrasound image panel, the bladder neck is the target structure of interest. The labels *B* and *R* refer to the bladder and the rectal balloon, respectively

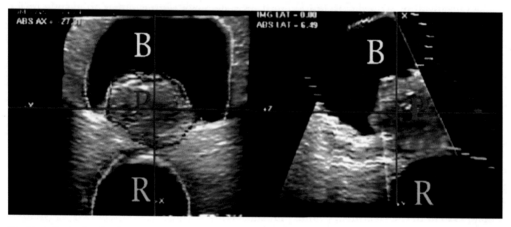

Fig. 12.7. Target localization using optical-guided 3D ultrasound for the treatment of a patient with intact prostate. As shown in the ultrasound image panel, the prostate is the target structure of interest. The labels *B*, *P*, and *R* refer to the bladder, the prostate, and the rectal balloon, respectively

and pelvic lesions (RYKEN et al. 2001; MEEKS et al. 2003). In general, the clinical use of optical-guided 3D ultrasound image guidance proceeds as follows: prior to CT scanning, the patient is immobilized using a custom vacuum cushion as is commonly used in radiation therapy (Vac-Loc, Med-Tec, Inc., Orange City, Iowa). The CT is acquired with the patient immobilized in the same position that will be used during the radiotherapy treatment in order to maintain a generally consistent position of mobile anatomy. The CT images are transferred to the treatment planning system, where the tumor volume and normal structures of interest are delineated. A treatment plan is then designed to conform the prescription dose closely to the planning target volume (PTV) while minimizing the dose to the nearby normal structures, using either 3D conformal radiotherapy or intensity modulated radiotherapy.

On each day of the treatment, the patient is placed in the same immobilization cushion that was used during CT scanning. The patient is initially set up relative to isocenter using conventional laser alignment. A 3D ultrasound volume is then acquired and reconstructed on the workstation. The target volume and critical structure outlines, as delineated on the planning CT scans, are overlaid on the acquired ultrasound volume in relation to the linear accelerator isocenter. The contours determined from the CT scans are then manipulated until they align with the anatomic structures on the ultrasound images (cf. Fig. 12.8). The amount of movement required to align the contours with the ultrasound images determines the magnitude of the target misregistration with isocenter based on conventional setup techniques. The target is then placed at the isocenter by tracking a fiducial array attached to the treatment couch, which allows precise translation from the initial treatment room laser setup position to the 3D ultrasound determined setup position. Once the shifts indicated by optically guided 3D ultrasound target localization system have been made, treatment proceeds as planned.

However, the question becomes: What is the dosimetric impact on the treatment plan due to shifting the patient daily using 3D ultrasound target localization? ORTON and TOMÉ (2004) have studied the effects of such daily shifts on dose distributions in the prostate PTV, rectal wall, and bladder wall for intensity-modulated radiotherapy for a ten-patient cohort. The shifts in their study were based on daily ultrasound imaging using an optical-guided 3D-ul-

trasound target localization system; however, their results are applicable to daily shifts derived using other methods. To investigate how these shifts affect dose distributions and predicted outcomes, they generated treatment plans for three cases: (a) the initial preplan, which represents the ideal case in which no shifts are necessary; (b) a postplan incorporating each day's actual shifts; and (c) a postplan in which no shifts were made but the internal organs were moved by the amounts indicated by daily 3D US imaging. They found that when daily shifts were made, doses to the target, rectal wall, and bladder wall are virtually identical to those in the preplan; however, when the indicated shifts were not carried out dose distributions degraded as shown in Fig. 12.9. For a typical patient, PTV-EUDs are 99.7% of the preplan PTV-EUD for the postplan with shifts and 92.7% of the preplan PTV-EUD for the postplan without shifts.

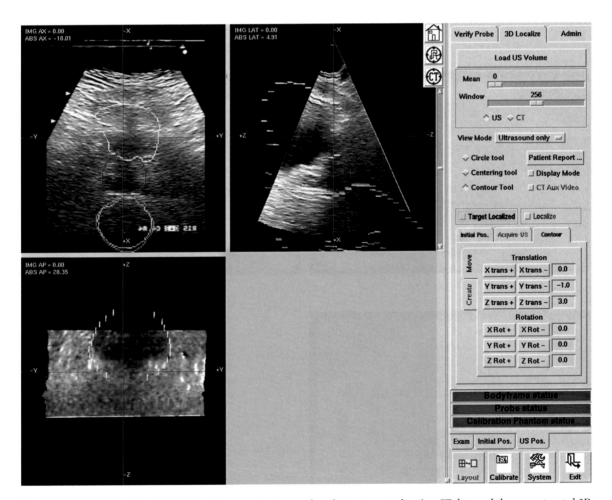

Fig. 12.8. Graphical interface which allows the operator to correlate the treatment-planning CT data and the reconstructed 3D ultrasound volumes to each other. Contours of the target structures and the organs at risk generated during the treatment-planning process are overlaid onto the 3D ultrasound data set. The operator can shift the contours in all three views until a satisfactory match between contours and depicted ultrasound anatomy is achieved

In their study an evenly spaced seven-beam arrangement around the patient was used, so differences in SSD and depth for each beam were largely offset by beams entering on the opposite side of the patient. It is important to note that other beam geometries might not yield such good results. Also, daily differences in organ shapes were not modeled in their study. Nonetheless, this study illustrates that delivery of IMRT without adequate target localization can induce cold spots in the target that can potentially compromise local tumor control probability, especially if smaller PTV margins are used (TOMÉ and FOWLER 2002).

Whereas ultrasound localization can provide high-precision target localization, it is important to note that the clinical use of ultrasound images in radiation therapy requires training and skill from the user. It is unlikely that users with no training in the interpretation of ultrasound images will gener-

ate results that improve target localization accuracy. To illustrate this, we have undertaken a pilot study at the University of Iowa to determine the impact of training in the interpretation of 3D ultrasound sound images on the localization of the prostate. In this pilot study nine different users independent of each other retrospectively registered 15 patient data sets. Four of the users had approximately 1 year of 3D ultrasound interpretation and localization experience, whereas the other five users had been trained in the use of 3D ultrasound localization system but had no experience in ultrasound interpretation. Results of this pilot study are shown in Table 12.3 and indicate that the standard deviation among untrained users is of the order of the average required shift. Among those trained in 3D ultrasound interpretation, however, the results are consistent and indicate that significant increases in target localization using daily 3D ultrasound localization can be obtained.

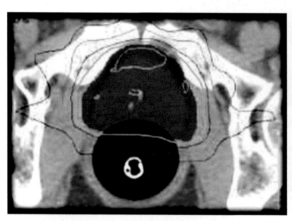

a   Theoretical composite treatment plan

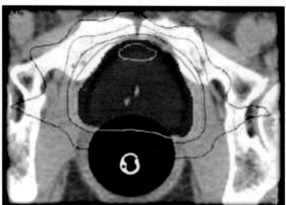

b   Composite Treatment plan in which the indicated shifts were applied

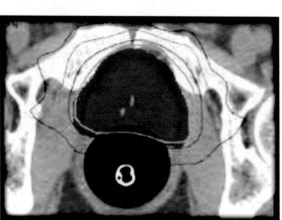

c   Composite Treatment plan in which the indicated organ shifts were not applied

Fig. 12.9a–c. a The ideal isodose distribution that would be delivered if internal organ motion and setup variation were absent. **b** Isodose distribution that is delivered when one corrects for organ motion and setup variation daily using 3D ultrasound imaging. **c** Isodose distribution that would result if one did not correct for these variations in setup. The 76 Gy (*green*), 70 Gy (*sky blue*), 65 Gy (*purple*), 54 Gy (*dark green*), and 44 Gy (*dark purple*) isodose lines are shown

Table 12.3 User variability in prostate localization using ultrasound guidance. *AP* anteroposterior

	AP (mm)	Lateral (mm)	Axial (mm)
Average shift	3.4	2.7	4.5
Standard deviation (four trained users)	1.2	0.9	1.4
Standard deviation (five untrained users)	3.6	1.5	2.9

References

Bouchet LG, Meeks SL, Goodchild G et al. (2001) Calibration of three-dimensional ultrasound images for image-guided radiation therapy. Phys Med Biol 46:559–577

Bouchet LG, Meeks SL, Bova FJ et al. (2002) 3D ultrasound image guidance for high precision extracranial radiosurgery and radiotherapy. Radiosurgery 4:262–278

Chinnaiyan P, Tomé WA, Patel R et al. (2003) 3D-ultrasound guided radiation therapy in the post-prostatectomy setting. Technol Cancer Res Treat 2:455–458

Horn BKP (1987) Closed-form solution of absolute orientation using unit quaternions. J Opt Soc Am 4:629–642

Lattanzi J, McNeeley S, Pinover W et al. (1999) A comparison of daily CT localization to a daily ultrasound-based system in prostate cancer. Int J Radiat Oncol Biol Phys 43:719–725

Lattanzi J, McNeeley S, Hanlon A et al. (2000) Ultrasound-based stereotactic guidance of precision conformal external beam radiation therapy in clinically localized prostate cancer. Urology 55:73–78

Meeks SL, Buatti JM, Bouchet LG et al. (2003) Ultrasound guided extracranial radiosurgery: technique and application. Int J Radiat Oncol Biol Phys 55:1092–1101

Orton NP, Tomé WA (2004) The impact of daily shifts on prostate IMRT dose distributions. Med. Phys. 31 2845–2848.

Patel RR, Orton NP, Tomé WA et al. (2003) Rectal dose-sparing with a balloon catheter and ultrasound localization in conformal radiation therapy for prostate cancer. Radiother Oncol 67:285–294

Ryken TC, Meeks SL, Buatti JM et al. (2001) Ultrasonic guidance for spinal extracranial radiosurgery: technique and application for metastatic spinal lesions. Neurosurg Focus 11:8

Shoup TE, Mistree F (1987) Optimization methods with applications for personal computers,1st edn. Prentice-Hall, Englewood Cliffs, New Jersey

Tomé WA, Fowler JF (2002) On cold spots in tumor subvolumes. Med Phys 29:1590–1598

Tomé WA, Meeks SL, Orton NP et al. (2002) Commissioning and quality assurance of an optically guided 3D ultrasound target localization system for radiotherapy. Med Phys 29:1781–1788

Troccaz J, Laieb N, Vassal P et al. (1995) Patient setup optimization for external conformal radiotherapy. J Imaging Guided Surg 1:113–120

Yang CC, Ting JY, Markoe A et al. (1999) A comparison of 3D data correlation methods for fractionated stereotactic radiotherapy. Int J Radiat Oncol Biol Phys 43:663–670

13.2
Definition of Target Volume

Both ICRU report 50 from 1993 and ICRU report 62 from 1999 standardised the nomenclature used for three-dimensional conformal treatment planning and thus gave the community of radiation oncologists a consistent language and guidelines for image-based target volume delineation. The following terms were defined: gross tumour volume (GTV); the clinical target volume (CTV); the internal target volume (ITV); the planning target volume (PTV); the treated volume; and the irradiated volume (Fig. 13.1).

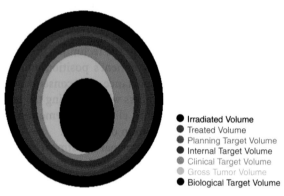

● Irradiated Volume
● Treated Volume
● Planning Target Volume
● Internal Target Volume
● Clinical Target Volume
● Gross Tumor Volume
● Biological Target Volume

Fig. 13.1. Concepts used in target volume definition for radiation treatment.

13.2.1
Gross Target Volume

The gross tumour volume (GTV) is the macroscopic (gross) extent of the tumour as determined by radiological and clinical investigations (palpation, inspection). The GTV-primary (GTV-P) defines the area of the primary tumour and GTV-nodal (GTV-N) the macroscopically involved lymph nodes. The GTV is obtained by summarising the area outlined by the radiation oncologist in each section, multiplied by the thickness of each section. The extension of the GTV is of major importance for the treatment strategy: in many cases the gross tumour tissue is irradiated with higher doses, as it is encompassed within the boost volume. The GTV can represent the total volume of the primary tumour, the macroscopic residual tumour tissue after partial tumour resection or the region of recurrence after either surgical, radiation or chemotherapeutic treatment.

The delineation of the GTV is usually based on data obtained from CT and MRI. In general, tumour tissue shows contrast enhancement, has a different density (CT) or intensity (MRI) compared with the normal tis-

sue and is surrounded by perifocal oedema. Based on these radiological characteristics, the radiation oncologist has to outline on each section the areas of gross tumour tissue. The demarcation of the GTV needs profound radiological knowledge. The technique used for the CT or MRI investigation and the window and level settings have to be appropriate for the anatomical region. The volume of the gross tumour tissue visualised on CT or MRI should correspond to the volume of the actual macroscopic tumour extension.

A major problem when delineating residual tumour tissue in surgically treated patients is the recognition and differentiation of changes caused by surgery itself. But also contrast enhancement, oedema and hyper- or hypodensity (intensity) of the tissue can cause difficulties, ultimately resulting in inaccurate GTV delineation.

Sometimes it is extremely difficult or even impossible to delineate the gross tumour mass when using conventional radiological investigation techniques such as CT and MRI. An example is the visualisation of prostate cancer with current imaging methods (CT, MRI, sonography). In this case, for three-dimensional treatment planning, the radiation oncologist delineates the prostate encompassing the GTV and the clinical target volume (CTV).

An accurate delineation of tumour tissue should focus the irradiation dose on the GTV and spare surrounding normal tissue. New investigative techniques, such as positron emission tomography PET, SPECT and MRS, visualise tumour tissue with a higher specificity. It seems likely that these techniques will help in the future to delineate tumour tissue with higher precision. The possibility to integrate these "biological" investigative techniques in GTV definition is discussed in the section "Biological Target Volume".

13.2.2
Clinical Target Volume

The GTV, together with the surrounding microscopic tumour infiltration, constitutes the primary clinical target volume CTV (CTV-P). It is important to mention that the definition of the CTV-P also includes the tumour bed, which has to be irradiated after a complete macroscopic tumour resection, in both R0 (complete microscopic resection) and R1 (microscopic residual tumour on the margin of the tumour bed) situation. Moreover, for CTV-P definition, anatomical tumour characteristics have to be considered, such as the likelihood of perineural and perivascular extension or tumour spread along anatomical borders. As the margins between CTV and GTV are not homogenous, they have to be adjusted to the probable microscopic tumour spread. The CTV-nodal (CTV-N) defines the assumed microscopic lymphatic tumour spread, which has also to be included in the radiation treatment planning.

13.2.3
Internal Target Volume

The internal target volume (ITV) is a term introduced by the ICRU report 62. The ITV encompasses the GTV/CTV plus internal margins to the GTV/CTV, caused by possible physiological movements of organs and tumour, due to respiration, pulsation, filling of the rectum, or variations of tumour size and shape, etc. It is defined in relation to internal reference points, most often rigid bone structures, in an internal patient coordinate system. Observation of internal margins is difficult, in many cases even impossible. Examples for reducing internal margins are the fixation of the rectum with a balloon during irradiation of the prostate or body fixation and administration of oxygen to reduce respiration movements during stereotactic radiotherapy.

13.2.4
Planning Target Volume

The planning target volume (PTV) incorporates the GTV/CTV plus margins due to uncertainties of patient setup and beam adjustment; therefore, these margins consider the inaccuracy in the geometrical location of the GTV/CTV in the irradiated space, due to variations in patient positioning during radiotherapy and organ motility. The PTV can be compared with an envelope and has to be treated with the same irradiation dose as the GTV/CTV. Movements of the GTV/CTV within this envelope should not change the delivered radiation dose to the GTC and CTV. The setup margins are not uniform. The radiation oncologist should define the margins together with the radiation physicist, taking into consideration the possible inaccuracy in patient and beam positioning. Significant reduction of setup margins can be obtained by applying patient fixation and re-positioning techniques, as used, for example, in stereotactic radiotherapy. Beam positioning uncertainties should be subject of a specific quality control program of each treatment unit. Consequently, the volume of normal tissue surrounding the tumour and included in high irradiation dose areas can be considerably

reduced and the dose applied to the GTV/CTV can be escalated.

13.2.5
Treated Volume

The treated volume is the volume of tumour and surrounding normal tissue included in an isodose surface representing the irradiation dose proposed for the treatment. As a rule this corresponds to the 95% isodose. Ideally, the treated volume should correspond to the PTV; however, in many cases the treated volume exceeds the PTV. The coherence of an irradiation plan can be described/illustrated as the correlation between PTV and treated volume.

13.2.6
Irradiated Volume

The irradiated volume is a volume included in an isodose surface with a possible biological impact on the normal tissue encompassed in this volume; therefore, the irradiated volume depends on the selected isodose curve and the normal tissue surrounding the tumour. The choice of an isodose surface depends on the end point defined for possible side effects in normal tissue. For grade-III and grade-IV side effects, the isodoses will correspond to higher doses, whereas low-dose areas are significant for the risk of carcinogenesis.

13.3
Definition of Organs at Risk

Organs at risk (ICRU report 50), also known as critical structures, are anatomical structures with important functional properties located in the vicinity of the target volume. They have to be considered in the treatment planning, since irradiation can cause pathological changes in normal tissue, with irreversible functional consequences. The critical structures must be outlined in the planning process and the dose applied to these areas has to be quantified by either visualising the isodose distribution or by means of dose-volume histograms (DVH). It is essential to include the tolerance dose of the organs at risk in the treatment strategy.

The ICRU report 62 furthermore added the term "planning organs at risk volume" (PRV). This term takes into account that the organs at risk are in the majority mobile structures; therefore, a surrounding margin is added to the organs at risk in order to compensate for geometric uncertainties.

With regard to histopathological properties of tissue, organs at risk can be classified as serial, parallel and serial-parallel (KÄLLMANN et al. 1992; JACKSON and KUTCHER 1993). Serial organs can lose their complete functionality even if only a small volume of the organ receives a dose above the tolerance limit. Best known examples are spinal cord, optical nerve and optical chiasm. In contrast, parallel organs are damaged only if a larger volume is included in the irradiation region, e.g. lung and kidney; however, in many cases these two models are combined in a serial-parallel organ configuration. Side effects occur due to various pathological mechanisms, size of the irradiated volume and the maximal dose applied to the organ. A classical example for a serial-parallel organ is the heart: the coronary arteries are a *parallel* and the myocardium a *serial organ.*

The organs at risk can be located at a distance from the PTV, close to the PTV or incorporated within the tumour tissue and thus be in the PTV. These three situations have to be considered in the treatment planning. Organs with a low tolerance to irradiation (lens, gonads) have to be outlined even if they are not located in the immediate vicinity of the PTV. For organs situated close to the PTV a plan with a very deep dose fall towards the critical structures has to be designed. If the organs at risk are encompassed within the PTV, it is crucial to achieve a homogeneous dose distribution, and to consider the D_{max} and the isodose distribution in the PTV.

13.4
New Concepts in Target Volume Definition: Biological Target Volume

In recent years, new methods for tumour visualisation have begun to have a major impact on radiation oncology. Techniques such as PET, SPECT and MRS permit the visualisation of molecular biological pathways in tumours; thus, additional information about metabolism, physiology and molecular biology of tumour tissue can be obtained. This new class of images, showing specific biological events, seems to complement the anatomical information from traditional radiological investigations. Accordingly, in addition to the terms GTV, CTV, PTV, etc., LING et al. (2000) proposed the terms "biological target volume" (BTV) and "multidimensional conformal radiotherapy" (MD-CRT).

Thus far, biological imaging has not yet been integrated into radiation treatment planning; however, several trials indicate that biological imaging could have a significant impact on the development of new treatment strategies in radiation oncology. These trials show that PET, SPECT or MRS might be helpful in obtaining more specific answers to the three essential questions given in the sections that follow.

13.4.1
Where is the Tumour Located and Where Are the (Macroscopic) Tumour Margins?

The co-registration of biological and anatomical imaging techniques seems to improve the delineation of viable tumour tissue compared with CT or MRI alone. Biological imaging enables the definition of a target volume on the basis of biological processes. For some cases this results in clearer differentiation between tumour and normal tissue as compared with CT and MRI alone; however, anatomical imaging will continue to be the basis of treatment planning because of its higher resolution. Incorporation of biological imaging into the tumour volume definition should only be done in tumours with increased sensitivity and specificity to biological investigation techniques.

We discuss three clinical entities: lung cancer, head and neck cancer, prostate cancer and brain gliomas as biological imaging proved to be helpful in target volume delineation.

13.4.1.1
Lung Cancer

The role of fluoro-deoxyglucose (FDG)-PET in radiation treatment planning of lung cancer has been thoroughly investigated in a total of 415 patients in 11 trials (Table 13.1; GROSU et al. 2005c). These investigations compared FDG-PET data with CT data. The CT/PET image fusion methods were used in five trials. One study used the integrated PET/CT system. All these studies suggested that FDG-PET adds essential information to the CT with significant consequences on GTV, CTV and PTV definition. The percentage of cases presenting significant changes in tumour volume after the integration of FDG-PET investigation in the radiation treatment planning ranged from 21 to 100%. Ten studies pointed out the significant implications of FDG-PET in staging lymph node involvement. These findings are supported by data in the literature, showing an advantage of FDG-PET over CT in lung cancer

Table 13.1 Impact of FDG-PET in gross target volume (*GTV*) and planning target volume (*PTV*) delineation in lung cancer. *PET* positron emission tomography

Reference	No. of patients	Change of GTV and PTV post-PET (*n*)	Increase of GTV and PTV using PET	Decrease of GTV and PTV using PET	Comments
HERBERT et al. (1996)	20	GTV: 7 of 20 (35%)	GTV 3 of 20 (15%)	GTV 4 of 20 (20%)	PET may be useful for delineation of lung cancer
KIFFER et al. (1998)	15	GTV: 7 of 15 (47%); PTV: 4 of 15 (27%)	GTV and PTV: 4 of 15 (27%)		PET defects positive lymph nodes, not useful in tumour delineation
MUNLEY et al. (1999)	35	PTV: 12 of 35 (34%)	PTV: 12 of 35 (34%)	PET target smaller than CT not evaluated	PET complements CT information
NESTLE et al. (1999)	34, stage IIIA–IV	Change of field size (cm^2) in 12 (35%); median 19.3%	Increase of field size 9 (26%)	Decrease of field size: 3 (9%)	Change of field size in patients with dys- or atelectasis
VANUYTSEL et al. (2000)	73 (N+), stage IIA–IIIB	GTV: 45 of 73 (62%)	GTV: 16 of 73 (22%) 11 patients with pathology; 1 patient unnecessary; 4 patients insufficient	GTV: 29 of 73 (40%); 25 patients pathology; 1 patient inappropriate; 3 patients insufficient	PET data vs pathology: 36 (49%)=pathology; 2 (3%) inappropriate; 7 (10%) insufficient
	First 10 for whom PET–CT–GTV <CT–GTV			PTV 29±18% (cc); max. 66%, min. 12%; V lung (20 Gy) 27±18%; max. 59%, min. 8%	Assessment of lymph node infiltration by PET improves radiation treatment planning
MACMANUS et al. (2001)	153 stage IA–IIIB, unresectable candidates for radical RT after conventional staging	GTV: 22 of 102 (21%)	GTV: 22 of 102 (21%); inclusion of structures previously considered uninvolved by tumour	GTV: 16 of 102 (15%); exclusion of atelectasis and lymph nodes	Post-PET stage but not pre-PET stage was significantly associated with survival
GIRAUD et al (2001)	12	GTV, PTV: 5 of 12 (42%)			4 of 12 lymph nodes; 1 of 12 atelectasis and distant meta
MAH et al. (2002)	30, stage IA–IIIB	GTV: 5 of 23 (22%); FDG-avid lymph nodes	PTV: 30–76% of cases (varied between the three physicians)	PTV: 24–70% of cases (varied between the three physicians)	Addition of PET does lower physician variation in PTV delineation; PET-significant alterations to patient management and PTV
ERDI et al. (2002)	11, N+	PTV: 11 of 11 (100%)	PTV: 7 of 11 (36%); 19% (5–46%) cc; detection of lymph nodes	PTV: 4 of 11 18% (2–48%) cc; exclusion of atelectasis trimming the target volume to spare critical structure	PET improves GTV and PTV definition
CIERNIK et al. (2003)	6	GTV: 100%	GTV: 1 of 6 (17%)	GTV: 4 of 6 (67%)	PET/CT improves GTV delineation
BRADLEY et al. (2004)	26	PTV: 14 of 24 (58%)	11 of 24 (46%); 10 lymph nodes, 1 tumour	3 of 24 (12%); tumour vs atelectasis	PET improves diagnosis of lymph nodes and atelectasis

staging. The PET may help to differentiate between viable tumour tissue and associated atelectasis; however, it shows limitations when secondary inflammation is present. As GTV delineation is a fundamental step in radiation treatment planning, the use of combined CT/PET imaging for all dose escalation studies with either conformal radiotherapy or IMRT has been recommended by RTOG as a standard method in lung cancer (CHAPMAN et al. 2003).

13.4.1.2
Head and Neck Cancer

A significant number of studies have demonstrated that FDG-PET can be superior to CT and MRI in the detection of lymph nodes metastases, the identification of unknown primary cancer and in the detection of viable tumour tissue after treatment.

RAHN et al. (1998) studied 34 patients with histologically confirmed squamous cell carcinoma of the head and neck by performing FDG-PET prior to treatment planning in addition to conventional staging procedures, including CT, MRI and ultrasound. The integration of FDG-PET in radiation treatment planning led to significant changes in the radiation field and dose in 9 of 22 (44%) patients with a primary tumour and in 7 of 12 (58%) patients with tumour recurrence. This was due mainly to the inclusion of lymph nodes metastases detected by FDG-PET. In a recent study, NISHIOKA et al. (2002) showed that the integration of FDG-PET in radiation treatment planning might also cause a reduction in the size of the radiation fields. The GTV for primary tumour was not altered by image fusion in 19 of 21 (90%) patients. Of the 9 patients with nasopharynx cancer, the GTV was enlarged by 49% in only 1 patient and decreased by 45% in 1 patient. In 15 of 21 (71%) patients the tumour-free FDG-PET detection allowed normal tissue to be spared. Mainly, parotid glands were spared and, thus, xerostomia was avoided. The authors concluded that the image fusion between FDG-PET and MRI/CT was useful in GTV, CTV and PTV determination, both for encompassing the whole tumour area in the irradiation field and for sparing of normal tissue. The integrated PET/CT was used for GTV and PTV definition in 12 patients with head and neck tumours and compared with CT alone. The GTV increased by 25% or more due to FDG-PET in 17% of these cases and was reduced by 25 in 33% of the patients. The corresponding change in PTV was approximately 20%; however, this study did not integrate MRI in the analysis, which has a higher sensitivity than CT (CIERNIK et al. 2003).

In conclusion, the value of FDG-PET for radiation treatment planning is still under investigation. In some cases, FDG-PET visualised tumour infiltration better than CT or MRI alone. The FDG-PET could also play an important role in the definition of the boost volume for radiation therapy; however, the relatively high FDG uptake in inflammation areas could sometimes lead to false-positive results.

13.4.1.3
Prostate Cancer

Choline and citrate metabolism within cytosol and the extracellular space were investigated in prostate cancer using H-MRS. Clinical trials analysing tumour location and extent within the prostate, extra-capsular spread and cancer aggressiveness in pre-prostatectomy patients have indicated that the metabolic information provided by H-MRS combined with the anatomical information provided by MRI can significantly improve the tumour diagnosis and the outline of tumour extension (MIZOWAKI et al. 2002; MUELLER-LISSE et al. 2001; WEFER et al. 2000).

COAKLEY et al. (2002) evaluated endorectal MRI and 3D MRS in 37 patients before prostatectomy and correlated the tumour volumes measured with MRI and H-MRS with the true tumour volume measured after prostatectomy. Measurements of tumour volume with MRI, MRS and a combination of both were all positively correlated with histopathological volume (Pearson's correlation coefficient of 0.49, 0.59 and 0.55, respectively), but only measurements with 3D MRS and a combination of MRI and MRS were statistically significant ($p<0.05$).

The integration of H-MRS in brachytherapy treatment planning for target volume definition in patients with organ-confined but aggressive prostatic cancer could improve the tumour control probability (ZAIDER et al. 2000).

13.4.1.4
Brain Gliomas

Although only preliminary data are available, the presented literature indicates that amino-acid PET, SPECT and H-MRS, in addition to conventional morphological imaging, are superior to the exclusive use of either MRI or CT in the visualisation of vital tumour extension in gliomas (Table 13.2).

Our group investigated the value of amino-acid PET and SPECT in GTV, PTV and boost volume (BV) definition for radiation treatment planning of brain gliomas. We demonstrated that I-123-alpha-methyltyrosine (IMT)-SPECT and L-(methyl-11C) methionine (MET)-PET offer significant additional infor-

Table 13.2 Impact of biological imaging: PET, SPECT and MRS in target volume delineation of gliomas. *High G* high-grade gliomas

Reference	No. of patients	Diagnosis	Biological imaging	Additional information to MRI and CT
VOGES (1997)	46	Low+high G	MET-PET, FDG-PET	Yes
JULOW (2000)	13	High G	MET-PET, SPECT ?	Yes
GROSS (1998)	18	High G	FDG-PET	No
NUUTINEN (2000)	11	Low G	MET-PET	Yes
GROSU (2000)	30	Low+high G	IMT-SPECT	Yes
GRAVES (2000)	36	High G	H-MRS	Yes
PIRZKALL (2001)	40	High G	H-MRS	Yes
PIRZKALL (2004)	30	High G	H-MRS	Yes
GROSU (2002)	66	High G	IMT-SPECT	Yes
GROSU (2005)	44	High G	MET-PET	Yes

mation concerning tumour extension in high-grade gliomas, compared with anatomical imaging (CT and MRI) alone (GROSU et al. 2000, 2002, 2003, 2005a,c). The MRS studies led to similar results (PIRZKALL et al. 2001, 2004). A current analysis of an amino-acid SPECT- or PET-planned subgroup of patients with recurrent gliomas suggests that the integration of amino-acid PET or SPECT in target volume definition might contribute to an improved outcome (Fig. 13.2; GROSU et al. 2005b).

13.4.2
Which Relevant Biological Properties of the Tumour Could Represent an Appropriate Biological Target for Radiation Therapy?

Tumour hypoxia (Fig. 13.3) is an unfavourable prognostic indicator in cancer as it can be linked to aggressive growth, metastasis and poor response to treatment (MOLLS and VAUPEL 2000; MOLLS 2001). In a clinical pilot study, CHAO et al. (2001) demonstrated the feasibility of [^{60}Cu]ATSM-PET guided radiotherapy planning in head and neck cancer patients; however, the tumour-to-background ratio in hypoxic tumour tissues did not dramatically differ from previous studies using [^{18}F]FMISO and other nitroimidazole compounds. The authors reported the integration of the hypoxia tracer Cu-ATSM in radiation treatment planning for patients with locally advanced head and neck cancer treated with IMRT. They developed a CT/PET fusion image based on external markers. The GTV outline was based on radiological and PET findings. Within the GTV, regions with a Cu-ATSM uptake twice that of the contralateral normal neck muscle were selected and outlined as hypoxic

GTV (hGTV). The IMRT plan delivered 80 Gy in 35 fractions to the ATSM-enriched tumour subvolume (hGTV) and the GTV received simultaneously 70 Gy in 35 fractions, keeping the radiation dose to the parotid glands below 30 Gy.

More recently, DEHDASHTI et al. (2003) were the first to demonstrate a negative predictive value of enhanced Cu-ATSM uptake for the response to treatment in 14 cervical cancer patients. RISCHIN et al. (2001) used ^{18}F-misonidazole scans to detect hypoxia in patients with T3/4, N2/3 head and neck tumours treated with tirapazamine, cisplatin and radiation therapy. Fourteen of 15 patients were hypoxia positive at the beginning of the treatment, but only one patient had detectable hypoxia at the end of radiochemotherapy.

By imaging either hypoxia (RISCHIN et al. 2001) with ^{18}F-misonidazole or [^{60}Cu]ATSM (CHAO et al. 2001), angiogenesis with ^{18}F-labelled RDG-containing glycopeptide and PET (HAUBNER et al. 2001), proliferation with fluorine-labelled thymidine analogue 3'-deoxy-3'-[^{18}F]-fluorothymidine (FLT) and PET (WAGNER et al. 2003), or apoptosis with a radio-labelled recombinant Annexin V and SPECT (BELHOCINE et al. 2004), different areas within a tumour can be identified and individually targeted. This approach, closely related to the IMRT technique, has been named "dose painting". The IMRT combined with a treatment plan based on biological imaging could be used for individualised, i.e. customised, radiation therapy. The biological process visualised by the tracer needs to be specified; therefore, the selected BTV should be named after the respective tracer (i.e. BTV$_{(FDG-PET)}$, BTV$_{(FAZA-PET)}$; GROSU et al. 2005c).

This novel treatment approach will generate a new set of problems and questions such as: What are the dynamics of the visualised biological processes? How

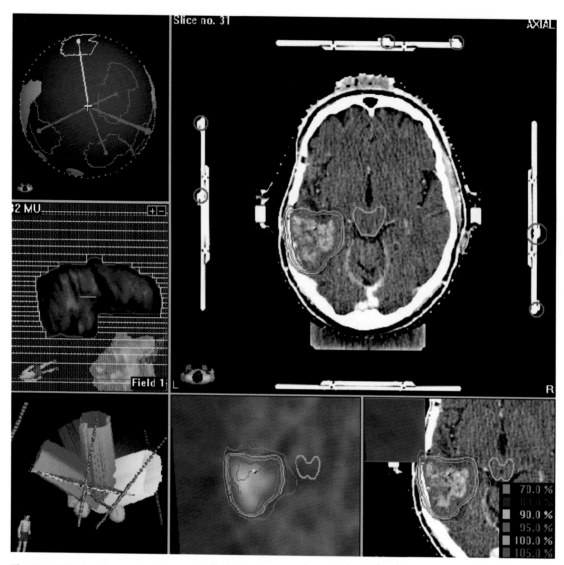

Fig. 13.2. MET-PET data were integrated in the planning system and were co-registered with the CT and MRI data using an automatically, intensity based image fusion method – Brain LAB (Grosu et al. 2005b)

many biological investigations are necessary during treatment and when? Which radiation treatment schedules have to be applied?

Prospective clinical trials and experimental studies should supply the answers to these questions.

13.4.3
What is the Biological Tumour Response to Therapy?

Biological imaging could be used to evaluate the response of a tumour to different therapeutic interventions. The usefulness of PET for monitoring patients treated with chemotherapy has been documented in several studies. Although evaluating the response af-

ter radiochemotherapy is often difficult due to treatment-induced inflammatory tissue changes, preliminary data for lung (WEBER et al. 2003; MACMANUS et al. 2003; CHOI 2002), oesophageal (FLAMEN et al. 2002) and cervical cancer (GRIGSBY et al. 2003) suggest that the decrease of FDG uptake after treatment correlates with histological tumour remission and longer survival; however, it still has to be clarified at which time points PET imaging should be performed, which tracer is to be used and which criteria can be used for the definition of a tumour response in PET. Theoretically, the integration of biological imaging in treatment monitoring could have significant consequences for future radiation treatment planning. This could result in changes of target volume/boost vol-

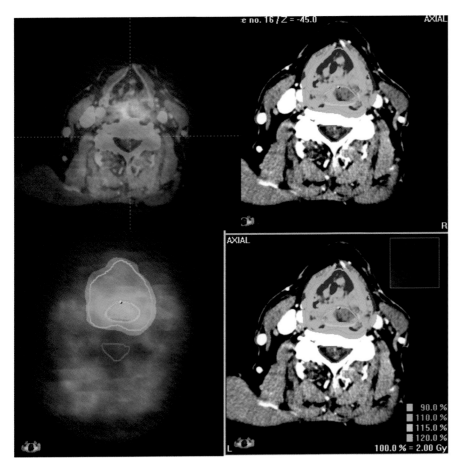

Fig. 13.3. Imaging of hypoxia in a patient with a cT4 cN2 cM0 G2 laryngeal cancer. [18F]Azomycin arabinoside [18F]FAZA-PET/CT image fusion. Using IMRT the hypoxic area is encompassed in the high dose area (BTV, orange) and the CTV is encompassed in the lower dose area - green. (Courtesy of Dr. M. Piert, Nuclear Medicine Department, Technical University Munich, Germany)

ume delineation during radiation therapy; however, this still has to be analysed in clinical studies.

In summary, biological imaging allows the visualisation of fundamental biological processes in malignancies. To date, the suggested superiority of biological strategies for treatment planning has not yet been sufficiently demonstrated; therefore, before the BTV definition can be generally recommended, further clinical studies based on integrated PET, SPECT or MRS need to be conducted. These studies must compare the outcome of biologically based treatment regimes with conventional treatment regimes.

13.5
Conclusion

Target volume definition is an interactive process. Based on radiological (and biological) imaging, the radiation oncologist has to outline the GTV, CTV, ITV and PTV and BTV. In this process, a lot of medical and technological aspects have to be considered. The criteria for GTV, CTV, etc. definition are often not exactly standardised, and this leads, in many cases, to variability between clinicians; however, exactly defined imaging criteria, imaging with high sensitivity and specificity for tumour tissue and special training could lead to a higher consensus in target volume delineation and, consequently, to lower differences between clinicians. It must be emphasised, however, that further verification studies and cost-benefit analyses are needed before biological target definition can become a stably integrated part of target volume definition.

The ICRU report 50 from 1993 and the ICRU report 62 from 1999 defining the anatomically based terms CTV, GTV and PTV must still be considered as the gold standard in radiation treatment planning; however, further advances in technology concerning signal resolution and development of new tracers with higher sensitivity and specificity will induce a shift of paradigms away from the anatomically based target volume definition towards biologically based treatment strategies. New concepts and treatment strategies should be defined based on these new investigation methods, and the standards in radiation treatment planning – in a continuous, evolutionary process – will have to integrate new imaging methods in an attempt to finally achieve the ultimate goal of cancer cure.

References

Belhocine T, Steinmetz N, Li C et al. (2004) The imaging of apoptosis with the radiolabeled annexin V: optimal timing for clinical feasibility. Technol Cancer Res Treat 3:23–32

Bradley J, Thorstad WL, Mutic S et al. (2004) Impact of FDG–PET on radiation therapy volume delineation in NSCLC. Int J Radiat Oncol Biol Phys 59:78–86

Chao KS, Bosch WR, Mutic S et al. (2001) A novel approach to overcome hypoxic tumour resistance: Cu-ATSM-guided intensity-modulated radiation therapy. Int J Radiat Oncol Biol Phys 49:1171–1182

Chapman JD, Bradley JD, Eary JF et al. (2003) Molecular (functional) imaging for radiotherapy applications: an RTOG symposium. Int J Radiat Oncol Biol Phys 55:294–301

Choi NC, Fischman AJ, Niemierko A et al. (2002) Dose-response relationship between probability of pathologic tumour control and glucose metabolic rate measured with FDG PET after preoperative chemoradiotherapy in locally advanced non-small-cell lung cancer. Int J Radiat Oncol Biol Phys 54:1024–1035

Ciernik IF, Dizendorf E, Baumert BG et al. (2003) Radiation treatment planning with an integrated positron emission and computer tomography (PET/CT): a feasibility study. Int J Radiat Oncol Biol Phys 57:853–863

Coakley FV, Kurhanewicz J, Lu Y et al. (2002) Prostate cancer tumour volume: measurement with endorectal MR and MR spectroscopic imaging. Radiology 223:91–97

Dehdashti F, Grigsby PW, Mintun MA et al. (2003) Assessing tumour hypoxia in cervical cancer by positron emission tomography with 60Cu-ATSM: relationship to therapeutic response: a preliminary report. Int J Radiat Oncol Biol Phys 55:1233–1238

Erdi YE, Rosenzweig K, Erdi AK et al. (2002) Radiotherapy treatment planning for patients with non-small cell lung cancer using positron emission tomography (PET). Radiother Oncol 62:51–60

Flamen P, Van Cutsem E, Lerut A et al. (2002) Positron emission tomography for assessment of the response to induction radiochemotherapy in locally advanced oesophageal cancer. Ann Oncol 13:361–368

Giraud P, Grahek D, Montravers F et al. (2001) CT and (18)F-deoxyglucose (FDG) image fusion for optimization of conformal radiotherapy of lung cancers. Int J Radiat Oncol Biol Phys 49:1249–1257

Graves EE, Nelson SJ, Vigneron DB et al. (2000) A preliminary study of the prognostic value of proton magnetic resonance spectroscopic imaging in gamma knife radiosurgery of recurrent malignant gliomas. Neurosurgery 46:319–326

Grigsby PW, Siegel BA, Dehdashti F et al. (2003) Posttherapy surveillance monitoring of cervical cancer by FDG–PET. Int J Radiat Oncol Biol Phys 55:907–913

Gross MW, Weber WA, Feldmann HJ et al. (1998) The value of F-18-fluorodeoxyglucose PET for the 3-D radiation treatment planning of malignant gliomas. Int J Radiat Oncol Biol Phys 41:989–995

Grosu AL, Weber WA, Feldmann HJ et al. (2000) First experience with I-123-Alpha-Methyl-Tyrosine SPECT in the 3-D radiation treatment planning of brain gliomas. Int J Radiat Oncol Biol Phys 47:517–527

Grosu AL, Feldmann HJ, Dick S et al. (2002) Implications of IMT-SPECT for postoperative radiation treatment planning in patients with gliomas. Int J Radiat Oncol Biol Phys 54:842–854

Grosu AL, Lachner R, Wiedenmann N et al. (2003) Validation of a method for automatic fusion of CT- and C11-methionine-PET datasets of the brain for stereotactic radiotherapy using a LINAC. First clinical experience. Int J Radiat Oncol Biol Phys 56:1450–1463

Grosu AL, Weber AW, Riedel E et al. (2005a) L-(Methyl-11C) methionine positron emission tomography for target delineation in resected high grade gliomas before radiation therapy. Int J Radiat Oncol Biol Phys 63:64–74

Grosu AL, Weber WA, Franz M et al. (2005b) Re-irradiation of recurrent high grade gliomas using amino-acids–PET(SPECT)/CT/MRI image fusion to determine gross tumour volume for stereotactic fractionated radiotherapy. Int J Radiat Oncol Biol Phys (in press)

Grosu AL, Piert M, Weber WA et al. (2005c) Positron emission tomography in target volume definition for radiation treatment planning. Strahlenther Onkol 181:483–499

Haubner R, Wester HJ, Weber WA et al. (2001) Noninvasive imaging of alpha(v)beta3 integrin expression using 18F-labeled RGD-containing glycopeptide and positron emission tomography. Cancer Res 61:1781–1785

Hebert ME, Lowe VJ, Hoffman JM et al. (1996) Positron emission tomography in the pretreatment evaluation and follow-up of non-small cell lung cancer patients treated with radiotherapy: preliminary findings. Am J Clin Oncol 19:416–421

ICRU 50 (1993) Prescribing, recording and reporting photon beam therapy. ICRU report no. 50. ICRU, Bethesda, Maryland

ICRU 62 (1999) Prescribing, recording and reporting photon beam therapy. ICRU report no. 62 (supplement to ICRU report no. 50). ICRU, Bethesda, Maryland

Jackson A, Kutcher GJ (1993) Probability of radiation-induced complications for normal tissue with parallel architecture subject to non-uniform irradiation. Med Phys 20:621–625

Julow J, Major T, Emri M et al. (2000) The application of image fusion in stereotactic brachytherapy of brain tumours. Acta Neurochir (Wien) 142:1253–1258

Källmann P, Ägren A, Brahme A (1992) Tumour and normal tissue responses to fractionated nonuniform dose delivery. Int J Radiat Oncol Biol Phys 62:249–262

Kiffer JD, Berlangieri SU, Scott AM et al. (1998) The contribution of 18F-fluoro-2-deoxy-glucose positron emission tomographic imaging to radiotherapy planning in lung cancer. Lung Cancer 19:167–177

Ling CC, Humm J, Larson S et al. (2000) Towards multidimensional radiotherapy (MD-CRT): biological imaging and biological conformality. Int J Radiat Oncol Biol Phys 47:551–560

MacManus MP, Hicks RJ, Ball DL et al. (2001) F-18 fluorodeoxyglucose positron emission tomography staging in radical radiotherapy candidates with nonsmall cell lung carcinoma: powerful correlation with survival and high impact on treatment. Cancer 92:886–895

MacManus MP, Hicks RJ, Matthews JP et al. (2003) Positron emission tomography is superior to computed tomography scanning for response assessment after radical radiotherapy or chemoradiotherapy in patients with non-small-cell lung cancer. J Clin Oncol 21:1285–1292

Mah K, Caldwell CB, Ung YC et al. (2002) The impact of (18)FDG–PET on target and critical organs in CT-based treatment planning of patients with poorly defined non-small-cell lung carcinoma: a prospective study. Int J Radiat Oncol Biol Phys 52:339–350

Mizowaki T, Cohen GN, Fung AY et al. (2002) Towards integrating functional imaging in the treatment of prostate cancer with radiation: the registration of the MR spectroscopy imaging to ultrasound/CT images and its implementation in treatment planning. Int J Radiat Oncol Biol Phys 54:1558–1564

Molls M (2001) Tumor oxygenation and treatment outcome. In: Bokemeyer C, Ludwig H (eds) ESO scientific updates, vol 6: Anemia in cancer. Elsevier, Amsterdam, pp 175–187

Molls M, Vaupel P (2000) The impact of the tumor environment on experimental and clinical radiation oncology and other therapeutic modalities. In: Molls M, Vaupel P (eds) Blood perfusion and microenvironment of human tumors: implications for clinical radiooncology. Springer, Berlin Heidelberg New York, pp 1–3

Mueller-Lisse UG, Vigneron DB, Hricak H et al. (2001) Localized prostate cancer: effect of hormone deprivation therapy measured by using combined three-dimensional 1H MR spectroscopy and MR imaging: clinicopathologic case-controlled study. Radiology 221:380–390

Munley MT, Marks LB, Scarfone C et al. (1999) Multimodality nuclear medicine imaging in three-dimensional radiation treatment planning for lung cancer: challenges and prospects. Lung Cancer 23:105–114

Nestle U et al. (1999) 18F-deoxyglucose positron emission tomography (FDG-PET) for the planning of radiotherapy in lung cancer: high impact in patients with atelectasis. Int J Radiat Oncol Biol Phys 44(3):593-7.

Nishioka T, Shiga T, Shirato H et al. (2002) Image fusion between 18FDG-PET and MRI/CT for radiotherapy planning of oropharyngeal and nasopharyngeal carcinomas. Int J Radiat Oncol Biol Phys 53:1051–1057

Nuutinen J, Sonninen P, Lehikoinen P et al. (2000) Radiotherapy treatment planning and long-term follow-up with [(11)C]methionine PET in patients with low-grade astrocytoma. Int J Radiat Oncol Biol Phys 48:43–52

Pirzkall A, McKnight TR, Graves EE et al. (2001) MR-spectroscopy guided target delineation for high-grade gliomas. Int J Radiat Oncol Biol Phys 50:915–928

Pirzkall A, Li X, Oh J et al. (2004) 3D MRSI for resected high-grade gliomas before RT: tumour extent according to metabolic activity in relation to MRI. Int J Radiat Oncol Biol Phys 59:126–137

Rahn AN, Baum RP, Adamietz IA et al. (1998) Value of 18F fluorodeoxyglucose positron emission tomography in radiotherapy planning of head–neck tumours. Strahlenther Onkol 174:358–364 [in German]

Rischin D, Peters L, Hicks R et al. (2001) Phase I trial of concurrent tirapazamine, cisplatin, and radiotherapy in patients with advanced head and neck cancer. J Clin Oncol 19:535–542

Vanuytsel LJ, Vansteenkiste JF, Stroobants SG et al. (2000) The impact of (18)F-fluoro-2-deoxy-D-glucose positron emission tomography (FDG–PET) lymph node staging on the radiation treatment volumes in patients with non-small cell lung cancer. Radiother Oncol

Voges J, Herholz K, Holzer T et al. (1997) 11C-methionine and 18F-2-fluorodeoxyglucose positron emission tomography: a tool for diagnosis of cerebral glioma and monitoring after brachytherapy with 125-I seeds. Stereotact Funct Neurosurg 69:129–135

Wagner M, Seitz U, Buck A et al. (2003) 3′-[18F]fluoro-3′-deoxythymidine ([18F]-FLT) as positron emission tomography tracer for imaging proliferation in a murine B-cell lymphoma model and in the human disease. Cancer Res 63:2681–2687

Weber WA, Petersen V, Schmidt B et al. (2003) Positron emission tomography in non-small-cell lung cancer: prediction of response to chemotherapy by quantitative assessment of glucose use. J Clin Oncol 21:2651–2657

Wefer AE, Hricak H, Vigneron DB et al. (2000) Sextant localization of prostate cancer: comparison of sextant biopsy, magnetic resonance imaging and magnetic resonance spectroscopic imaging with step section histology. J Urol 164:400–404

Zaider M, Zelefsky MJ, Lee EK et al. (2000) Treatment planning for prostate implants using magnetic-resonance spectroscopy imaging. Int J Radiat Oncol Biol Phys 47:1085–1096

14 Virtual Therapy Simulation

Rolf Bendl

CONTENTS

14.1 Introduction

Therapy simulation is widely used and has a long tradition in radiotherapy. A therapy simulator is an X-ray imaging device with the same geometric properties as the linear accelerator used for therapy. This device allows the acquisition of images with the radiation source located in the same position as during therapy. This way it is used to determine and verify patient setup. Furthermore, therapy simulators are used to determine irradiation directions.

Modern fast computer technology and dedicated software tools allow the simulation and examination of a large variety of problems of real life in a virtual reality. By means of those tools complex courses and tasks can be simulated and optimised in advance. It was an early insight in radiotherapy planning that those simulations can improve treatment planning and hopefully the treatment itself markedly; therefore, the core of all modern treatment planning systems are programs and algorithms which allow interactive simulation of the visible parts of the treatment, e.g. determination of irradiation directions and beam portals, as well as simulation of the physical dose deposition in the patient's tissue (SHEROUSE and CHANEY 1991). Since the term "*therapy simulation*" has already been used, this kind of simulation is called "*virtual therapy simulation*". Often "*virtual therapy simulation*" is also used as synonym for radiotherapy planning.

The goal of virtual therapy simulation is to find the optimal treatment plan for an individual patient by testing, evaluating and optimising alternative treatment strategies before treatment starts. In this chapter the most basic concepts and tools used for simulating the "visible" parts of treatment are explained. For more information about pre-calculation of dose distributions refer to Chap. 15.

The idea of conformal radiotherapy is to concentrate the therapeutic dose precisely to the shape of the tumour and to shield surrounding healthy tissue as best as possible. This way side effects can be reduced, while simultaneously the opportunity to escalate dose in the target volume can be increased which can result in a better tumour control probability.

One of the two most important techniques to concentrate dose to target volume is the precise adjustment of beam shapes to the shape of the tumour, to spare surrounding organs at risk. The second technique is the determination of treatment techniques which consist of series of different beams from different directions. These beams will superimpose in the target volume and sum up there to the desired dose level, whereas dose in regions which are hit only by one or a small number of beams can be kept below tissue-dependent tolerance levels. Virtual simulators basically supply interactive graphical tools to facilitate those complex tasks.

R. BENDL, PhD
Abteilung Medizinische Physik in der Strahlentherapie, Deutsches Krebsforschungszentrum, Im Neuenheimer Feld 280, 69120 Heidelberg, Germany

14.2
Modelling

For virtual simulation a three-dimensional model of the patient's anatomy is required. This model is based on at least one or more three-dimensional image series. Two-dimensional tomograms can be combined to form such a three-dimensional "image cube".

Since modern dose calculation algorithms refer to "Hounsfield" units to calculate the dose deposition, at least one series acquired with a CT scanner is necessary, but consideration of other image modalities (MRI, PET, SPECT) can help in precise delineation of anatomy and in outlining of the target volume. Since CT imaging is a robust technique and images are considered to be geometrically correct, they are of particular relevance and treatment planning starts with the acquisition of a planning CT series. To use information of image modalities simultaneously, it is necessary to establish well-defined relationships between the volume elements of the different image sequences. That means they must be registered with the planning CT series.

Since all subsequent steps of therapy planning, optimisation and evaluation are based on that three-dimensional patient model, the delineation of anatomy and the definition of the target volume is one of the most crucial steps in treatment planning. While that step is time-consuming and difficult, image segmentation is an inevitable pre-requisite or component of radiotherapy planning, since treatment plan selection is based on quantitative evaluations of dose distributions and information on how much dose is applied to organs at risk and the target volume. Without delineation of therapy-relevant structures it would not be possible to generate answers to those questions. While delineation of normal anatomy can be supported by various interactive, semi-automatic or automatic segmentation tools, definition of target volumes cannot be automated reasonably, since target volumes do not include only visible tumour regions but surrounding suspicious areas and security margins as well to compensate for organ movement and uncertainties in patient positioning. Therefore, the target volume spans over image regions with various visually perceptible types of tissue and therefore automatic algorithms considering only image intensities normally fail. For more information about registration and segmentation refer to Chaps. 3–5.

Based on segmentation results it is possible to create a three-dimensional model of the individual anatomy which can be presented and utilized in various views during treatment planning.

14.3
The Radiotherapy Planning Cycle

In conventional forward treatment planning the definition of treatment parameters is an iterative, trial-and-error process. The therapist starts with the definition of one or more alternative treatment strategies, then he has to start dose calculations based on the defined strategies. Subsequently, the expected dose distributions can be analysed and compared. If plans do not meet specific individual constraints, the treatment parameters are modified until an acceptable plan is found (Fig. 14.1).

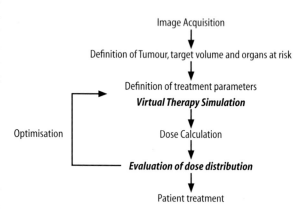

Fig. 14.1. Radiotherapy planning cycle

14.4
Graphical Tools for Virtual Simulation

To supply a virtual environment which allows the definition of all treatment parameters and the examination of consequences to the expected dose distributions, modern three-dimensional planning systems offer a variety of different graphical tools. Besides the availability of three-dimensional images, the interactivity of these scenes is of particular relevance. The possibility of getting direct feedback on how defined settings will influence the treatment plan plays an important role and enables users to generate highly individualized treatment plans in a short time. Due to the extensive increase of graphical performance of currently available computer hardware, it is possible to display very complex scenes without any delay and this way the necessary interactivity can be assured.

14.4.1
Beam's Eye View: Selection of Irradiation Directions and Beam Portals

The most important tool for selecting appropriate irradiation directions and for defining beam portals is the beam's eye view (Fig. 14.2; GOITEIN et al. 1983). In the beam's eye view the three-dimensional model of the patient's anatomy is presented from the position of the radiation source. In regarding this scene it immediately becomes clear which structures are enclosed by the current beam, since the positions of the beam-limiting devices are integrated, too; therefore, an interactive beam's eye view is the best tool to select appropriate irradiation directions.

Subsequently, the beam's eye view can be used to adjust the shape of the beam to the shape of the target volume. Depending on the used radiation type, several possibilities for transcription of these definitions should be available. For example, irregular beam shaping in conformal radiotherapy with photons normally is done by using multi-leaf collimators; therefore, the defined leaf positions should be integrated in the beam's eye view , too. On using electrons beam shaping is mostly done by applicators or individual blocks. Necessary tools for defining and displaying these components should also be available.

14.4.2
Observer's View: Optimal Combination of Irradiation Directions

The observer's view (Fig. 14.3) was designed to help treatment planners in combining several beams applied from different irradiation directions. These three-dimensional scenes show the same patient model as the beam's eye view from an arbitrary point of view (BENDL et al. 1993). In addition, the shapes of all defined beams are integrated, since one of the most important criteria in combining beams is to minimize that sub-volume where the single beams overlap. In this scene the degree of overlap can easily be perceived. In addition, the observer's view gives a fast impression of the complete treatment strategy and is therefore a good base for information exchange between the involved staff. Last, but not least, is the observer's view a valuable tool for qualitative evaluation of dose distributions. By means of integrated dose information the overall quality of treatment plans can be examined and hot spots on organs at risk or cold spots on the target volume can be detected easily.

During definition of irradiation directions the therapist must consider whether the selected directions can be transferred to the existing treatment

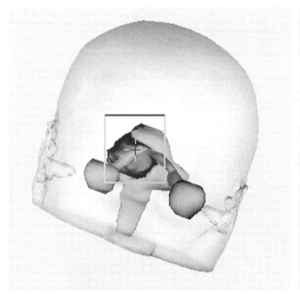
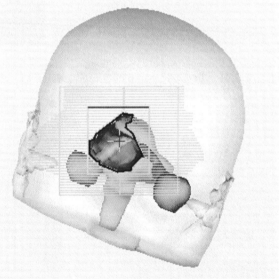

Fig. 14.2. Beam's eye view, three-dimensional patient model. *Red*: target volume; *green*: brain stem; *violet*: eyeballs. *Left*: a conventional rectangular beam portal which encloses a large volume of healthy tissue. *Right*: with multi-leaf collimators it is possible to form irregular-shaped portals and to spare healthy tissue

unit. This aspect is of particular relevance, especially when using non-coplanar irradiation directions. Depending on the accelerator specifications and the selected accessories (multi-leaf collimators, applicators, etc.) it might be possible that some desired directions cannot be realised because the selected gantry and couch angles would lead to a collision of gantry and couch or even with the patient. These "collision zones" depend on the position of the target point in the patient and cannot be determined universally. Since the patient model consists only of that part of the patient viewed by the scanner during image acquisition, it is not possible to establish a reliable automatic collision detection procedure without supplementary information. An approved method is therefore the presentation of a three-dimensional scene of the treatment unit where the patient model is integrated correctly. In this scene, all degrees of freedom of the linear accelerator (Linac; Fig. 14.4) and Couch can be inspected visually. Since therapists know their patients, and are normally able to estimate the extension of a patient quite well, they can determine visually which directions can be used without any problems and which directions must be verified with the patient at the accelerator prior to the first treatment fraction.

a   b

Fig. 14.3a,b. Observer's view. **a** The sub-volume where all three beams overlap can be identified easily. **b** Display of a 14 technique with 14 individually shaped beams, designed for stereotactic single dose therapy

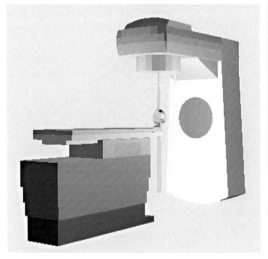
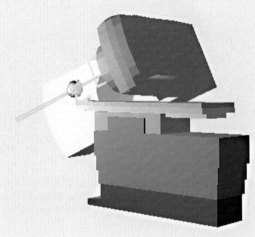

Fig. 14.4. Linac view: presentation of treatment situation

14.4.3
Multi-planar Reconstructions of Tomographic Image Series

Besides artificial three-dimensional scenes, it is necessary to project selected irradiation directions and beam shapes onto the slices of the planning CT, too, to allow the examination of selected parameters on the originally acquired patient data. Of course, it is helpful to have not only projections on the initially acquired transversal slices but at least on multi-planar reconstructions in the other two main directions (sagittal, coronal). The selection of oblique sections might be helpful during examination of dose distributions. Sometimes an additional gain of information can be achieved by projecting beams on images of other modalities, but this is usually considered to be of minor priority.

14.4.4
Digital Reconstructed Radiographs

Generation and presentation of digital reconstructed radiographs (DRRs) is a virtual simulation of real treatment simulators or portal imaging devices. Based on the information in the acquired CT image series it is possible to calculate artificial X-Ray images from arbitrary directions comparable to those mentioned above (GALVIN et al. 1995).

During X-Ray image acquisition, X-Rays are absorbed differentially according to tissue density. Structures with high density (bones) absorb more intensity of the ray than soft tissue, which means that the incident radiation on the film after having passed bony structures is low, resulting in a low optical density on the film (bright). Water, fat muscle tissue or air absorb only a small amount (or none) of the radiation. Rays running only through soft tissue keep a higher intensity which results in a high optical density on the film (dark or black).

To simulate these behaviours ray-tracing algorithms are used (see also Chap. 4.4.2). Starting from the position of the X-ray source, the algorithm traces a bundle of beams through an image cube, until a specified projection plane (i.e. the film position) is reached (SIDDON 1985). Each ray passes through different elements of the data cube (volume elements or voxel). Depending on how the values of the voxels passed through contribute to the final value of the ray in the imaging plane, different kinds of images can be produced (Fig. 14.5).

If the absorption of a ray's intensity is considered to be proportional to the density of a given voxel, the total absorption is proportional to the sum of the voxel densities along the ray. To generate DRRs, the voxel intensities must simply be summed (Fig. 14.5a). If an algorithm considers only the maximum voxel values along a ray and projects that value onto the image plane, maximum intensity projections (MIPs) can be generated. Applying such an algorithm to ap-

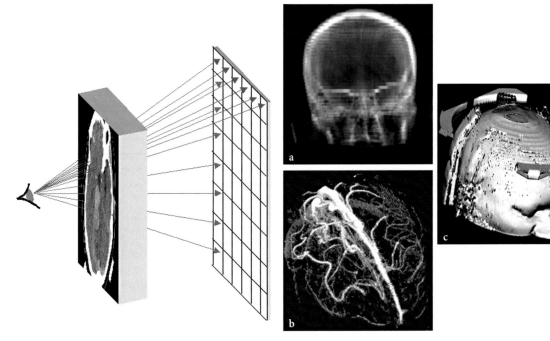

Fig. 14.5a–c. Principle of ray tracing. **a** Digital reconstructed radiographs. **b** Surface rendering. **c** Maximum intensity projection

propriate image cubes (CT or MR angiographies), the blood vessel system can be visualised (Fig. 14.5b). Adding some constraints to the ray-tracing algorithm, e.g. the ray tracing should finish when a voxel with a particular intensity arrives, the surface of anatomical structures can be detected and visualized. As a prerequisite, the intensity of those structures must have good contrast with respect to the surrounding tissue. In this way, the patient's surface and bony structures can easily be visualised (Fig. 14.5c). Images generated with surface-rendering methods from the position of the radiation source show the same geometrical relations as the beam's eye view images explained above.

For radiotherapy, DRRs and MIPs are especially important. The MIPs can be used for detecting malformations of blood vessels (AVMs). The DRRs can help during patient positioning and positioning control. Based on the specified treatment parameters and the known geometry of the X-ray imaging systems, DRRs can be pre-calculated. To verify patient positioning X-ray images can be acquired prior or during treatment. They can be compared to the calculated DRRs, deviations between both images can be calculated automatically and it is possible to compensate for displacements by correcting the treatment position to the planned one.

On comparing DRRs with portal images generated with the beam of the treatment unit it is necessary to take care of different imaging energies. The CT images are normally acquired with an imaging energy in the lower kV range, the energy used for irradiation of photons is in the range of 6–20 MV. Mega-voltage images usually show a much lower contrast, and it is difficult to perceive soft tissue structures. Due to those differences, a comparison of artificial images with verification images is often difficult, too. A solution could be the consideration of different mean image energies (MOHAN and CHEN 1985) already on generation of the DRRs.

14.5
Tools for Evaluating Dose Distributions

An optimal virtual simulator would not only present the result of geometrical changes (beam directions, shapes, etc.) immediately, but also the consequences on the expected dose distribution in real-time. Admittedly, the immediate feedback can be assured only on the geometrical aspects of treatment parameters. Up to now it is still not possible to display the consequences of changed settings on dose distributions in real-time on standard computer hardware.

Nevertheless, the possibility of examination and evaluation of dose distributions should be an integral part of a virtual simulator, since only the resulting dose distribution can answer the question, whether the selected treatment strategy considers all individual constraints sufficiently.

There is a large variety on methods to display dose information. By integrating dose distributions into the three-dimensional patient model, presented in the observer's view, a fast impression about the global dose distribution can be achieved. This integration can be done with different techniques. Figure 14.6 shows two examples. On the left side dose information is integrated as iso-dose ribbons, which means that sub-volume which receives a dose higher than a given level is enclosed by some yellow ribbons. This way a planner can immediately verify whether the target volume is enclosed completely by the therapeutic dose level. On the right side, the surface dose is displayed. That means the surfaces of all defined structures are coloured according to the expected dose. Black and blue areas do not receive any or only a small amount of dose; in red areas the dose is equal to the desired therapeutic dose level. This way, hot spots on organs at risk (here, for example, on the chiasm and on the right optical nerve) or cold spots in the target volume (here in the topmost region) can be detected very easily.

Of course variations of these techniques are common, too. For example, some treatment planning systems offer the possibility to display dose clouds as semi-transparent surfaces or are using more advanced display techniques such as fog.

While those 3D scenes give a very fast impression of dose distributions, they are not sufficient to get precise information about the dose distribution inside of an organ at risk or in the target volume, since the patient model is reduced to the surface of anatomical structures and dose clouds. Therefore, most of the available radiotherapy simulators and planning systems supply possibilities to project dose information onto the initially acquired images. Again, various different techniques are possible. The most common techniques are the display of dose distributions as iso-dose lines, and the integration of dose information using colour-wash techniques (see Fig. 14.8: comparison of concurrent treatment plans.). An important feature is the ability to present dose not only on the original CT image slices but on multi-planar reconstructions and oblique sections, too. The possibility to project dose not only on the planning CT but also on other available series of other imaging modalities is desirable.

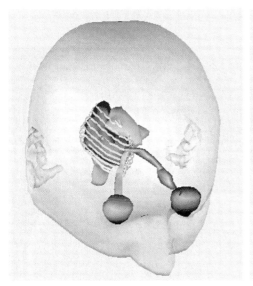
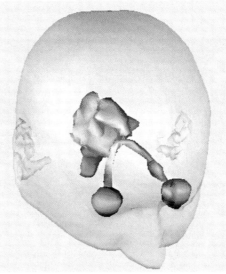

Fig. 14.6. Integration of dose information into observer's view. *Left*: display as isodose ribbons; *right*: display of surface dose. Dose distribution calculated based on beam configuration displayed in Fig. 14.3a

Besides those qualitative evaluation tools, a quantitative analysis of dose distributions should be possible, since the complex shape of three-dimensional dose distributions makes it difficult to compare concurrent plans. Calculation of quantitative parameters enables objective measurements, allows the perception of under- and overdosed regions and supports the evaluation of the dose homogeneity.

Dose-volume histograms (DVHs; Fig. 14.7) are a simple and the most accepted way of displaying information about three-dimensional dose distributions. The DVHs are usually displayed as cumulative histograms, showing the fraction of the total volume of a particular structure receiving doses up to a given value (CHEN et al.1987; DRZYMALA et al. 1991). Since the information of the complex three-dimensional dose distribution is reduced to a group of simple two-dimensional graphs, comparison of concurrent plans is facilitated. The loss of spatial information can be compensated for by qualitative display techniques mentioned above.

The examination of additional statistical parameters, such as minimum, maximum and mean dose, variance, and number of voxels below or above certain levels can give additional information.

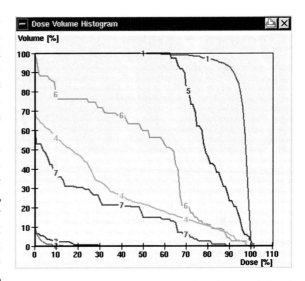

Fig. 14.7. Dose volume histogram. Dose distribution calculated based on beam configuration displayed in Fig. 14.3a

14.5.1
Comparison of Treatment Plans

During the radiotherapy planning cycle not only tools for defining a single treatment plan are needed, but tools for comparing alternative treatment strategies are desirable as well. During the iterative enhancement of treatment plans differences in dose distributions decrease, and often it is difficult to recognize resulting changes in dose distributions; therefore, a planning system or a virtual simulator should offer possibilities to compare dose distributions of concurrent treatment plans and special display techniques to enhance small deviations in dose distributions. Figure 14.8 shows an example how differences in dose distributions can be presented and emphasised. In the upper row a dose distribution is shown

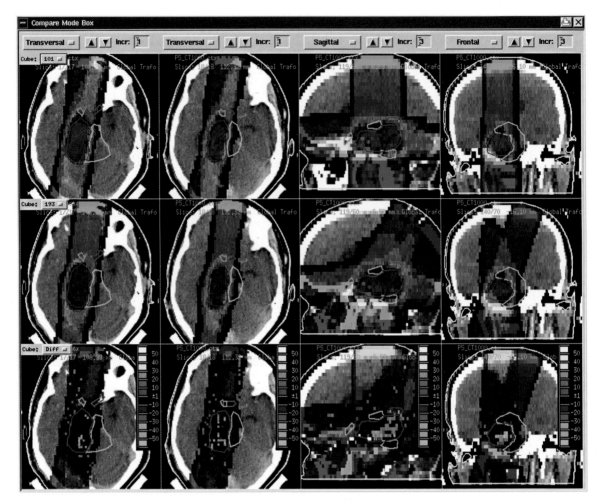

Fig. 14.8 Comparison of concurrent treatment plans. *Upper row:* colour-wash display of dose distribution of plan based on beam configuration of Fig. 14.3a. *Centre row:* dose of an alternative plan with four beams. *Bottom row:* display of difference dose

generated based on beam configuration displayed in Fig. 14.8, left. In the second row a dose distribution is displayed of a plan where a fourth beam was added. The third row presents the difference dose. Regions where the first one shows a higher dose are coloured in red and yellow, and those regions where the second plan lets to higher doses are coloured in blue and green. Besides these qualitative comparisons, methods for quantitative analysis, e.g. simultaneous display of DVHs and generation of difference DVHs, are favourable.

References

Bendl R, Pross J, Schlegel W (1993) VIRTUOS - A program for VIRTUAL radiotherapy simulation. In: Lemke HU et al. (eds) Proc Int Symp CAR 93. Springer, Berlin Heidelberg New York, pp 676–682

Chen GTY, Pelizzari CA, Spelbring DR, Awan A (1987) Evaluation of treatment plans using dose volume histograms. Front Radiat Ther Oncol 21:44–57

Drzymala RE, Mohan R, Brewster L, Chu J, Goitein M, Harms W, Urie M (1991) Dose-volume histograms. Int J Radiat Oncol Biol Phys 21:71–78

Galvin JM, Sims C, Dominiak G, Cooper JS (1995) The use of digitally reconstructed radiographs for three-dimensional treatment planning and CT-simulation. Int J Radiat Oncol Biol Phys 31:935–942

Goitein M, Abrams M, Rowell D, Pollari H, Wiles J (1983) Multi dimensional treatment planning: II. Beam's eye view, back-projection and projection through CT-sections. Int J Radiat Oncol Biol Phys 9:789–797

Mohan R, Chen C (1985) Energy and angular distributions of photons from medical linear accelerators. Med Phys 12:592–597

Sherouse GW, Chaney EL (1991) The portable virtual simulator. Int J Radiat Oncol Biol Phys 21:475–482

Siddon RL (1985) Fast calculation of the exact radiological path for a three-dimensional CT array. Med Phys 12:252–255

15 Dose Calculation Algorithms

Uwe Oelfke and Christian Scholz

CONTENTS

15.1 Introduction

The accurate and fast calculation of a 3D dose distribution within the patient is one of the most central procedures in modern radiation oncology. It creates the only reliable and verifiable link between the chosen treatment parameters, and the observed clinical outcome for a specified treatment technique, i.e., the prescribed dose level for the tumor, the number of therapeutic beams, their angles of incidence, and a set of intensity amplitudes – obtained by a careful treatment plan optimization – result in a distribution of absorbed dose which is the primary physical quantity available for an analysis of the achieved clinical

U. OELFKE, PhD
Deutsches Krebsforschungszentrum, Im Neuenheimer Feld 280, 69120 Heidelberg, Germany
C. SCHOLZ, PhD
Siemens Medical Solutions, Oncology Care Systems, Hans-Bunte-Str. 19, 69123 Heidelberg, Germany

effects of this specific treatment. The twofold application of dose calculation algorithms in radiation oncology practice, firstly for the plan optimization in the treatment planning process, and secondly for the retrospective analysis of the correlation between treatment parameters and clinical outcome, defines two mutually conflicting goals of the respective dose algorithms. Firstly, the dose calculation has to be fast such that the treatment planning process can be completed in clinically acceptable time frames, and secondly, the result of the dose calculation has to be sufficiently accurate so that the establishment of correlations between delivered dose and clinical effects remains reliable and meaningful. The conflict between "high speed" and "high accuracy" is one of the crucial challenges for the development of modern dose calculation algorithms.

The accuracy of the dose calculation algorithms becomes a problem only for very heterogeneous tissues, where a very detailed modeling of the energy transport in the patient is required. The prediction of the dose around air cavities, e.g., often encountered for tumors adjacent to the paranasal sinuses or for solid tumors embedded in lung tissue, is intricate and time-consuming. Almost all new developments related to dose algorithms are specifically concentrating on these or equivalent areas of tissue heterogeneities, whereas for the majority of clinical cases with almost homogeneous tissues existing, simple calculation methods can be reliably applied.

Naturally, dose algorithms for high-energy photon beams were first developed for the ultimate "homogeneous" patient – a patient completely consisting of water (SCHOKNECHT 1967). Measurements of a set of generic dose functions, e.g., tissue air ratios, tissue phantom ratios, output factors, and off-axis ratios, are measured in a water phantom for a set of regular treatment fields under reference conditions. The dose within a patient is then calculated by extrapolating these measurements to the specific chosen treatment fields and by the application of various correction algorithms, e.g., for the inclusion of missing tissues at the patient surface or the approximate consideration

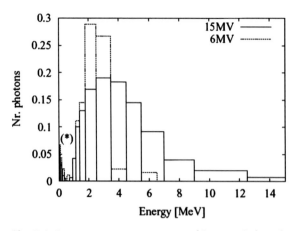

Fig. 15.2. Energy spectra to generate multi-energetic kernels for 6- and 15-MV photon beams. Energy spectra of a Siemens Primus linear accelerator obtained by a phenomenological fit of depth dose curves (SCHOLZ et al. 2003b) are shown.

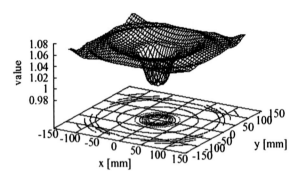

Fig. 15.3. Primary fluence matrix. In this figure a 2D primary fluence matrix obtained by 1D measurements along the diagonals of square treatment fields in air is shown. It is normalized to 1.0 at the origin. The increase of fluence with the distance to the center is induced by the shape of the flattening filter inside the treatment head, which improves the dose profile at the isocenter plane.

different technical procedures can be applied, such as those described, for instance, by SCHOLZ et al. (2003b).

15.4
Dose Calculation in Homogeneous Media

15.4.1
TERMA

We first consider the dose deposition of a mono-energetic, infinitely narrow photon beam in z-direction of energy E and initial fluence Φ in a homogeneous medium, which is naturally chosen to be water for

all applications in radiotherapy. The energy fluence Ψ of the primary photons is to a first approximation determined by the linear photon absorption coefficient $\mu(E)$ in water, i.e., at the interaction point of the primary photons $\vec{r}$ one gets:

$$\Psi(\vec{r}) = \Phi(\vec{r}_{\perp},0) \, E \, e^{-\mu(E)}$$

where $\vec{r}_{\perp}$ denotes the coordinates perpendicular to the beam direction. The rate of the primary photon interactions in the medium determines the TERMA $T(\vec{r})$, i.e., the total energy per unit mass released by a radiation field interacting with a medium of density ρ at a certain point $\vec{r}$:

$$T(\vec{r}) = \frac{\mu}{\rho}(\vec{r})\Psi(\vec{r})$$

This locally released energy of the radiation field is subsequently available for a further transport emerging from the interaction point $\vec{r}$, which is usually described by the concept of dose kernels.

15.4.2
Dose Kernels: Point-Spread Kernel and Pencil Beam

It is common to use two elementary dose kernels for model-based algorithms (MOHAN et al. 1986; BORTFELD et al. 1993; MACKIE et al. 1988; AHNESJÖ 1989). The most elementary kernel $k(\vec{r}, \vec{r}', E)$, the so-called point-spread kernel, indicates the distribution of absorbed energy in water at the coordinate $\vec{r}$ which is created by interactions of primary photons of energy E at the coordinate $\vec{r}'$ (see Fig. 15.4). The elemental mono-energetic dose deposition kernels can be taken from Monte Carlo simulations, e.g., from MACKIE et al. (1988).

The second class of dose kernels and most commonly used in current treatment planning systems is the pencil beam. A pencil-beam kernel is obtained through the integration of all point-spread kernels along an infinite ray of photons in the medium as indicated in Fig. 15.5. It is evident that the pencil-beam kernel uses the more condensed information about the dose in water along the central kernel axis, i.e., it provides a coarser sampling of the physical processes than the point-spread kernel and it is therefore harder to adapt the dose calculations based on pencil-beam kernels to regions with intricate tissue inhomogeneities. On the other hand, pencil-beam kernels bear the obvious advantage of reduced dose calculation times. The first pencil-beam-type dose calcula-

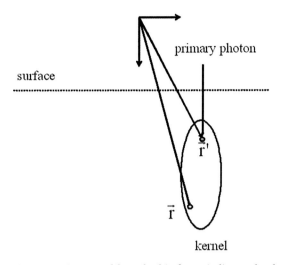

Fig. 15.4. Point-spread kernel. This figure indicates the dose distribution at the points $\vec{r}$ deposited by primary photons that interact at the $\vec{r}'$. In clinical mega-voltage photon beams the macroscopic extent of this dose kernel is given mainly by the range of secondary electrons, typically several centimeters

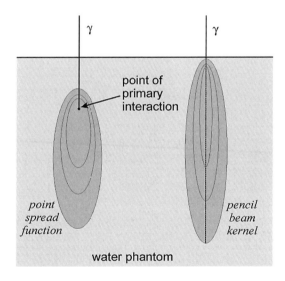

Fig. 15.5. Point kernels and pencil-beam kernels. In this figure on the left a point kernel or so-called point-spread function is pictured. It describes the energy spread around a single interaction point of primary unscattered photons. By the integration of this kernel inside the patient along an infinite thin ray a so-called pencil-beam kernel is derived, which is displayed on the right-hand side

tion was introduced by MOHAN et al. (1986). Pencil-beam kernels can be either created again with Monte Carlo simulations or can be directly derived from a few standard measurements, e.g., output factors of regular fields and tissue phantom ratios. The technical details of this approach are given by BORTFELD et al. (1993).

15.4.3
Superposition and Convolution Algorithms

Finally, the two components of TERMA and dose kernels are combined to achieve the accurate calculation of absorbed dose. The most general approach of model-based dose calculation algorithms is the superposition method (MACKIE et al. 1985; AHNESJÖ 1989; SCHOLZ et al. 2003b). It generates the dose delivered at a point $\vec{r}$ by superimposing the dose contributions from all dose kernels $k(\vec{r}, \vec{r}', E)$ of the defined energy spectrum originating from all primary interaction points $\vec{r}'$ and weighs their contributions with the respective TERMA, i.e.,

$$D(\vec{r}) = \int dE' \int d^3r' \; T(\vec{r}', E') \; k(\vec{r}, \vec{r}', E').$$

The superposition approach certainly is a too sophisticated method to be applied for a dose calculation in homogenous media, but it can be used very well for dose calculations in regions of interest with tissue inhomogeneities.

A reduction of the computational effort required for the superposition approach, is achieved, if one assumes that the shape of the dose kernel is translational invariant. Then, the kernel $k(\vec{r}, \vec{r}', E)$ is only a function of the distance between the interaction $\vec{r}'$ point and the coordinate $\vec{r}$ where the dose is measured such that the superposition formula reduces to the well known convolution approach, i.e.,

$$D(\vec{r}) = \int dE' \int d^3r' \; T(\vec{r}', E') \; k(|\vec{r} - \vec{r}'|, E').$$

The calculation times of the convolution algorithm are reduced dramatically compared with the more accurate superposition methods.

The application of a pencil-beam kernel, usually employed with convolution algorithms, leads to a further substantial reduction of calculation times, e.g., with a simple single value decomposition of the kernel the dose calculation can be reduced to a few 2D convolutions (BORTFELD et al. 1993), so that a 3D dose calculation for a conventional treatment plan can be accomplished in seconds.

15.5
Accounting for Tissue Inhomogeneities

In order to calculate the dose in regions with tissue inhomogeneities, the dose calculation has to account for the variations of electron densities derived from

CT scans. The electron densities influence the dose calculation in two aspects:

1. The local TERMA depends on the path of the primary photons through the patient to their interaction point.
2. The energy distribution around the primary interaction point described by the dose kernels is also influenced by variations of the respective electron densities. The accurate calculation of the TERMA is accomplished by ray tracing the pathway of a photon to its interaction point. For this process the absorption rate of the photons along a considered trajectory is scaled by the ratio of the electron densities of the encountered media to the electron density of water. More important and more difficult to deal with is the influence of electron density variations on the dose kernels. This problem and some related practical aspects are briefly discussed below.

15.5.1
Pencil Beam and Pathlength Scaling

The pencil beam is a dose kernel describing the 3D dose distribution of an infinitely narrow mono-energetic photon beam in water. The individual interaction points of the photons are all assumed to be on the central axis of the pencil beam. Here tissue inhomogeneities are accounted for by the same pathlength scaling with relative electron densities between tissue and water as applied for the calculation of the TERMA. The values of the pencil-beam kernel in water are used according to the radiological pathlength calculated along the central axis of the pencil beam. Electron variations perpendicular to the pencil-beam axis are not accounted for, i.e., this inhomogeneity correction assumes a slab geometry of tissue inhomogeneities, which are represented by the values of the electron densities along the pencil-beam axis.

the energy transport through the kernel from the interaction point $\vec{r}'$ to the dose point $\vec{r}$ occours along a straight line, i.e., within the kernel one also introduces an "internal" ray tracing. Along each internal ray the contribution of the kernel is scaled with the average electron density encountered along the line connecting $\vec{r}'$ and $\vec{r}$. This leads to the so-called density-scaled dose deposition kernels.

As an example we show in Fig. 15.6 a hypothetical dose kernel in water together with a density-scaled kernel including some tissue inhomogeneities. It can be clearly seen that the kernel extends further from the interaction point if the internal energy transport encounters a medium of an electron smaller than water on its way to the dose deposition point. The dimensions of the scaled dose kernel are shrinking in comparison with the original dose kernel in water if higher electron densities represent an additional obstacle for the energy transport within the dose kernel.

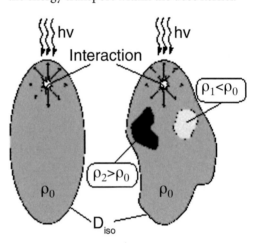

Fig. 15.6. Isodose curves of density-scaled kernels. Here the density-scaling of the Superposition kernel is displayed. The range of the energy spread behind tissue inhomogeneities is adapted to the radiological distance between the interaction point at $\vec{r}'$ and the dose point $\vec{r}$. This leads to a deformation of the isodose lines behind inhomogeneities as shown in the right part of the figure

15.5.2
Density Scaling of Point-Spread Kernels

For the superposition algorithm the most accurate sampling of the primary interaction points within the patient is required. Consequently, this method also applies the most accurate technique of inhomogeneity corrections for the dose kernel. According to O'Connor's theorem, MOHAN et al (1986) devised a density scaling method which is applied directly to the point-spread kernel. Basically, it is assumed that

15.5.3
Collapsed Cone and Kernel Tilting

The problem of the superposition approach are its long computation times, which may still take hours for a complicated IMRT case even if performed with state-of-the-art hardware technology (SCHULZE 1995); therefore, various approximations have been suggested for accelerating the dose calculation with the superposition technique. One method introduced by AHNESJÖ (1989) is the collapsed cone-beam technique which refers to a specific internal sampling of the dose kernels. Furthermore, it seems to be important that the axis of the point-spread kernels be aligned with the original photon rays employed for the determination of the TERMA within the patient (AHNESJÖ 1989; SCHOLZ et al. 2003b; SHARPE and BATTISTA 1993). The neglect of the required kernel tilting saves considerable calculation time; however, it also can introduce significant dose errors (SHARPE and BATTISTA 1993; LIU et al. 1997).

Another effect which might have to be considered is the hardening of the photon-energy spectrum. Works of LIU et al. (1997) and METCALFE et al. (1990) showed that these effects are usually minimal in routine clinical practice.

15.5.4
A Simple Example:
Doses at Cork–Water Interfaces

In order to demonstrate the sensitivity of the various types of dose algorithms to inhomogeneous media, we show the results of calculations performed for a simple phantom geometry. A 4-cm slab of cork was placed at a depth of 6 cm into a water phantom. The irradiation geometry of a 6-MV photon field was fixed as a 3×3 cm² open field with a source-to-skin distance SSD=95 cm. This field size was chosen since the measurements could still be performed reliably and because a significant effect on the dose patterns was anticipated. The 3D dose distributions for this simple geometry were calculated with three different algorithms: a pencil beam, a superposition, and a Monte Carlo dose calculation algorithm (SCHOLZ 2004). All applied algorithms were proven to give equivalent results for 3D dose distributions in water for various regular and irregularly shaped fields.

Firstly, we show the result of the 2D lateral dose distributions on a central slice of the phantom in Fig. 15.7. It is clearly indicated in Fig. 15.7a that the pencil beam almost completely misses the effect of lateral scattering within the cork slab. The result of the superposition algorithm, displayed in Fig. 15.7b, accounts for most of the lateral scattering of the secondary electrons as indicated by the extended 10% isodose line in cork. The effect of energy transport through secondary electrons is even more pronounced if the results of the Monte Carlo calculation, shown in Fig. 15.7c, are considered.

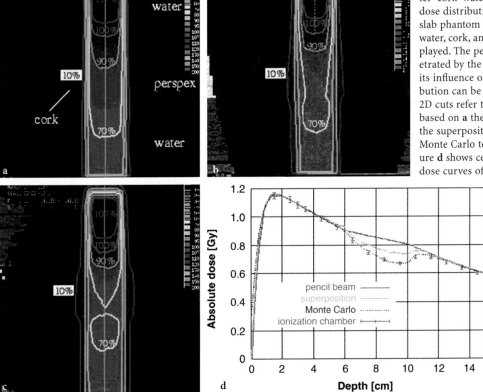

Fig. 15.7a–d. Dose distribution of a 6-MV beam through water–cork–water slabs. Here the dose distributions through a slab phantom consisting of solid water, cork, and perspex are displayed. The perspex is not penetrated by the beam directly, so its influence on the dose distribution can be neglected. **a–c** The 2D cuts refer to dose calculation based on **a** the pencil beam, **b** the superposition, and **c** the Monte Carlo technique. The figure **d** shows central-axis-depth dose curves of all three dose distributions compared with ionization chamber measurements

These results are consistent with the respective depth-dose curves shown in Fig 15.7d. The enhanced lateral scattering of secondary electrons in the low-density medium reduces the dose values on the central-beam axis, an effect which is almost completely missed by the pencil-beam calculation resulting in a severe overestimation of the respective dose of up to 12% of the maximum dose. This discrepancy is substantially reduced by the superposition algorithm. Only the Monte Carlo calculation was able to reproduce the experimental data with an accuracy of 2%, which was the estimated accuracy of the measurement.

While these simple phantom experiments may reveal some generic features of the discussed dose calculation algorithms, it is not clear to what extent the various algorithms generate dose differences which are of clinical relevance.

In the final section of this review we address this issue briefly by considering the influence of different dose calculation engines on IMRT dose optimization for clinical cases with abundant severe tissue inhomogeneities.

15.6
Dose Calculations and IMRT Optimization

The role of advanced high-energy photon dose calculations for iterative IMRT treatment planning is still under investigation. Since conventional algorithms, such as pencil-beam methods with poor consideration of inhomogeneities, could produce substantial deviations in media different to water, the quality of intensity-modulated treatment plans generated through those dose calculation methods was recently reviewed by different groups (JERAJ et al. 2002; SIEBERS et al. 2001, 2002; SCHOLZ et al. 2003a).

Particularly two aspects are important referring to dose calculation in IMRT: firstly, one has to assess the accuracy of the respective algorithm as required for the determination and evaluation of the final dose pattern originating from a given set of fluence amplitudes. Deviations from a reference dose calculation can be called *systematic errors*. Secondly, the influence of the dose algorithm in the optimization process itself should be analyzed, since deviations in the dose calculation induces differences in bixel intensities in the fluence maps, which is usually referred to as *convergence error*. In order to demonstrate the nature of these two errors, we consider one of the most sensitive clinical cases with respect to dose calculations, the irradiation of small, solid lesions embedded in lung tissue.

15.6.1
The Systematic Error

As mentioned previously, the systematic error indicates the difference in dose which arises from the application of different dose calculation algorithms for a given set of fluence matrices. A solid lung tumor of about 30-mm diameter located in the left lung is totally surrounded by low-density lung tissue. It is irradiated with five coplanar, intensity-modulated beams. As indicated by the dose distribution on a transversal CT slice of the patient in Fig. 15.8a, the original optimization based on the pencil-beam dose calculation seems to cover well the PTV by the prescribed dose of 63 Gy (100% isodose); however, the recalculation with the superposition algorithm shown in Fig. 15.8b reveals that only a small volume of the target is encompassed by the 60 Gy isodose (95% isodose) and that a severe underdosage of about 13 Gy in mean dose is observed for the PTV.

The systematic error induced by the pencil-beam algorithm demonstrates that this dose calculation method severely overestimates the dose inside the tumor. For the considered case the respective underdosage of the tumor shown by the superposition algorithm is based on the fact that for high-energy photon fields the range of many secondary electrons is larger than the radius of the tumor volume shown above; thus, there is not enough material to absorb the total number of secondary electrons inside the target region. The remaining electrons simply are scattered into the low-density lung tissue. These findings are also well represented by the dose-volume histogram shown in Fig. 15.8d.

15.6.2
The Convergence Error

The convergence error is defined by the dose difference which arises from the application of two different fluence matrices – obtained by inverse planning with two different dose algorithms – for which the same reference dose calculation is applied. For our example we compare the dose distributions in Fig. 15.8b, where the pencil-beam algorithm was used for the generation of the fluence matrices, and the dose distribution in Fig. 15.8c, where the superposition algorithm was employed for the optimization. Once the fluence matrices were established, the final dose distributions were obtained by a calculation with the superposition algorithm. As clearly indicated by the fluence matrices shown in Fig. 15.9, the optimiza-

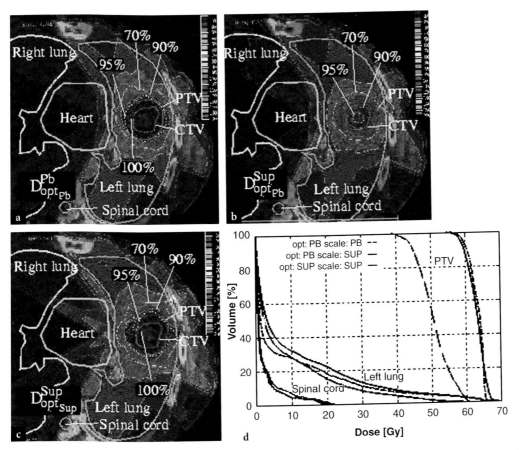

Fig. 15.8a–d. Absolute dose distributions for a lung patient. a–c A transversal slice of the 3D dose distribution in a lung tumor patient is shown exemplarily. a Pencil-beam (*PB*) optimization plus PB recalculation. b Pencil-beam optimization plus superposition recalculation. c Superposition optimization plus superposition recalculation. The 100% isodose line represents the prescribed dose of 63 Gy. The dose-volume histograms of the planning target volume (*PTV*), the left lung, and the spinal cord for the different plans are displayed in d. One clearly recognizes the reduction of the mean PTV dose by about 13 Gy produced by the original optimization with a pencil beam algorithm which was recalculated with superposition. Compared with this, the superposition-optimized plan reaches the prescribed mean dose of 63 Gy in the target volume at the expense of a slightly higher irradiation of the left lung

tion with the superposition algorithm compensates the obvious underdosage of the PTV created by the fluence generated with the pencil-beam method. The fluence matrix on the right-hand side of Fig. 15.9, obtained with the superposition algorithm, enhances the bixel weights at the periphery of the projected tumor such that an adequate dose coverage of the PTV is guaranteed. This fact is also well reflected by the respective values of the dose-volume histogram shown in Fig. 15.8d.

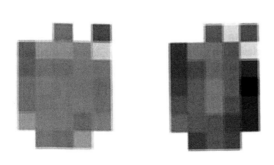

Fig. 15.9. Fluence matrices of the pencil beam (*left*) and superposition-optimized plan (*right*). In these images the bixel intensities of one beam of both plans for the lung tumor case shown in Fig. 15.8 are displayed. The enhanced fluence values in the peripheral regions of the plan optimized with the superposition algorithm are clearly recognized

15.7
Conclusion

Dose calculation algorithms play a central role for the clinical practice of radiation therapy. They form

the volume should receive more than a dose of D_{max}". They can be visualized as a barrier with a corner at the point D_{max}, V_{max} on the DVH plot. The use of multiple DVH constraints may also be indicated in some cases (CAROL et al. 1997), and DVH constraints can be used for the target volume as well.

One general advantage of physical criteria, such as dose-volume criteria, is that, because they are simply and clearly defined, they can be used easily in clinical protocols. Such IMRT protocols using a combination of various physical dose criteria have been developed and distributed by the Radiation Therapy Oncology Group (RTOG, www.rtog.org). A well-known example is their protocol 022 for IMRT of oropharyngeal cancer.

A potential mathematical problem with dose-volume constraints is that they are non-convex (DEASY 1997). As a consequence of this, optimization based on DVH constraints may get trapped in local minima; however, it has been shown that this is mainly a theoretical problem, which seems to be of little relevance in practical optimization (WU and MOHAN 2002; LLACER et al. 2003).

The final physical dose criterion that we discuss is the *equivalent uniform dose* (EUD). It is defined as the uniform dose that would create the same biological effect in a specific organ as the actual non-uniform dose distribution (BRAHME 1984; NIEMIERKO 1997). Because it involves the biological effect, it is often interpreted as a biological rather than physical dose criterion; however, according to a more recent definition by NIEMIERKO (1999) the EUD is simply the generalized mean (*a*-norm) of the physical dose distribution:

$$EUD = \left(\sum_i v_i \cdot d_i^a \right)^{1/a}$$

where v_i is the volume of voxel i divided by the total volume of the organ. With this definition and with $a = 1$ the EUD is in fact the mean dose and with $a = \infty$ it equals the maximum dose. Of course, the value of a is organ specific. Values between 1 and infinity represent organs with a mixture between parallel and serial structures. It is interesting to note that the value of a can be derived from the well-known power-law relationship of the tolerance dose TD as a function of the relative treated volume v:

$$TD(v) = \frac{TD(1)}{v^n}.$$

One finds that $a = 1/n$.

For target volumes, the value of a is negative. The beauty of this approach is its simplicity and general-ity. All organs can be characterized by just one parameter. The EUD has recently been implemented into experimental IMRT optimization systems (THIEKE et al. 2002; WU et al. 2002). A convenient feature of the EUD is that it is a convex function of the dose distribution, and therefore of the beam intensities, for $a \geq 1$ and for $a \leq 0$, i.e., for almost all organs. This facilitates optimization and makes the use of projection onto convex sets (POCS) methods possible (CHO et al. 1998; THIEKE et al. 2002).

17.2.1.2
Physical and Technical IMRT Delivery Criteria

The most obvious physical limitation of this type is that the *beam intensities must not be allowed to take negative values*. This does not require any further discussion.

Another important aspect of optimized treatment plans is that they have to be deliverable in an acceptable amount of time. Also, small errors in the delivery should not affect the resulting dose distribution too much, i.e., the plan should be robust. Both of these requirements can be fulfilled with intensity maps that are smooth. If there are multiple solutions to the optimization problem, which is frequently the case, one should obviously choose the solution with the smoothest (easiest to deliver) intensity maps. If one has the choice between a highly conformal plan with extremely complex intensity maps, and a plan that is almost as good but much easier and safer to deliver, one would probably choose the latter plan. Basically there are two ways to ensure smooth intensity maps. First, one can smooth the maps using a filter. Median filters can be used for this purpose (WEBB et al. 1998; KESSEN et al. 2000). Secondly, one can add a smoothing term to the objective function (ALBER and NÜSSLIN 2000).

17.2.1.3
The Objective Function

The criteria (or "costlets") mentioned in the previous subsections must somehow be combined to form the total objective function. An alternative is to use only one criterion in the objective function and set all other criteria as constraints. The most common approach is to add some criteria using weight factors that reflect the importance of the different criteria. Multiplication of costlets has also been suggested (FRAASS 2002). A typical approach is to add the deviations from the dose prescription (tumor) and maximum doses (critical structures) as defined in subsection 2.1.1 using organ-specific weight factors u_k and w_k. According to this definition the optimum treatment is the one with the smallest overall deviation from the prescription. In the plan steering process (the "human iteration loop") it is not a priori clear by how much to change the weights to achieve the desired effect on the dose distribution. Suitable weight factors have to be determined by trial and error, which can be quite time-consuming and is discussed further below. Some researchers argued that optimized "inverse" IMRT planning is just replacing the conventional manual trial-and-error search to find the best beam parameters by a manual trial-and-error search for suitable weight factors and constraints. While there may be some truth to this statement, it is also true that IMRT allows one to deliver highly conformal concave dose distributions that are impossible to achieve with conventional 3D techniques.

17.2.2
Variables to Be Optimized

After determining the objective function and the set of constraints, the next question is about the variables to be optimized. Even with the most powerful modern computers it is impossible to optimize all treatment parameters (variables). In practice, a subset of the treatment parameters, including, for example, the beam angles, need to be manually pre-selected.

17.2.2.1
Intensity Maps

In IMRT the main variables to be optimized are obviously the intensity maps for each beam. Each beam is typically subdivided into beam elements (bixels) of 5×5 to 10×10 mm^2. The intensity (fluence) for each of the bixels is optimized. The total number of bixels for all beams is typically of the order of 1000–10,000. Because there is no way to deliver intensity-modulated photon beams directly with a linac, the intensity maps are then converted to a series of multileaf collimator (MLC) shapes (segments) in a second almost independent step, which is called leaf sequencing. Of course, there has to be some link between optimization and sequencing. For example, the optimizer must know the leaf width of the MLC and should use that as the bixel size in one dimension. More thorough approaches for the consideration of delivery constraints in intensity map optimization have also been suggested (CHO and MARKS 2000; ALBER and NÜSSLIN 2001).

Besides the common two-step approach, it has also been suggested to avoid the intermediate step of using intensity maps altogether and directly optimize MLC shapes (apertures) and their weights. This approach has been suggested by DENEVE et al. (1996), among others. The MLC shapes can be determined manually based on the geometry (anatomy) of the problem, or they can be directly optimized together with the weights of the segments. The latter method has recently been published (SHEPARD et al. 2002). The direct optimization of MLC shapes and weights is mathematically a difficult, non-convex problem.

17.2.2.2
Number of Beams

The question of how many intensity-modulated beams should be used is highly relevant for the practical delivery of IMRT. As a consequence of the analogy between image reconstruction in computed tomography (CT) and inverse radiotherapy planning, the early theoretical approaches to inverse planning assumed a very high number of coplanar beams (BRAHME et al. 1982; CORMACK and CORMACK 1987). Later it was recognized that quite acceptable results can also be achieved with a moderate number of beams. In fact, some publications claim that one generally does not need more than three intensity-modulated beams to obtain results that can hardly be improved any more (SÖDERSTRÖM and BRAHME 1995). These issues have been discussed frequently in the literature (BRAHME 1993; BRAHME 1994; MACKIE et al. 1994; MOHAN and LING 1995; MOHAN and WANG 1996; SÖDERSTRÖM and BRAHME 1996).

In principle, it is clear that the higher the number of beams, the higher is the dose conformation potential; however, the incremental improvement of the conformity of the total dose distribution diminishes as more beams are added. The real question about the "optimum" number of beams in IMRT is therefore: What is the number of beams beyond which one does not see any practically relevant improvement of the treatment plan? Several authors have found independently that it is hardly ever necessary to use much more than ten intensity-modulated beams to achieve results that are close to optimum (BORTFELD et al. 1990; WEBB 1992; SÖDERSTRÖM and BRAHME 1995; STEIN et al. 1997).

On the other extreme, with a number of beams as small as three or four, the conformity of the resulting dose distribution is considerably reduced when compared with a plan incorporating ten beams; however, thanks to intensity modulation, it is still possible to imprint the desired shape on at least one isodose curve, e.g., the 80% isodose. This finding is related to image reconstruction of homogeneous objects using only three or four projections (NATTERER 1986); hence, in cases where the tolerance of the critical structures is not too low as compared with the required target dose, one may get away with very few intensity-modulated beams (SÖDERSTRÖM and BRAHME 1995).

17.2.2.3
Beam Angles

The question about the optimum beam angles is related to the number of beams. Clearly, if very many beams (more than ten) are used, they can be placed at evenly spaced angular intervals and there will be no need to optimize beam orientations. Even with a moderate number of beams of the order of seven or nine, one may often use evenly spaced beams without compromising the dose distribution (BORTFELD and SCHLEGEL 1993); however, it has been shown that this is case dependent, and complex cases, such as head and neck, sometimes benefit from beam orientation optimization even for nine or more beams (PUGACHEV et al. 2001). With very few beams, such as four or less, very careful and time-consuming optimization of beam orientations is always essential; otherwise, one will not be able to achieve acceptable results.

It should be noted that optimal orientations of intensity-modulated beams are generally different from those of uniform beams: in IMRT it is not generally necessary and often not even advantageous to avoid beam directions through organs at risk, because these can be spared by reducing the intensity for the corresponding rays (STEIN et al. 1997). Also, for similar reasons, non-coplanar beams are rarely used in IMRT. Another point worth mentioning is that parallel opposed beams should be avoided in IMRT, such that, for evenly spaced beams, the number of beams should be odd. The reason for this is that a parallel opposed beam adds much less beam-shaping potential than a slightly angled beam. Clearly, if the attenuation were zero, parallel-opposed beams would be completely useless.

17.2.2.4
Number of Intensity Levels

Most IMRT planning methods assume a *continuous* modulation of the intensity. Several investigations have shown that promising results can be achieved with step-like beam profiles as well (BORTFELD et al. 1994; GUSTAFSSON et al. 1994; DENEVE et al. 1996). In fact, using a moderate number of stair steps with five to seven "intensity levels" in each beam profile, the results are almost as good as with continuous modulation (KELLER-REICHENBECHER et al. 1998). Consequently, it is not necessary to go to a fully dynamic treatment mode to perform IMRT with an MLC. The IMRT can instead be realized in a "step-and-shoot" mode, i.e., by the successive delivery of a number of static MLC-shaped beam segments from each direction of incidence. The total number of beam segments or "subfields" to be delivered in this way is of the order of 100. Modern treatment machines can deliver such a sequence of subfields automatically and quickly. A comparison of the features of dynamic vs step and shoot IMRT has been published by CHUI et al. (2001).

17.2.2.5
Beam Energy

The choice of the beam energy is less critical in IMRT than in conventional radiotherapy. In fact, it was suggested that very low energies around 1 MeV or less suffice in IMRT. The reason is once again that in IMRT one tends to spread the beams more evenly around the patient, and the depth-dose fall-off is therefore not very relevant.

17.2.3
Optimization Algorithms

The objective function as a function of the variables to be optimized, in combination with the optimization constraints, defines the optimization problem. As for its solution, many mathematical algorithms have been developed to solve optimization problems of various kinds. The algorithm of choice depends on the type of the problem (linear/non-linear, convex/non-convex). We cannot describe these algorithms in any detail in this chapter. The interested reader is referred to other reviews (BORTFELD 1999; SHEPARD et al. 1999).

The most common approach in commercial IMRT optimization systems is the non-linear (often quadratic) problem formulation. The algorithm used for finding the solution is usually a variant of the "gradient" technique. It converges rapidly and can be applied to a wide range of optimization problems; however, in the non-linear case one does not usually let the algorithm converge to the numerical optimum, but stops after an acceptable number of iteration steps. Statements about the proximity to the true optimal solution are more difficult to make than in the linear case, or impossible.

There is no doubt that in general optimization the biggest experience exists in the field of linear optimization algorithms ("linear programming"). Even though many of the IMRT objectives and constraints discussed above are non-linear, a linear model or its variants can approximate the problem sufficiently well to yield useful solutions (LANGER et al. 1996; ROMEIJN et al. 2003). An advantage of the linear problem formulation is that the optimality of the solution can be proven.

17.3
Inverse Planning in Practice

17.3.1
General Approach

Most planning systems used in clinical practice presently were not designed from ground up for intensity-modulated radiotherapy. Instead, the IMRT optimization engine is usually implemented as a separate program that communicates with the conventional three-dimensional forward treatment planning system (TPS). Since many steps of the complete planning process are the same for both forward and inverse planned treatments, this is not necessarily a disadvantage.

A general approach to radiotherapy planning, including IMRT, can be formulated as follows:

Step 1. The contours of target structures and organs at risk are outlined in the TPS (using the appropriate imaging modalities such as CT, MRI, and PET).

Step 2. In some cases, at this stage it might be unclear whether conventional unmodulated fields are sufficient or whether IMRT is necessary to achieve a satisfying dose distribution. In these cases one should first try to find a good treatment plan with open fields by varying number, direction, and weight of the beams and their individual shape defined by the MLC leafs. E.g., for intracranial meningiomas a 3D-planned, conventional treatment with uniform fields usually is sufficient. But some meningiomas are located directly next to several critical structures, such as the optical nerves and the brain stem, so it may turn out that curative target doses with open fields would lead to unacceptable doses in those organs at risk, and IMRT should be used instead.

There are also clinical cases where even complex shaped targets can safely be treated with conventional techniques by skillfully combining different radiation modalities (photons and electrons), wedges, individual blocks, and non-coplanar beam angles (YAJNIK et al. 2003); however, the planning and delivery of those treatments might be even more time-consuming than an IMRT treatment with a comparable dose distribution. So even from an economic point of view there can be an indication for IMRT. But of course, in general IMRT is more time-consuming and more expensive than conventional radiotherapy and should only be applied if a clinical benefit for the patient can be expected.

Step 3. From here on we assume that open fields did not lead to a satisfying dose distribution or the geometry of the planning problem obviously is too complex for open field treatment. The IMRT module has to be started from within the TPS. It imports the CT data, the beam configuration, and the organ contours from the TPS.

Step 4. Now the parameters needed for the optimization have to be defined, namely the dose prescription and weight factors for each structure. Figure 17.1 shows the input window of the inverse planning program KonRad (distributed by Siemens OCS). Please note that a priority is assigned to structures which have some overlap with other structures. In the region of overlap the optimization engine is using the settings of the structure with the higher priority.

Step 5. Then the optimization is carried out. Depending on the specific case and on the program used, this process takes from less than a minute up to hours. In the end the inverse planning program presents the resulting treatment plan. It is evaluated with regard to the dose distribution in the target structures and the organs at risk using DVHs and full 3D dose information in form of isodoses or color-wash displays. If the actual treatment plan is not acceptable in one or more parts, the optimization has to be restarted with modified start parameters. The steering parameters are both the dose constraints and the weight factors. In the case of step-and-shoot delivery, the planner also has to check the number of segments of the plan. Too many segments can lead to unacceptable delivery times, making it necessary to reduce the number of intensity levels or to activate profile smoothing.

HUNT et al. (2002) have systematically investigated the interplay of the optimization parameters for a simple test case consisting of a concave-shaped target structure (PTV) around an organ at risk (normal tissue, NT) shown in Fig. 17.2a. The evaluation crite-

Langer M, Brown R, Urie M, Leong J, Stracher M, Shapiro J (1990) Large scale optimization of beam weights under dose-volume restrictions. Int J Radiat Oncol Biol Phys 18:887–893

Langer M, Morrill S, Brown R, Lee O, Lane R (1996) A comparison of mixed integer programming and fast simulated annealing for optimizing beam weights in radiation therapy. Med Phys 23:957–964

Langer M, Lee EK, Deasy JO, Rardin RL, Deye JA (2003) Operations research applied to radiotherapy, an NCI-NSF-sponsored workshop 7–9 February, 2002. Int J Radiat Oncol Biol Phys 57:762–768

Lauve A, Morris M, Schmidt-Ullrich R, Wu Q, Mohan R, Abayomi O, Buck D, Holdford D, Dawson K, Dinardo L, Reiter E (2004) Simultaneous integrated boost intensity-modulated radiotherapy for locally advanced head-and-neck squamous cell carcinomas. Part II: clinical results. Int J Radiat Oncol Biol Phys 60:374–387

Levegrun S, Jackson A, Zelefsky MJ, Skwarchuk MW, Venkatraman ES, Schlegel W, Fuks Z, Leibel SA, Ling CC (2001) Fitting tumor control probability models to biopsy outcome after three-dimensional conformal radiation therapy of prostate cancer: pitfalls in deducing radiobiologic parameters for tumors from clinical data. Int J Radiat Oncol Biol Phys 51:1064–1080

Llacer J, Deasy JO, Portfeld TR, Solberg TD, Promberger C (2003) Absence of multiple local minima effects in intensity modulated optimization with dose-volume constraints. Phys Med Biol 48:183–210

Mackie R, Deasy J, Holmes T, Fowler J (1994) Optimization of radiation therapy and the development of multileaf collimation. Int J Radiat Oncol Biol Phys 28:784–785

McDonald SC, Rubin P (1977) Optimization of external beam radiation therapy. Int J Radiat Oncol Biol Phys 2:307–317

Mohan R, Ling CC (1995) When becometh less more? Int J Radiat Oncol Biol Phys 33:235–237

Mohan R, Wang XH (1996) Physical vs biological objectives for treatment plan optimization. Radiother Oncol 40:186–187

Mohan R, Wu Q, Manning M, Schmidt-Ullrich R (2000) Radiobiological considerations in the design of fractionation strategies for intensity-modulated radiation therapy of head and neck cancers. Int J Radiat Oncol Biol Phys 46:619–630

Natterer F (1986) The mathematics of computerized tomography. Teubner, Stuttgart

Niemierko A (1992) Random search algorithm (RONSC) for the optimization of radiation therapy with both physical and biological endpoints and constraints. Int J Radiat Oncol Biol Phys 23:89–98

Niemierko A (1997) Reporting and analyzing dose distributions: a concept of equivalent uniform dose. Med Phys 24:103–110

Niemierko A (1999) A generalized concept of equivalent uniform dose (EUD). Med Phys 26:1100

Pugachev A, Li JG, Boyer AL, Hancock SL, Le QT, Donaldson SS, Xing L (2001) Role of beam orientation optimization in intensity-modulated radiation therapy. Int J Radiat Oncol Biol Phys 50:551–560

Redpath AT, Vickery BL, Wright DH (1975) A set of fortran subroutines for optimizing radiotherapy plans. Comput Programs Biomed 5:158–164

Romeijn HE, Ahuja RK, Dempsey JF, Kumar A, Li JG (2003) A novel linear programming approach to fluence map optimization for intensity modulated radiation therapy treatment planning. Phys Med Biol 48:3521–3542

Shepard DM, Earl MA, Li XA, Naqvi S, C Yu (2002) Direct aperture optimization: a turn key solution for step-and-shoot IMRT. Med Phys 29:1007–1018

Shepard DM, Ferris MC, Olivera GH, Mackie TR (1999) Optimizing the delivery of radiation therapy to cancer patients. SIAM Rev 41:721–733

Söderström S, Brahme A (1995) Which is the most suitable number of photon beam portals in coplanar radiation therapy. Int J Radiat Oncol Biol Phys 33:151–159

Söderström S, Brahme A (1996) Small is beautiful: and often enough. Int J Radiat Oncol Biol Phys 34:757–758

Spirou SV, Chui C-S (1998) A gradient inverse planning algorithm with dose-volume constraints. Med Phys 25:321–333

Starkschall G, Pollack A, Stevens CW (2000) Using a dose-volume feasibility search algorithm for radiation treatment planning. Thirteenth International Conference on the Use of Computers in Radiation therapy. Springer, Berlin Heidelberg New York

Stein J, Mohan R, Wang XH, Bortfeld T, Wu Q, Preiser K, Ling CC, Schlegel W (1997) Number and orientation of beams in intensity-modulated radiation treatments. Med Phys 24:149–160

Thieke C (2003) Multicriteria optimization in inverse radiotherapy planning. Doctor of Natural Sciences dissertation. Combined Faculties for the Natural Sciences and for Mathematics, Ruperto-Carola University of Heidelberg

Thieke C, Bortfeld T, Niemierko A (2002) Direct Consideration of EUD Constraints in IMRT Optimization. Med Phys 29:1283

Webb S (1989) Optimisation of conformal radiotherapy dose distributions by simulated annealing. Phys Med Biol 34:1349–1370

Webb S (1992) Optimization by simulated annealing of three-dimensional, conformaltreatment planning for radiation fields defined by a multileaf collimator:II. Inclusion of the two-dimensional modulation of the X-ray intensity. Phys Med Biol 37:1689–1704

Webb S, Convery DJ, Evans PM (1998) Inverse planning with constraints to generate smoothed intensity-modulated beams. Phys Med Biol 43:2785–2794

Wu Q, Mohan R (2002) Multiple local minima in IMRT optimization based on dose-volume criteria. Med Phys 29:1514–1527

Wu Q, Mohan R, Niemierko A, Schmidt-Ullrich R (2002) Optimization of intensity-modulated radiotherapy plans based on the equivalent uniform dose. Int J Radiat Oncol Biol Phys 52:224–235

Yajnik S, Rosenzweig KE, Mychalczak B, Krug L, Flores R, Hong L, Rusch VW (2003) Hemithoracic radiation after extrapleural pneumonectomy for malignant pleural mesothelioma. Int J Radiat Oncol Biol Phys 56:1319–1326

18 Biological Models in Treatment Planning

CHRISTIAN P. KARGER

CONTENTS

18.1
Basic Parameters of Radiation Tolerance

18.1.1
Tumour Control and Normal Tissue Complications

The aim of radiotherapy is to give sufficent dose to the tumour to achieve local control without introducing severe complications in the surrounding

C. P. KARGER, PhD
Abteilung Medizinische Physik in der Strahlentherapie, Deutsches Krebsforschungszentrum, Im Neuenheimer Feld 280, 69120 Heidelberg, Germany

normal tissue. These conflicting aims can be quantitatively described by dose-response curves for the tumour and normal tissue, respectively (Fig. 18.1). With increasing dose to the tumour, the tumour control probability (TCP) also increases. Dose escalation, however, also rises the normal tissue complication probability (NTCP), which frequently is the limiting factor in clinical situations. In the region between both curves (denoted as "therapeutic window"), the probability of tumour control without normal tissue complications reaches a maximum at the optimum dose D_{opt}. If type and probability of the related complications are not acceptable, however, this optimum dose may not be feasible to be applied in clinical situations and the probability for tumour control will therefore be even lower.

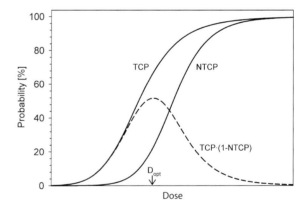

Fig. 18.1 Dose-response curve for tumour control probability (*TCP*) and normal tissue complication probability (*NTCP*). The maximum probability for tumour control without normal tissue complications (*dashed line*) is reached at the optimum dose (D_{opt})

18.1.2
Determination of Dose-Response Curves

In experimental or clinical situations, dose-response data are obtained in terms of incidence rates x/n (x out of n subjects show the selected endpoint) at several dose levels. As incidence rates are binomially distributed and show a large spread for a small number of subjects, n, an analytical curve is adjusted to the data using a maximum likelihood fit. Although several parameterisations may be used to describe dose-response curves mathematically (KÄLLMAN et al. 1992; NIEMIERKO and GOITEIN 1991), the following one is frequently used because of its simplicity:

$$P(D) = \frac{1}{1 + \left(\dfrac{D_{50}}{D}\right)^k} \qquad (1)$$

P(D) gives the expectation value of the probability that the selected end point occurs at the dose level D. The parameter D_{50} (often referred to as tolerance dose, TD_{50}, or equivalent dose, ED_{50}) is the dose at which an effect probability of 50% is expected. The parameter k is related to the slope of the dose-response curve.

Most statistical software packages supply the logistic formula for parameterisation of dose-response curves, which is not equivalent to Eq. (1). The parameters D_{50} and k of Eq. (1) as well as their standard errors, however, may be calculated from the fit of the logistic function, if ln(D) instead of the dose D is used as independent variable (KARGER and HARTMANN 2001).

Tolerance doses for normal tissue complications may be determined with reasonable accuracy in animal experiments. The question of whether these data also apply to humans, however, remains an intrinsic problem, and tolerance doses from animal experiments are generally not believed to be directly transferable to humans. In humans, on the other hand, the parameter D_{50} usually cannot be determined as such high complication rates are usually prevented by clinical experience. For clinical applications, quantities such as D_5 (dose leading to 5% complication probability) are more relevant. From such low complication probabilities, however, it is nearly impossible to determine the slope of the curve.

EMAMI et al. (1991) published a compilation of data for D_5 and D_{50} for a variety of normal tissues. Although the authors stressed the limited accuracy of these data, no significant improvements in knowledge was achieved up to now and most articles still refer to these data.

18.1.3
Implicit Dependencies of Dose-Response Curves

Figure 18.1 suggests that the effect probability is solely a function of the radiation dose. The dose-response curves, however, implicitly depend on several biological and physical parameters. At first, the biological end point for the tissue response has to be specified, including the time after radiation used for follow-up. For normal tissue complications, the dose-response curves will be different for early and late effects and even within one type of complication the curves may differ, depending on the exact end point definition and the method of investigation. In the same way, the curve for the tumour may depend on the definition of tumour control, e.g. whether local control or remission of the tumour is regarded. In the clinical situation, relevant end points have to be selected, e.g. radiological tumour control and early or late complications which are definitely to be prevented.

Besides the definition of the biological end point, the dose-response curves are strongly influenced by various treatment parameters, most of them being of physical nature. The most important parameters are given by the time pattern of the applied dose, the volume of irradiated normal tissue and the radiation quality used for treatment. Since the very beginning of radiotherapy treatments, it was the aim of many investigations to optimise the physical treatment parameters to improve clinical outcome, i.e. to open the "therapeutic window" between the dose-response curves for tumour control and normal tissue complications (Fig. 18.1).

18.2
Aims of Biological Models

The aim of biological models is to predict the radiation response of biological systems. While early approaches focused on modelling the radiation response for different fractionation schemes, newer developments attempt to model effect probabilities (TCP and NTCP), their volume dependence and the relative biological efficiency (RBE) of radiation with high linear energy transfer (LET).

In treatment planning, biological models may be applied with different intentions:

1. Transfer of one treatment regime to a biologically iso-effective new regime or new radiation modality without predicting absolute values for TCP or NTCP.

2. Calculation of TCP or NTCP values to compare either competing treatment plans for an individual patient or different treatment techniques for a specified clinical application. In this case, the TCP/

NTCT values are not expected to be completely correct in their absolute values, but it is believed that they can be used as rationale to prefer one treatment plan (or technique) over another.

3. Prediction of absolute TCP or NTCP values for individual patients.
4. Integration of TCP/NTCP models into the cost function of the dose optimisation algorithm to generate biologically optimised treatment plans.

18.3
Fractionation Regime and Treatment Time

Among the various influence factors on TCP and NTCP four factors have always been considered to be most important. These factors are denoted as the four "R"s of radiotherapy (WITHERS 1992; THAMES et al. 1989) and include repair of sublethal cell damage, repopulation of tumour cells, redistribution of cells over different cell cycle phases with different radio-sensitivity and reoxygenation of radio-resistant hypoxic tumours after beginning of the radiotherapy course. Repair and repopulation have been identified as the most important factors with respect to radiation response and were therefore subject of early models.

18.3.1
Historical Models

Repair and repopulation are related to the fractionation regime, i.e. to the dose per fraction, the number of fractions and the overall treatment time. Historically, several models have been developed on the basis of skin data (BARENDSEN 1982; ULMER 1986) to describe iso-effect relations for different treatment regimes. Examples for these models are the nominal standard dose (NSD) model and its derivates denoted as the partial tolerance (PT), the time, dose, fractionation (TDF) and the cumulative radiation effect (CRE) model. These models, however, were criticized for several reasons (BARENDSEN 1982; FOWLER 1984) and should not be used anymore. The most significant problem with the NSD model probably is the time factor. The underlying assumption is that repopulation directly starts at the beginning of the radiotherapy course. As radiation injury becomes manifest only if the cell attempts division, the amount of cell loss depends on the turnover of the target cells. It is known that proliferation is not significant

until some time after radiation (BARENDSEN 1982; FOWLER 1984, 1989, 1992; THAMES 1982). While this so-called kick-off time is in the range of a few weeks for early reactions, it may last up to several months or longer for late reacting tissues; therefore, late radiation damage is not influenced by the overall treatment time, i.e. no time factor has to be considered for late effects (FOWLER 1989). Instead, the dose per fraction, rather than the fraction number, was identified as leading factor.

18.3.2
The Linear-Quadratic Model

The linear-quadratic model was introduced to replace the former iso-effect relations for different fractionation regimes (BARENDSEN 1982; FOWLER 1984, 1989, 1992; WITHERS 1992). Using this model, the survival fraction of cells irradiated with a single dose d is described by

$$SF = e^{-(\alpha d + \beta d^2)}. \tag{2}$$

α and β are parameters measuring the amount of lethal and sub-lethal cell damage, respectively. In a logarithmic representation, the survival curve shows an initial linear decrease at low doses followed by a "shoulder" for which the bending is determined by the ratio α/β (Fig. 18.2).

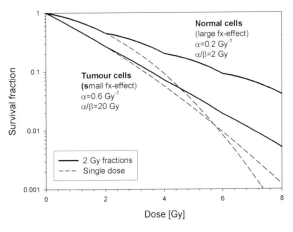

Fig. 18.2. Cell survival curves for tumour and late-reacting normal tissue. The curves are displayed for singe and fractionated treatment, respectively. For single doses the curves for tumour and normal cells intersect each other, leading to lower survival fractions for the normal cells. For a fractionated treatment, however, the survival fraction for tumour cells always remains below the one for normal cells. A fractionated treatment therefore spares late responding normal tissue

The smaller α/β is, the more pronounced the shoulder of the survival curve will be. If the time between two fractions exceeds the minimum of about 6 h (FOWLER 1989), the sub-lethal damage can be repaired and the shape of the cell survival curve will be reproduced for the subsequent fraction (Fig. 18.2). After n fractions of equal doses, d, the survival fraction will then be $e^{-n(\alpha d+\beta d^2)}$. As the survival level is determined by $n(\alpha d+\beta d^2)$, a biologically effective dose (also termed as extrapolated response dose, ERD) may be defined (FOWLER 1989, 1992; PRASAD 1992) by the equation

$$BED = nd(1+\frac{d}{\alpha/\beta}). \tag{3}$$

Two fractionation regimes having the same BED are considered to be iso-effective. As a consequence, one fractionation regime may be converted to another iso-effective regime using the relation

$$n_2 d_2 = n_1 d_1 \cdot \frac{\alpha/\beta+d_1}{\alpha/\beta+d_2} \tag{4}$$

n_1, n_2, and d_1, d_2 are the number of fractions and the doses per fraction for both regimes, respectively. It is a great advantage of the linear-quadratic model that only the ratio α/β, rather than the absolute values of α and β are required in Eq. (4).

Values for α/β can be measured for in vitro as well as for in vivo systems. Although the same cell type may be involved, the α/β-values may be different as cells in intact tissue are not expected to respond independently from their physiological environment. As in vivo settings do not allow to determine the fraction of surviving cells, α/β has to be measured by equating the BED of two different fractionation regimes according to Eq. (3). The required iso-effective total doses may be obtained from the dose-response curve of the respective regime (e.g. the tolerance doses D_{50}). The resulting equation can then be resolved for α/β. Alternatively, the inverse of the total dose $(nd)^{-1}$ may be plotted against the dose per fraction d (referred to as Douglas-Fowler plot) for more than two iso-effective fractionation regimes (BARENDSEN 1982; FOWLER 1984; THAMES 1982). According to Eq. (3), α/β may then be determined from a linear regression to the data.

The value for α/β does not only depend on the irradiated type of tissue, but also on the considered biological end point. Values for α/β are given in the literature for various tissues and end points (BARENDSEN 1982; FOWLER 1984, 1989; THAMES et al. 1989; WITHERS 1992). Typical α/β-values are around 3 Gy for late reactions and around 10 Gy for early re-

actions. Most tumours show α/β-values of 10 Gy or more. As a consequence, late reacting tissue can be spared by fractionated treatment relative to the tumour response, while the fractionation effect is small or may even be neglected for early reacting tissues. For these tissues the overall treatment time is more important.

Although attempts have been made to derive Eq. (2) as an approximation from mechanistic considerations (GILBERT 1980), the linear-quadratic model is mostly regarded as an empirical parameterisation. The linear-quadratic model has been validated in the dose range of about 2–8 Gy per fraction (FOWLER 1984, 1989; WITHERS 1986, 1992). At higher doses some experimental cell survival curves asymptotically approach to a purely exponential shape and deviations to the linear-quadratic model have been seen. This behaviour may be explained by multi-hit killing from accumulation of multiple sub-lethal events (WITHERS 1992). Although there are other parameterisations for cell survival curves (e.g. the two-component model), which account for the purely exponential shape at high doses, the linear-quadratic model is usually preferred since the derived iso-effect relation of Eq. (4) depends only on the single parameter α/β (WITHERS 1992). Because of the limitation of the linear-quadratic model at high doses, Eq. (4) should not be used to transfer a fractionated treatment to an iso-effective single dose treatment, although it has previously been used as approximation (LARSON et al. 1993; PRASAD 1992; FLICKINGER et al. 1990).

18.3.3
Extensions to the Linear-Quadratic Model

Several extensions have been developed for the linear-quadratic model to account for additional dependencies of the radiation response. As repopulation may play an important role for early reacting tissues, a time factor has been introduced:

$$SF = e^{-(\alpha d+\beta d^2)+\gamma(T-T_k)} \tag{5},$$

where γ is related to the average doubling time T_p of the cells by $\gamma = \ln2/T_p$ and T_k is the kick-off time until proliferation starts (FOWLER 1989, 1992). For late effects the time factor can be neglected and in this case Eq. (5) reduces to Eq. (2). Equation (5) leads to a modified expression for the BED given by

$$BED = nd(1+\frac{d}{\alpha/\beta}) - \gamma\frac{(T-T_k)}{\alpha}. \tag{6}$$

The apparent disadvantage of this extended model is that the absolute values of α is required which is not always available with sufficient accuracy. In addition, the values of the two parameters γ and T_k have to be known.

As there is some evidence that α and β depend on the cell cycle phase, the linear-quadratic model was extended to account for this heterogeneity (SCHULTHEISS 1987). To do so, three additional parameters are introduced, describing the spread and the correlation of the parameters α and β. In an other approach it was attempted to give a complete description for repair, repopulation, redistribution and re-oxygenation (BRENNER 1995). This model also needs five input parameters.

A special situation is given by the application of brachytherapy techniques, since the irradiation is performed either continuously over several days or in a fractionated fashion, where the time between the fractions is in the order of one hour. In both cases, the sub-lethal damage may not completely be repaired. Several investigations have been performed to model the radiation response for incomplete repair (BRENNER and HALL 1991; DALE 1986; DALE et al. 1988; LING and CHUI 1993; NILSSON et al. 1990; POP et al. 1996; THAMES 1985). As a result, the cell survival curve for incomplete repair is given as

$$SF = e^{-(\alpha d + G\beta d^2)}, \tag{7}$$

where G is a factor calculated from the temporal characteristics of the dose distribution (BRENNER and HALL 1991). In the case of complete repair, G is equal to 1 and Eq. (7) reduces to Eq. (2). For continuous radiation G is given by

$$G = 2\left(\frac{t_0}{T}\right)^2\left[\frac{T}{t_0} - 1 + e^{-\frac{T}{t_0}}\right], \tag{8}$$

where T and t_0 are the total time of irradiation and the characteristic repair time, respectively (BRENNER and HALL 1991; POP et al. 1996). t_0 is related to the repair half time $T_{1/2}$ by $T_{1/2} = t_0 \cdot \ln 2$. For a general fractionation regime of n equal fractions, each of duration T_F and period Δt between fractions, the factor G may be calculated according to BRENNER 1991:

$$G = \frac{2}{(nT_F)^2}\left[nt_0T_F + nt_0^2\left(1 - \frac{1}{x}\right) + \right.$$
$$\left. t_0^2\left(x + \frac{1}{x} - 2\right)\left\{\frac{n}{1-y} - \frac{y(1-y^N)}{(1-y)^2}\right\}\right]. \tag{9}$$

In Eq. (9) the abbreviations $x = e^{-T_F/t_0}$ and $y = e^{-\Delta t/t_0}$ have been used. For a standard fractionation regime with $T_F \ll t_0$ and $\Delta t \gg t_0$, G reduces to 1/N and Eq. (7) then will be reduced to $SF = e^{-n(\alpha d + \beta d^2)}$, which is the product of the survival fractions of n biologically independent fractions.

18.4
NTCP Models

The NTCP models aim to describe the complication probability in normal tissues in terms of dose-response curves. As there is extensive evidence, that the radiation response of normal tissue depends on the amount of irradiated normal tissue (BURMAN et al. 1991; COHEN 1982; EMAMI et al. 1991; FLICKINGER et al. 1990; SCHULTHEISS 1983; WITHERS et al. 1988), the irradiated volume is included as an additional important parameter. The extent of the volume effect is dependent on the architecture of the respective tissue and several models have been proposed. While some of them are only of phenomenological nature, others include more basic bio-statistical principles.

18.4.1
The Lyman-Kutcher Model

A four-parameter model was proposed by LYMAN (LYMAN 1985). In this model, the complication probability P(D,v) for a uniform irradiation of a normal tissue volume V with a dose D is given by (BURMAN et al. 1991; BURMAN 2002; LYMAN 1985; KUTCHER 1996):

$$P(D,v) = \frac{1}{\sqrt{2\pi}}\int_{-\infty}^{t} e^{-\frac{x^2}{2}} dx \tag{10a}$$

$$t = \frac{1}{m}\left(\frac{D}{TD_{50}(v)} - 1\right) \tag{10b}$$

$$TD_{50}(v) = TD_{50}(1) \cdot v^{-n} \tag{10c}$$

$$v = \frac{V}{V_{ref}}. \tag{10d}$$

The four parameters of the model are given by TD_{50}, m, n and V_{ref}, which have to be adjusted to clinical data for each tissue type using a specified biological end point. $TD_{50}(v)$ is the tolerance dose for the fractional volume v, m is related to the slope of the dose-response curve, n describes the volume effect and V_{ref} is the reference volume to which the fractional volume refers to. V_{ref} may be chosen as the whole organ or as a part of it. Equation (10c) relates the tolerance doses of the partial volume v to that of the reference volume (v=1).

EMAMI et al. (1991) published tolerance doses for various tissues and fractional volumes which were derived from a literature search and from clinical experience. The authors considered the uncertainty of these tolerance doses to be rather high. Subsequently, the model parameters of Eq. (10) were adjusted to fit these tolerance data (BURMAN et al. 1991). As there were only few tolerance doses for each organ, the parameters were fit "by eye" rather than by using statistical methods; therefore, and due to the uncertainty of the underlying data, the derived model parameters have to be treated with great caution. Although some of the tolerance doses were refined later (BURMAN 2002), most of them have remained unchanged up to now.

In clinical practice, the normal tissue will not be uniformly irradiated as assumed by the Lyman model; therefore, the model was extended by introducing histogram-reduction algorithms, which transform the multi-step dose volume histogram obtained for a specific treatment plan into a biologically iso-effective single-step histogram, i.e. a non-uniform irradiation is transformed in an biologically iso-effective uniform irradiation. Two different types of reduction algorithms have been proposed which lead to similar although not identical NTCP-values:

The first one (LYMAN and WOLBARST 1987, 1989) replaces the two rightmost bins (at doses D_n and D_{n-1} and volumes V_n and V_{n-1}) of the cumulative histogram by a single bin at dose D'_{n-1} and Volume V_{n-1}. The dose D'_{n-1} is calculated such that the new histogram has the same NTCP according to Eq. (10). This procedure is iterated until a single-step histogram is achieved which corresponds to a homogeneous irradiation of the reference volume (v=1) with a dose D_1 for which the Lyman model can directly be applied.

The second algorithm (KUTCHER and BURMAN 1989; KUTCHER et al. 1991) transforms the initial multi-step histogram (having the maximum dose D_{max}) to a biologically iso-effective single-step histogram with an effective volume V_{eff} at the dose D_{max}. For this approach, a volume effect according to Eq. (10c) is assumed. The single-step histogram then corresponds to a homogeneous irradiation of the fractional volume $v_{eff}=V_{eff}/V_{ref}$ and the NTCP is then calculated by Eq. (10).

18.4.2
The Critical Element Model

The critical element model (NIEMIERKO and GOITEIN 1991; SCHULTHEISS 1983; WOLBARST 1984) assumes that an organ consists of a number of identical functional subunits (FSU; WITHERS et al. 1988), each of them responding independently to radiation. The term "critical element" means that it is additionally assumed that a complication occurs, if a single FSU is inactivated (NIEMIERKO and GOITEIN 1991). The critical element model is expected to describe the radiation response for organs such as spinal cord, brain or bowel.

If $P(D,v)$ is the complication probability that a dose D to the fractional volume v will produce a complication, $[1- P(D,v)]$ is the probability that no complication occurs. If a whole organ consisting of N equal-sized compartments (each of volume v=1/N) is uniformly irradiated with a dose D, the probability that the organ escapes injury $P(D,1)$ is given by the product of the probabilities that each sub-volume escapes injury. $P(D,v)$ can then be expressed by (SCHULTHEISS 1983):

$$P(D,v) = 1 - [1 - P(D,1)]^v \qquad (11)$$

Equation (11) may easily be generalized to non-uniform dose distributions {D} by replacing the right side by $1 - \prod_i [1 - P(D_i,1)]^{\Delta v_i}$.

As the size of the product is strongly affected by the smallest factor, the size of the complication probability $P(D,v)$ is governed by large values of $P(D_i,1)$, i.e. by the highest doses of {D_i}.

It follows from Eq. (11) that the dose-response curve for any partial volume irradiation can be calculated if the dose-response curve for the whole organ is known. No specific dose-response model has to be assumed.

It is a characteristic feature of the critical element model that the dose-volume iso-effect curve is determined solely by the slope parameter (k in Eq. (1) and m in Eq. (10b), respectively) of the dose-response curve for the whole organ (NIEMIERKO and GOITEIN 1991). In contrast to this, the volume dependence in the Lyman model (Eq. (10c)) uses the additional parameter n, which can be selected independently from the slope parameter m. That means that in general, the Lyman model describes a tissue architecture different from the one of the critical element model.

In the approximation of P(D,1)<<1 (i.e. small doses, $D<<D_{50}$), however, Eq. (11) yields $P(D,v)=v \cdot P(D,1)$. If in addition the dose-response model of Eq. (1) is taken in the same approximation one obtains $P(D,1)=(D/D)^k$. From these two relations, a dose-volume iso-effect relation can be derived (SCHULTHEISS 1983):

$$D_v = v^{-1/k} \cdot D_1 \qquad (12)$$

This relation is of the same structure as Eq. (10c). This means that the Lyman model is able to describe the dose-volume relation of tissue with critical element structure only for small complication probabilities (NIEMIERKO and GOITEIN 1991). It also has to be pointed out that the histogram reduction methods (LYMAN and WOLBARST 1987, 1989; KUTCHER and BURMAN 1989; KUTCHER et al. 1991) of the Lyman model implicitly make use of Eq. (10c) which is in general not valid for tissues with critical element structure. In this case the algorithms have to be adapted according to NIEMIERKO and GOITEIN (1991).

The two dose-volume relations of Eqs. (11) and (12) were tested in animal experiments for the spinal cord and the brain, which both are considered to be of the critical element architecture (NIEMIERKO and GOITEIN 1991; SCHULTHEISS 1983). As expected, Eq. (11) was found to give a better description of the data. The critical element model was also used to calculate the complication probabilities and dose-volume iso-effect relations for radiosurgery treatments of the brain (LAX and KARLSSON 1996; FLICKINGER 1989; FLICKINGER et al. 1990).

Although the publication of SCHULTHEISS (1983) does not explicitly use the term "FSU", the described model comprises all characteristic features of the critical element model. A more theoretical approach was presented by WOLBARST (1984), using the dose-response curve of a single FSU as starting point to model the radiation response of the entire organ.

18.4.3
The Critical Volume Model

The critical volume model describes tissues, where the FSUs of an organ are assumed to be arranged in a parallel fashion (JACKSON et al. 1993; NIEMIERKO and GOITEIN 1993a; WOLBARST et al. 1982; YORKE et al. 1993). In contrast to the critical element model, an inactivation of a single FSU will not lead to a

complication in the organ as the organ function will be maintained by the remaining FSUs. If more than a critical number of FSUs will be inactivated, however, a complication will occur. This especially means that the organ tolerates any dose as long as the number of affected FSUs is below this threshold. The critical volume model is expected to describe the complication probabilities of organs such as the lung, kidney, liver, or parotid glands.

If an organ is assumed to consist of N parallel organized and independently responding FSUs, the probability that more than M FSUs are inactivated by an uniform dose D is given by (NIEMIERKO and GOITEIN 1993a):

$$P(D) = \sum_{k=M+1}^{N} \binom{N}{k} \cdot P_{FSU}^k (1-P_{FSU})^{N-k}, \qquad (13)$$

where P_{FSU} is the dose-dependent probability for inactivating a single FSU. Equation 13 may also be generalized for inhomogeneous dose distributions (NIEMIERKO and GOITEIN 1993a; JACKSON et al. 1993; YORKE et al. 1993) leading to the concept of integral responding tissues (WOLBARST et al. 1982). As a consequence, the radiation response of parallel organized tissues should be governed by the mean rather than by the maximum doses as found for tissues of critical element structure.

For the special case of M=0, the critical volume model reduces to the critical element model (see previous section). In this case Eq. (13) can be written as:

$$P(D) = 1-(1-P_{FSU})^N. \qquad (14)$$

Another special case is given by M=N, which means that all FSUs have to be inactivated to produce a complication. In this case, Eq. (13) yields:

$$P(D) = P_{FSU}^N. \qquad (15)$$

An example for this situation is a tumour where the FSU is identified with a single clonogenic cell. P_{FSU} then is the probability of inactivating one cell and P(D) is the probability of controlling the tumour, i.e. that all N clonogenic cells of the tumour are inactivated; therefore, with respect to the end-point tumour control, the critical volume model can be applied. With respect to the end-point tumour recurrence, however, tumours behave according to the critical element model. This can be seen by substituting the survival probability of the FSU by $S_{FSU}=1-P_{FSU}$ and the recurrence probability for the tumour $S(D)=1-P(D)$ into Eq. 15 which then results in an expression, which is formally identical to Eq. 14:

parameter t_0) have to be known in their absolute values, which may be difficult for an in vivo setting. Predictions of these extensions have to be treated with ever greater caution.

18.7.2
NTCP and TCP Models

The clinical application of NTCP models has significantly improved the understanding of the volume dependence of normal tissue response to radiotherapy. For clinical applications, mainly the phenomenological model of Lyman has been applied using the tolerance data provided by EMAMI et al. (1991) and the fit parameters of BURMAN et al. (1991). Although the uncertainty of these data was already stressed by the authors, these data are still used as reference in most of the recent literature. Since these tolerance data have been published, almost no attempt was made to refine this data base and it is not likely that this situation will improve in the near future. One reason for this may be the fact that the Lyman model requires input data from uniform partial-volume irradiations, which were less frequently applied with the upcoming of 3D-conformal radiotherapy. If this is the actual reason, the historical data published by EMAMI et al. (1991) may be the best data which can be achieved. Prediction of absolute NTCP values are therefore problematic, and the use of such absolute NTCP values as only criteria for clinical decisions is currently not warranted.

Although extensions to the Lyman model have been proposed, they are usually not applied to clinical data. As these extensions need additional biological parameters, the uncertainty of all parameters in the model increases and it is not expected that the description of clinical data will be improved.

The NTCP models are frequently applied to comparative planning studies and it is argued that this is justified as the models are known to give a correct qualitative description of the radiation response and only the ranking of NTCP values is considered. Although this kind of application contains weaker demands to the models, one major problem persists: the uncertainty of the predictions due to the uncertainty of the model parameters is mostly not specified quantitatively and the question arises as to whether a difference in NTCP values for different treatment plans (or techniques) may be regarded as significant.

The NTCP models based on more radiobiological principles may be applied for improving the principal understanding of normal tissue response to radiation. As these models contain more parameters than the Lyman model, they are usually not applied to clinical data.

In principal, the restrictions to NTCP models apply also to TCP models. For TCP models, however, the situation is even more complicated, since parameters such as proliferation, oxygenation and angiogenesis are much more heterogeneous for tumours than for normal tissues. Moreover, as these parameters may change under radiotherapy, the clinical application of TCP models is difficult. Nevertheless, the models may be applied to improve the understanding of the tumour response to radiation and its interaction with accompanying influence factors.

Although TCP/NTCP models have been developed and implemented into the cost functions of dose optimisation algorithms (BRAHME 2001), clinically applied treatment plans continue to be optimised in terms of physical dose because of the intrinsic uncertainties of the biological models. AMOLS et al. (1997) discussed an approach to optimise treatment plans on the basis of NTCP- and TCP predictions and some additional parameters describing the risk acceptance of individual patients and physicians. This integrated approach to biological plan optimisation appears to be far away from application in clinical reality.

18.7.3
RBE of High-LET Radiation

The RBE models for high-LET radiation take an exceptional position in the field of biological modelling as most clinical experience is based on photon therapy. For the application of high-LET radiation, the RBE must be considered somehow to make use of this experience. In this context, RBE models are the only biological models which are routinely applied in clinical practice and it is out of discussion that the application is necessary for the optimisation of treatment plans.

The RBE variations for protons are small and there are currently no clinical indications that a more detailed RBE model than the constant factor of 1.1 is needed. For heavy ions, however, the RBE varies between much larger values and moreover depends on several physical as well as biological parameters. Current models describe the main characteristics of these dependencies and enable the safe application of heavy ion therapy. Similar to other biological models, intrinsic uncertainties are involved in the predictions of these models. This uncertainty has to be kept in mind, if the model is introduced into clinical application. This especially means that the prescribed dose has to be selected very carefully and the full potential of heavy ions has to be determined in dose escalation studies as it is currently done at the HIMAC facility in Japan (TSUJII et al. 2002).

18.8
Conclusion

Several biological models have been developed. Although these models give a correct description of the main characteristics of the radiation response, great caution has to be taken if these models are to be applied to patients.

While the linear-quadratic model provides a good description of experimental settings, a larger uncertainty is involved in the prediction of iso-effects for clinical applications. The more advanced NTCP and TCP models should only be applied for relative, rather than absolute, predictions of effect probabilities. When using relative values, the uncertainty of the predictions should be considered to decide whether a detected difference is really significant. As TCP/NTCP models are currently not completely validated, integration of these models into the cost function of the dose optimisation algorithm is not warranted. Whether it is possible to arrive at fully biologically optimised treatment plans for photon therapy has to be investigated by further research.

In this context, the clinical application of heavy charged particles plays an exceptional role as biological optimisation is routinely performed and an adequate RBE model is an essential prerequisite. The applied RBE model may still contain some degree of uncertainty which has to be considered carefully at treatment plan assessment and dose prescription.

References

Amols HI, Zaider M, Mayes MK et al. (1997) Physician/patient-driven risk assignment in radiation oncology: reality or fancy? In J Radiat Oncol Biol Phys 38:455–461

Barendsen GW (1982) Dose fractionation, dose rate and iso-effect relationships for normal tissue responses. Int J Radiat Oncol Biol Phys 8:1981–1997

Borkenstein K, Levegrün S, Peschke P (2004) Modeling and computer simulations of tumor growth and tumor response to radiotherapy. Radiat Res 162:71–83

Brahme A (2001) Individualizing cancer treatment: biological optimization models in treatment planning and delivery. Int J Radiat Oncol Biol Phys 49:327–337

Brenner DJ, Hall JH (1991) Conditions for the equivalence of continuous to pulsed low dose rate brachytherapy. Int J Radiat Oncol Biol Phys 20:181–190

Brenner DJ, Hlatky LR, Hahnfeldt PJ et al. (1995) A convenient extension of the linear-quadratic model to include redistribution and reoxygenation. Int J Radiat Oncol Biol Phys 32:379–390

Burman C (2002) Fitting of tissue tolerance data to analytic function: improving the therapeutic ratio. Front Radiat Ther Oncol 37:151–162

Burman C, Kutcher GJ, Emami B et al. (1991) Fitting normal tissue tolerance data to an analytic function. Int J Radiat Oncol Biol Phys 21:123–135

Cohen L (1982) The tissue volume factor in radiation oncology. Int J Radiat Oncol Biol Phys 8:1771–1774

Dale RG (1986) The application of the linear-quadratic model to fractionated radiotherapy when there is incomplete normal tissue recovery between fractions, and possible implication for treatments involving multiple fractions per day. Br J Radiol 59:919–927

Dale RG, Huczkowski J, Trott KR (1988) Possible dose rate dependence of recovery kinetics as deduced from a preliminary analysis of the effects of fractionated irradiations at varying dose rates. Br J Radiol 61:153–157

Emami B, Lyman J, Brouwn A et al. (1991) Tolerance of normal tissue to therapeutic irradiation. Int J Radiat Oncol Biol Phys 21:109–122

Flickinger JC (1989) An integrated logistic formula for prediction of complication from radiosurgery. Int J Radiat Oncol Biol Phys 17:879–885

Flickinger JC, Schell MC, Larson D (1990) Estimation of complications for linear accelerator radiosurgery with the integrated logistic formula. Int J Radiat Oncol Biol Phys 19:143–148

Fowler JF (1984) What next in fractionated radiotherapy? Br J Cancer 49 (Suppl VI):285–300

Fowler JF (1989) The linear-quadratic formula and progress in fractionated radiotherapy. Br J Radiol 62:679–694

Fowler JF (1992) Brief summary of radiobiological principles in fractionated radiotherapy. Semin Radiat Oncol 2:16–21

Gilbert CW, Hendry JH, Major D (1980) The approximation in the formulation for survival $S=\exp-(\alpha D+\beta D^2)$. Int J Radiat Biol 37:469–471

Haberer T, Becher W, Schardt D at al. (1993) Magnetic scanning system for heavy ion therapy. Nucl Instrum Meth A330:296–305

Jackson A, Kutscher GJ, Yorke ED (1993) Probability of radiation induced complications for normal tissues with parallel architecture subject to non-uniform irradiation. Med Phys 20:613–625

Källman P, Agren A, Brahme A (1992) Tumour and normal tissue responses to fractionated non-uniform dose delivery. Int J Radiat Biol 62:249–262

Kanai T, Furusawa Y, Fukutsu K et al. (1997) Irradiation of mixed beam and design of spread-out Bragg peak for heavy-ion radiotherapy. Radiat Res 147:78–85

Kanai T, Endo M, Minohara S et al. (1999) Biophysical characteristics of HIMAC clinical irradiation system for heavy-ion radiation therapy. Int J Radiat Oncol Biol Phys 44:201–210

Karger CP, Hartmann GH (2001) Determination of tolerance dose uncertainties and optimal design of dose response experiments with small animal numbers. Strahlenther Onkol 177:37–42

Kraft G (2000) Tumortherapy with heavy charged particles. Prog Part Nucl Phys 45:S473–S544

Kraft G, Scholz M, Bechthold U (1999) Tumor therapy and track structure. Radiat Environ Biophys 38:229–237

Krämer M, Scholz M (2000) Treatment planning for heavy-ion radiotherapy: calculation and optimization of biologically effective dose. Phys Med Biol 45:3319–3330

Krämer M, Weyrather WK, Scholz M (2003) The increased relative biological efficiency of heavy charged particles: from radiobiology to treatment planning. Technol Cancer Res Treat 2:427–436

Kutcher GJ (1996) Quantitative plan evaluation: TCP/NTCP models. Front Radiat Ther Oncol 29:67–80

Kutcher GJ, Burman C (1989) Calculation of complication probability factors for non-uniform normal tissue irradiation: the effective volume model. Int J Radiat Oncol Biol Phys 16:1623–1630

Kutcher GJ, Burman C, Brewster L (1991) Histogram reduction method for calculating complication probabilities for three-dimensional treatment planning evaluations. Int J Radiat Oncol Biol Phys 21:137–146

Larson DA, Flickinger JC, Loeffler JS (1993) The radiobiology of radiosurgery. Int J Radiat Oncol Biol Phys 25:557–561

Lax I, Karlsson B (1996) Prediction of complications in gamma knife radiosurgery of ateriovenous malformations. Acta Oncol 35:49–55

Ling CC, Chui CS (1993) Stereotactic treatment of brain tumors with radioactive implants or external photon beams: radiobiophysical aspects. Radiother Oncol 26:11–18

Lyman JT (1985) Complication probability as assessed from dose-volume-histograms. Radiat Res 104:S13–S19

Lyman JT, Wolbarst AB (1987) Optimization of radiation therapy III: a method of assessing complication probabilities from dose volume histograms. Int J Radiat Oncol Biol Phys 13:103–109

Lyman JT, Wolbarst AB (1989) Optimization of radiation therapy IV: a dose volume histogram reduction algorithm. Int J Radiat Oncol Biol Phys 17:433–436

Niemierko A (1998) Radiobiological models of tissue response to radiation in treatment planning systems. Tumori 84:140–143

Niemierko A, Goitein M (1991) Calculation of normal tissue complication probability and dose-volume histogram reduction schemes for tissue with critical element architecture. Radiother Oncol 20:166–176

Niemierko A, Goitein M (1993a) Modelling of normal tissue response to radiation: the critical volume model. Int J Radiat Oncol Biol Phys 25:135–145

Niemierko A, Goitein M (1993b) Implementation of a model for estimating tumor control probability for an inhomogeneously irradiated tumor. Radiother Oncol 29:140–147

Nilsson P, Thames HD, Joiner MC (1990) A generalized formulation of the incomplete-repair model for cell survival and tissue response to fractionated low dose-rate irradiation. Int J Radiat Biol 57:127–142

Paganetti H (2003) Significance and implementation of RBE variations in proton beam therapy. Technol Cancer Res Treat 2:413–426

Pop LAM, van den Broek JFCM, Visser AG, van der Kogel AJ (1996) Constraints in the use of repair half time and mathematical modelling for the clinical application of HDR and PDR treatment schedules as an alternative for LDR brachytherapy. Radiother Oncol 38:153–162

Prasad SC (1992) Linear quadratic model and biologically equivalent dose for single fraction treatments. Med Dosim 17:101–102

Roberts SA, Hendry JH (1998) A realistic closed-form radiobiological model of clonical tumor-control data incorporating intertumor heterogeneity. Int J Radiat Oncol Biol Phys 41:689–699

Sanchez-Nieto B, Nahum AE (1999) The delta-TCP concept: a clinically useful measure of tumor control probability. Phys Med Biol 44:369–380

Scholz M, Kraft G (1994) Calculation of heavy ion inactivation probabilities based on track structure, X-ray sensitivity and target size. Radiat Proton Dosim 52:29–33

Scholz M, Kellerer AM, Kraft-Weyrather G et al. (1997) Computation of cell survival in heavy ion beams for therapy. The model and its approximation. Radiat Environ Biophys 36:59–66

Schultheiss TE, Orton CG, Peck RA (1983) Models in radiotherapy: volume effects. Med Phys 10:410–415

Schultheiss TE, Zagars GK, Peters LJ (1987) An explanatory hypothesis for early- and late-effect parameter values in the LQ model. Radiother Oncol 9:241–248

Thames HD (1985) An "incomplete-repair" model for survival after fractionated and continuous irradiations. Int J Radiat Biol 47:319–339

Thames HD, Bentzen SM, Turesson I et al. (1989) Fractionation parameters for human tissues and tumors. Int J Radiat Biol 56:701–710

Thames HD, Withers HR, Peters LJ et al. (1982) Changes in early and late responses with altered dose fractionation: implications for dose survival relationships. Int J Radiat Oncol Biol Phys 8:219–226

Tsujii H, Morita S, Miyamoto T et al. (2002) Experiences of carbon ion radiotherapy at NIRS. In: Kogelnik HD, Lukas P, Sedlmayer F (eds) Progress in radio-oncology, vol 7. Monduzzi Editore, Bologna, pp 393–405

Ulmer W (1986) Some aspects of the chronological dose distribution in the radiobiology and radiotherapy. Strahlenther Onkol 162:374–385

Van Vliet-Vroegindeweij C, Wheeler F, Stecher-Rasmussen F et al. (2001) Microdosimetry model for Boron neutron capture therapy. Part II. Theoretical estimation of the effectiveness function and surviving fractions. Radiat Res 155:498–502

Wambersie A, Menzel HG (1993) RBE in fast neutron therapy and boron neutron capture therapy. A useful concept or a misuse. Strahlenther Onkol 169:57–64

Webb S, Nahum AE (1993) A model for calculating tumor control probability in radiotherapy including the effects of inhomogeneous distributions of dose and clonogenic cells. Phys Med Biol 38:653–666

Wilkens JJ, Oelfke U (2003) Analytical linear energy transfer calculations for proton therapy. Med Phys 30:806–815

Withers HR (1986) Predicting late normal tissue responses. Int J Radiat Oncol Biol Phys 12:693–698

Withers HR (1992) Biologic basis of radiation therapy. In: Perez CA, Brady LW, (eds) Principles and practice of radiation oncology, 2nd edn. Lippincott, Philadelphia, pp 64–96

Withers HR, Taylor JMG, Maciejewski B (1988) Treatment volume and tissue tolerance. Int J Radiat Oncol Biol Phys 14:751–759

Wolbarst AB (1984) Optimization of radiation therapy. Part

II. The critical-voxel model. Int J Radiat Oncol Biol Phys 10:741–745

Wolbarst AB, Chin LM, Svenson GK (1982) Optimization of radiation therapy: integral-response of a model biological system. Int J Radiat Oncol Biol Phys 8:1761–1769

Yashkin PN, Silin DI, Zolotov VA et al. (1995) Relative biological effectiveness of proton medical beam at Moscow synchrotron determined by the Chinese hamster cells assay. Int J Radiat Oncol Biol Phys 31:535–540

Yorke ED, Kutscher GJ, Jackson A et al. (1993) Probability of radiation induced complications in normal tissues with parallel architecture under conditions of uniform whole or partial organ irradiation. Radiother Oncol 26:226–237

Zamenhof R, Redmond E, Solares G et al. (1996) Monte-Carlo based treatment planning for Boron neutron capture therapy using custom designed models automatically generated from CT data. Int J Radiat Oncol Biol Phys 35:383–397

gle catheter or applicator. This is the case for the "standard treatments" using simple standard applicators as in the case of a cylinder applicator for the postoperative intracavitary brachytherapy of corpus uteri carcinomas (KRIEGER et al. 1996; BALTAS et al. 1999).

The treatment delivery itself is then based on pre-existing standard plans with isodose distribution documentation. Due to the rigidity of such kind of applicators, the main item/challenge here is to check the correct placement of the catheter in the patient.

The 2D treatment planning procedure is mainly applicator oriented/based. When the placement of the applicator is validated using simple X-rays and is found to be at the adequate position, then the dose delivery to the anatomy around the applicator can be assumed as appropriate for such kind of simple geometries and catheter/source configurations.

19.3
3D Treatment Planning

Here the target and organ at risk localization as well as the catheter reconstruction are based on 3D methods using modern imaging modalities. The same is valid for the dose calculation and evaluation.

A common procedure, at least in the past for gynaecological and other intracavitary applications, was based on two or more X-ray films, which are mainly used for the 3D reconstruction of the used catheters or applicators. For the intracavitary brachytherapy of the primary cervix carcinoma a set of discrete anatomical points has been and is continuously being used for documenting the dose distribution to the patient anatomy. These points have been selected in a way that they can be identified on X-ray films when a specific geometry is applied (ICRU 1985; HERBORT et al. 1993). This method of reconstruction is called projectional reconstruction method (PRM; TSALPATOUROS et al. 1997; BALTAS et al. 1997; BALTAS et al. 2000). Due to the fact that PRM is of limited practicability with reference to the definition of anatomical volumes such as PTV and OARs, PRM can be considered an intermediate, 2.5D, treatment planning method, where the catheters and the dose calculations are realized in the 3D space but only a limited correlation of this distribution to the anatomy can be achieved.

Figure 19.1 demonstrates the two localization radiographs used for the treatment planning of a brachytherapy cervix implant using the ring applicator.

Due to the missing correlation between anatomy and dosimetry when using conventional X-ray radiographs, it is presently common to use 3D sectional imaging such as CT, MR or ultrasound (US) for treatment planning purposes. This is becoming increasingly more popular and tends to replace the traditional methods, at least in the western world. In fact, the establishment of brachytherapy as first-choice treatment for early stages of prostate cancer, where US imaging for the pre- and intraoperative planning

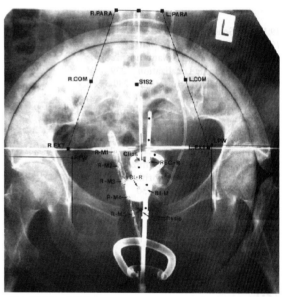

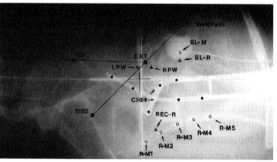

Fig. 19.1a,b. Demonstration of the use of projectional reconstruction method (*PRM*) for the treatment planning of a brachytherapy treatment of cancer of the cervix using the ring applicator. Here orthogonal radiographs are used. **a** Anterior–posterior (*AP*) view. **b** Lateral view. The applicator and the Foley catheter used to obtain the bladder reference point (ICRU 1985) are clearly seen. On both radiographs the reference points regarding the organs at risk – bladder (*BL-R*) and rectum (*REC-R*) – and the reference points related to bony structures (pelvic wall: *RPW, LPW*; lymphatic trapezoid: *PARA, COM* and *EXT*) are also shown. The measurement probes for rectum (*R-M1–R-M5*) and for bladder (*BL-M*) as well as the clips used to check the position of the applicator with respect to portion and the position of the central shielding block during the external beam radiotherapy are also seen. (From HERBORT et al. 1993)

and needle insertion, as well as the CT imaging for the post-planning, are mandatory for an effective and safe treatment, gave rise to developments in the field of imaging-based treatment planning which is also of benefit for all other brachytherapy applications.

The use of 3D imaging enables an anatomy-adapted implantation and anatomy-based treatment planning and optimization in brachytherapy (TSALPATOUROS et al. 1997; BALTAS et al. 1997; ZAMBOGLOU et al. 1998; KOLOTAS et al. 1999a; KOLOTAS et al. 1999b; BALTAS et al. 2000; KOLOTAS et al. 2000).

In addition CT, MR and US are currently used for guidance during needle insertion, offering through this a high degree of safety and intra-implantation approval of the needle position relative to the anatomy (ZAMBOGLOU et al. 1998; KOLOTAS et al. 1999a; KOLOTAS et al. 1999b; KOLOTAS et al. 2000).

Table 19.1 presents an overview of the different imaging modalities regarding their role and possibilities for treatment planning in brachytherapy.

Herein the different steps of the 3D imaging-based treatment planning in modern brachytherapy is addressed in detail.

19.3.1
Anatomy Localization

One or more imaging modalities can be included for the delineation of the patients anatomy, GTV, CTV, PTV and organs at risk (OARs) that have to be considered either for the preparation of the implant (pre-planning) or for the planning of brachytherapy delivery when all catheters are already placed (post-planning).

Here the standard tools, known as the external beam planning systems, are also used for effective and accurate 3D delineation of tissues and organs. Figures 19.2–19.4 demonstrate the tissue delineation for MRI-based pre-planning, 3D US-based intraoperative pre-planning and CT-based post-planning of a prostate monotherapy implant, respectively.

For a more accurate delineation especially of localization in soft tissues, such as gynaecological tumours, brain tumours and perhaps prostate, MRI imaging (pre-application) can be considered fused with CT or US imaging used for the implantation procedure itself.

19.3.2
Catheter Localization

The greatest benefit when using 3D imaging modalities such as CT, MRI or US for the localization and reconstruction of catheters is that there is no need of identifying and matching of catheters describing markers or points on two different projections, as is the case for the PRM method (TSALPATOUROS et al. 1997; MILICKOVIC et al. 2000a; MILICKOVIC et al. 2000b). The PRM methods are man-power intensive and require the use of special X-ray visible markers that are placed within the catheters in order to make them visible for the reconstruction.

The available technology enables effective and fast catheter reconstruction using CT imaging and currently US imaging using automatic reconstruction tools (MILICKOVIC et al. 2000a; GIANNOULI et al. 2000). Such a kind of technology makes the reconstruction procedure user independent and increases the reliability of the brachytherapy method.

Figures 19.5 and 19.6 demonstrate the results of the 3D US-based and CT-based reconstruction, respectively, of the realized implant with 19 catheters for the prostate cancer case shown in Figs. 19.2–19.4.

The 3D US imaging has been used for the intraoperative, live, planning and irradiation, whereas the CT imaging has been used for the treatment planning of the second fraction with that implant.

The in-plane resolution of the US imaging is as high as some tenths of a millimetre, whereas for the CT imaging the in-plane resolution is about 0.5 mm. The limited resolution when using CT imaging in the sagittal and coronal planes results from the inter-slice

Table 19.1. Usability of the different available imaging modalities with regard to the localization of anatomy, catheters or applicators, and their classification according to their availability, speed and ability to offer live and interactive imaging for navigation and guidance of catheter insertion

Imaging modality	Anatomy	Catheter/applicators	Availability	Speed	Live/interactive
Conventional X-ray	+	+++++	+++++	+++++	+++++
CT	+++++	++++	+++	+++	+
MRI	+++++	++	+	++	+
US	+++	+++	+++++	+++++	+++++
3D US	+++	+++	++	++++	+++

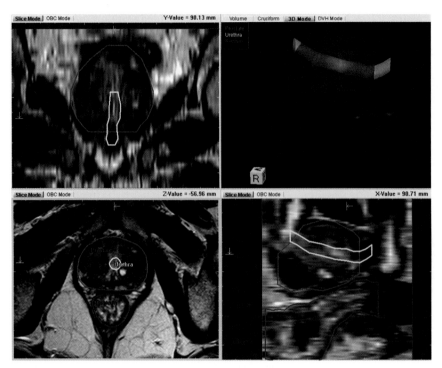

Fig. 19.2. Example of an anatomy delineation for the pre-planning of a high dose rate (HDR) monotherapy implant using MR imaging (T2-weighted imaging) and the SWIFT treatment planning system (Nucletron B.V., Veenendaal, The Netherlands). *Upper left:* coronal view. *Lower left:* axial image. *Lower right:* sagittal view. *Upper right:* 3D view of the planning target volume (PTV; *red*), urethra (*yellow*) and rectum (*purple*)

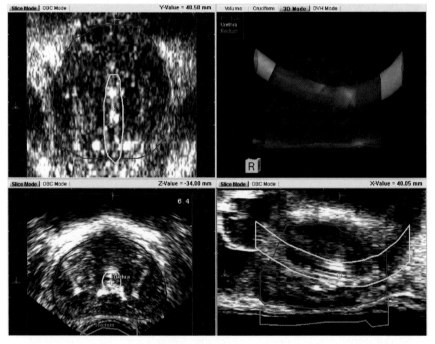

Fig. 19.3. Example of an anatomy delineation for the intraoperative pre-planning for the HDR monotherapy implant of the prostate cancer case of Fig. 2, using 3D US imaging and the SWIFT treatment planning system (Nucletron B.V., Veenendaal, The Netherlands). *Upper left:* coronal view. *Lower left:* axial image. *Lower right:* sagittal view. *Upper right:* 3D view of the PTV (*red*), urethra (*yellow*) and rectum (*purple*). The benefit of using US imaging for identifying the apexal prostate limits is clearly demonstrated in the sagittal view. This 3D imaging and anatomy model was used for the creation of an intraoperative pre-plan for the catheter placement

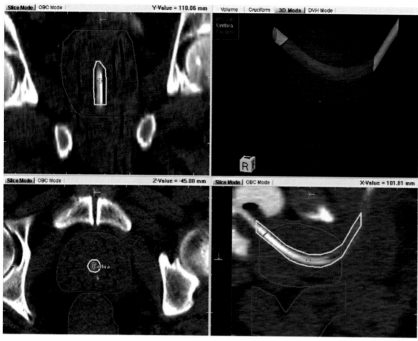

Fig. 19.4. Example of the anatomy delineation for the post-planning of an HDR monotherapy implant for the second brachytherapy fraction and for the prostate cancer case of Fig. 19.2, using CT imaging and the SWIFT treatment planning system (Nucletron B.V., Veenendaal, The Netherlands). *Upper left:* coronal view. *Lower left:* axial image. *Lower right:* sagittal view. *Upper right:* 3D view of the PTV (*red*), urethra (*yellow*) and rectum (*purple*). Contrast media has been used for an adequate visualization of the bladder and the urethra. The difficulty of identifying the apexal prostate limits when using CT imaging is demonstrated in the sagittal view. The 19 implanted catheters (*black holes* or *curves*) are also clearly identified on all images

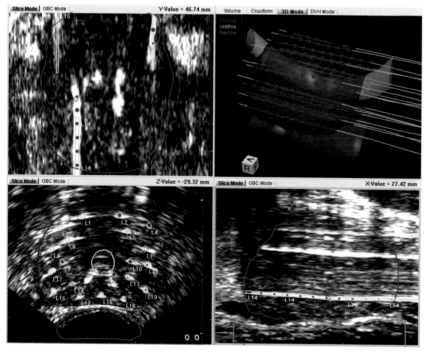

Fig. 19.5. Catheter reconstruction for the intraoperative live planning of the HDR monotherapy implant of the prostate cancer case of Fig. 19.2, using 3D US imaging. *Upper left:* coronal view. *Lower left:* axial image. *Lower right:* sagittal view. *Upper right:* 3D view of the PTV (*red*), urethra (*yellow*) and rectum (*purple*), and of the catheters (*yellow lines*) with the automatically selected appropriate source steps in these (*red circles*). All 19 implanted catheters (*white surfaces* and *curves*) are also clearly identified on all images

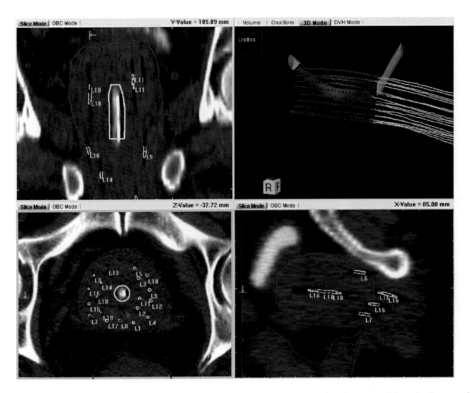

Fig. 19.6. Catheter reconstruction for the post-operative planning for the second brachytherapy fraction and for the HDR monotherapy implant of the prostate cancer case of Fig. 19.2, using CT imaging. *Upper left:* coronal view. *Lower left:* axial image. *Lower right:* sagittal view. *Upper right:* 3D view of the PTV (*red*), urethra (*yellow*) and rectum (*purple*), and of the catheters (*yellow lines*) with the automatically selected appropriate source steps in these (*red circles*). All 19 implanted catheters (*black holes* and *curves*) are also clearly identified on all images

distance. In the example of Figs. 19.4 and 19.6 this was 3.0 mm. Generally, the total accuracy that can be achieved with CT or MR imaging is half the slice thickness with the precondition that slice thickness equals the inter-slice distance (no gap). It is recommended to use for the reconstruction of not straight (metallic) catheters a slice thickness and inter-slice distance of 3 mm, achieving in this way an accuracy as high as 1.5 mm. When using axial MR imaging, then usually the longitudinal image distance is 5.0 mm resulting in an accuracy of 2.5 mm, which makes MRI for catheter reconstruction in several cases of limited interest. This can be overcome if non-axial MRI can be incorporated in the reconstruction procedure.

In contrast to CT, in 3D US imaging the inter-plane distance is as low as 1.0 mm (Figs. 19.3, 19.5), resulting thus in an accuracy better than 1.0 mm (actually of ca. 0.5 mm). Another benefit of US imaging for the reconstruction of catheters is the possibility to combine 3D volume reconstruction with live 2D imaging, offering in this way the possibility of an interactive reconstruction.

19.3.3
Dose Calculation

Although several national protocols exist for the dose calculation around brachytherapy sources, the protocol proposed and established by the American Association of Physicists in Medicine (AAPM), Task Group 43, and published in 1995 (NATH et al. 1995), has been widely accepted and builds the standard protocol that the majority of vendors of treatment planning systems in brachytherapy are following. Even if this was primarily focused to low dose rate (LDR) sources (in the original publication it was explicitly mentioned that high activity sources and iridium wires were beyond the scope of that report), the TG 43 formalism has been widely used and virtually internationally accepted also for high dose rate (HDR) iridium sources used in remote afterloading systems

The TG 43 formalism is a consistent, and a simple to implement, formalism based on a small number of parameters/quantities that can be easily extracted from Monte Carlo (MC) calculated dose rate distributions around the sources in a water-equivalent medium.

The basic concept of the TG 43 dosimetry protocol is to derive dosimetry parameters for calculating dose rates or dose values directly from measured or MC calculated dose distributions around the sources in water or water-equivalent medium. This increases the accuracy of the calculations to be carried out in the clinic, which are always for water medium and not in free space. Furthermore, this method avoids the use of any term of activity (apparent or contained) that has led to significant discrepancies in the past.

19.3.3.1
The TG 43 Dosimetry Protocol

Figure 19.7 summarizes the geometry and coordinate definitions used in the TG 43 dosimetry protocol.

The dose rate $\dot{D}(r,\theta)$ at a point P around a source having cylindrical coordinates (r,θ) relative to the source coordinate system is given according to that protocol by:

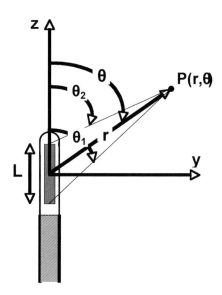

Fig. 19.7. The geometry and the definitions used for the TG 43 protocol. A source of an active length, L, the encapsulation geometry and the guidance wire is shown. This is the usual configuration of an HDR iridium source. The origin of the coordinate system is positioned at the centre of the active core of the source. The z-axis is along the tip of the source. A cylindrical symmetry for the activity distribution within the core is here assumed. The point of interest, P, is at a radial distance, r, from the origin and has a polar angle coordinate, θ, in the cylindrical coordinate system

$$\dot{D}(r,\theta) = S_K \cdot \Lambda \cdot \frac{G(r,\theta)}{G(r_0,\theta_0)} \cdot g(r) \cdot F(r,\theta) \tag{1}$$

where S_K is the air kerma strength of the source, Λ is the dose rate constant, $G(r,\theta)$ is the geometry function, $g(r)$ is the radial dose function and $F(r,\theta)$ is the anisotropy function.

Air Kerma Strength

The air kerma strength (S_k) replaces the previous commonly used quantity *apparent activity* A_{app} and describes the strength of the brachytherapy source. S_k is defined as the product of air kerma rate in free space at the distance of calibration of the source d $\dot{K}(d)$ and the square of that distance, d^2:

$$S_K = \dot{K}(d) \cdot d^2 \tag{2}$$

The calibration must be performed at a distance, d, defined along the transverse bisector of the source, $r=d$ and $\theta=\pi/2$ in Fig. 19.7, that is large enough so that the source can be considered as a point source. For direct measurements of $\dot{K}(d)$ using an in-air setup the possible attenuation of the radiation in air has to be considered. According to the above definition, S_k accounts also for the scattering and attenuation of the radiation which occurred in the source core and source encapsulation. The reference calibration distance for $\dot{K}(d)$ is common use is 1 m. TG 43 recommends as a unit for reporting the air kerma strength S_k of a source, $\mu Gy.m^2.h^{-1}$, and denotes this by the symbol U: $1\ U = 1\ \mu Gy.m^2.h^{-1} = 1\ cGy.cm^2.h^{-1}$

Dose Rate Constant

The dose rate constant Λ is defined according to:

$$\Lambda = \frac{\dot{D}(r_0,\theta_0)}{S_K} \tag{3}$$

where $r_0=1.0$ cm and $\theta_0=\pi/2$. For the definition of the polar coordinates r and θ see Fig. 7.

Geometry Function G(r,θ)

$G(r,\theta)$ is the geometry function describing the effect of the spatial distribution of the activity in the source volume on the dose distribution and is given by:

$$G(r,\theta) = G(\vec{r}) = \frac{\int\limits_{source} \frac{\rho(\vec{r}')}{|\vec{r}-\vec{r}'|^2} dV'}{\int\limits_{source} \rho(\vec{r}')dV'} \tag{4}$$

where $\rho(\mathbf{r}')$ is the activity per unit volume at a point $\mathbf{r}'$ inside the source and dV' is an infinitesimal volume element located at the same position. This factor reduces to:

$$G(r,q) = G(r) = \frac{1}{r^2} \text{ for a point source} \qquad (5)$$

$$G(r,\theta) = \frac{\theta_2 - \theta_1}{L \cdot r \cdot \sin\theta} \text{ for a finite line source} \qquad (6)$$

Here L is the active length of the source and the angles θ_1 and θ_2 are illustrated in Fig. 19.7.

Reference Point of Dose Calculations

The reference point is that for the formalism chosen to be the point lying on the transverse bisector of the source at a distance of 1 cm from its centre: expressed in polar coordinates as defined in Fig. 19.7, i.e. $(r_0, \theta_0) = (1\text{cm}, \pi/2)$.

Radial Dose Function

$g(r)$ is the radial dose function and is defined as:

$$g(r) = \left[\frac{G(r_0, \theta_0)}{G(r, \theta_0)} \right] \cdot \left[\frac{\dot{D}(r, \theta_0)}{\dot{D}(r_0, \theta_0)} \right] \qquad (7)$$

Anisotropy Function

Finally, $F(r,\theta)$ is the anisotropy function defined as:

$$F(r,\theta) = \left[\frac{G(r, \theta_0)}{G(r, \theta)} \right] \cdot \left[\frac{\dot{D}(r, \theta)}{\dot{D}(r, \theta_0)} \right] \qquad (8)$$

Anisotropy Factor

Because of the difficulty in determining the orientation of the implanted seeds, post-implant dosimetry for low dose rate permanent implants is based on the point source approximation using the anisotropy factor $\varphi_{an}(r)$:

$$\phi_{an}(r) = \left[\frac{1}{2 \cdot \dot{D}(r, \theta_0)} \right] \cdot \int_0^\pi \dot{D}(r, \theta) \cdot \sin\theta \cdot d\theta =$$

$$\left[\frac{1}{2 \cdot G(r, \theta_0)} \right] \cdot \int_0^\pi F(r, \theta) \cdot G(r, \theta) \cdot \sin\theta \cdot d\theta \qquad (9)$$

This is uncommon for the case of HDR iridium-192-based brachytherapy where the TG 43 formalism as given in Eq. (1) with the line source approximation described in Eq. (6) is used.

Anisotropy Constant

Using a $1/r^2$ weighted-average of anisotropy factors, for r>1 cm, the distance independent anisotropy factor φ_{an} is calculated using the equation:

$$\varphi_{an} = \frac{\sum_i \frac{\varphi_{an}(r_i)}{r_i^2}}{\sum_i \frac{1}{r_i^2}} \qquad (10)$$

In the literature the TG 43 parameter values and functions for the common used seeds or HDR iridium sources can be found.

Recently AAPM has updated the TG 43 protocol for low-activity seeds (RIVARD et al. 2004), where it is recommended to use separately radial dose functions $g(r)$ and anisotropy functions $F(r,\theta)$ as well as anisotropy factors $\varphi_{an}(r)$ and anisotropy constants φ_{an} in dependence on the geometry factor is used; point (see Eq. (5)) or line (see Eq. (6)) source approximation. This report contains the corresponding tables for all factors and functions for the most common seeds according to the new formulation.

Although the TG 43 formalism offers a stable platform for calculation of dose or dose rate distributions in brachytherapy, it can be easily seen from Eq. (1) that the TG 43 formalism is actually a 2D model.

Tissue inhomogeneities and bounded geometries are not considered by this formalism. The effects of the presence of inhomogeneities and the variable dimensions of patient-specific anatomy are ignored. The MC simulation would be the only accurate solution to the aforementioned deficiencies based on actual patient anatomical data. That is, however, still too time-consuming to be incorporated in a clinical environment in spite of promising correlated simulation techniques (HEDTJÄRN et al. 2002); therefore, kernel superposition methods (WILLIAMSON and BAKER 1991; CARLSSON and AHNESJÖ 2000; CARLSSON TEDGREN and AHNESJÖ 2003) and analytical models (WILLIAMSON et al. 1993; KIROV and WILLIAMSON 1997; DASKALOV et al. 1998) have been employed and tested in a variety of geometries, thus opening the way of handling tissue and shielding material inhomogeneities and bounded patient geometries.

A simpler analytical dosimetry model (ANAGNOSTOPOULOS et al. 2003) based on the primary and scatter separation technique (RUSSELL and AHNESJÖ 1996; WILLIAMSON 1996) was published and evaluated in patient-equivalent phantom geometries (PANTELIS et al. 2004; ANAGNOSTOPOULOS et al. 2004). The kernel superposition as well as the analytical dose calculation models announce in this way the future of 2.5 and real 3D dose calculation in brachytherapy.

In the following a short description of this recently proposed simple analytical dose calculation model that has been shown to describe adequately the dose distribution in inhomogeneous tissue environments is given.

19.3.3.2
The Analytical Dose Calculation Model

According to the analytical dose rate calculation formalism proposed in the work of Anagnostopoulos et al. (2003) the dose rate per unit air kerma strength, S_K, in a homogeneous tissue medium surrounding a real ^{192}Ir source can be calculated using the following equation (Anagnostopoulos et al. 2003; Pantelis et al. 2004):

$$\left[\frac{\dot{D}_{medium}(r,\theta)}{S_K} \right]_{real} = \tag{11}$$

$$\left(\frac{\mu_{en}}{\rho} \right)_{air}^{medium} e^{-\mu_{med}\cdot r}\left(1+SPR_{water}(\rho_{medium}r)\right)G(r,\theta)F(r,\theta)$$

where r is the radial distance, $\left(\mu_{en}/\rho\right)_{air}^{medium}$ is the effective mass energy absorption coefficient ratio of the medium of interest to air, μ_{medium} is the effective linear attenuation coefficient of the medium, SPR_{water} is the scatter to primary dose rate ratio calculated in water medium and μ_{medium} is the density of the medium.

The effective mass energy absorption coefficient ratio, $\left(\mu_{en}/\rho\right)_{air}^{medium}$ and the effective linear attenuation coefficient, μ_{medium}, are calculated by weighting over the primary ^{192}Ir photon spectrum, while the scatter to primary dose rate ratios for water medium (Russell and Ahnesjö 1996; Williamson 1996) is calculated using the polynomial fitted function (Anagnostopoulos et al. 2003; Pantelis et al. 2004):

$$SPR_{water}(\rho r)=0.123\,(\rho r)+0.005\,(\rho r)^2 \tag{12}$$

that can accurately calculate (within 1%) the SPR_{water} values for density-scaled distances of $\rho r \leq 10$ g cm^{-2}. For the general application of Eq. (11) for every homogeneous medium, changing from homogeneous water to a different homogeneous medium would necessitate MC calculated $SPR_{medium}(r)$ results thus reducing the versatility of an analytical model; how-

ever, due to the range of the ^{192}Ir energies and the consequent predominance of incoherent scattering (Anagnostopoulos et al. 2003), $SPR_{medium}(r)$ results for tissue materials are in good agreement with that of water when plotted vs distance scaled for the corresponding density (i.e. in units of grams per square centimetre). This is shown in Fig. 19.8 where MC calculated (Briesmeister 2000) $SPR_{bone}(r)$ ratios of cortical bone are also plotted vs distance from a point ^{192}Ir source multiplied by the corresponding material density (1.92 g/cm^3 for bone) and an overall good agreement within 1–5% may be observed (Anagnostopoulos et al. 2003).

In this equation, $G(r,\theta)$ is the geometry function of the source accounting for the spatial distribution of radioactivity. $F(r,\theta)$ is the anisotropy function accounting for the anisotropy of dose distribution around the source.

In order to account for the presence of different inhomogeneous materials along the path connecting the source and a dose point in a patient anatomy-equivalent phantom, Eq. (11) was generalized to:

$$\frac{\dot{D}_{medium}(r,\theta)}{S_K} = \tag{13}$$

$$\left(\frac{\mu_{en}}{\rho} \right)_{air}^{medium} e^{-\sum_i \mu_{medium\,i}\cdot r_i}\left(1+SPR_{water}(\rho_i,r_i)\right)G(r,\theta)F(r,\theta)$$

where i is the index of every material transversed along the connecting path of the source point to

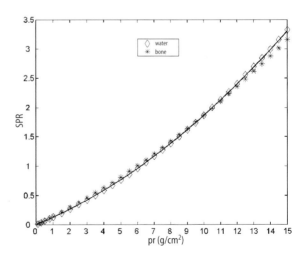

Fig. 19.8. Scatter to primary (*SPR*) dose rate ratio results for unbounded, homogenous water and bone phantoms calculated with MC simulations are plotted vs density-scaled distance, ρr, in units of grams per square centimetre. In the same figure a polynomial fit on water SPR values SPR(ρr)=0.123(ρr)+0.005 (ρr)^2 is also presented

the dose calculation point. The SPR in water for the scaled distance is parameterized according to Eq. (12), where ⌡r is the sum of the mass density scaled path lengths inside the inhomogeneities along the radius connecting the source with the dose calculation point.

The effect of patient inhomogeneities surrounding the oesophagus on the dosimetry planning of an upper thoracic oesophageal ^{192}Ir HDR brachytherapy treatment was studied (ANAGNOSTOPOULOS et al. 2004) and the analytical dose calculation model of Eq. (13) was found to correct for the presence of tissue inhomogeneities as it is evident in Fig. 19.9, where dose calculations with the analytical model are compared with corresponding results from the MCNPX Monte Carlo code (HENDRICKS et al. 2002) as well as with corresponding calculations by a contemporary treatment planning system software featuring a full TG-43 dose calculation algorithm (PLATO BPS v. 14.2.4, Nucletron B.V, The Netherlands) in terms of isodose contours. The presence of patient inhomogeneities had no effect on the delivery of the planned dose distribution to the PTV; however, regarding the OARs, the common practice of current treatment planning systems to consider the patient geometry as

a homogeneous water medium leads to a dose overestimation of up to 13% to the spinal cord and an underestimation of up to 15% to the bone of sternum (ANAGNOSTOPOULOS et al. 2004). These discrepancies correspond to the dose region of about 5–10% of the prescribed dose and are only significant in case that brachytherapy is used as a boost to external beam therapy.

19.3.4
Dose Optimization

The objectives of brachytherapy treatment planning are to deliver a sufficiently high dose in the cancerous tissue and to protect the surrounding normal tissue (NT) and OARs from excessive radiation. The problem is to determine the position and number of source dwell positions (SDPs), number of catheters and the dwell times, such that the obtained dose distribution is as close as possible to the desired dose distribution. Additionally, the stability of solutions can be considered with respect to possible movements of the SDPs. The planning includes clinical constraints such as a realistic range of catheters as

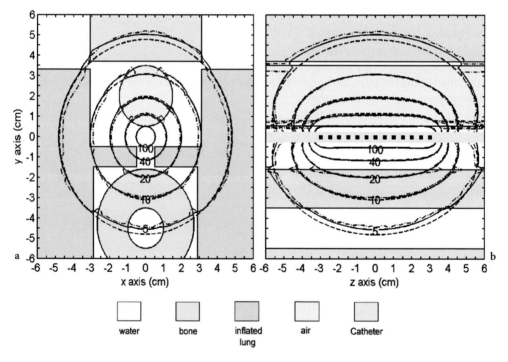

Fig. 19.9a,b. Percentage isodose contours calculated with the PLATO BPS v. 14.2.4 (- - -), the Monte Carlo (- · - · -) and the analytical model (—) in the inhomogeneous patient-equivalent geometry of an upper thoracic oesophageal ^{192}Ir HDR brachytherapy treatment (ANAGNOSTOPOULOS et al. 2004). The 100% isodose contour encompasses the cylindrical-shaped oesophageal PTV and is not altered due to the presence of the surrounding tissue inhomogeneities. The results in **a** are plotted on the central transversal plane (z=0 cm, adjacent to the central CT slice), whereas in **b** the same results are plotted on the sagittal plane containing the catheter inserted inside the oesophagus (x=0 cm)

well as their positions and orientations. The determination of an optimal number of catheters is a very important aspect of treatment planning, as a reduction of the number of catheters simplifies the treatment plan in terms of time and complexity. It also reduces the possibility of treatment errors and is less invasive for the patient.

As analytical solutions cannot be determined, the solution is obtained by inverse optimization or inverse planning. The term "inverse planning" is used considering this as the opposite of the forward problem, i.e. the determination of the dose distribution for a specific set of SDPs and dwell times. If the positions and number of catheters and the SDPs are given after the implantation of the catheters, we term the process "post-planning". Then, the optimization process to obtain an optimal dose distribution is called "dose optimization". Dose optimization can be considered as a special type of inverse optimization where the positions and number of catheters and the SDPs are fixed.

Inverse planning has to consider many objectives and is thus a multiobjective (MO) optimization problem (MIETTINEN 1999). We have a set of competing objectives. Increasing the dose in the PTV will increase the dose outside the PTV and in the OARs. A trade-off between the objectives exists as we never have a situation in which all the objectives can be in the best possible way satisfied simultaneously. One solution of this MO problem is to convert it into a specific single objective (SO) problem by combining the various objective functions with different weights into a single objective function. Optimization and analysis of the solutions are repeated with different sets of weights until a satisfactory solution is obtained as the optimal weights are a priori unknown. In MO optimization a representative set of all possible so-called non-dominated solutions is obtained and the best solution is selected from this set. The optimization and decision processes are decoupled. The set provides a coherent global view of the trade-offs between the objectives necessary to select the best possible solution, whereas the SO approach is a trial-and-error method in which optimization and decision processes are coupled.

19.3.4.1
Optimization Objectives

An ideal dose function $D(\mathbf{r})$ with a constant dose equal to the prescription dose, D_{ref}, inside the PTV and 0 outside is physically impossible since radiation cannot be confined to the PTV only as some part of the radiation has to traverse the OARs and the surrounding NT. Out of all possible dose distributions

the problem is to obtain an optimal dose distribution without any a priori knowledge of the physical restrictions. Optimality requires quantifying the quality of a dose distribution. A natural measure quantifying the similarity of a dose distribution at N sampling points with dose values, d_i, to the corresponding optimal dose values, d_i^*, is a distance measure. A common measure is the L_p norm:

$$L_p = \left(\sum_{i=1}^{N} (d_i - d_i^*)^p \right)^{\frac{1}{p}} \qquad (14)$$

For $p=2$, i.e. L_2 we have the Euclidean distance.

The treatment planning problem is transformed into an optimization problem by introducing as an objective the minimization of the distance between the ideal and the achievable dose distribution. These objectives can be expressed in general by the objective functions $f_L(\mathbf{x})$ and $f_H(\mathbf{x})$:

$$f_L(\mathbf{x}) = \frac{1}{N} \sum_{i=1}^{N} \Theta(D_L - d_i(\mathbf{x}))(D_L - d_i(\mathbf{x}))^p,$$

$$f_H(\mathbf{x}) = \frac{1}{N} \sum_{i=1}^{N} \Theta(d_i(\mathbf{x}) - D_H)(d_i(\mathbf{x}) - D_H)^p \qquad (15)$$

where $d_i(\mathbf{x})$ is the dose at the ith sampling point that depends on parameters $\mathbf{x}$ such as dwell times, p is a parameter defining the distance norm, N the number of sampling points, D_L and D_H the low and high dose limits; these are used if dose values above D_L and below D_H are to be ignored expressed by the step function $\Theta(\mathbf{x})$.

The difference between various dosimetric based objective functions is the norm used for defining the distance between the ideal and actual dose distribution, on how the violation is penalized and what dose normalization is applied. For $p=2$ we have the quadratic-type or variance-like set of objective functions. Specific objectives of this type were used by MILICKOVIC et al. (2002) including an objective for the dose distribution of sampling points on the PTV surface that results in an objective value that is correlated with the so-called conformity index used by LAHANAS et al. (1999) directly as an objective. The objective functions require that the SDPs are all inside the PTV. In the case of SDPs outside the PTV additional or modified objective functions are required. For $p=1$, a linear form, results were presented by LESSARD and POULIOT (2001). For $p=0$ (LAHANAS et al. 2003a) we have DVH-based objectives as the DVH value at the dose, D_H, is given by

$$DVH(D_H) = \frac{100}{N} \sum_{i=1}^{N} \Theta(d_i - D_H) \qquad (16)$$

$$c_1 = \frac{PTV_{ref}}{PTV} \qquad\qquad c_2 = \frac{PTV_{ref}}{V_{ref}}$$

$$c_3 = \prod_{i=1}^{N_{OAR}} \left(1 - \frac{V_{OAR}^i \ (D > D_{crit}^i)}{V_{OAR}^i}\right)$$

where the coefficient c_1 is the fraction of the PTV (PTV_{ref}) that receives dose values of at least D_{ref}. The coefficient c_2 is the fraction of the reference iso-dose volume V_{ref} that is within the PTV (see also Fig. 19.14).

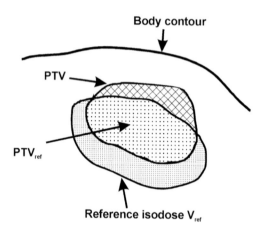

Fig. 19.14. Volumes necessary for computation of the conformal index COIN

V_{OAR}^i is the volume of the ith OAR and $V_{OAR}^i(D > D_{crit}^i)$ is the volume of the ith OAR that receives a dose that exceeds the critical dose value D_{crit}^i defined for that OAR. The product in the equation for c_3 is calculated for all N_{OAR} OARs included in the treatment planning.

In the case where an OAR receives a dose, D, above the critical value defined for that structure, the conformity index will be reduced by a fraction that is proportional to the volume that exceeds this limit. The ideal situation is COIN=c_1=c_2=c_3=1.

The COIN assumes in this form that the PTV, the OARs and the surrounding normal tissue are of the same importance.

When the COIN value is calculated for every dose value, D, according to Eq. (17), then the conformity distribution or the COIN histogram is calculated.

Figure 19.15 demonstrates the conformity distribution for the implant shown in Fig. 19.12 and for the solution having the maximum COIN value. A good implant is that where the maximum COIN value is observed exactly at or near the reference dose V_{ref} (100%).

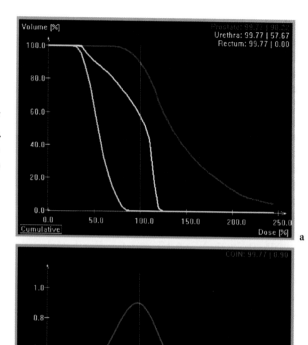

Fig. 19.15a,b. Evaluation results for the implant shown in Fig. 19.12. Here the solution with the highest maximum COIN value is selected. The critical dose values or dose limits for the OARs urethra und rectum were 125 and 85% of the reference dose (100%). **a** The cumulative DVHs for PTV, and the OARs urethra and rectum. **b** The conformity distribution (COIN distribution) calculated according to Eq. (17) and based on the above DVHs including both OARs demonstrates a COIN value of 0.90 for the 100% reference isodose line. The maximum COIN value is 0.91 and it is observed for the 98% isodose value that is very close to the reference dose for that implant and treatment plan selected

References

Anagnostopoulos G, Baltas D, Karaiskos P, Pantelis E, Papagiannis P, Sakelliou L (2003) An analytical model as a step towards accounting for inhomogeneities and bounded geometries in ^{192}Ir brachytherapy treatment planning. Phys Med Biol 48:1625–1647

Anagnostopoulos G, Baltas D, Pantelis E, Papagiannis P, Sakelliou L (2004) The effect of patient inhomogeneities in oesophageal ^{192}Ir HDR brachytherapy: a Monte Carlo and analytical dosimetry study. Phys Med Biol 49:2675–2685

Anderson LA (1986) "Natural" volume-dose histogramm for brachytherapy. Med Phys 13:898–903

Ash D, Flynn A, Battermann J, de Reijke T, Lavagnini P, Blank

L (2000) ESTRO/EAU/EORTC recommendations on permanent seed implantation for localized prostate cancer. Radiother Oncol 57:315–321

Baltas D, Lapukins Z, Dotzel W, Herbort K, Voith C, Lietz C, Filbry E, Kober B (1994) Review of treatment of oesophageal cancer with special reference to Darmstadt experience. In: Bruggmoser G, Mould RF (eds) Brachytherapy review, Freiburg Oncology Series monograph no. 1. Proc German Brachytherapy Conference, Freiburg, pp 244–270

Baltas D, Tsalpatouros A, Kolotas C, Ioannidis G, Geramani K, Koutsouris D, Uzunoglou N, Zamboglou N (1997) PROMETHEUS: Three-dimensional CT based software or localization and reconstruction in HDR brachytherapy. In: Zamboglou N, Flentje M (eds) Onlologische Seminare lokoregionaler Therapie, no. 2. New developments in interstitial remote controlled brachytherapy, pp 60–76

Baltas D, Kolotas C, Geramani K, Mould RF, Ioannidis G, Kekchidi M, Zamboglou N (1998) A conformal index (COIN) to evaluate implant quality and dose specification in brachytherapy. Int J Radiat Oncol Biol Phys 40:515–525

Baltas D, Kneschaurek P, Krieger H (1999) DGMP Report no. 14 Leitlinien in der Radioonkologie P3, Dosisspezifikation in der HDR-Brachytherapie

Baltas D, Milickovic N, Giannouli S, Lahanas M, Kolotas C, Zamboglou N (2000) New tools of brachytherapy based on three-dimensional imaging. Front Radiat Ther Oncol 34:59–70

Briesmeister JF (ed) (2000) MCNP^{TM} – a general Monte Carlo N-particle transport code: version 4C report LA-13709-M. Los Alamos National Laboratory, Los Alamos, New Mexico

Carlsson AK, Ahnesjö A (2000) Point kernels and superposition methods for scatter dose calculations in brachytherapy. Phys Med Biol 45:357–382

Carlsson Tedgren AK, Ahnesjö A (2003) Accounting for high Z shields in brachytherapy using collapsed cone superposition for scatter dose calculation. Med Phys 30:2206–2217

Daskalov GM, Kirov AS, Williamson JF (1998) Analytical approach to heterogeneity correction factor calculation for brachytherapy. Med Phys 25:722–735

Giannouli S, Baltas D, Milickovic N, Lahanas M, Kolotas C, Zamboglou N, Uzunoglou N (2000) Autoactivation of source dwell positions for HDR brachytherapy treatment planning. Med Phys 27:2517–2520

Hedtjärn H, Carlsson GA, Williamson JF (2002) Accelerated Monte Carlo based dose calculations for brachytherapy planning using correlated sampling Phys Med Biol 47:351–376

Hendricks JS, McKinney GW, Waters LS, Hughes HG (2002) MCNPX user's manual, version 2.4.0, Report LA CP 02-408. Los Alamos National Laboratory, Los Alamos, New Mexico

Herbort K, Baltas D, Kober B (1993) Documentation and reporting of brachytherapy. Brachytherapy in Germany. Proc German Brachytherapy Conference, Köln, pp 226–247

ICRU report 38 (1985) Dose and volume specification for reporting intracavitary therapy in gynaecology. International Commission on Radiation Units and Measurements

ICRU report 58 (1997) Dose and volume specification for reporting interstitial therapy. International Commission on Radiation Units and Measurements

Kirov AS, Williamson JF (1997) Two-dimensional scatter integration method for brachytherapy dose calculations in 3D geometry. Phys Med Biol 42:2119–2135

Kolotas C, Baltas D, Zamboglou N (1999a) CT-Based Interstitial HDR brachytherapy. Strahlenther Onkol 9:175:419–427

Kolotas C, Birn G, Baltas D, Rogge B, Ulrich P, Zamboglou N (1999b) CT guided interstitial high dose rate brachytherapy for recurrent malignant gliomas. Br J Radiol 72:805–808

Kolotas C, Baltas D, Strassmann G, Martin T, Hey-Koch S, Milickovic N, Zamboglou N (2000) Interstitial HDR brachytherapy using the same philosophy as in external beam therapy: analysis of 205 CT guided implants. J Brachyther Int 16:103–109

Krieger H, Baltas D, Kneschaurek B (1996) Vorschlag zur Dosis- und Volumenspezifikation und Dokumentation in der HDR-Brachytherapie. Strahlenther Onkol 172:527–542

Lahanas M, Baltas D (2003) Are dose calculations during dose optimization in brachytherapy necessary? Med Phys 30:2368–2375

Lahanas M, Baltas D, Zamboglou N (1999) Anatomy-based three-dimensional dose optimization in brachytherapy using multiobjective genetic algorithms. Med Phys 26:1904–1918

Lahanas M, Baltas D, Zamboglou N (2003a) A hybrid evolutionary algorithm for multi-objective anatomy-based dose optimization in high-dose-rate brachytherapy. Phys Med Biol 48:399–415

Lahanas M, Baltas D, Giannouli S (2003b) Global convergence analysis of fast multiobjective gradient based dose optimization algorithms for high-dose-rate brachytherapy. Phys Med Biol 48:599–617

Lessard E, Pouliot J (2001) Inverse planning anatomy-based dose optimisation for HDR-brachytherapy of the prostate using fast simulated annealing and dedicated objective functions. Med Phys 28:773–779

Miettinen K M (1999) Nonlinear multiobjective optimisation. Kluwer, Boston

Milickovic N, Giannouli S, Baltas D, Lahanas M, Kolotas C, Zamboglou N, Uzunoglu N (2000a) Catheter autoreconstruction in computed tomography based brachytherapy treatment planning. Med Phys 27:1047–1057

Milickovic N, Baltas D, Giannouli S, Lahanas M, Zamboglou N (2000b) CT imaging based digitally reconstructed radiographs and its application in brachytherapy. Phys Med Biol 45:2787–2800

Milickovic N, Lahanas M, Papagiannopoulou M, Zamboglou N, Baltas D (2002) Multiobjective anatomy-based dose optimisation for HDR-brachytherapy with constraint free deterministic algorithms. Phys Med Biol 47:2263–2280

Nag S, Beyer D, Friedland J, Grimm P, Nath R (1999) American Brachytherapy Society (ABS) Recommendations for Transperineal Permanent Brachytherapy of Prostate Cancer. Int J Radiat Oncol Biol Phys 44:789–799

Nag S, Cano E, Demanes J, Puthawala A, Vikram B (2001) The American Brachytherapy Society recommendations for high-dose-rate brachytherapy for head-and-neck carcinoma. Int J Radiat Oncol Biol Phys 50:1190–1198

Nath R, Anderson L, Luxton G, Weaver K, Williamson JF, Meigooni AS (1995) Dosimetry of interstitial brachytherapy sources: recommendations of the AAPM Radiation Therapy Committee Task Group 43. Med Phys 22:209–234

Pantelis E, Papagiannis P, Anagnostopoulos G, Baltas D, Karaiskos P, Sandilos P, Sakelliou L (2004) Evaluation of a TG-43

20 Beam Delivery in 3D Conformal Radiotherapy Using Multi-Leaf Collimators

W. Schlegel, K.H. Grosser, P. Häring, and B. Rhein

CONTENTS

20.1
Conformal Treatment Techniques

Conformational radiation therapy (CRT) was introduced in the early 1960s by radiation oncologist S. Takahashi, who came up with many ideas of how to concentrate the dose to the target volume using various forms of axial transverse tomography and rotating multi-leaf collimators (MLC; Takahashi 1965).

Three-dimensional conformal radiotherapy (3D CRT) is an extension of CRT by the inclusion of 3D treatment planning and can be considered to be one of the most important advances in treating patients with malignant disease. It is performed in nearly all modern radiotherapy units.

The goal of 3D CRT is the delivery of a high radiation dose which is precisely conformed to the target volume while keeping normal tissue complications at a minimum.

W. Schlegel, PhD; P. Häring, PhD; B. Rhein, PhD
Abteilung Medizinische Physik in der Strahlentherapie,
Deutsches Krebsforschungszentrum, Im Neuenheimer Feld
280, 69120 Heidelberg, Germany
K. H. Grosser, PhD
Radiologische Klinik der Universität Heidelberg,
Abteilung Strahlentherapie, Im Neuenheimer Feld 400,
59120 Heidelberg, Germany

The preconditions which have to be fulfilled in order to achieve conformal dose distributions are discussed in the preceding chapters (Chaps. 2–12) on imaging and treatment planning (Chaps. 13–19). In summary, it can be said that first of all detailed diagnostic imaging information has to be available from a variety of sources including conventional X-ray imaging, CT, MRI and PET in order to be able to define the target volume and the organs at risk with sufficient accuracy. Furthermore, a 3D computerized treatment planning system and an exact and reproducible patient positioning system have to be used. If these boundary conditions are fulfilled, conformal irradiation techniques can be used optimally.

In general, the attainable dose conformity in conventional conformal radiation therapy depends on the boundary conditions described in Table 20.1. As is seen from this table, there are many approaches to conformal therapy using sophisticated irradiation techniques with multiple isocentric beam irradiations, irregularly shaped fields (either using cerrobend blocks or MLCs), and computer-controlled dynamic techniques).

An important step in 3D CRT was the introduction of irregularly shaped irradiation fields, made of metal blocks from alloys with low melting points (in radiotherapy often called "cerrobend" blocks). Individually shaped irregular fields realized by cerrobend blocking turned out to be time-consuming and expensive; therefore, great progress in conformal radiotherapy was achieved by the development and application of MLCs. The dose distributions achieved with MLCs turned out to be equivalent to conformal blocks; however, the cost of conformal radiation therapy could be minimized and the flexibility significantly enhanced by using the new MLC technology (Adams et al. 1999; Foroudi et al. 2000).

The MLCs are beam-shaping devices that consist of two opposing banks of attenuating leaves, each of which can be positioned independently. The leaves can either be moved manually or driven by motors to such positions that, seen from the "beam's eye view" of the irradiation source, the collimator opening corresponds to the shape of the tumor (Fig. 20.1). The leaf settings are usually defined within the virtual

Table 20.1 Irradiation techniques for conformal radiotherapy. *MLC* multi-leaf collimator

Physical parameter	Impact on conformity	Scorecard	Drawbacks
No. of beam incidents	Better conformity can be achieved with a higher number of irradiating directions	++	Higher complexity, longer planning time, normal tissue dose becomes possibly larger, longer irradiation time
Optimization of beam directions	Higher conformity possible	++	Higher complexity, longer planning time; in case of non-coplanar beam directions: longer treatment time
Optimization of beam energy (photons)	Higher conformity possible	+	
Application of an MLC	Higher conformity possible	+++	
Width of the MLC leaves	Small leaf width enables a better field adjustment and therefore better conformity	++	Higher complexity, longer planning time
More than one target point at the same time	In cases where more than one target volume is treated simultaneously, conformity might be higher	+	Higher complexity, longer planning time, sometimes a lower homogeneity
Moving beam irradiations; computer-controlled dynamic radiotherapy	Depends on shape of target volume	+	Longer irradiation time, complex verification and quality assurance

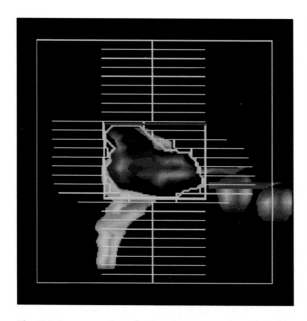

Fig. 20.1 Beams eye view of a planning target volume (PTV) in the brain, together with organs at risk (*green:* brain stem; *red:* eyes; *blue:* optic nerves) and MLC setting (*yellow*)

20.2
Multi-Leaf Collimators

Multi-leaf collimators permit the quick and flexible adjustment of the irradiation fields to the tumor shape and the shape of the organs at risk. Though already proposed by TAKAHASHI in 1960, it took nearly 25 years before the first commercial computer controlled MLCs appeared on the market. This was due to the fact that MLCs are mechanical devices with high mechanical complexity, and they have to fulfill very rigid technical, dosimetric, and safety constraints. Detailed reviews of the history and performance of MLCs for 3D CRT are given by WEBB (1993, 1997, 2000). The use of MLCs for static or dynamic IMRT is discussed in more detail in another work by WEBB (2005).

This chapter describes briefly the general design and performance of the MLCs as they are currently being used in routine applications for 3D CRT.

20.2.1
Geometrical and Mechanical Properties

The most important technical parameters (Fig. 20.2) which characterize the performance of an MLC are mechanical and geometrical properties such as:
1. The maximum field size
2. The leaf width
3. Maximum overtravel

therapy simulation program of 3D treatment planning (see Chap. 14; BOESECKE et al. 1988, 1991; ÉSIK et al. 1991).

There were different other designs of MLCs with up to six leaf banks (TOPOLNJAK et al. 2004), which, however, up to now have not played an important role in the clinical practice of 3D CRT.

4. Interdigitation
5. Configuration of the MLC with respect to the collimator jaws

For MLCs which are used for IMRT, other important parameters are also the minimum and maximum leaf speed and the precision of leaf positioning. Other aspects are of course the complexity of calibration and the overall efforts for maintenance.

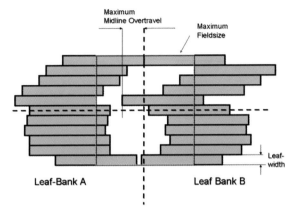

Fig. 20.2 Multi-leaf collimator with the most relevant mechanical parameters

Maximum Field Size

Two kinds of MLCs are employed presently: those for medium-sized and large fields of up to 40×40 cm², which are implemented in the gantry of linacs; and "add-on-MLCs" for small field sizes (often called mini- or micro-MLCs) which can be attached to the accessory holder of the treatment head, and, for example, used in conjunction with stereotactic conformal radiotherapy. Mini- and micro-MLCs have characteristic maximum field sizes of about 10×10 cm². Maximum field sizes depend for some MLCs (see Tables 20.2, 20.3) from the maximum overtravel: when the maximum overtravel is used, maximum field size will become smaller, because the whole leaf banks have to be shifted in order to achieve complete overtravel.

Leaf Width

MLCs integrated into the linac head. Computer-controlled MLCs integrated into the head of the accelerator usually have a spatial resolution of 0.5–1 cm in the isocenter plane, perpendicular to the leaf-motion direction, and a positioning accuracy in the range of 1 mm in the direction of the motion.

The leaf width (measured in the isocenter plane) should be adapted to the size and complexity of the target volumes. Maybe an effective leaf width of 10 mm is completely sufficient in case of prostate cancer; however, in the case of a small target volume located around the spinal cord, 10 mm is too large! A leaf width of 5 mm is presently considered to be a good compromise.

Table 20.2 Commercial integrated MLCs

Manufacturer	Product name	Leaf width at isocenter (mm)	Midline over-travel (cm)	No. of leaves	Maximum field size (cm²)	Focusing properties	Remarks
Elekta-1	Integrated MLC	10	12.5	40×2	40×40	Single focusing	
Elekta-2	Beam modulator	4	11	40×2	16×22	Single focusing	
Siemens-1	3D MLC	10	10	29×2	40×40[b]	Double focusing	
Siemens-2	Optifocus	10	10	41×2	40×40	Double focusing	
Siemens-3	160 MLC	5	20[a]	80×2	40×40	Single focusing	Announced for 2006
Varian-1	Millennium MLC-52	10	20[a]	26×2	26×40	Single focusing	
Varian-2	Millennium MLC-80	10	20[a]	40×2	40×40	Single focusing	
Varian-3	Millennium MLC-120	Central 20 cm of field: 5 mm; outer 20 cm of field: 10 mm	20[a]	60×2	40×40	Single focusing	

[a]Requires movement of the complete leaf bank and leads to reduced maximum field sizes

[b]The Siemens 3D MLC consists of 2×27 inner leaves with 1-cm leaf width and two outer leaves with 6.5-cm leaf width

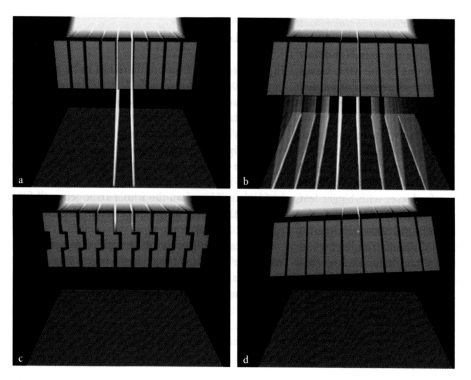

Fig. 20.10a-d. Various MLC concepts to prevent leakage **a** unfocused without correction, **b** focused without correction, **c** with tongue and groove, and **d** with flipped collimator

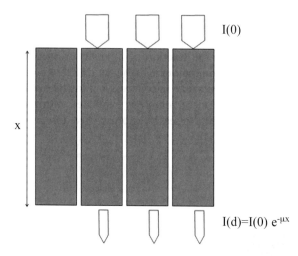

Fig. 20.11 Leaf transmission for leaves with height X and transmission coefficient

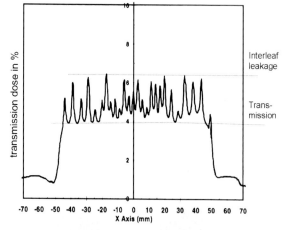

Fig. 20.12 Dose profile measured with dosimetric film under a completely closed accessory-type MLC. Leaf transmission and interleaf leakage can be detected

At a rough guess, the transmitted intensity is twice the original physical level. If the transmission is, for example, 2%, the maximum of the transmitted intensity mounts up to 4% at some positions. Especially for MLCs used in IMRT, transmission as well as interleaf leakage should therefore be kept as low as possible.

Restrictions for leakage radiation of MLCs are given in IEC (1998): If the MLC is covered by rectangular jaws, which are automatically adjusted to the MLC shape, leakage radiation must be below 5% of an open 10×10-cm² field; otherwise, maximum leakage should be less than 2% and average leakage less than 0.5%.

20.2.3
Operating Modes

There are in principle two different operating modes for MLCs for 3D CRT, depending on whether the leaves are moving when the beam is on (dynamic mode) or the beam is shut off (static mode).

Although modern MLCs in principle have dynamic properties, they are still commonly applied in static treatment techniques. The real potential of MLCs in 3D CRT will be demonstrated in the future, when, for example, dynamic arc treatment techniques with MLCs will be more widely accepted and implemented to perform dynamic field shaping.

The two modes (static mode=step-and-shoot technique; dynamic mode=sliding-window technique) currently play a bigger role in IMRT delivery (see Chap. 23). In IMRT dynamic delivery is in many cases more time efficient than step and shoot, but the step-and-shoot technique is less complex and quality assurance may be easier to perform. Chui et al. (2001) compared the delivery of IMRT by dynamic and static techniques.

20.3
Commercial MLCs

20.3.1
Linac-Integrated MLCs

The major manufacturers of commercial MLCs integrated into the irradiation head of a linear accelerator are the companies Elekta, Siemens, and Varian. Galvin (1999) provides a useful review with tables of MLC properties. Huq et al. (2002) compared the MLCs of all three manufacturers using precisely the same criteria and experimental methods. The result of this investigation was that there is no clear superiority of one vendor compared with the others: the different designs and configurations of the MLCs are leading to a balance of advantages and disadvantages and there was no clear superiority of one MLC compared with the others. An updated list of MLC characteristics is given in Table 20.2.

20.3.2
Accessory-Type MLCs

There are a couple of companies manufacturing and distributing accessory MLCs, which are especially suited to treat small target volumes in conjunction with stereotactic irradiation techniques. Bortfeld et al. (1999) have provided an overview on the characteristics and performances of these mini- and micro-MLCs. The specifications of these add-on collimators are summarized in Table 20.3. In summary, MLCs of this type are closer positioned to the patient's surface and have smaller leaf widths. As a consequence, mini- and micro-MLCs have much smaller penumbras and produce dose distributions with higher conformity; however, they are restricted to the treatment of small target volumes.

20.4
The Limits of Conventional Conformal Radiation Therapy

Complex-shaped target volumes close to radio-sensitive organs remain a challenge for the radiotherapist. With MLCs and conventional irradiation techniques conformal and homogeneous dose distributions cannot be obtained in all cases. This is particularly true for concave-shaped target volumes. In Chaps. 17 and 23 (Inverse planning and IMRT) of this book it is shown that as the result of a superposition of several irradiation segments with homogeneous intensity, a concave-shaped dose distribution is produced. The MLCs to produce these IMRT field segments have to fulfill very special criteria concerning leaf speed, accuracy, and reproducibility of leaf positioning, overtravel, transmission, and leakage. The specific requirements of MLCs in IMRT are discussed and analyzed in much more detail by Schlegel and Mahr (2000) and Webb (2005).

References

Adams EJ, Cosgrove VP, Shepherd SF, Warrington AP, Bedford JL, Mubata CD, Bidmead AM, Brada M (1999) Comparison of a multi-leaf collimator with conformal blocks for the delivery of stereotactically guided conformal radiotherapy. Radiother Oncol 51:205–209

Boesecke R, Doll J, Bauer B, Schlegel W, Pastyr O, Lorenz WJ (1988) Treatment planning for conformation therapy using a multi-leaf collimator. Strahlenther Onkol 164:151–154

Boesecke R, Becker G, Alandt K, Pastyr O, Doll J, Schlegel W, Lorenz WJ (1991) Modification of three-dimensional treatment planning system for the use of multi-leaf collimators in conformation radiotherapy. Radiother Oncol 21:261–268

Bortfeld T, Schlegel W, Höver KH (1999) Mini- and micro-multileaf collimators. Med Phys 26:1094

Bortfeld T, Oelfke U, Nill S (2000) What is the optimum leaf width of a multileaf collimator? Med Phys 27:2494–2502

Butson MJ, Yu PK, Cheung T (2003) Rounded end multi-leaf penumbral measurements with radiochromic film. Phys Med Biol 48:N247–N252

Chui CS, Chan MF, Yorke E, Spirou S, Ling CC (2001) Delivery of intensity modulated radiation therapy with a conventional multileaf collimator: comparison of dynamic and segmental methods. Med Phys 28 2441–2449

Debus J, Engenhart-Cabillic R, Holz FG, Pastyr O, Rhein B, Bortfeld T (1997) Stereotactic precision radiotherapy in the treatment of intraocular malignancies with a micro-multileaf collimator. Front Radiat Ther Oncol 30:39–46

Ésik O, Buerkelbach J, Boesecke R, Schlegel W, Németh G, Lorenz WJ (1991) Three-dimensional photon radiotherapy planning for laryngeal and hypopharyngeal cancers. 2. Conformation treatment planning using a multileaf collimator. Radiother Oncol 20:238–244

Foroudi F, Lapsley H, Manderson C, Yeghiaian-Alvandi R (2000) Cost-minimization analysis: radiation treatment with and without a multi-leaf collimator. Int J Radiat Oncol Biol Phys 47:1443–1448

Föller M, Bortfeld T, Hädinger U, Häring P, Schulze C, Seeber S, Richter J, Schlegel W (1998) Beam shaping with a computer controllled micro-multileaf-collimator. Radiother Oncol 48 (Suppl 1):S76

Galvin JM (1999) The multileaf collimator: a complete guide. Proc AAPM Annual Meeting

Hartmann GH, Föhlisch F (2002) Dosimetric characterization of a new miniature multileaf collimator. Phys Med Biol 47: N171–N177

Hartmann GH, Pastyr O, Echner G, Hädinger U, Seeber S, Juschka J, Richter J, Schlegel W (2002) A new design for a mid-sized multileaf collimator (MLC). Radiother Oncol 64:S214

Huq MS, Das IJ, Steinberg T, Galvin JM (2002) A dosimetric comparison of various multileaf collimators. Phys Med Biol 47 N159–N170

IEC (1998) IEC 601-2-1, 2nd edn

Nill S, Tücking T, Münter M, Oelfke U (2002) IMRT with MLCs of different leaf widths: a comparison of achievable dose distributions. Radiother Oncol 64 (Suppl 1):107–108

Pastyr O, Echner G, Hartmann GH, Richter J, Schlegel W (2001) Dynamic edge focussing: a new MLC-design to deliver IMRT with a double focussing high precision multi-leaf collimator. Radiother Oncol 61:S24

Schlegel W, Pastyr O, Bortfeld T, Becker G, Schad L, Gademann G, Lorenz WJ (1992) Computer systems and mechanical tools for stereotactically guided conformation therapy with linear accelerators. Int J Radiat Oncol Biol Phys 24:781–787

Schlegel W, Pastyr O, Bortfeld T, Gademann G, Menke M, Maier Borst W (1993) Stereotactically guided fractionated radiotherapy: technical aspects. Radiother Oncol 29:197–204

Schlegel W, Pastyr O, Kubesch R, Stein J, Diemer T, Höver KH, Rhein B (1997) A computer controlled micro-multileaf-collimator for stereotactic conformal radiotherapy. In: Leavitt DD (ed) Proc XIIth Int Conf on the Use of Computers in Radiation Therapy. Medical Physics Publishing, Madison, Wisconsin, pp 79–82

Takahashi S (1965) Conformation radiotherapy: rotation techniques as applied to radiography and radiotherapy of cancer. Acta Radiol (Suppl 242):1–142

Topolnjak R, van der Heide UA, Raaymakers BW, Kotte AN, Welleweerd J, Lagendijk JJ (2004) A six-bank multi-leaf system for high precision shaping of large fields. Phys Med Biol 49:2645–2656

Webb S (ed) (1993) The physics of three-dimensional radiation therapy: conformal radiotherapy, radiosurgery and treatment planning. IOP Publishing, Bristol, UK

Webb S (ed) (1997) The physics of conformal radiotherapy: advances in technology IOP Publishing, Bristol, UK

Webb S (ed) (2000) Intensity modulated radiation therapy. IOP Publishing, Bristol, UK

Webb S (ed) (2005) Intensity modulated radiation therapy. Developing physics and clinical implementation. IOP Publishing, Bristol, UK.

21 Stereotactic Radiotherapy/Radiosurgery

Anca-Ligia Grosu, Peter Kneschaurek, and Wolfgang Schlegel

CONTENTS

21.1
Introduction

Stereotactic radiotherapy dates back more than 50 years; however, this form of treatment has entered the domain of radiation oncology only in the past 10–15 years. Initially an exotic technique, stereotactic radiotherapy has become an established, widespread treatment approach, characterized by the delivery of the irradiation with a very high geometrical precision. The method is especially used for benign and malignant cranial tumors but has lately been adapted to the body.

Stereotactic coordinates are used in the methodology of different irradiation techniques: implantation of seeds; Bragg-Peak irradiation; irradiation with gamma knife (Co60) or with linear accelerator (LINAC or X-Knife). This chapter focuses on the description of the technique of stereotactic tele-therapy (percutaneous stereotactic radiotherapy) with LINAC, also known as stereotactic convergent beam irradiation. This is the most widespread irradiation technique using stereotactic coordinates.

21.2
Definition

Stereotaxy (*stereo* + *taxis* – Greek, orientation in space) is a method which defines a point in the patient's body by using an external three-dimensional coordinate system which is rigidly attached to the patient. Stereotactic radiotherapy uses this technique to position a target reference point, defined in the tumor, in the isocenter of the radiation machine (LINAC, gamma knife, etc.) with a high accuracy. This results in a highly precise delivery of the radiation dose to an exactly defined target (tumor) volume. The aim is to encompass the target volume in the high-dose area and, by means of a steep dose gradient, to spare the surrounding normal tissue. This allows the definition of the planning target volume (PTV) without, or with only a very small (1–2 mm), safety margin around the gross tumor volume (GTV) or the clinical target volume (CTV). Stereotactic radiotherapy is performed in various treatment techniques, including gamma-knife units using 60 Co photons, heavy charged particles, or neutron beams units and modified LINAC units.

Treatment can be administered in a single fraction (radiosurgery, RS) or as multiple fractions (stereotactic fractionated radiotherapy, SFR). Some authors use the term stereotactic radiotherapy for the fraction-

A.-L. Grosu MD; P. Kneschaurek, PhD
Department of Radiation Oncology, Klinikum rechts der Isar, Technical University, Ismaningerstrasse 22, 81675 Munich, Germany
W. Schlegel, PhD
Deutsches Krebsforschungszentrum, Im Neuenheimer Feld 280, 69121 Heidelberg, Germany

ated delivery of the radiation; however, we consider it as a general term for all the irradiation treatments which are characterized by the integration of stereo-tactic coordinates in the treatment planning and delivery, including the RS and the SFR.

The intention of RS is to produce enough cell kill within the target volume in a single fraction in order to eradicate the tumor. This single high irradiation dose can produce considerable side effects in normal tissue located close to the tumor or within the target volume. The SFR combines the precision of target localization and dose application of RS with the radiobiological advantage of fractionated radiotherapy, i.e., breaking the total dose into smaller parts and thus allowing repair of DNA damage in normal tissue during the time between fractions. Time intervals of more than 6 h between fractions can significantly reduce the risk of side effects in normal tissue (HALL and BRENNER 1993; SCHLEGEL et al. 1993).

21.3
Historical Background

The first one to combine stereotactic methodology – which up to that point was used only in neurosurgery – with radiation therapy was the Swedish neurosurgeon Lars Leksell. He had the idea that X-ray could be used to treat certain neurological functional diseases instead of stereotactically guided needles used in neurosurgery. Leksel performed the first treatment in 1951, at the Karolinska Institute, and called the new therapy approach radiosurgery (RS). The patient was fixed in a stereotactic ring and irradiated with a precisely guided roentgen tube using 200-kV Röntgenbremsstrahlung (LEKSELL 1951). Bragg-peak studies with protons were begun in Uppsala, Boston, and Berkeley (KJELLBERG et al. 1968; LARSSON et al. 1958; LAWRENCE et al. 1962). In Berkeley Bragg-peak RS using helium ion beams was also developed (LYMAN et al. 1977). Leksel continued his work and built the first isotope radiation machine, in 1968, the gamma knife.

The gamma-knife unit consists of 201 Co-60 sources arranged on the surface of a hemispherical shell, each aiming at an isocenter in a uniform distance of 40 cm. Each source is collimated by a fixed primary collimator and subsequently by a helmet, which consists of 201 collimators to which the patient and frame are attached and which is docked with the primary collimator array at the time of treatment. The helmets define an approximately spherical dose distribution at the isocenter with nominal diameters

of 4, 8, 14, or 18 mm. The physical point of interest inside the patient is positioned at the isocenter of this distribution before treatment, and, if required, a sequence of treatment at different isocenters, with possibly different diameter helmets, are used to produce a conformal dose distribution. Multiple isocenters treatments are required when the target shape is irregular. By combining several spherical dose distributions of smaller diameter in the appropriate location, it is possible to create a sum of isodose distributions similar to the target volume. As a result the dose to the target itself is inhomogeneous, with larger number of isocenters leading to higher mean target doses. To date, the radiobiological consequences of dose inhomogeneity in tumor tissue cannot be evaluated yet, but the advantages of dose escalation in the tumor and dose reduction in the normal tissue are undisputed (LINDQUIST et al. 1995; VERHEY and SMITH 1995).

The stereotactic radiation therapy with LINAC started in the early 1980s: the Swedish physicist Larsson proposed to use the LINAC instead Co 60 or protons (LARSSON et al. 1974). The first published reports on clinical use of LINAC came from Buenos Aires (BETTI and DERECHINSKY 1983), Heidelberg (HARTMANN et al. 1985), and Vicenza (COLOMBO et al. 1985), the authors of which all had developed the concept to deliver a single-dose radiation with convergent non-coplanar dynamic irradiation. Many clinical investigators made use of this technology all over the world in a very short time, showing, in comparison with the gamma-knife RS, comparable clinical results (ENGENHART et al. 1989, 1992; MEHTA et al. 1995; BECKER et al. 1996; DEBUS et al. 1999). Further progress was made in LINAC stereotactic radiotherapy due to the development of non-invasive, replaceable head-fixation systems, which allow the implementation of the dose fractionation (SCHLEGEL et al. 1992, 1993; STÄRK et al. 1997). Another important development is the micro-multileaf collimator which permits the adoption of the irradiated area to irregular-shaped target volumes (SCHLEGEL et al. 1997) and the extension of the stereotactic irradiation technique to the head and neck (GADEMANN et al. 1993) and the body (HERFARTH et al. 2000).

21.4
Stereotactic Coordinates

The key idea of all stereotactic radiation therapy techniques is the use of stereotactic coordinates to define the volume which has to be irradiated three dimensionally; therefore, target volume and anatomical structures are localized in the space defined by the stereotactic coordinates system. Computed tomography (CT) or magnetic resonance imaging (MRI) are used for defining the anatomical structures which are of crucial importance for radiation therapy. The stereotactic localizer which is imaged simultaneously allows transfer of the image coordinates into the stereotactic coordinates system. During the planning of the radiation therapy a target point in the stereotactic space is defined. Before onset of radiation the patient is positioned in such a way that this target point is placed in the isocenter of the radiation machine with the help of the stereotactic positioning system.

In general, the stereotactic coordinates are a cartesian three-dimensional coordinates system attached to the stereotactic frame in a rigid relationship. The origin of the stereotactic coordinates system is generally in the center of the volume defined by the stereotactic frame: the x and y axes correspond to the lateral and frontal side of the frame and the z axis to the cranio-caudal direction (Fig. 21.1).

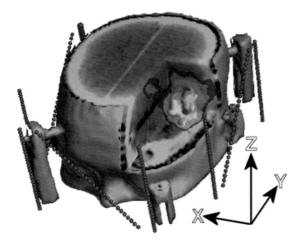

Fig. 21.1 The stereotactic coordinates define three-dimensionally the space which has to be irradiated

21.5
Method of Stereotactic Irradiation Treatment

The main steps in the planning and delivering of stereotactic irradiation treatment are:
1. Rigid application of the stereotactic frame to the patient
2. Imaging (CT, MRI, angiography) of the patient with the frame and localizer attached to the frame

3. Treatment planning
4. Positioning of the patient for the stereotactic radiation therapy
5. Delivery of the irradiation
6. Quality assurance

21.5.1
Stereotactic Frame

Stereotactic radiotherapy is based on the rigid connection of the stereotactic frame to the patient during CT, MRI, and angiography imaging (Figs. 21.2, 21.3). The stereotactic frame is the base for the fixation of the other stereotactic elements (localizer and positioner) and for the definition of the origin (point 0) of

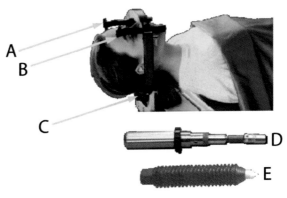

Fig. 21.2 Head-ring fixation. A Carbon fiber posts. B, E Artifact-free fixation pins with ceramic tips. C Robust and lightweight frame. D Two torque wrenches for pressure adjustment of pins

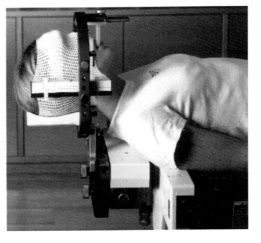

Fig. 21.3 Non-invasive repeat head fixation with the mask system and upper jaw support

the stereotactic coordinates. The constant geometrical relationship between the stereotactic frame and anatomical structures, including the PTV, is realized by fixation of the frame to the patient. During the whole treatment procedure, from the performance of the stereotactic imaging to the delivery of the irradiation treatment, the stereotactic frame must not be removed from the patient. In case of relocatable frames it must be assured that the position of the patient is exactly the same relative to the frame after reapplication of the relocatable frame.

For the treatment of cranial lesions by RS the frame system is neurosurgically fixed onto the patient's skull (Fig. 21.2). For SFR the head is fixed non-invasively in a relocatable thermoplastic mask attached on the stereotactic frame (Fig. 21.3).

There are different stereotactic frame systems described in detail in the literature: the BRW system (BROWN 1979; HEILBRUN et al. 1983); the CRW system (COULDWELL and APUZZO 1990; SPIEGELMANN and FRIEDMAN 1991); the Leksell system (LEKSELL and JERNBERG 1980; LEKSELL et al. 1985); the Leibinger-Fischer system (RIECHERT and MUNDINGER 1955; STURM et al. 1983); and the BrainLAB system (STÄRK et al. 1997; AUER et al. 2002). Each system is different with regard to material of the stereotactic frame, design, and connection with the localizer and positioner and accuracy of repositioning (KORTMANN et al. 1999).

21.5.2
Imaging for Stereotactic Irradiation Treatment

Imaging is used in stereotactic radiotherapy for: (a) localization and positioning; (b) definition of target volume and organs at risk; and (c) calculation and 3D representation of the isodose distribution.

Most stereotactic systems use CT for localization. During the CT investigation the localizer is attached to the frame (Fig. 21.4). The localizer is a box with CT-compatible fiducial markers on each plane, which are visualized on CT on each scan; thus, the localizer defines the link between the stereotactic coordinates and the imaging coordinates, so that for any point in the imaging the 3D stereotactic coordinates can be determined. The stereotactic frame, the patient fixation system, and the localizer form a fix unit. They have to be compatible with the radiological investigations and offer an accurate visualization of the tumor and of the critical structures, without artifacts. The use of MRI for localization and positioning needs a high homogeneity of the magnetic field to avoid spatial distortions artifacts, which could disturb the geometrical correlation between the stereotactic coordinates system and imaging coordinates system.

Non-invasive imaging techniques are a central component of treatment planning in radiation oncology. The information gained from different imaging modalities is usually of a complementary nature: (a)

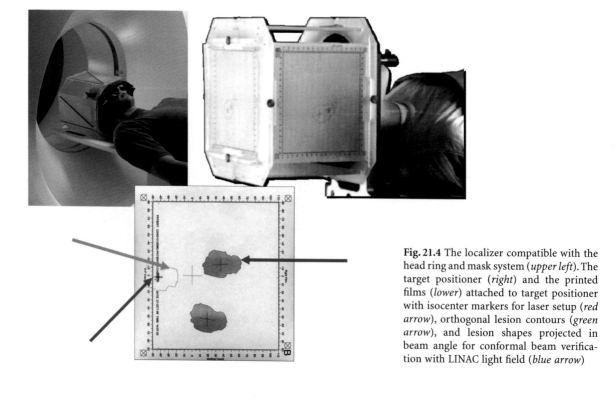

Fig. 21.4 The localizer compatible with the head ring and mask system (*upper left*). The target positioner (*right*) and the printed films (*lower*) attached to target positioner with isocenter markers for laser setup (*red arrow*), orthogonal lesion contours (*green arrow*), and lesion shapes projected in beam angle for conformal beam verification with LINAC light field (*blue arrow*)

MRI describes the anatomical structures of soft tissue with a high accuracy; (b) CT is important for the delineation of bone and soft tissue; (c) positron emission tomography (PET) and single photon computed emission tomography (SPECT) offer additional information about tumor extension and biology; and (d) angiography is essential for the visualization of the arterio-venous malformations; thus, the definition of tumor extension and critical structures is characterized by the correct integration of multiple different investigational tools, by using specialized image fusion software (Figs. 21.5, 21.6; GROSU et al. 2003).

The calculation and 3D representation of the isodose distribution is discussed in the next chapter.

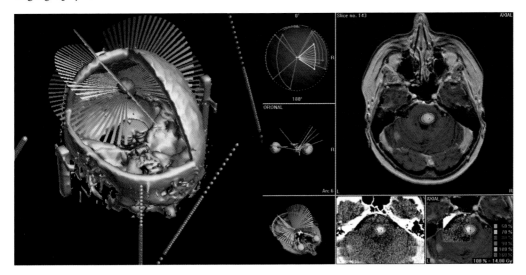

Fig. 21.5 The images *left* show the rotational stereotactic technique using six radiation arcs and conical collimators. The images *right* present the dose distribution for the radiosurgery treatment of a brain stem metastases on MRI (*upper*) and the CT/MRI image fusion (*lower*)

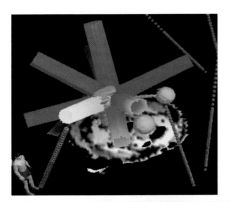

Fig. 21.6 Stereotactic irradiation using eight static fields and micro-multileaf collimator in a patient with acoustic neuroma treated with stereotactic fractionated radiotherapy. The MRI/CT image fusion and the conformal dose distribution is shown in the two lower images

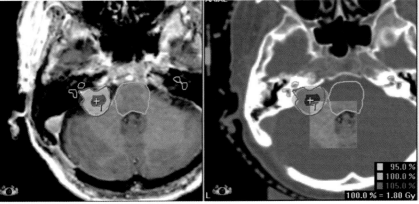

21.5.3
Treatment Planning

21.5.3.1
Definition of Target Volume and Organs at Risk

The delineation of the target volume is a complex interactive process in which all the information of the imaging tools and the clinical information (operation, histopathology, other treatment approaches, etc.) are considered. The tumor-specific morphology, the growth pattern of the tumor, and the anatomical relationship to the normal tissue are essential parameters in defining the target volume.

Of major importance for the stereotactic radiation therapy is the delineation of the organs at risk. All the organs at risk which may get significant dose have to be delineated.

21.5.3.2
Definition of the Stereotactic Target Point

The target point is the point in the target volume that must be positioned with exact precision in the isocenter of the LINAC. The position of the target point can be defined interactively. One or more target points can be defined. In stereotactic planning programs the coordinates of the target points are related in such way that the resulting dose distribution meets the clinical requirements. The planning system outputs the position of these points in stereotactic coordinate. Prior to therapy, these coordinates will be used to correctly position the patient. This is performed with a positioner, a device attached to the stereotactic frame, which allows the connection of the stereotactic coordinate system to the room coordinate system, where the isocenter of the treatment device is defined (Fig. 21.4).

21.5.3.3
Planning of the Radiation Technique

The stereotactic radiation is characterized by a very steep dose fall-off on the margin of the target volume. The steep dose gradient is achieved by the use of appropriate collimators and a multitude of radiation directions.

Stereotactic Collimators. Tertiary stereotactic collimators for circular or oval target volumes are attached to the tray holder of the LINAC. The diameter of the irradiated area is defined by the size of the circular collimators and varies usually between 1 and 35 mm (Fig. 21.7). For irregular target volumes

different individual apertures may be used, but the production and the use thereof is very cumbersome.

Only recently have micro-multileaf collimators become available (Fig. 21.8). The beam shape can be selected by computer or by hand. In this way the contours of the irradiation field can be adjusted individually to the tumor shape. Micro-multileaf collimators, in comparison with the traditional multi-leaf collimators, have the advantage of a decreased leaf width and therefore optimized the resolution (between 1 and 3 mm). Computer-controlled micro-multileaf

Fig. 21.7 Set of eight conical collimators

Fig. 21.8 Micro-multileaf collimator: 3-mm fine leaves at center; 4.5-mm intermediate leaves; and 5.5-mm outside leaves. Maximum field size is 10×10 cm

Hamilton AJ, Lulu BA, Fosmire H et al. (1996) LINAC-based spinal stereotactic radiosurgery. Stereotact Funct Neurosurg 66:1–9

Hanley J, Debois MM, Mah D et al. (1999a) Deep inspiration breath-hold technique for lung tumors: the potential value of target immobilization and reduced lung density in dose escalation. Int J Radiat Oncol Biol Phys 45:603–611

Hanley J, Debois MM, Mah M et al. (1999b) Deep inspiration breath-hold technique for lung tumors: the potential value of target immobilization and reduced lung density in dose escalation. Int J Radiat Oncol Biol Phys 45:603–611

Hara R, Itami J, Kondo T et al. (2002) Stereotactic single high dose irradiation of lung tumors under resperitory gating. Radiother Oncol 63:159–163

Hartmann GH, Schlegel W, Sturm V et al. (1985) Cerebral radiation surgery using moving field irradiation at a linear accelerator facility. Int J Radiat Oncol Biol Phys 11:1185–1192

Hayakawa K, Mitsuhashi N, Saito Y (1999) Limited field irradiation for medically inoperable patients with peripheral stage I non-small cell lung cancer. Lung Cancer 26:137–142

Herfarth KK, Debus J (2003) Stereotactic radiation therapy of liver tumors. Radiother Oncol 68 (Suppl 1):S45

Herfarth KK, Debus J, Lohr F et al. (2000a) Stereotactic single dose radiation treatment of tumors in the lung. Radiology 217 (Suppl):148

Herfarth KK, Debus J, Lohr F et al. (2000b) Extracranial stereotactic radiation therapy: set-up accuracy of patients treated for liver metastases. Int J Radiat Oncol Biol Phys 46:329–335

Herfarth KK, Pirzkall A, Lohr F et al. (2000c) Erste Erfahrungen mit einem nicht-invasiven Patientenfixierungssystem für die stereotaktische Strahlentherapie der Prostata. Strahlenther Onkol 176:217–222

Herfarth KK, Debus J, Lohr F et al. (2001a) Stereotactic single dose radiation therapy of liver tumors: results of a phase I/II trial. J Clin Oncol 19:164–170

Herfarth KK, Debus J, Lohr F et al. (2001b) Stereotaktische Bestrahlung von Lebermetastasen. Radiologe 41:64–68

Herfarth KK, Hof H, Bahner ML et al. (2003a) Assessment of focal liver reaction by multiphasic CT after stereotactic single-dose radiotherapy of liver tumors. Int J Radiat Oncol Biol Phys 57:444–451

Herfarth KK, Münter MW, Debus J (2003b) Strahlentherapeutische Behandlung von Lebermetastasen. Chir Gastroenterol 19:359–363

Hof H, Herfarth KK, Munter M et al. (2003a) The use of the multislice CT for the determination of respiratory lung tumor movement in stereotactic single-dose irradiation. Strahlenther Onkol 179:542–547

Hof H, Herfarth KK, Munter M et al. (2003b) Stereotactic single-dose radiotherapy of stage I non-small-cell lung cancer (NSCLC). Int J Radiat Oncol Biol Phys 56:335–341

Hof H, Herfarth K, Wannenmacher M et al. (2004) Stereotaktische Bestrahlung von Lungentumoren: Erfahrungen mit einem Einzeitkonzept. Strahlenther Onkol 180 (Suppl 1):17

International Registry of Lung Metastases (1997) Long-term results of metastasectomy: prognostic analyses based on 5206 cases. J Thorac Cardiovasc Surg 113:37–49

Koong AC, Le QT, Ho A et al. (2004) Phase I study of stereotactic radiosurgery in patients with locally advanced pancreatic cancer. Int J Radiat Oncol Biol Phys 58:1017–1021

Lax I, Blomgren H, Näslund I et al. (1994) Stereotactic radiotherapy of malignancies in the abdomen. Acta Oncol 33:677–683

Leksell L (1951) The stereotactic method and radiosurgery of the brain. Acta Chir Scand 102:316–319

Lohr F, Debus J, Frank C et al. (1999) Noninvasive patient fixation for extracranial stereotactic radiotherapy. Int J Radiat Oncol Biol Phys 45:521–527

Mageras GS, Yorke E (2004) Deep inspiration breath hold and respiratory gating strategies for reducing organ motion in radiation treatment. Semin Radiat Oncol 14:65–75

Milker-Zabel S, Zabel A, Thilmann C et al. (2003) Clinical results of retreatment of vertebral bone metastases by stereotactic conformal radiotherapy and intensity-modulated radiotherapy. Int J Radiat Oncol Biol Phys 55:162–167

Minohara S, Kanai T, Endo M et al. (2000) Respiratory gated irradiation system for heavy-ion radiotherapy. Int J Radiat Oncol Biol Phys 47:1097–1103

Nagata Y, Negoro Y, Aoki T et al. (2002) Clinical outcomes of 3D conformal hypofractionated single high-dose radiotherapy for one or two lung tumors using a stereotactic body frame. Int J Radiat Oncol Biol Phys 52:1041–1046

Nakagawa K, Aoki Y, Tago M et al. (2000) Megavoltage CT-assisted stereotactic radiosurgery for thoracic tumors: original reseach in the treatment of thoracic neoplasms. Int J Radiat Oncol Biol Phys 48:449–457

Onishi H, Kuriyama K, Komiyama T et al. (2003) CT evaluation of patient deep inspiration self-breath-holding: How precisely can patients reproduce the tumor position in the absence of respiratory monitoring devices? Med Phys 30:1183–1187

Onishi H, Araki T, Shirato H et al. (2004a) Stereotactic hypofractionated high-dose irradiation for stage I nonsmall cell lung carcinoma: clinical outcomes in 245 subjects in a Japanese multiinstitutional study. Cancer 101:1623–1631

Onishi H, Kuriyama K, Komiyama T et al. (2004b) Clinical outcomes of stereotactic radiotherapy for stage I non-small cell lung cancer using a novel irradiation technique: patient self-controlled breath-hold and beam switching using a combination of linear accelerator and CT scanner. Lung Cancer 45:45–55

Pirzkall A, Debus J, Lohr F et al. (1998) Radiosurgery alone or in combination with whole-brain radiotherapy for brain metastases. J Clin Oncol 16:3563–3569

Rosenzweig KE, Hanley J, Mah D et al. (2000) The deep inspiration breath-hold technique in the treatment of inoperable non-small-cell lung cancer. Int J Radiat Oncol Biol Phys 48:81–87

Ryu S, Fang Yin F, Rock J et al. (2003) Image-guided and intensity-modulated radiosurgery for patients with spinal metastasis. Cancer 97:2013–2018

Starkschall G, Forster KM, Kitamura K et al. (2004) Correlation of gross tumor volume excursion with potential benefits of respiratory gating. Int J Radiat Oncol Biol Phys 60:1291–1297

Takeda T, Takeda A, Kunieda E et al. (2004) Radiation injury after hypofractionated stereotactic radiotherapy for peripheral small lung tumors: serial changes on CT. Am J Roentgenol 182:1123–1128

Thilmann C, Schulz-Ertner D, Zabel A et al. (2002) Intensity-modulated radiotherapy of sacral chordoma: a case report and a comparison with stereotactic conformal radiotherapy. Acta Oncol 41:395–399

Timmerman R, Papiez L, McGarry R et al. (2003) Extracranial stereotactic radioablation: results of a phase I study in medically inoperable stage I non-small cell lung cancer. Chest 124:1946–1955

Uematsu M, Shioda A, Tahara K et al. (1998) Focal, high dose, and fractionated modified stereotactic radiation therapy for lung carcinoma patients. Cancer 82:1062–1070

Warner MA, Warner ME, Buck CF et al. (1988) Clinical efficacy of high frequency jet ventilation during extracorporal shock wave lithotripsy of renal and ureteral calculi: a comparison with conventional mechanical ventilation. J Urol 139:486–487

Willemart S, Nicaise N, Struyven J et al. (2000) Acute radiation-induced hepatic injury: evaluation by triphasic contrast enhanced helical CT. Br J Radiol 73:544–546

Wong J, Sharpe M, Jaffray D et al. (1999) The use of active breathing control (ABC) to reduce margin for breathing motion. Int J Radiat Oncol Biol Phys 44:911–919

Wulf J, Hädinger U, Oppitz U et al. (2000) Stereotactic radiotherapy of extracranial targets: CT-simulation and accuracy of treatment in the stereotactic body frame. Radiother Oncol 57:225–236

Wulf J, Hädinger U, Oppitz U et al. (2001) Stereotactic radiotherapy of targets in the lung and liver. Strahlenther Onkol 177:645–655

Wulf J, Haedinger U, Oppitz U et al. (2004) Stereotactic radiotherapy of primary lung cancer and pulmonary metastases: a non-invasive treatment approach in medically inoperable patients. Int J Radiat Oncol Biol Phys 60:186–196

Yin F, Kim JG, Haughton C et al. (2001) Extracranial radiosurgery: immobilizing liver motion in dogs using high-frequency jet ventilation and total intravenous anesthesia. Int J Radiat Oncol Biol Phys 49:211–216

Zierhut D, Bettscheider C, Schubert K et al. (2001) Radiation therapy of stage I and II non-small cell lung cancer (NSCLC). Lung Cancer 34 (Suppl) 3:S39–S43

23 X-IMRT

Simeon Nill, Ralf Hinderer, and Uwe Oelfke

CONTENTS

23.1
Introduction

The IMRT dose delivery techniques are the second cornerstone of X-IMRT in addition to the concepts of inverse treatment planning and optimization (discussed in Chap. 17). The result of an inverse treatment plan usually consists of ideal photon-fluence distributions for each selected beam angle. In this chapter we discuss on a very fundamental level the main concepts and strategies for the actual delivery of these fluence patterns to a patient. This very brief summary of some of the interesting techniques cannot be complete and very detailed. The interested reader who aims to explore the touched topics in more depth is referred to the standard literature, e.g., Webb (2001) or (Palta and Mackie 2003).

The main focus in this chapter is on the discussion of the most prominent dose delivery techniques with linac-integrated multi-leaf collimators (MLCs). The

S. Nill, PhD; R. Hinderer, PhD; U. Oelfke, PhD
Abteilung Medizinische Physik in der Strahlentherapie, Deutsches Krebsforschungszentrum, Im Neuenheimer Feld 280, 69120 Heidelberg, Germany

basic concepts employed for the two standard methods – the "step-and-shoot" approach and the "dynamic" dose delivery – are briefly reviewed before the new concept of "helical tomography" is introduced. Finally, a short section describes the dose delivery technique with individually designed compensators.

23.2
MLC-based IMRT Delivery

Most modern linear accelerators are equipped with MLCs, which were originally designed to replace the use of lead blocks to shield normal tissue during the treatment (Biggs et al. 1991; Boesecke et al. 1991; Jordan and Williams 1994). In the next section we first describe the main IMRT relevant characteristics of a typical MLC including its dosimetric properties. Then, various dose delivery techniques are briefly described. Based on the MLC computer-control system, two main dose delivery methods are distinguished: (a) the "step-and-shoot" approach; and (b) the "dynamic" dose delivery concept. A very detailed description of both techniques with references to the original papers can be found in the excellent book by Webb (2001). Besides these two basic IMRT delivery concepts with linac-integrated MLCs, we also briefly discuss some examples of "high-resolution" IMRT with add-on mini- or micro-MLCs.

23.2.1
MLC Design

23.2.1.1
IMRT-Relevant MLC Data

Multi-leaf collimators and their design and function for 3D conformal radiotherapy (3D CRT) are described in detail in Chap. 20, together with their geometrical and mechanical characteristics, see also Galvin et al. (19939.

To deliver intensity-modulated fields (IMRT) the following geometric characteristics are of most importance:

1. Leaf width. The leaf width (see Chap. 20.2.1) determines one dimension of the achievable spatial resolution of the fluence modulation. The spatial resolution of the fluence in the direction of the leaves' movement is only restricted by the positioning accuracy of the MLC. For the delivery of IMRT fields a possible large overtravel range of the different leaves (10–15 cm) is often required to irradiate extended target volumes.
2. Maximum leaf speed. The leaf speed is especially important for the "dynamic" delivery of intensity-modulated fields. A typical leaf speed is of the order of 2–4 cm/s (WEBB 2001; HUG 2002).
3. The maximum field size of the MLC. For IMRT applications, the maximum field size is not the same as the maximum open field which can be achieved, but it is mainly defined by the maximum overtravel.
4. Maximum overtravel. Maximum overtravel (see Chap. 20.2.1) is the distance of how far a leaf can move over the midline of the MLC.
5. Leakage and transmission. The leakage and transmission (see Chaps. 20.2.2.2, 20.2.2.3) play an important role specifically for the delivery of IMRT treatments because the dose delivery often requires a substantial number of monitor units (MU); these are applied where most areas of the treatment field are covered with closed leaves, i.e., the leakage radiation received by these areas is significantly enhanced.
6. Leaf-positioning accuracy. One of the most important properties of an MLC with respect to IMRT is the leaf-positioning accuracy. For IMRT delivery the leaf-positioning accuracy is even more important than for most conventional treatment techniques due to the large number of small field components for an average IMRT treatment. For a typical fluence grid with a resolution of 10 mm in both directions a leaf-positioning error of 1 mm has to be considered as a large error. This is due to the characteristic output factor curve with a very steep gradient for small field sizes. An error of 1 mm for the leaf position can cause a dose error of up to approximately 10% irradiating a 10×10-mm field and therefore the maximum leaf-positioning error should not be larger than 0.5 mm. It can also be concluded that the required leaf-positioning accuracy depends on the fluence grid resolution of the fluence maps.

23.2.2
Step-and-Shoot Dose Delivery

As result of the inverse planning process, described in Chap. 17, one receives the "ideal" fluence maps for each incident beam direction. In most treatment planning systems these ideal intensity maps have to be converted into field segments to be delivered with the MLC, a process that is called the "sequencing" of the fluence. In this section we shortly describe the "step-and-shoot" dose delivery method and discuss briefly the two basic sequencing techniques.

23.2.2.1
The Step-and-Shoot Technique

The "step-and-shoot" technique of IMRT dose delivery (BORTFELD et al. 1994) is a straightforward extension of the conventional multiple-field irradiation technique. The "step-and-shoot" approach superimposes the dose delivered by a number of irregularly shaped and partially overlapping treatment fields, often called subfields or segments. For each segment a well-defined number of monitor units is delivered. Then, the beam is turned off while the leaves of the MLC move to the positions required by the next IMRT segment. After the verification and record system (V&R) has validated the new leaf positions, the beam is turned on and the dose is delivered for this segment. This process is repeated for all segments per incident beam angle and all beam directions (see Fig. 23.1).

23.2.2.2
Leaf-Sequencing Algorithms

In order to describe the basic features of leaf-sequencing algorithms we introduce the following terms: a fluence map is defined on a 2D fixed grid covering the respective beam aperture. It is usually divided into discrete, quadratic elements called "bixels." The fluence map assigns one beam intensity to each bixel. A 1D line of intensities for all bixels created by a single leaf pair is called a channel. The width of each channel coincides with the leaf width of the MLC. The goal of the sequencing process is to decompose the fluence maps into a number of field components or subfields.

The first step of the sequencing process is the stratification of the continuous fluence profiles provided by the inverse planning process, i.e., each line of intensity for a specific channel is forced to take on only a few discrete fluence levels (Fig. 23.2). For each

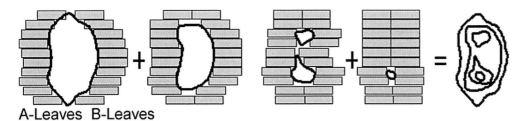

A-Leaves B-Leaves

Fig. 23.1. The basic idea of the "step-and-shoot" approach is to deliver an intensity-modulated beam as a superposition of a set of irregularly shaped, partially overlapping field components

Fig. 23.2. Stratification of a continuous fluence distribution into a fluence map with a number of discrete levels

fluence level the MLC leaves shape and deliver a beam segment. In general, not all levels may be deliverable without an additional step. One problem is the appearance of spatial "holes" for a fluence pattern generated by the optimization process (e.g., two pyramids next to each other with zero fluence in between). For these cases each level has to be divided into separately deliverable subfields. This is one constraint which, in addition to other MLC hardware constraints, must be taken into account during the process of calculating the required leaf positions.

The two most prominent concepts are the "close-in" and "sweep" technique. In Fig. 23.3 we show an example of the "close-in" technique. For this example the fluence map was divided into three levels. The typical numbers of levels for clinical cases is in the range of five to ten. In the first step the left leaf is moved towards the first positive gradient of the fluence pattern while the right leaf is moving towards the first negative fluence gradient. This defines the first segment which is then delivered. Next, the two leaves move to the next positive and negative gradients of the respective fluence map to define the next subfield followed by its delivery. This procedure is repeated until the whole fluence map is delivered.

We briefly describe the basic algorithm of the "sweep" technique for the 1D example of a fluence distribution given in Fig. 23.4. For this fluence map,

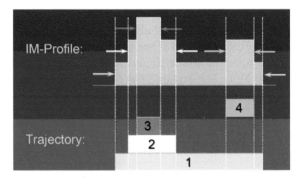

Fig. 23.3. Decomposition of a one-dimensional fluence map into deliverable segments using the "close-in" technique

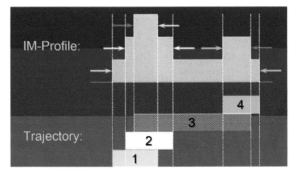

Fig. 23.4. Decomposition of a one-dimensional fluence map into deliverable segments using the "sweep" technique

one can clearly see that the "sweep" approach is more complicated than the "close-in" technique. For the "sweep" technique the left leaf again probes all positive fluence gradients from the left side while the right leaf first moves to the negative fluence gradient located most to the left. Then both leaves move in the same direction while again left and right leaf stop at the respective positive and negative fluence gradients. With this technique the treatment time is reduced compared with the "close-in" technique since the leaves are always moving only in one direction.

Most inverse treatment planning programs have a build-in leaf-sequencing algorithm. The total number of segments depends on the complexity of the fluence maps, the number of beams, and other technical factors. Since the total treatment time depends linearly on the number of segments, a significant effort is put into optimizing the sequencing algorithm to find the best solution in terms of treatment time. Up to now no optimal sequencer has been found which creates the best possible solutions for all clinical cases (XIA and VERHEY 1998; SIOCHI 1999).

23.2.2.3
Hardware Constraints: Matchlines

For real 2D fluence distributions the sequencer must not only calculate the leaf positions but also must take into account the geometric and dosimetric properties of the MLC. For example, the leaf-end design and the focusing properties, which can lead to significant dose artifacts if not taken care of, must be taken into account. Two of these artifacts are described below.

The first dose artifact is called the tongue-and-groove effect. To reduce the interleaf leakage some MLCs are using a tongue-and-groove design (Fig. 23.5; see also Fig. 20.8 in Chap. 20). If a large field perpendicular to the leaf motion direction is divided into two subfields, an underdosage at the matchline of the two treatment fields is observed.

This can be explained by the fact that at the border of the two subfields the beam attenuation through the leaves is slightly different due to the tongue-and-groove design (see Fig. 23.5). This effect can cause an underdosage of up to 20%. Most sequencers at present are capable of minimizing this effect by reducing the number of segments which can lead to this dose artifact (KAMATH et al. 2004).

The second dose artifact can be caused by transversal matchlines. If two segments have a joint borderline perpendicular to the leaf motion direction, stripes of underdosage can occur. These underdosages are an effect of an incomplete compensation for the penumbras of the two fields. Possible reasons are the finite thickness of the leaves, the extended radiation source, and scatter effects. This effect leads to the same order of underdosage as the tongue-and-groove effect and can be compensated for by adjusting the leaf positions of the involved segments.

The main differences between conventional techniques and the IMRT "step-and-shoot" approach are the increased number of fields, a longer treatment time, and that very small fields are delivered. In addition, the number of monitor units per field is low compared with conventional treatment techniques and a higher total number of monitor units are delivered per fraction. Especially the low number of monitor units per segment is of great importance. It must be validated that the linear accelerator is capable of delivering such small monitor units so that 10×3 MU equals the dose delivered with 30 MU. If this is not the case, a large additional error is introduced.

23.2.3
Dynamic Dose Delivery

For the "dynamic" delivery technique (DMLC; Fig. 23.6) the intensity modulation is achieved by an individual variation of the velocities of the moving

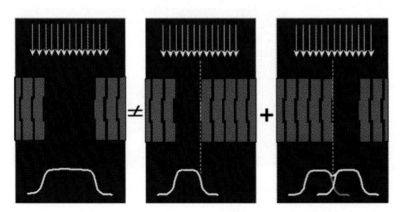

Fig. 23.5. The tongue-and-groove underdose effect is caused by the tongue-and-groove leaf design to suppress interleaf leakage

leaves, i.e., the treatment can be realized without interrupting the treatment beam. The clinical application of this technique was pioneered at the Memorial Sloan Kettering Center in New York (LoSasso et al. 1998; Chui et al. 2001; Zelefsky et al. 2002). The "dynamic" MLC sequencing problem has been solved in various ways (see Webb 1997, and references therein).

One way to illustrate its basic features is to consider the "dynamic" mode as an extension of the "sweep" mode of the static dose delivery as illustrated in Fig. 23.7. Firstly, the intensity profile in Fig. 23.7a is converted into positive intensity differences that are equal to the intensities to be delivered (see Fig. 23.7b). It can be easily seen that the differences of the upper and lower profile in Fig. 23.7b equals the intensity profile in Fig. 23.7a. If the discrete bixels in Fig. 23.7b are made smaller and smaller as shown in Fig. 23.7c and d, the representation of the continuous intensity

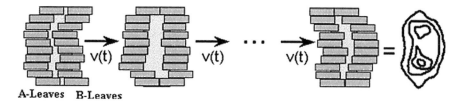

A-Leaves   B-Leaves

Fig. 23.6 Principle of dynamic IMRT delivery

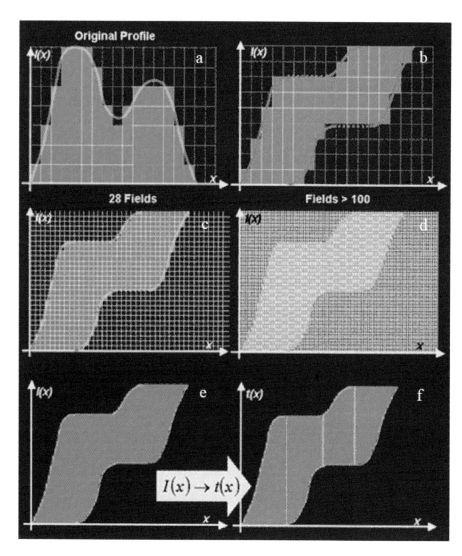

Fig. 23.7a–f Transition from leaf-sweep "step-and-shoot" approach to the "dynamic" approach (if the number of intensity levels converges to infinity)

profile in Fig. 23.7e is achieved. Next, the difference of the two introduced upper and lower profiles is interpreted as irradiation time for the locations on the x-axis. This is equivalent to the required intensity profile if one assumes a constant dose rate for the dose delivery process. In the t(x) representation, shown in Fig. 23.7f, the lower profile can be viewed as the trajectory of the "leading" leaf and the upper profile corresponds to the trajectory of the "trailing" leaf, i.e., the inverse of the derivative of these profiles represent the required velocity profile of the leaves. For example, at the beginning of the irradiation with closed leaves, the leading leave has to move with an "infinite" velocity to the position of the first intensity maximum before the trailing leaf shapes the required intensity profile. As a final step, the trajectories are adjusted to the maximal available speed of the leaves. With this approach the leaf trajectories can be changed in such a way that still the same fluence profiles are delivered (Spirou and Chui 1994; Stein et al. 1994; Ma et al. 1998; Convery and Webb 1998; Boyer and Yu 1999).

The treatment time to deliver calculated leaf trajectories is approximately the time the leaves need to move from the leftmost side to the rightmost side plus the delivery time to irradiate the leaf sweep decomposition. The irradiation time is defined mostly by the complexity of the fluence patterns and therefore given by the treatment time of the intensity channel which takes the longest time to be delivered. The total treatment time for "dynamic" technique is mostly shorter than the treatment time to deliver the same fluence pattern with any "step-and-shoot" approach (Ma et al. 1999).

In addition to the technical properties required for an MLC to be used in static mode, the "dynamic" delivery technique requires that the leaf speed be controllable with a very high accuracy and that the calibration process for leaf speed and leaf position must be easily possible (LoSasso 1998). The leaf-position accuracy is not only important to create the correct aperture for small fields but any leaf-positioning error may lead to a dose error even for large fields (Zygmanski et al. 2003); therefore, strict quality assurance for both parameters is needed.

For the "dynamic" delivery process matchlines perpendicular to the direction of the leaf movement do not pose a problem. The tongue-and-groove effect, on the other hand, basically leads to the same problem as for the static delivery process. Using a method called leaf synchronization the effect could eliminated from the delivery process, but this could lead to a longer treatment time (Webb et al. 1996; Van Stanvoort et al. 1996).

Comparison of Static and Dynamic Delivery Process

One often discussed question is which delivery technique is superior to the other (Palta and Mackie 2003). Both techniques have their advantages and disadvantages. The following section discusses only the main aspects for the comparison. From a technical point of view the "dynamic" approach is more complex than the "step-and-shoot" approach. For the "dynamic" process the leaves are moving while the beam is turned on and therefore a very accurate control of the leaf positions, leaf speed, and dose rate at the same time must be achieved. On the other hand, a shorter delivery time is the main advantage of this technique. The static approach allows an easier verification process and is viewed as a natural extension of established conventional dose delivery techniques. Plan comparisons of static and DMLC optimized plans have shown that the differences for the achievable dose distribution for the target volume are less then a few percent (Palta and Mackie 2003). Up to now both techniques have been successfully implemented clinically (Zelefsky et al. 2002; Eisbruch 2001; Lee et al. 2002; Münter et al. 2002; Thilmann et al. 2002; Thilmann et al. 2004) and both techniques are still improving (Chui et al. 2001).

23.2.4
Improved Spatial Resolution: Add-on MLCs

The spatial resolution currently achievable with internal MLCs is in the range of 5–10 mm. It was shown by Bortfeld et al. (2000) that for a 6-MV photon beam the optimal leaf width according to basic physics is in the range of 1.5–2 mm. A number of MLCs with smaller leaf widths in the range of 1.6–5.5 mm were developed to generate more conformal dose distributions (Chap. 20.3.2; Cosgrove et al. 1999; Meeks et al. 1999; Xia et al. 1999; Hartmann and Föhlisch 2002). For intensity-modulated radiation therapy (IMRT) the effect of the leaf width on the physical dose distribution for a MLC with 5- and 10-mm leaf width was recently evaluated by Fiveash et al (2002) and for even smaller leaves by Nill et al. (2005). The result of these studies was that for most clinical cases the spatial resolution of the internal MLCs is good enough to create acceptable 3D dose distributions, but for very complex geometries of the anatomy where for instance the organ at risk is in close proximity to the target volume, the 3D dose distribution achievable with an improved spatial resolution is significantly better. The disadvantages with add-On MLC approach are the reduced clearance between the MLC and the patient, the limited field size and, due to the increased spatial resolution, the increased number of total monitor units to be delivered.

23.3
Helical Tomotherapy Delivery

The basic idea of tomotherapy is that the fluence modulation of a radiation beam could be achieved by leaves, which move rapidly in and out of a fan beam; therefore, a binary MLC was developed to temporally modulate a fan beam. To achieve a dose distribution highly conformal to the target, an irradiation from many directions is needed. One possible solution is a continuously rotating linear accelerator with the binary MLC attached. To cover the whole target volume, the radiation must be delivered in a slice-by-slice fashion. This technique is analogous to a CT scanner. The combination of the word tomography and radiation therapy leads to the expression tomotherapy (MACKIE et al. 1993).

One realization of the principle of tomotherapy is helical tomotherapy. Here, a linear accelerator is mounted on a ring gantry (see Fig. 23.8). The binary MLC is attached downstream to the linear accelerator. Technically, this is the fusion of a helical CT scanner and a linear accelerator. During the continuous rotation of the linear accelerator around the patient the couch is moved continuously in longitudinal direction through the bore of the gantry. From the patient's point of view the radiation is delivered in a helical fashion (MACKIE et al. 1993; MACKIE et al. 1999; OLIVERA et al. 1999). This is the main difference between helical tomotherapy and another implementation of tomotherapy, sequential tomotherapy, as it is employed by the Peacock system (CAROL 1996). In sequential tomotherapy the patient couch is in a fixed position during the rotation of the linear accelerator. After each revolution, the couch is indexed in longitudinal direction by the width of the fan beam. Compared with helical tomotherapy, this procedure leads to a prolonged treatment time.

The binary MLC of the helical tomotherapy machine consists of 64 interdigitated leaves with 32 leaves on each side. The motion of the leaves is performed by a pneumatic system. Due to the high air pressure, the leaves can open or close the radiation field within only 40 ms. The leaves are made of an alloy consisting of 95% tungsten. A single leaf has a height of about 10 cm and projects to a width of 0.625 cm at isocenter. A tongue-and-groove design is employed to reduce the interleaf leakage. The modulation of the beam fluence is achieved by varying the leaf opening time (PALTA and MACKIE 2003).

The inverse treatment planning for helical tomotherapy is done by approximating the rotational de-

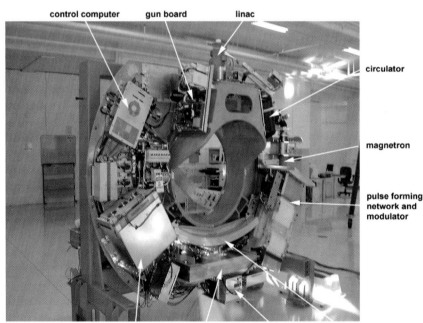

Fig. 23.8 Helical tomotherapy machine and its components mounted on a ring gantry. Some of the components, such as the control computer, high-voltage power supply, and data acquisition system are required for the operation of the detector rather than the beam delivery system. (Courtesy T.R. Mackie, University of Wisconsin, and G.H. Olivera, TomoTherapy Inc.)

livery with a series of discrete beams (MACKIE et al. 1995; PALTA and MACKIE 2003). Typically, one revolution is divided into a series of up to 51 beam projections. During the optimization process the rotation time, the helical pitch (the ratio of the longitudinal distance traveled per rotation to the slice thickness), and the configurations of the leaves for all gantry and couch positions are determined. Typical values for the pitch are in the range of 0.2–0.5 with a typical slice thickness between 2.5 and 5.0 cm. A common rotation time is 20 s. Consequently, the total treatment time is equal to 20 s multiplied by the number of rotations required to treat the whole target volume.

23.4
Compensator-Based IMRT Delivery

Not all linear accelerators are equipped with an MLC. An alternative to an MLC is the use physical fluence attenuators called compensators (ELLIS et al. 1959; MAGERAS et al. 1991; JIANG 1998; CHANG et al. 2004). A physical compensator is made of a material which absorbs radiation and is mounted in the accessory tray of the linear accelerator. The compensators consist of bixels which are filled with different thickness of a photon absorbing alloy, which determines the fluence modulation. The thickness map over the beam aperture is calculated by a computer program based on the result of the optimization process (see Fig. 23.9). To mount and manufacture compensators special equipment, such as a furnace, to melt the suitable compensator material, a milling machine to produce the moulds (see Fig. 23.10) and a special holder to mount the compensator to the accessory tray are needed. A typical material for compensator is the alloy MCP96.

Since each compensator represents one individual fluence distribution, a new compensator must be manufactured for each field of a patient after the treatment planning process. Another important issue is the time needed to replace the compensators during the daily field by field treatment. For each field a technician has to walk into the treatment room and change the compensator in the accessory tray. Another problem of the compensator approach is the achievable range of fluence modulations (CHANG et al. 2004). Despite these obvious disadvantages, there are a number of important advantages of compensator-based IMRT. As already mentioned with compensators, IMRT can be performed on even older linacs and the spatial resolution of the fluence modulation

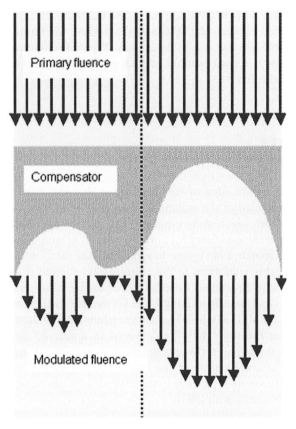

Fig. 23.9 Physical compensators offer a very simple way to create IMFs by means of individually tailored absorbers

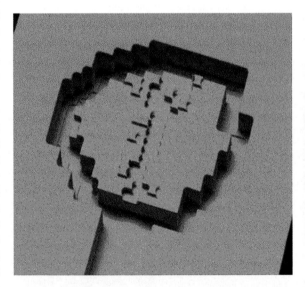

Fig. 23.10 A mould of a compensator mould processed with a milling machine. The material is a plastic called "obomodulan." The indicated cavities in the mould will be filled with a photon attenuation alloy

can be very high, e.g., <5 mm. Another important advantage of compensators is that the total monitor units delivered for a complete IMRT plan is only slightly increased compared with the monitor units used for the delivery of a conventional treatment plan. This might be an important fact in the ongoing discussion about secondary radiation-induced tumors due to the higher exposure of normal tissue with low doses for MLC-based IMRT delivery (HALL and WUU 2003).

In conclusion, IMRT with compensators is feasible, but the effort to produce compensators can be very high and special equipment is needed; however, there seems to be advantages which can justify the application of compensators (CHANG 2004).

References

Biggs P, Capalucci J, Russell M (1991) Comparison of the penumbra between focused and nondivergent blocks: implications for multileaf collimators. Med Phys 18:753–758

Boesecke R, Becker G, Alandt K, Pastyr O, Doll J, Sclegel W, Lorenz WJ (1991) Modification of a three-dimensional treatment planning system for the use of multileaf collimators in conformation readiotherapy. Radiother Oncol 21:261–268

Bortfeld T, Boyer AL, Schlegel W, Kahler DL, Waldron TJ (1994) Realization and verification of three-dimensional conformal radiotherapy with modulated fields. Int J Radiat Oncol Biol Phys 30:899–908

Bortfeld T, Oelfke U, Nill S (2000) What is the optimum leaf width of a multifleaf collimator? Med Phys 27:2494–2502

Boyer Al, Yu CX (1999) Intensity-modulated radiation therapy with dynamic multileaf collimators. Semin Radiat Oncol 9:48–59

Carol M, Grant WH III, Pavord D, Eddy P, Targovnik HS, Butler B, Woo S, Figura J, Onufrey V, Grossman R, Selkar R (1996) Initial clinical experience with the Peacock intensity modulation of a 3-D conformal radiation therapy system. Stereotact Funct Neurosurg 66:30–34

Chang SX, Cullip TJ, Deschesne KM, Miller EP, Rosenman JG (2004) Compensators: an alternative IMRT delivery technique. J Appl Clin Med Phys 5:15–36

Convery DI, Webb S (1998) Generation of discrete beam-intensity modulation by dynamic multileaf collimation under minimum leaf separation constraints. Phys Med Biol 43:2521–2538

Cosgrove V, Jahn U, Pfaender M, Bauer S, Budach V, Wurm R (1999) Commissioning of a micro multi-leaf collimator and planning system for stereotactic radiosurgery. Radiother Oncol 50:325–336

Chui CS, Chanf MF, Yorke E, Spirou S, Ling CC (2001) Delivery of intensity-modulated radiation therapy with a conventional multileaf collimator: comparison of dynamic and segmental methods. Med Phys 28:2441–2449

Eisbruch A, Kim HM, Terrell JE, Marsh LH, Dawson LA, Ship JA (2001) Xerostomia and its predictors following parotid-sparing irradiation of head-and-neck cancer. Int J Radiat Oncol Biol Phys 50:695–704

Ellis F, Hall EJ, Oliver R (1959) A compensator for variations in tissue thickness for high energy beams. Br J Radiol 32:421–422

Fiveash JB, Murshed H, Duan J, Hyatt M, Caranto J, Bonner JA, Popple RA (2002) Effect of multileaf collimator leaf width on physical dose distributions in the treatment of CNS and head and neck neoplasms with intensity modulated radiation therapy. Med Phys 29:1116–1119

Hall EJ, Wuu CS (2003) Radiation-induced second cancers: the impact of 3D-CRT and IMRT. Int J Radiat Oncol Biol Phys 56:83–88

Hartmann GH, Föhlisch F (2002) Dosimetric characterization of a new miniature multileaf collimator. Phys Med Biol 47:N171–N177

Hug MS, Das IJ, Steinberg T, Galvin JM (2002) A dosimetric comparison of various multileaf collimators. Phys Med Biol 47:N159–N170

Jiang SB (1998) On modulator design for photon beam intensity-modulated conformal therapy. Med Phys 25:668–675

Jordan TF, Williams PC (1994) The design and performance characteristics of a multileaf collimator. Phys Med Biol 39:231–251

Kamath S, Sahni S, Ranka S, Li J, Palta J (2004) Comparison of step-and-shoot leaf sequencing algorithms that eliminate tongue-and-groove effect. Phys Med Biol 49:3137–3143

Lee N, Xia P, Quivey JM, Sultanem K, Poon I, Akazawa C, Akazawa P, Weinberg, Fu KK (2002) Intensity modulated radiotherapy in the treatment of nasopharyngeal carcinoma: an update of the UCSF experience. Int J Radiat Oncol Biol Phys 53:12–22

LoSasso T, Chui CS, Ling CC (1998) Physical and dosimetric aspects of a multileaf collimation system used in the dynamic mode for implementing intensity modulated radiotherapy. Med Phys 25:1919–1927

Ma A, Boyer AL, Xing L, Ma CM (1998) An optimized leaf-setting algorithm for beam intensity modulation sung dynamic multileaf collimator. Phys Med Biol 43:1629–1643

Ma L, Boyer A, Ma CM, Xing L (1999) Synchronizing dynamic multileaf collimators for producing two-dimensional intensity-modulated fields with minimum beam delivery time. Int J Radiat Oncol Biol Phys 44:1147–1154

Mackie TR, Holmes T, Swerdloff S, Reckwerdt P, Deasy JO, Yang J, Paliwal B, Kinsella T (1993) Tomotherapy: a new concept for the delivery of conformal radiotherapy using dynamic collimation. Med Phys 20:1709–1719

Mackie TR, Holmes TW, Reckwerdt PJ, Yang J (1995) Tomotherapy: optimized planning and delivery of radiation therapy. Int J Imaging Syst Tech 6:43–55

Mackie TR, Balog J, Ruchala K, Shepard D, Aldridge S, Fitchard E, Reckwerdt PJ, Olivera G, McNutt TR, Metha M (1999) Tomotherapy. Semin Radiat Oncol 9:108–117

Mageras GS, Mohan R, Burman C, Barset GD, Kutcher GJ (1991) Compensators for three dimensional treatment planning. Med Phys 18:133–140

Meeks S, Bova F, Kim S, Tome W, Buatti J, Friedman W (1999) Dosimetric characteristics of a double-focused miniature multileaf collimator. Med Phys 26:729–733

Münter MW, Debus J, Hof H, Nill S, Haring P, Bortfeld T, Wannenmacher M (2002) Inverse treatment planning and stereotactic intensity-modulated radiation therapy (IMRT) of the tumor and lymph node levels for nasopharyngeal carcinomas. Description of treatment technique, plan comparison, and case study. Strahlenther Onkol 178:517–523

Nill S, Tücking T, Münter MW, Oelfke U (2005) Intensity modulated radiation therapy with multi leaf collimators of different leaf widths: a comparison of achievable dose distributions. Radiol Oncol75:106–111

Olivera GH, Shepard DM, Ruchala K, Aldridge JS, Kapatoes JM, Fitchard EE, Reckwerdt PJ, Fang G, Balog J, Zachman J and Mackie TR (1999) Tomotherapy. In: Van Dyk J (ed) Modern technology of radiation oncology. Medical Physics Publishing, Madison, Wisconsin

Palta J, Mackie TR (2003) (eds) Medical physics monograph 29:51–75

Siochi A (1999) Minimizing static intensity modulation delivery time using an intensity solid paradigm. Int J Radiat Oncol Biol Phys 43:671–680

Spirou S, Chui C (1994) Generation of arbitrary intensity profiles by dynamic jaws or multileaf collimators. Med Phys 21:1031–1041

Stein J, Bortfeld T, Dorschel B, Schlegel W (1994) Dynamic X-ray compensation for conformal radiotherapy by means of multileaf collimator. Radiother Oncol 32:163–173

Thilmann C, Schulz-Ertner D, Zabel A, Herfarth KK, Wannenmacher M, Debus J (2002) Intensity-modulated radiotherapy of sacral chordoma: a case report and a comparison with stereotactic conformal radiotherapy. Acta Oncol 41:395–399

Thilmann C, Sroka-Perez G, Krempien R, Hoess A, Wannenmacher M, Debus J (2004) Inversely planned intensity modulated radiotherapy of the breast including the internal mammary chain: a plan comparison study. Technol Cancer Res Treat 3:69–75

Van Stanvoort J, Heijmen B (1996) Dynamic multileaf collimation without tongue-and-groove underdose effects. Phys Med Biol 41:2091–2105

Webb S (1997) The physics of conformal radiotherapy. Advances in technology. Institute of Physics Publishing, Bristol

Webb S (2001) Intensity-modulated radiation therapy. Institute of Physics, Publishing, Bristol

Webb S, Bortfeld T, Stein J, Convery D (1996) The effect of stair-step leaf transmission on the "tongue-and-groove problem" in dynamic radiotherapy with multileaf collimator. Phys Med Biol 42, 595–202

Xia P, Verhey LJ (1998) Multileaf collimator leaf sequencing algorithm for intensity modulated beams with multiple static segments. Med Phys 25:1424–1434

Xia P, Geis P, Xing L (1999) Physical characteristics of a miniature multileaf collimator. Med Phys 26:65–70

Zelefsky M, Fuks Z, Hunt M, Yamada Y, Marion C, Ling C, Amalos H Venkatramen ES, Leibel SA (2002) High-dose intensity modulated radiation therapy for prostate cancer: early toxicity and biochemical outcome in 772 patients. Int J Radiat Oncol Biol Phys 53:1111–1116

Zygmanski P, Kung JH, Jiang SB, Chin L (2003) Dependence of fluence errors in dynamic IMRT on leaf-positional errors varying with time and leaf number. Med Phys 30:2736–2349

24 Control of Breathing Motion: Techniques and Models (Gated Radiotherapy)

Timothy D. Solberg, N.M. Wink, S.E. Tenn, S. Kriminski, G.D. Hugo, and N. Agazaryan

CONTENTS

T. D. Solberg, PhD
Department of Radiation Oncology, University of Nebraska Medical Center, 987521 Nebraska Medical Center, Omaha, NE 68198-7521, USA
T. D. Solberg, PhD; N. M. Wink, MS; S. E. Tenn, MS; S. Kriminski, PhD; N. Agazaryan, PhD
Department of Radiation Oncology, Suite B265, David Geffen School of Medicine at UCLA, 200 UCLA Medical Plaza, Los Angeles, CA 90096-6951, USA
G.D. Hugo, PhD
Department of Radiation Oncology, William Beaumont Hospital, 3601 W. Thirteen Mile Road, Royal Oak, MI 48073, USA

This work was supported in part by grant # 03-028-01-CCE from the American Cancer Society

24.1 Introduction

It is well known throughout the cancer community that a high degree of tumor control and ultimately cure can be achieved through the administration of appropriate doses of ionizing radiation. Throughout the traditional radiotherapy experience, however, administration of a lethal tumor dose has routinely been limited by the tolerance of nearby healthy tissues that cannot be adequately excluded from the radiation field. New methodologies for radiation delivery, such as intensity-modulated radiotherapy (IMRT) and body radiosurgery, offer the possibility for further dose escalation in order to optimally avoid critical structures and reduce treatment-associated morbidity. With these new delivery techniques, however, the practical and very real problems associated with patient positioning, target localization, and respiratory motion take on added importance. It is disadvantageous and perhaps counterproductive to deliver a very exacting dose distribution when there is uncertainty in the location of the anatomical structures of interest. The IMRT Collaborative Working Group (2001), sponsored by the NCI, recognizes the hazards involved in applying a highly focused radiotherapy approach to moving targets: "application of IMRT to sites that are susceptible to breathing motion should be limited until proper accommodation of motion uncertainties is included." Thus, for continued clinical gains in the practice of radiotherapy, management of breathing motion is essential.

24.2 Clinical Rationale for Management of Respiratory Motion

Recent clinical gains in a number of tumor sites have been truly remarkable. In prostate cancer, for example, 5-year survival has increased from 43% for the 5-year period ended in 1954 to 99% for the 5-year

(DIBH) technique in the treatment of lung tumors (HANLEY et al. 1999; MAH et al. 2000; ROSENZWEIG et al. 2000). Two main benefits of DIBH are the reduction of the lung density and reduction of tumor motion. The DIBH technique involves verbally coaching the patient to the same, reproducible deep inspiration level. Although the DIBH technique appears feasible for clinical implementation, the technique requires comprehensive quality assurance procedures to ensure accurate treatment delivery.

Active breathing control, first introduced by WONG et al. (1999) at William Beaumont Hospital, reproducibly facilitates breath-hold without requiring the patient to reach maximum inspiration capacity (FRAZIER et al. 2004; REMOUCHAMPS et al. 2003a–c; STROMBERG et al. 2000). Elekta has since commercialized the method under trademark Active Breathing Coordinator. The method has been also implemented at Mount Vernon Cancer Centre (Northwood, Middlesex, UK; WILSON et al. 2003). The ABC apparatus can be used to suspend breathing at any pre-determined position along the normal breathing cycle. The system makes use of digital spirometer to measure the respiratory trace. The spirometer actively controls a balloon valve, consequently taking control over the patients breath-hold. In the example shown in Fig. 24.4, the patient was initially brought to quiet tidal breathing and then verbally coached to perform a slow deep inspiration, a slow deep expiration, and a second slow deep inspiration followed by breath-hold. Treatment planning and delivery can then be performed at identical ABC conditions with minimal margin for breathing motion. It is even possible to use the ABC method with standard operation of the accelerator to deliver therapy with a short beam on

time, where tumor and organ positions are immobilized before the beam is turned on and not released until the beam is turned off.

24.5.4
Free-Breathing Gating Techniques

Active gating is another available option of external beam radiotherapy applied to regions affected by intra-fraction motion. Respiratory gating involves the administration of radiation within a particular window of the patient's breathing cycle as demonstrated in Fig. 24.5. This method requires monitoring patient's respiratory motion using either an external or internal fiducial markers. To date, the only commercially available respiratory gating systems are the Varian RPM system (described in section 24.3.3) and the BrainLAB ExacTracGating system.

A main obstacle for active gating is the difficulty associated with real-time detection of the moving target location during treatment. The RPM gating system uses an external IR marker as a surrogate for tumor motion and may not necessarily correlate with internal motion (TSUNASHIMA et al. 2004). In contrast, the ExacTracGating system combines the capability to track externally placed markers in three dimensions with a digital radiography system capable of locating internally placed fiducials such as gold BBs. While the paradigm is similar to the Cyberknife, the linac does not move dynamically and the imaging system does not try to find a tumor trajectory. Instead, the system assumes that the target position is correlated to the position of the surface markers during a given window in the respiratory cycle. This

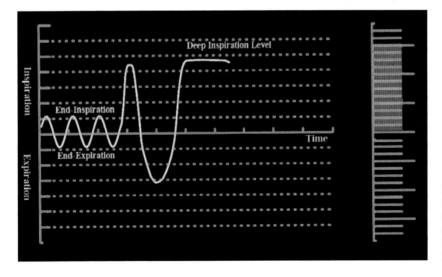

Fig. 24.4 Representation of spirometry tracings from the modified slow vital capacity maneuver and the deep inspiration breath-hold. (From ROSENZWEIG et al. 2000)

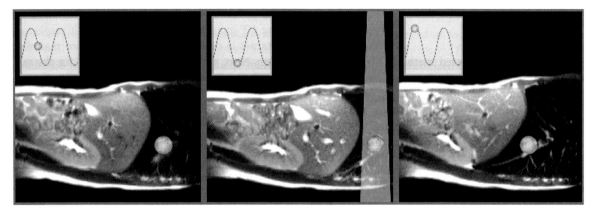

Fig. 24.5 Respiratory gating involves administration of radiation within a particular window of the patient's breathing cycle

assumption simplifies the gating procedure since no tumor trajectories need to be calculated or related to external motion. Clinical validation is needed to test the validity of this assumption.

ExacTracGating works by detecting the patient's breathing pattern using an IR camera system with reflecting markers attached to the skin. Once it has detected a breathing pattern, the system can trigger each of two isocentrically aimed diagnostic X-ray units at a user-defined point in the breathing cycle. From these two images the 3D location of an internally placed radio-opaque fiducial marker can be determined. The patient can then be moved into the correct treatment position based on the position of the internal markers. This has the advantage over positioning via external fiducials in that implanted markers are in close proximity to the actual target and are more likely to indicate its true position. The system also has the capability to trigger the linear accelerator at the same point of the breathing cycle that the localization X-rays were acquired. Assuming that the target position remains directly related to the position of the patient's surface, the target will be hit each time it arrives at that same point in the breathing cycle. At least two groups (VEDAM et al. 2003; MAGERAS et al. 2001) have found a good correlation between the position of externally placed markers and the position of the diaphragm; however, the assumption has been questioned by others (OZHASOGLU et al. 2002; CHEN et al. 2001). Therefore, the ExacTracGating system has the ability to intermittently verify that the fiducials are at the correct location during the gate period using its radiography system. This system has been developed in close collaboration with academic centers (HUGO et al. 2002, 2003; VERELLEN et al. 2003).

Several groups have constructed moving phantoms to investigate the characteristics of an active gating approach (HUGO et al. 2002; DUAN et al. 2003; JIANG et al. 2003). Examples of these devices are shown in Fig. 24.6. The dosimetric aspects of active gating are highlighted in section 24.6.

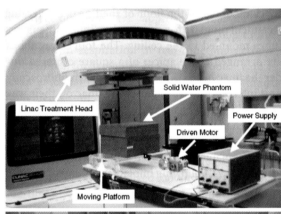

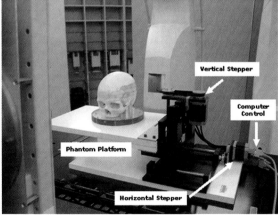

Fig. 24.6 Platforms designed by JIANG et al. (2003; *top*) and HUGO et al. (2002; *bottom*) in which dosimetric studies on moving objects can be performed

motion had been previously obtained, and that image-guided and adaptive radiotherapy would be optimally facilitated using a two-source system.

24.6
Dosimetric Consequences of Respiratory Motion

The subject of gated operation of modern radiotherapy treatment devices, and the dosimetric characteristics under such conditions, is one that has received little attention in the literature. The first clinical investigations were performed in Japan using a commercial linear accelerator (ML15MDX, Mitsubishi, Osaka, Japan; Ohara et al. 1989). A mask was used to detect ventilation and provide a respiration curve that was used to gate the linac. Beam stability, characterized by output, symmetry, and uniformity, was evaluated for gated and non-gated operation. Variations in measured dose of up to 5% were observed, although the monitor units delivered were quite low (≤10) compared with those typically required in clinical practice. Additionally, deviations as large as 4.5% in cross- and in-plane uniformity were reported. Subsequently, the same group reported on the gated operation of a 500-MeV proton linac located operated at the University of Tsukuba, Japan (Ohara et al. 1989; Inada et al. 1992).

Kubo and Hill (1996) performed a similar investigation using a commercial electron/photon linac (Varian Medical Systems, Palo Alto, Calif.). Several sensors were evaluated for their ability to monitor and accurately reproduce a human respiration pattern. These sensors were electronically interfaced with the linac for subsequent dosimetric characterization. The beam was gated by tapping the gun delay circuit controlling the grid on the electron source. With 200 monitor units (MU) delivered at a rate of 240 MU/min, absolute dose differed by less than 0.1% when gating was performed with a duty cycle of 50%. Similarly, differences in symmetry were less than 1.2%. In addition to the obvious fact that different linacs were used, the fact that these results are significantly better than those reported by Ohara et al. (1989) is likely due to the small number of monitor units used in the earlier study. Ramsey et al. (1999) performed a subsequent investigation in which several delivery parameters were systematically evaluated. Dosimetric parameters were measured as monitor units were varied from 10 to 100 in five segments of 2–20 MU pulses and with inter-pulse delays from

0.0 to 4.2 s. The dose rate remained fixed at 320 MU/min. Results were quite similar to those of Kubo and Hill (1996), with maximum deviations of 0.8, 1.9, and 0.8% in output, flatness and symmetry, respectively. The largest deviations were observed when the smallest number of monitor units was used.

Solberg et al. (2000) reported on the operation of a commercial gating implementation on a Novalis linear accelerator (BrainLAB AG, Heimstetten, Germany). Similar to the Varian units described previously, gating was achieved through a three-pin connector that provides access to a 12-V DC signal to the MHOLDOFF/status bit on the console backplane. A beam inhibit is triggered whenever this bit is low. This bit toggles the initial position interlock (IPSN) interlock on the accelerator controller which subsequently halts both radiation and, in the case of dynamic treatment, leaf motion. Gating was performed under software control through a personal computer interfaced to the three-pin connector. In the study, monitor units were varied from 25 to 200 at gating frequencies from 0.2 to 1.0 Hz (at a constant 50% duty cycle). Two-dimensional dose maps were obtained using a 20×20-cm^2 amorphous-silicon imaging flat-panel array (Scanditronix Medical/Wellhöfer Dosimetrie, Uppsala, Sweden) consisting of 256×256 detector elements. Frames were acquired every 80 ms and integrated to obtain an absolute dose map. Through the full dose-rate range (160–800 MU/min), no clinically significant differences were observed in output, flatness, or symmetry under gated operation.

Like the Varian unit, injection current on the Siemens ONCOR AvantGarde accelerator (Siemens Medical Solutions, Concord, Calif.) is regulated by a triode grid; however, in order to avoid in the temperature variations within the linac components, neither electron nor microwave generation are interrupted during the beam-off state under gated operation. Instead, the beam is interrupted by injecting electrons asynchronously with the microwave generation. With this mechanism, energy dissipation, and therefore temperature variation in the linac components during gated operation, remains the same as during standard operation.

An investigation of the dosimetry characteristics under gated operation has been performed recently on the ONCOR accelerator at our institution. Ionization readings were recorded from a 0.6-cm^3 chamber placed at a depth of 5 cm in solid water. Twenty-five to 200 monitor units were delivered in non-gated operation and at gating frequencies ranging from 0.0625 to 0.50 Hz. A 50% duty cycle and dose rate of 300 MU/min were used. As with previous investigations, the largest deviations were observed when the smallest number of monitor units was used in combination with the high gating frequencies. With the exception of the 0.5 Hz/25 MU combination, output variations were all less than 0.5% (see Fig. 24.9).

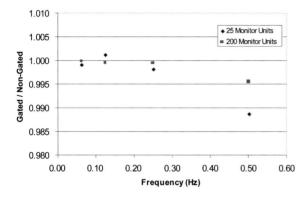

Fig. 24.9 Ratio of ionization chamber readings for the ONCOR accelerator under gated and non-gated operation

The dosimetric consequences of respiratory motion are likely to be most significant when dynamic delivery techniques, such as virtual wedge or intensity modulated radiotherapy (IMRT), are employed. These effects were first quantified in a theoretical study performed by YU et al. (1998). In their investigation, the authors suggested that under clinically realistic conditions, the interplay between the motion-of-field defining devices, such as multileaf collimators (MLC) and target motion due to respiration, could produce intensity variations in the target anatomy as large as 480%. The authors pointed out, however, that the largest variations occurred in small spatial regions and would be significantly reduced through the random smoothing process that occurs in fractionated delivery.

BORTFELD et al. (2002) expanded the investigation of the role of fractionation on the IMRT delivery/respiratory motion interplay. They concluded that intrafraction effects rendered variations in the expected dose essentially independent of treatment technique.

Experimental investigations of the dosimetric variation associated with dynamic delivery and respiratory motion have been reported by several groups. KUBO and WANG (2000) evaluated the ability of a Varian 2100C equipped with an 80-leaf MLC to deliver dynamic wedge and IMRT fields under gated operation. Several gating patterns were tested: one with an alternating 2-s beam on/beam off, another with an alternating 10-s beam on/beam off to simulate a breath-hold technique, and a third set

to gate the beam during on the exhalation phase of a realistic respiratory cycle. In the case of dynamic wedge delivery, dosimetric differences between gated and non-gated operation were minimal except at the field edges. Similarly, gating had a negligible effect on IMRT delivery.

In contrast, SOLBERG et al. (2000) observed significant dosimetric variations in gated wedged and intensity-modulated fields delivered through dynamic MLC, although the largest errors occurred in small spatial regions. The magnitude of these discrepancies was found to increase as smaller numbers of total monitors units were used and at higher gating frequencies (Fig. 24.10). A more in-depth investigation of these effects was performed recently DUAN et al. (2003), who quantified discrepancies between gated and non-gated delivery in a stationary phantom using both absolute dose and 2D dose distribution measurements. Output differences of up to 3.7% were observed in gated dynamic wedge fields as measured with a 0.6-cm^3 ionization chamber. Additionally, discrepancies were observed in 2D dose distribution measurements. Dosimetric integrity under gated operation was found to correlate closely with dose rate, leaf speed, and frequency of beam interruptions. This was attributed primarily to the lag in leaf motion during dynamic delivery. Observations corresponded reasonably well with those of SOLBERG et al. (2000) shown in Fig. 24.9.

At least three groups have experimentally investigated the dosimetric consequences of respiratory motion. HUGO et al. (2002) investigated the ability of a prototype gating system to reproduce non-gated IMRT dose distributions. A phantom capable of simultaneous 2D movement (see Fig. 24.6) was set in motion using an analytical liver motion function obtained from the literature (LUJAN et al. 1999). Under IMRT delivery, considerable dosimetric errors were observed between nonmoving and moving conditions (Fig. 24.11). Gating was found to substantially reduce these errors.

DUAN et al. (2003) performed similar experiments using film oriented perpendicular to the beam axis, in a phantom capable of 1D motion in the cranial–caudal direction. The Varian RPM system was used to generate gating signals. When respiratory gating was applied to the moving phantom (using a 25% duty cycle), differences between moving and stationary measurements were reduced from >10 mm distance errors and 5% dose errors to <1 mm and 1%.

Finally, JIANG et al. (2003) used an ionization chamber in solid water phantom placed on a moving platform to characterize errors in IMRT delivery.

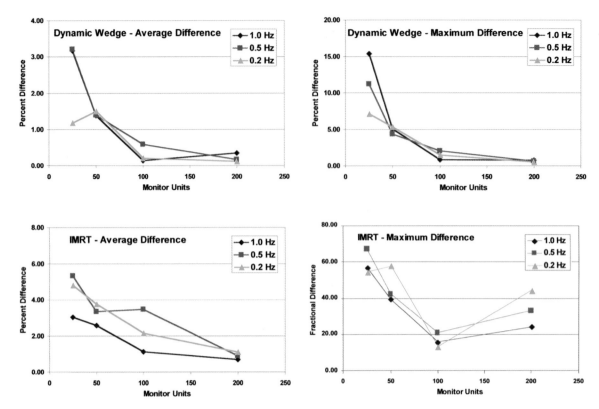

Fig. 24.10 Average and maximum deviations of the 2D dose maps obtained in a plane perpendicular to the beam axis. Wedged fields were created by opening a leaf gap, and a sliding window technique was used for intensity-modulated radiotherapy (IMRT) delivery

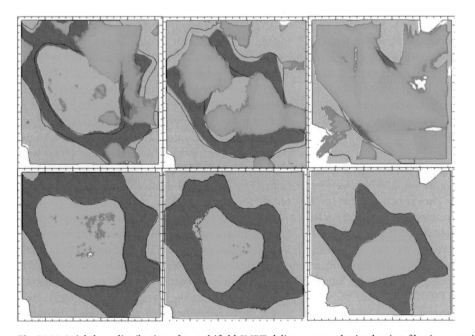

Fig. 24.11 Axial dose distributions for multifield IMRT delivery were obtained using film in a moving phantom under non-gated (*top*) and gated (*bottom*) delivery. The measured 80, 50, and 20% isodose regions are shown in color-wash superimposed on results from treatment planning calculations (*solid lines*). Characterization was performed on the central axis (*left*), and 1 and 2 cm off axis (*center* and *right*, respectively). Regions where the acceptability criteria were exceeded, as measured by the multidimensional Gamma parameter (Low et al. 1998), are shown in the *dark green overlay*

Motion was constrained to 1D sinusoidal movement, with an adjustable period and fixed amplitude of 2 cm. While point dose errors of up to 30% for single and up to 18% for multiple fields were observed, the authors suggested that fractionated delivery would significantly reduce the magnitude of these errors. Nevertheless, they pointed out that further study was needed to characterize the effects in normal/critical tissue outside of the target volume.

24.7
Conclusion

For continued clinical gains in the practice of radio-therapy, management of breathing motion is essential. The problem of organ motion in radiotherapy is complex; thus, interventions to reduce organ-motion-related uncertainties require effort, expertise, and collaboration from many disciplines. The application of image-guidance techniques, i.e., image-guided radiotherapy, will play an increasing important role in developing new and improved delivery techniques, i.e., adaptive radiotherapy. With some anecdotal clinical evidence and many potentially beneficial but unproven technologies under development and on the horizon, it is essential to place equal emphasis on the planning and implementation of prospective clinical trials.

References

Adler JR Jr, Chang SD, Murphy MJ et al. (1997) The Cyberknife: a frameless robotic system for radiosurgery. Stereotact Funct Neurosurg 69:124–128

Adler JR Jr, Murphy MJ, Chang SD et al. (1999) Image-guided robotic radiosurgery. Neurosurgery 44:1299–1306

Allen AM, Siracuse KM, Hayman JA et al. (2004) Evaluation of the influence of breathing on the movement and modeling of lung tumors. Int J Radiat Oncol Biol Phys 58:1251–1257

Aruga T, Itami J, Aruga M et al. (2000) Target volume definition for upper abdominal irradiation using CT scans obtained during inhale and exhale phases. Int J Radiat Oncol Biol Phys 48:465–469

Axelsonn P, Johnsson R, Stromqvist B (1996) Mechanics of the external fixation test an the lumbar spine: a roentgen stereophotogrammetric analysis. Spine 21:330–333

Balter JM, Ten Haken RK, Lawrence TS et al. (1996) Uncertainties in CT-based treatment plans due to patient breathing. Int. J. Radiation Oncology Biol. Phys. 36:167–174

Balter JM, Lam KL, McGinn CJ et al. (1998) Improvement of CT-based treatment-planning models of abdominal tar-gets using static exhale imaging. Int. J. Radiation Oncology Biol. Phys. 41:939-943

Balter JM, Dawson LA, Kazanijian S et al. (2001) Determination of ventilatory liver movement via radiographic evaluation of diaphragm position, Int. J. Radiation Oncology Biol. Phys. 51:267–270

Baroni G, Ferrigno G, Orecchia R et al. (2000) Real-time three-dimensional motion analysis for patient positioning verification. Radiother Oncol 54:21-7

Berbeco RI, Jiang SB, Sharp GC et al. (2004) Integrated radiotherapy imaging system (IRIS): design considerations of tumour tracking with linac gantry-mounted diagnostic X-ray systems with flat-panel detectors. Phys Med Biol 49:243–255

Blomgren H, Lax I, Naslund I et al. (1995) Stereotactic high dose fraction radiation therapy of extracranial tumors using an accelerator. Clinical experience of the first thirty-one patients. Acta Oncol 34:861–870

Blomgren H, Lax I, Goranson H et al. (1999) Radiosurgery for tumors in the body: clinical experience using a new method. J Radiosurg 2:239–245

Bortfeld T, Jokivarsi K, Goitein M et al. (2002) Effects of intra-fraction motion on IMRT dose delivery: statistical analysis and simulation. Phys Med Biol 47:2203–2220

Bova FJ, Buatti JM, Friedman WA et al. (1997) The University of Florida frameless high-precision stereotactic radiotherapy system. Int J Radiat Oncol Biol Phys 38:875–882

Brugmans MJ, van der Horst A, Lebesque JV et al. (1999) Beam intensity modulation to reduce the field sizes for conformal irradiation of lung tumors: a dosimetric study. Int J Radiat Oncol Biol Phys 43:893–904

Carman AB, Milburn PD (1997) Conjugate imagery in the automated reproduction of three dimensional coordinates from two dimensional coordinate data. J Biomech 30:733–736

Chen QS, Weinhous MS, Deibel FC et al. (2001) Fluoroscopic study of tumor motion due to breathing: facilitating precise radiation therapy for lung cancer patients. Med Phys 28:1850–1856

De Koste JRV, Lagerwaard FJ, Schuchhard-Schipper RH et al. (2001) Dosimetric consequences of tumor mobility in radiotherapy of stage I non-small cell lung cancer: an analysis of data generated using "slow" CT scans. Radiother Oncol 61:93–99

De Koste JRV, Lagerwaard FJ, de Boer HCJ et al. (2003) Are multiple CT scans required for planning curative radiotherapy in lung tumors of the lower lobe? Int J Radiat Oncol Biol Phys 55:1394–1399

Dhanantwari AC, Stergiopoulos S, Iakovidis I (2001a) Correcting organ motion artifacts in X-ray CT medical imaging systems by adaptive processing. I. Theory Med Phys 28:1562–1576

Dhanantwari AC, Stergiopoulos S, Zamboglou N et al. (2001b) Correcting organ motion artifacts in X-ray CT systems based on tracking of motion phase by the spatial overlap correlator. II. Experimental study. Med Phys 28:1577–1596

Duan J, Shen S, Fiveash JB et al. (2003) Dosimetric effect of respiration-gated beam on IMRT delivery. Med Phys 30:2241–2252

Ekberg L, Holmberg O, Wittgren L et al. (1998) What margins should be added to the clinical target volume in radiotherapy treatment planning for lung cancer? Radiother Oncol 48:71–77

Farre R, Montserrat JM, Rotget M et al. (1998) Accuracy of thermistors and thermocouples as flow-measuring devices for detecting hypopnoeas. Eur Respir J 11:179–182

Frazier RC, Vicini FA, Sharpe MB et al. (2004) Impact of breathing motion on whole breast radiotherapy: a dosimetric analysis using active breathing control. Int J Radiat Oncol Biol Phys 58:1041–1047

Ford EC, Mageras GS, Yorke E et al. (2002) Evaluation of respiratory movement during gated radiotherapy using film and electronic portal imaging. Int J Radiat Oncol Biol Phys 52:522–531

Ford EC, Mageras GS, Yorke E et al. (2003) Respiration-correlated spiral CT: a method of measuring respiratory-induced anatomic motion for radiation treatment planning. Med Phys 30:88–97

George R, Keall P, Chung T et al. (2004) Dynamic multileaf collimation and three-dimensional verification in IMRT. In: Yi BY, Ahn SD, Choi EK, Ha SW (eds) The use of computers in radiation therapy. Jeong Publishing, Seoul, pp 437–441

Giraud P, De Rycke Y, Dubray B et al. (2001) Conformal radiotherapy (CRT) planning for lung cancer: analysis of intrathoracic organ motion during extreme phases of breathing. Int J Radiat Oncol Biol Phys 51:1081–1092

Hanley J, Debois MM, Mah D et al. (1999) Deep inspiration breath-hold technique for lung tumors: the potential value of target immobilization and reduced lung density in dose escalation. Int J Radiat Oncol Biol Phys 45:603–611

Herfarth KK, Debus J, Lohr F et al. (2000) Extracranial stereotactic radiation therapy: set-up accuracy of patients treated for liver metastases. Int J Radiat Oncol Biol Phys 46:329–335

Hugo GD, Agazaryan N, Solberg TD (2002) An evaluation of gating window size, delivery method, and composite field dosimetry of respiratory-gated IMRT. Med Phys 29:2517–2525

Hugo GD, Agazaryan N, Solberg TD (2003) The effects of tumor motion on planning and delivery of respiratory-gated IMRT. Med Phys 30:1052–1066

Inada T, Tsuji H, Hayakawa Y et al. (1992) Proton irradiation synchronized with respiratory cycle. Nippon Acta Radiol 52:1161–1167

Intensity Modulated Radiation Therapy Collaborative Working Group (2001) Intensity-modulated radiotherapy: current status and issues of interest. Int J Radiat Oncol Biol Phys 51:880–914

Jiang SB, Pope C, Al Jarrah MK et al. (2003) An experimental investigation on intra-fractional organ motion effects in lung IMRT treatments. Phys Med Biol 48:1773–1784

Johnson CA, Keltner JL, Krohn MA et al. (1979) Photogrammetry of the optic disc in glaucoma and ocular hypertension with simultaneous stereo photography. Invest Ophthal Vis Sci 18:1252–1263

Keall PJ, Kini VR, Vedam SS et al. (2001) Motion adaptive X-ray therapy: a feasibility study. Phys Med Biol 46:1–10

Keall PJ, Kini VR, Vedam SS et al. (2002) Potential radiotherapy improvements with respiratory gating. Australas Phys Eng Sci Med 25:1–6

Keall PJ, Starkschall G, Shukla H et al. (2004) Acquiring 4D thoracic CT scans using a multislice helical method. Phys Med Biol 49:2053–2067

Kearfoot K, Juang RJ, Marzke MW (1993) Implementation of digital stereo imaging for analysis of metaphyses and joints in skeletal collections. Med Biol Eng Computing 31:149–156

Kini VR, Keall PJ, Vedam SS et al. (2000) Preliminary results from a study of a respiratory motion tracking system: underestimation of target volume with conventional CT simulation. Int J Radiat Oncol Biol Phys 48:164

Kini VR, Vedam SS, Keall PJ et al. (2003) Patient training in respiratory-gated radiotherapy. Med Dosim 28:7–11

Kitamura K, Shirato H, Seppenwoolde Y et al. (2002a) Three-dimensional intrafractional movement of the prostate measured during real-time tumor-tracking radiotherapy in supine and prone treatment positions. Int J Radiat Oncol Biol Phys 53:1117–1123

Kitamura K, Shirato H, Seppenwoolde Y et al. (2002b) Three-dimensional intrafractional movement of prostate measured during real-time tumor-tracking radiotherapy in supine and prone treatment positions. Int J Radiat Oncol Biol Phys 53:1117–1123

Kitamura K, Shirato H, Shimizu S et al. (2002c) Registration accuracy and possible migration of internal fiducial gold marker implanted in prostate and liver treated with real-time tumor-tracking radiation therapy (RTRT). Radiother Oncol 62:275–281

Kitamura K, Shirato H, Seppenwoolde Y et al. (2003) Tumor location, cirrhosis, and surgical history contribute to tumor movement in the liver, as measured during stereotactic irradiation using a real-time tumor-tracking radiotherapy system. Int J Radiat Oncol Biol Phys 56:221–228

Ko Y, Yi B, Ahn S et al. (2004) Immobilization effect of air injected blanket for abdomen fixation. In: Yi BY, Ahn SD, Choi EK, Ha SW (eds) The use of computers in radiation therapy. Jeong Publishing, Seoul, pp 421–423

Kubo HD, Hill BC (1996) Respiration gated radiotherapy treatment: a technical study. Phys Med Biol 41:83–91

Kubo HD, Wang L (2000) Compatibility of Varian 2100C gated operations with enhanced dynamic wedge and IMRT dose delivery. Med Phys 27:1732–1738

Kubo HD, Len PM, Minohara S et al. (2000) Breathing-synchronized radiotherapy program at the University of California Davis Cancer Center. Med Phys 27:346–353

Kwa SLS, Lebesque JV, Theuws JCM et al. (1998) Radiation pneumonitis as a function of meal lung dose: an analysis of pooled data of 540 patients. Int J Radiat Oncol Biol Phys 42:1–9

Lagerwaard FJ, de Koste JRV, Nijssen-Visser MRJ et al. (2001) Multiple "slow" CT scans for incorporating lung tumor mobility in radiotherapy planning. Int J Radiat Oncol Biol Phys 51:932–937

Lawrence TS, Robertson JM, Anscher MS et al. (1995) Hepatic toxicity resulting from cancer treatment. Int J Radiat Oncol Biol Phys 31:1237–1248

Lax I, Blomgren H, Naslund I et al. (1994) Stereotactic radiotherapy of malignancies in the abdomen. Methodological aspects. Acta Oncol 33:677–683

Lindemann J, Leiacker R, Rettinger G et al. (2002) Nasal mucosal temperature during respiration. Clin Otolaryngol 27:135–139

Low DA, Harms WB, Mutic S et al. (1998) A technique for the quantitative evaluation of dose distributions. Med Phys 25:656–661

Low D, Wahab S, El Naqa I et al. (2003a) Four-dimensional

computed tomography using a 16-slice scanner. Med Phys 30:1506

Low D, Nystrom M, Kalinin E et al. (2003b) A method for the reconstruction of four-dimensional synchronized CT scans acquired during free breathing. Med Phys 30:1254–1263

Lu W, Mackie TR (2002) Tomographic motion detection and correction directly in sinogram space. Phys Med Biol 47:1267–1284

Lujan AE, Larsen EW, Balter JM et al. (1999) A method for incorporating organ motion due to breathing into 3D dose calculations. Med Phys 26:715–720

Mageras GS, Yorke E (2004) Deep inspiration breath hold and respiratory gating strategies for reducing organ motion in radiation treatment. Semin Radiat Oncol 14:65–75

Mageras GS, Yorke E, Rosenzweig K et al. (2001) Fluoroscopic evaluation of diaphragmatic motion reduction with a respiratory gated radiotherapy system. J Appl Clin Med Phys 2:191–200

Mah D, Hanley J, Rosenzweig KE et al. (2000) Technical aspects of the deep inspiration breath-hold technique in the treatment of thoracic cancer. Int J Radiat Oncol Biol Phys 48:1175–1185

Marks MK, South M, Carter BG (1995) Measurement of respiratory rate and timing using a nasal thermocouple. J Clin Monit 11:159–164

Martel MK, Ten Haken RK, Hazuka MB et al. (1999) Estimation of tumor control probability model parameters from 3-D dose distributions of non-small cell lung cancer patients. Lung Cancer 24:31–37

Mehta M, Scrimger R, Mackie R et al. (2001) A new approach to dose escalation in non-small-cell lung cancer. Int J Radiat Oncol Biol Phys 49:23–33

Menke M, Hirschfeld F, Mack T et al. (1994) Stereotactically guided fractionated radiotherapy: technical aspects. Int J Radiat Oncol Biol Phys 29:1147–1155

Minohara S, Kanai T, Endo M et al. (2000) Respiratory gated irradiation system for heavy-ion radiotherapy. Int J Radiat Oncol Biol Phys 47:1097–1103

Mori S, Endo M, Tsunoo T et al. (2004) Physical performance evaluation of a 256-slice CT-scanner for four-dimensional imaging. Med Phys 31:1348–1356

Morita K, Fuwa N, Suzuki M et al. (1997) Radical radiotherapy for medically inoperable non-small cell lung cancer in clinical stage I: a retrospective analysis of 149 patients. Radiother Oncol 42:31–36

Murphy MJ, Cox RS (1996) The accuracy of dose localization for an image-guided frameless radiosurgery system. Med Phys 23:2043–2049

Murphy MJ, Adler JR Jr, Bodduluri M et al. (2000) Image-guided radiosurgery for the spine and pancreas. Comput Aided Surg 5:278–288

Murphy MJ, Martin D, Whyte R et al. (2002) The effectiveness of breath-holding to stabilize lung and pancreas tumors during radiosurgery. Int J Radiat Oncol Biol Phys 53:475–482

Nagata Y, Negoro Y, Aoki T et al. (2003) Clinical outcomes of 3D conformal hypofractionated single high-dose radiotherapy for one or two lung tumors using a stereotactic body frame. Int J Radiat Oncol Biol Phys 52:1041–1046

Narayan S, Henning GT, Ten Haken RK et al. (2004) Results following treatment to doses of 92.4 or 102.9 Gy on a phase I dose escalation study for non-small cell lung cancer. Lung Cancer 44:79–88

Negoro Y, Nagata Y, Aoki T et al. (2001) The effectiveness of an immobilization device in conformal radiotherapy for lung tumor: reduction of respiratory tumor movement and evaluation of the daily setup accuracy. Int J Radiat Oncol Biol Phys 50:889–898

Neicu T, Shirato H, Seppenwoolde Y et al. (2003) Synchronized moving aperture radiation therapy (SMART): average tumour trajectory for lung patients. Phys Med Biol 48:587–598

Norman RG, Ahmed MM, Walsleben TA et al. (1997) Detection of respiratory events during NPSG: nasal cannula/pressure sensor versus thermistor. Sleep 20:1175–1184

Ohara K, Okumura T, Akisada M et al. (1989) Irradiation synchronized with respiration gate. Int J Radiat Oncol Biol Phys 17:853–857

Ozhasoglu C, Murphy MJ (2002) Issues in respiratory motion compensation during external-beam radiotherapy. Int J Radiat Oncol Biol Phys 52:1389–1399

Pan T, Lee T-Y, Rietzel E et al. (2004) 4D-CT imaging of a volume influenced by respiratory motion on multi-slice CT. Med Phys 31:333–340

Peterson B, Palmerud G (1996) Measurement of upper extremity orientation by video stereometry system. Med Biol Eng Computing 34:149–154

Ramsey CR, Cordrey IL, Oliver AL (1999) comparison of beam characteristics for gated and nongated clinical X-ray beams. Med Phys 26:2086–2091

Reed GB, Cox AJ (1966) The human liver after radiation injury. Am J Pathol 48:597–612

Remouchamps VM, Letts N, Vicini FA et al. (2003a) Initial clinical experience with moderate deep-inspiration breath hold using an active breathing control device in the treatment of patients with left-sided breast cancer using external beam radiation therapy. Int J Radiat Oncol Biol Phys 56:704–715

Remouchamps VM, Letts N, Yan D et al. (2003b) Three-dimensional evaluation of intra- and interfraction immobilization of lung and chest wall using active breathing control: a reproducibility study with breast cancer patients. Int J Radiat Oncol Biol Phys 57:968–978

Remouchamps VM, Vicini FA, Sharpe MB et al. (2003c) Significant reductions in heart and lung doses using deep inspiration breath hold with active breathing control and intensity-modulated radiation therapy for patients treated with locoregional breast irradiation. Int J Radiat Oncol Biol Phys 55:392–406

Ries LAG, Eisner MP, Kosary CL et al. (eds) (2004) NCI Surveillance, Epidemiology and End Results. SEER Cancer Statistics Review, 1975–2001, National Cancer Institute. Bethesda, Maryland

Rietzel E, Chen GT, Doppke KP et al. (2003) 4D computed tomography for treatment planning. Int J Radiat Oncol Biol Phys 57:S232–S233

Ritchie CJ, Hsieh J, Gard MF et al. (1994) Predictive respiratory gating: a new method to reduce motion artifacts on CT scans. Radiology 190:847–852

Robertson JM, Lawrence TS, Andrews JC et al. (1997a) Long-term results of hepatic artery fluorodeoxyuridine and conformal radiation therapy for primary hepatobiliary cancers. Int J Radiat Oncol Biol Phys 37:325–330

Robertson JM, Ten Haken RK, Hazuka MB et al. (1997b) Dose

escalation for non-small cell lung cancer using conformal radiation therapy. Int J Radiat Oncol Biol Phys 37:1079–1085

Roof KS, Fidias P, Lynch TJ et al. (2003) Radiation dose escalation in limited-stage small-cell lung cancer. Int J Radiat Oncol Biol Phys 57:701–708

Rosenzweig KE, Hanley J, Mah D et al. (2000) The deep inspiration breath-hold technique in the treatment of inoperable non-small-cell lung cancer. Int J Radiat Oncol Biol Phys 48:81–87

Ross C, Hussey DH, Pennington EC et al. (1990) Analysis of movement on intrathoracic neoplasm using ultrafast computerized tomography. Int J Radiat Oncol Biol Phys 18:671–677

Ruschin M, Sixel KE (2002) Integration of digital fluoroscopy with CT-based radiation therapy planning of lung tumors. Med Phys 29:1698–1709

Sang-Wook Lee S, Eun Kyung Choi EK, Heon Joo Park HJ et al. (2003) Stereotactic body frame based fractionated radiosurgery on consecutive days for primary or metastatic tumors in the lung. Lung Cancer 40:309–315

Sato M, Uematsu M, Yamamoto F et al. (1998) Feasibility of frameless stereotactic high-dose radiation therapy for primary or metastatic liver cancer. J Radiosurg 1:233–240

Schlegel W, Pastyr O, Bortfeld T et al. (1993) Stereotactically guided fractionated radiotherapy: technical aspects. Radiother Oncol 29:197–204

Schweikard A, Glosser G, Bodduluri M et al. (2000) Robotic motion compensation for respiratory movement during radiosurgery. Comput Aided Surg 5:263–277

Selvik G (1990) Roentgen stereophotogrammetric analysis. Acta Radiol 31:113–126

Seppenwoolde Y, Shirato H, Kitamura K et al. (2002) Precise and real-time measurement of 3D tumor motion in lung due to breathing and heartbeat, measured during radiotherapy. Int J Radiat Oncol Biol Phys 53:822–834

Shimizu S, Shirato H, Ogura S et al. (2001) Detection of lung tumor movement in real-time tumor-tracking radiotherapy. Int J Radiat Oncol Biol Phys 51:304–310

Shirato H, Harada T, Harabayashi T et al. (2003) Feasibility of insertion/implantation of 2.0-mm-diameter gold internal fiducial markers for precise setup and real-time tumor tracking in radiotherapy. Int J Radiat Oncol Biol Phys 56:240–247

Shirato H, Shimizu S, Kitamura K et al. (2000a) Four-dimensional treatment planning and fluoroscopic real-time tumor tracking radiotherapy for moving tumor. Int J Radiat Oncol Biol Phys 48:435–442

Shirato H, Shimizu S, Kunieda T et al. (2000b) Physical aspects of a real-time tumor-tracking system for gated radiotherapy. Int J Radiat Oncol Biol Phys 48:1187–1195

Sixel K, Ruschin M, Cheung PC (2001) Integration of digital fluoroscopy with CT simulation: patient specific planning target volumes. Int J Radiat Oncol Biol Phys 51 (S1):122–123

Sixel KE, Ruschin M, Tirona R et al. (2003) Digital fluoroscopy to quantify lung tumor motion: potential for patient-specific planning target volumes. Int J Radiat Oncol Biol Phys 57:717–723

Solberg TD, Paul TJ, Agazaryan N et al. (2000) Dosimetry of gated intensity modulated radiotherapy. In: Schlegel W, Bortfeld T (eds) The use of computers in radiation therapy. Springer, Berlin Heidelberg New York, pp 286–288

Stevens CW, Munden RF, Forster KM et al. (2001) Respiratory-driven lung tumor motion is independent of tumor size, tumor location, and pulmonary function. Int J Radiat Oncol Biol Phys 51:62–68

Stromberg JS, Sharpe MB, Kim LH et al. (2000) Active breathing control (ABC) for Hodgkin's disease: reduction in normal tissue irradiation with deep inspiration and implications for treatment. Int J Radiat Oncol Biol Phys 48:797–806

Stroom JC, Heijmen BJ (2002) Geometrical uncertainties, radiotherapy planning margins, and the ICRU-62 report. Radiother Oncol 64:75–83

Suit HD, Becht J, Leong J et al. (1988) Potential for improvement in radiation therapy. Int J Radiat Oncol Biol Phys 14:777–786

Taguchi K (2003) Temporal resolution and the evaluation of candidate algorithms for four-dimensional CT. Med Phys 30:640–650

Ten Haken RK, Balter JM, Marsh LH et al. (1997) Potential benefits of eliminating planning target volume expansions for patient breathing in the treatment of liver tumors. Int J Radiat Oncol Biol Phys 38:613–617

Tsunashima Y, Sakae T, Shioyama Y et al. (2004) Correlation between the respiratory waveform measured using a respiratory sensor and 3D tumor motion in gated radiotherapy. Int J Radiat Oncol Biol Phys. 60:951–958

Vedam SS, Keall PJ, Kini VR et al. (2001) Determining parameters for respiration-gated radiotherapy. Med Phys 28:2139–2146

Vedam SS, Keall PJ, Kini VR et al. (2003) Acquiring a four-dimensional computed tomography dataset using an external respiratory signal. Phys Med Biol 48:45–62

Vedam SS, Keall PJ, Docef A et al. (2004) Predicting respiratory motion for four-dimensional radiotherapy. Med Phys 31:2274–2283

Verellen D, Soete G, Linthout N et al. (2003) Quality assurance of a system for improved target localization and patient set-up that combines real-time infrared tracking and stereoscopic X-ray imaging. Radiother Oncol 67:129–141

Wagman R, Yorke E, Ford E et al. (2003) Respiratory gating for liver tumors: use in dose escalation. Int J Radiat Oncol Biol Phys 55:659–668

Wahab S, Low D, El Naqa I et al. (2003) Use of four-dimensional computed tomography in conformal therapy planning for lung cancer. Med Phys 30:1364

Wang LT, Solberg TD, Medin PM et al. (2001) Infrared patient positioning for stereotactic radiosurgery of extracranial tumors. Comput Biol Med 31:101–111

Waters K, Terzopoulos D (1992) The computer synthesis of expressive faces. Philos Trans R Soc London Series B Biol Sci 335:87–93

Whyte RI, Crownover R, Murphy MJ et al. (2003) Stereotactic radiosurgery for lung tumors: preliminary report of a phase I trial. Ann Thorac Surg 75:1097–1101

Willett CG, Linggood RM, Stracher MA et al. (1987) The effect of the respiratory cycle on mediastinal and lung dimensions in Hodgkin's disease. Implications for radiotherapy gated to respiration. Cancer 60:1232–1237

Willner J, Baier K, Caragiani E et al. (2002) Dose, volume, and tumor control predictions in primary radiotherapy of non-small-cell lung cancer. Int J Radiat Oncol Biol Phys 52:382–389

Wilson EM, Williams FJ, Lyn BE et al. (2003) Validation of

active breathing control in patients with non-small-cell lung cancer to be treated with CHARTWEL. Int J Radiat Oncol Biol Phys 57:864–874

Wong JW, Sharpe MB, Jaffray DA et al. (1997) The use of active breathing control (ABC) to minimize breathing motion during radiation therapy. Int J Radiat Oncol Biol Phys 39:164 (abstract)

Wong JW, Sharpe MB, Jaffray DA et al. (1999) The use of active breathing control (ABC) to reduce margin for breathing motion. Int J Radiat Oncol Biol Phys 44:911–919

Wulf J, Hädinger U, Oppitz U et al. (2001) Stereotactic radio-

therapy of targets in the lung and liver. Strahlenther Onkol 177:645–655

Xiong C, Sjoberg BJ, Sveider P et al. (1993) Problems in the timing of respiration with the nasal thermistor technique. J Am Soc Echocardiogr 6:210–216

Yu CX, Jaffray DA, Wong JW (1998) The effects of intra-fraction organ motion on the delivery of dynamic intensity modulation. Phys Med Biol 43:91–104

Zhang T, Jeraj R, Keller H et al. (2004) Treatment plan optimization incorporating respiratory motion. Med Phys 31:1576–1586

25 Image-Guided/Adaptive Radiotherapy

DI YAN

CONTENTS

25.1
Introduction

Adaptive radiotherapy is a treatment technique that can systematically improve its treatment plan in response to patient/organ temporal variations observed during the therapy process. Temporal variations in radiotherapy process can be either patient/organ

D. YAN, D.Sc.
Director, Clinical Physics Section, Department of Radiation Oncology, William Beaumont Hospital, Royal Oak, MI 48073-6769, USA

geometry or dose-response related. Examples of the former include inter- and intra-treatment variations of patient/organ shape and positions caused by patient setup, beam placement, and patient organ physiological motion and deformation. Examples of dose response characteristics include the variations of size and location of tumor hypoxic volume, the apparent tumor growth fraction, and normal tissue damage/repair kinetics. Furthermore, tumor and normal organ dose response also induce changes in tissue shape and positions.

This chapter focuses on the use of adaptive strategies to manage patient/organ shape and position related temporal variations; however, the concepts of adaptive radiotherapy can be extended to a much broader range, including the management in temporal variations of patient/organ biology.

It is generally accepted that temporal variations are the predominant sources of treatment uncertainty in conventional radiation treatment. Numerous imaging studies have demonstrated that substantial temporal variations of patient/organ shape and position could occur during a typical radiotherapy course (BRIERLEY et al. 1994; DAVIES et al. 1994; HALVERSON et al. 1991; MARKS and HAUS 1976; MOERLAND et al. 1994; NUYTTENS et al. 2001; ROESKE et al. 1995; ROSS et al. 1990). Consequently, the radiation dose delivered to the target and a critical normal organ adjacent to the target can significantly deviate from that calculated in the pre-treatment planning. This deviation causes a time-dependent, or temporal, variation in the organ-dose distribution, consisting of both dose variation per treatment fraction and cumulative dose variation in each subvolume of the organ, and results in major treatment uncertainties. These uncertainties induce fundamental obstacles to assuring treatment quality and understanding the normal tissue dose response, thereby hindering reliable treatment optimization.

Temporal variations in patient/organ shape and position during the radiotherapy course can be separated into a systematic component and a random component. The systematic component represents a consistent discrepancy between the patient/organ shape and position appearing in pre-treatment simulation/planning and that at treatment delivery; therefore, it is also called treatment preparation error (VAN HERK et al. 2000). The random component represents patient/organ shape and position variations between treatment deliveries; these are also referred to as treatment execution errors. Because there is almost always some random component in a temporal variation, observations achieved by imaging the patient repeatedly during the treatment course are essential to characterize the variations. The most important function of repeat imaging is to identify the systematic component of a temporal variation, and consequently eliminate its effect in the treatment.

Due to intrinsic temporal variations, targeting in radiotherapy process is, in principle, a four-dimensional (4D) problem, i.e., involving not only space but also time; therefore, it is an adaptive optimal control methodology ideally suited to manage this process. A general adaptive radiotherapy system consists of five basic components: (a) *treatment delivery* to deliver radiation dose to the patient based on a treatment plan; (b) *imaging/verification* to observe and verify patient/organ temporal variation before, during, and/or after a treatment delivery; (c) *estimation/evaluation* to estimate, based on image feedback, the parameters which can characterize the undergoing temporal variation process, and evaluate the corresponding treatment parameters, such as the cumulative dose, biological effective dose, TCP, NTCP, etc.; (d) *design of adaptive planning/adjustment* to design and update planning/adjustment parameters, as well as modify imaging, delivery, and adjustment sched-

ules, in response to the estimation and the evaluation; and (e) *adaptive planning/adjustment* to perform a 4D conformal or IMRT planning with using the planning parameters specified in the adaptive planning/adjustment design, and adjust treatment delivery accordingly. A typical adaptive radiotherapy system is illustrated in Fig. 25.1. In the standard textbook of adaptive control (ASTROM and WITTENMARK 1995), this system is called the self-tuning regulator (STR), indicating a system that can update its planning and control parameters automatically. Patient treatment in this system is initiated by a pre-treatment plan and resides within two feedback loops. The inner loop consists of treatment delivery, imaging/verification, and planning/adjustment, which have been designed primarily to perform online image-guided treatment adjustment. The planning/adjustment parameters are updated and modified most likely offline in the outer loop, which is composed of the imaging/verification, parameter estimation/evaluation for a temporal variation process, design of adaptive planning/adjustment, and adaptive planning/adjustment. In addition, the schedules of adaptive planning/adjustment, treatment delivery, and imaging/verification in the adaptive radiotherapy system are most likely pre-designed and specified in a clinical adaptive treatment protocol; however, these schedules can be modified and updated during the treatment based on new observation and estimation (the dashed lines in Fig. 25.1).

The adaptive radiotherapy system shown in Fig. 25.1 has a very rich configuration. Only a few potentials have been investigated thus far in radiotherapy which are outlined in this chapter as the exam-

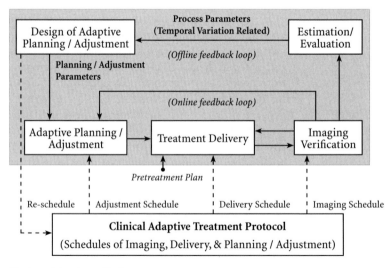

Fig. 25.1 Adaptive radiotherapy system

ples of clinical implementation. In this chapter, each key component in the adaptive radiotherapy system is described. In section 25.2, temporal variations of patient/organ shape and position are outlined and classified based on their characteristics. In section 25.3, X-ray imaging and verification techniques are described. Estimation and evaluation of temporal variation-related process parameters are introduced in section 25.4. In section 25.5 design and selection of control parameters for adaptive planning and adjustment are discussed. These parameters are directly used in the 4D planning/adjustment described in section 25.6. Finally, section 25.7 provides two typical adaptive treatment protocols that have been implemented or intends to be implemented in the clinic.

25.2
Temporal Variation of Patient/Organ Shape and Position

Temporal variation of an organ shape and position with respect to the radiation beams can be determined by the position displacement of subvolumes in the organ (V). For a given patient treatment, patient organ (target or normal structure) can be defined as a set of subvolumes or volume elements v, such that $V=\{v\}$.

The notation $\bar{x}_t(v) = \begin{bmatrix} x_t^{(1)}(v) \\ x_t^{(2)}(v) \\ x_t^{(3)}(v) \end{bmatrix} \in R^3$ indicates the three-

dimensional (3D) position vector of a subvolume v at a time instant t; therefore, shape and position variation of the organ of interest during the entire treatment period, T, is specified as

$$\bar{x}_t(v) = \bar{x}_r(v) + \bar{u}_t(v), \quad \forall v \in V ; t \in T \qquad (1)$$

where $\bar{x}_r(v) \in R^3$ is the subvolume position manifested on a pre-treatment CT image for treatment planning, and

$$\bar{u}_t(v) = \begin{bmatrix} u_t^{(1)}(v) \\ u_t^{(2)}(v) \\ u_t^{(3)}(v) \end{bmatrix} \in R^3$$

is the displacement vector of the subvolume at a time instant t.

Denoting T_i, $i=1, ..., n$ to be the time interval of dose delivery (<5 min) in each of the number n treatment fractions, then the organ shape and position variation represented by the subvolume displacements during the entire course of treatment delivery

can be modeled as a process of time, or a temporal variation process, as

$$\left\{ \bar{u}_t(v) \mid t \in \bigcup_{i=1}^{n} T_i \subset T \right\}, \forall v \in V \qquad (2)$$

On the other hand, the organ shape and position variation during each dose delivery can be modeled as

$$\left\{ \bar{u}_t(v) \mid t \in T_i \right\}, \forall v \in V ; i = 1,...,n \qquad (3)$$

It is clear that the processes [Exp. (3)] are subprocesses of the whole treatment process [Exp. (2)], and have been called intra-treatment process. Patient/organ shape and position variations have been classified into the inter-treatment variation, defined as $\bar{u}_t(v) = const$, $\forall t \in T_i$, and the intra-treatment variation where $\bar{u}_t(v)$ changes within T_i; however, both the variations most likely exist simultaneously during the treatment delivery and cannot be easily separated. Typical example of inter-treatment variation is the daily treatment setup error with respect to patient bony structure. Meanwhile, the typical example of intra-treatment variation is the patient respiration-induced organ motion.

Given an organ subvolume, the displacement sequence, denoted as a set of random vectors in Exps. (2) or (3), can be modeled as a random process within the time domain T of the treatment course or T_i of a treatment delivery. Using Eq. (1), subvolume displacement in the random process can be decomposed (YAN and LOCKMAN 2001) such that

$$\bar{u}_t(v) = \bar{\mu}_t(v) + \bar{\xi}_t(v), \quad \forall v \in V ; t \in T \qquad (4)$$

where $\bar{\mu}_t(v) = E[\bar{u}_t(v)]$ is the mean of the displacement or the mean of the random process, and $\bar{\xi}_t(v)$ is the random vector which has a zero mean but same shape of probability distribution of the displacement. The mean, by definition, is the systematic variation, and the standard deviation,

$$\bar{\sigma}_t(v) = \sqrt{E[\bar{\xi}_t(v)]^2} = \sqrt{E[\bar{u}_t(v) - \bar{\mu}_t(v)]^2},$$

is used to characterize the random variation $\bar{\xi}_t(v)$. In addition, the mean and the standard deviation have been proved to be the most important factors to influence treatment dose distribution; thus, they have been selected as the primary process parameters of temporal variation considered in the design of an adaptive treatment plan.

A temporal variation process can be a stationary random process if it has a constant mean during the treatment course, such as $\bar{\mu}_t(v) = \bar{\mu}(v)$, $\forall t \in T$, and a constant standard deviation, such as

$\bar{\sigma}_t(v) = \bar{\sigma}(v), \;\; \forall t \in T$. The condition of the constant standard deviation is slightly stronger than the formal definition of the stationary (wide sense) random process in the textbook (Wong 1983); however, it is more suitable for describing a temporal variation process in radiotherapy.

Followed by the above definition, a temporal variation process can be described using the stationary random process if its systematic variation and the standard deviation of the random variation are constants within the entire course of radiotherapy; otherwise, it is a non-stationary random process. Most of temporal variation process of patient/organ shape and position in radiotherapy can be considered as a stationary process. Examples of non-stationary process are most likely dose-response related, such as a process of organ displacement with its mean displacement drifted due to reopening of atelectasis lung, a process affected by organ filling that is changed by radiation dose, or a process with a normal organ adjacent to a shrinking target.

A subprocess of intra-treatment variation, $\left\{ \bar{u}_t(v) \mid t \in T_i \right\}$, can also be classified as the stationary and non-stationary. In this case, example of the stationary process could be related to patient respiration induced organ motion. On the other hand, example of the non-stationary process could be related to an organ-filling process such as intra-treatment bladder filling.

25.3
Imaging and Verification

Imaging (sampling) patient/organ shape and position frequently during the treatment course is the major means of verifying and characterizing anatomical variation in radiotherapy. Ideally, imaging should be performed with patient setup in treatment position and with a sampling schedule compatible with the frequency of the temporal variation considered. Commonly, imaging schedule in an adaptive radiotherapy is pre-designed in the treatment protocol based on specifications required for the estimation and evaluation of temporal variation process parameters, which are further discussed in section 25.4.

Three modes of X-ray imaging have been implemented in the radiotherapy clinic to observe patient anatomy-related temporal variation, which are radiographic, fluoroscopic, and volumetric CT imaging. In addition, 4D CT image can also be created (Ford et al. 2003; Sonke et al. 2003). Onboard imaging devices with partial or all three modes are commercially available, which include onboard MV or kV imager, in-room kV imager, CT on rail, tomotherapy unit with onboard MV CT, and onboard cone-beam kV CT.

25.3.1
Verification with Radiographic Imaging

Onboard MV radiographic imaging has been used to verify patient daily setup measured using the position of patient bony anatomy, or position of a region of interest with implanted radio-markers. Normally, the position error is determined using the rigid body registration between a daily treatment radiographic image and a reference radiographic image, most likely a digital reconstructed radiographic (DRR) image created in treatment planning. There have been numerous methods on 2D X-ray to 2D X-ray registration, which have been outlined in a survey paper (Antonie Maintz et al. 1998). Capability of using radiographic image for treatment verification has been extensively studied and is conclusive. It can be applied to determine bony anatomy position as a surrogate to verify patient setup position. In addition, it can also be used to locate implanted radio-markers' position as a surrogate to verify the position of a region of interest.

Patient/organ position displacement caused by a rigid body motion at the ith treatment delivery has been denoted using a vector of three translational parameters and a 3×3 matrix with three rotational parameters as $\bar{u}_t(v) = \bar{\Delta}_t(p) + R_t(p) \cdot \left[\bar{x}_r(v) - \bar{x}_r(p) \right], \;\; \forall v \in V; \; t \in T_i,$

where $\bar{\Delta}_t(p) = \begin{bmatrix} \delta_t^{(1)} \\ \delta_t^{(2)} \\ \delta_t^{(3)} \end{bmatrix}$ is the translational vector with

shift, $\delta_t^{(j)}$, along j^{th} axis determined with respect to a reference point p pre-defined on the reference image. $R_t(p) = R^{(1)} R^{(2)} R^{(3)}$ is the rotation matrix with respect to the same reference point and rotation around individual axis, such that

$$R^{(1)} = \begin{pmatrix} 1 & 0 & 0 \\ 0 & \cos\theta_t^{(1)} & -\sin\theta_t^{(1)} \\ 0 & \sin\theta_t^{(1)} & \cos\theta_t^{(1)} \end{pmatrix},$$

$$R^{(2)} = \begin{pmatrix} \cos\theta_t^{(2)} & 0 & \sin\theta_t^{(2)} \\ 0 & 1 & 0 \\ -\sin\theta_t^{(2)} & 0 & \cos\theta_t^{(2)} \end{pmatrix} \text{ and}$$

$$R^{(3)} = \begin{pmatrix} \cos\theta_t^{(3)} & -\sin\theta_t^{(3)} & 0 \\ \sin\theta_t^{(3)} & \cos\theta_t^{(3)} & 0 \\ 0 & 0 & 1 \end{pmatrix}$$

representing the subrotation matrix around j^{th} axis by an angle $\theta_t^{(j)}$. Since the displacements of all sub-volumes in a region of interest are uniquely determined by the translational vector and rotation matrix, only the six parameters, $\delta_t^{(j)}$; $\theta_t^{(j)}$, $j = 1, 2, 3$, are needed to determine patient/organ rigid body motion. Conventionally, the translational vector, $\bar{\Delta}_t$, $t = t_1, ..., t_n$, observed using a portal imaging device before, during, and/or after treatment delivery, have been used to represent the temporal variation of patient setup error, when rotation error in patient setup is insignificant.

25.3.2
Verification with Fluoroscopic Imaging

Fluoroscopy has been conventionally used to observe patient respiration-induced organ motion at a treatment simulator to guide target margin design in radiotherapy planning of lung cancer treatment. Recently, due to the availability of onboard kV imaging, it is being applied to verify intra-treatment organ motion induced by patient respiration (HUGO et al. 2004). This verification has been established by comparing the online portal fluoroscopy to the digital reconstructed fluoroscopy (DRF) created using the 4D CT image. Respiration-induced organ motion can be determined by tracking the motion of a landmark or a radio-marker implanted in or close to the organ of interest. Consequently, the frequency or the density of the motion can be derived by calculating the ratio of an accumulated time, within which the patient respiration-induced displacement is equal to a constant, versus the entire interval of breathing motion measurement (LUJAN et al. 1999).

Symbolically, the motion frequency or density function for a point of interest p can be calculated as

$$\varphi(p, c) = \frac{\left|\left\{\tau \mid \bar{u}_\tau(p) \equiv c, \forall \tau \in T_i\right\}\right|}{|T_i|},$$

where $\bar{u}_\tau(p), \tau \in T_i$ is the respiratory displacement of p measured using the fluoroscopic image within the time interval T_i. Figure 25.2 shows a typical time-position curve of patient breathing motion of a point of interest and its corresponding density function. In clinical practice, both the respiratory motion and its frequency are important for adaptive treatment design and planning to compensate for a patient respiration-induced temporal variation (LIANG et al. 2003). For treatment planning purpose, fluoroscopic image can be obtained in treatment position from either a simulator or an onboard kV imager; how-

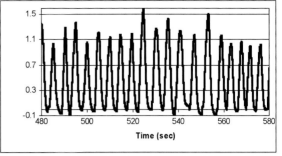

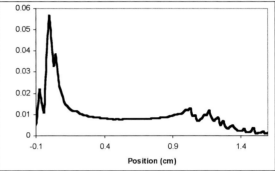

Fig. 25.2 A typical example of patient respiration-induced motion of a subvolume position and its corresponding position density distribution

ever, onboard fluoroscopy is preferred for verifying treatment delivery. Positions and frequency of points of interest, specifically the mean and the standard deviation of the displacement, measured from an online fluoroscopy, are compared with those pre-determined from the DRF created in the adaptive planning to verify the treatment quality.

25.3.3
Verification with CT Imaging

Volumetric CT has been the most useful imaging mode in verifying temporal variation of patient anatomy. Using this mode, the treatment dose in organs of interest could be constructed. The treatment plan can be designed in response to changes of patient/organ shape and position during the therapy course; however, due to overwhelming information contained in a 3D and 4D anatomical image, it also brings a great challenge in the applications of volumetric image feedback.

One of the most difficult tasks in applying volumetric image feedback in adaptive treatment planning is the image-based deformable organ registration. Unlike rigid body registration that has been well developed and discussed everywhere, deformable organ registration is quite immature. Methods of volumetric image-based deformable organ registration have been conventionally classified into two classes (ANTONIE MAINTZ et al. 1998): the segmentation-based registration method and the voxel property-based registration method. Segmentation-based registration utilizes the contours or surface of an organ of interest delineated from the reference image to elastically match the organ manifested on the second image (MCINERNEY and TERZOPOULOS 1996). On the other hand, voxel property-based registration method utilizes mutual information manifested in two images to perform the registration (PLUIM et al. 2003). Both registration methods, in principle, share a same problem on the interpretation of the rest of points of interest. Mathematically, this problem can be described as for given conditions $\{\bar{x}_r(v) \mid v \in V\}$ – the subvolume position of an organ of interest manifested on the reference image and $\{\bar{x}_t(v) = \bar{x}_r(v) + \bar{u}_t(v) \mid v \in \partial V \subset V\}$ – the boundary condition of surface points or mutual information, determining $\{\bar{x}_t(v) = \bar{x}_r(v) + \bar{u}_t(v) \mid v \in V - \partial V\}$ – the rest of subvolume positions manifested on the secondary image. Existing methods of interpretation are the finite element analysis that determines subvolume position based on the mechanical con-

stitutive equations and tissue elastic properties, and the direct interpretation of using a linear or a spline interpolation. Applications in radiotherapy include using the finite element method to perform CT image-based deformable organ registration for organs of interest in the prostate cancer treatment (YAN et al. 1999), the GYN cancer treatment (CHRISTENSEN et al. 2001) and the liver cancer treatment (BROCK et al. 2003). Deformable organ registration followed by volumetric image feedback provides the distribution of organ subvolume displacements (Fig. 25.3), which plays an important role in the adaptive or 4D planning; however, there is no clear answer thus far as to what degree of registration accuracy can be achieved utilizing each interpretation method and what is needed for an adaptive treatment planning.

Two types of sequential CT imaging have been applied in adaptive treatment planning. The first one has a longer elapse (day or days) of imaging (sampling) to primarily measure an inter-treatment temporal variation. Clinical applications of using multiple daily images have been limited to prostate cancer treatment (YAN et al. 2000), colon-rectal cancer treatment (NUYTTENS et al. 2002), and head and neck cancer treatment. The second sequential imag-

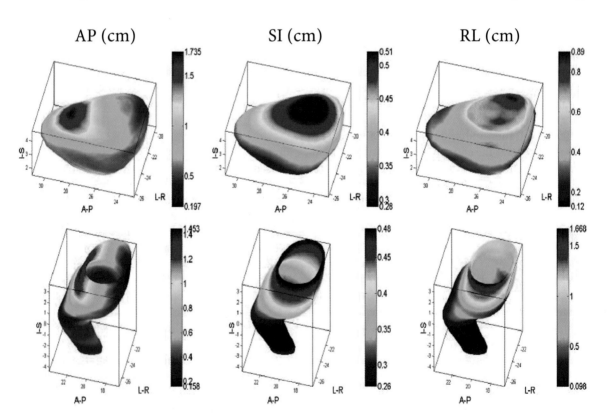

Fig. 25.3 A typical example of subvolume displacement distribution for a bladder wall and a rectal wall. The color map from *red* to *blue* indicates the range (large to small) of the standard deviation of each subvolume displacement in centimeters

ing has been aimed to detect organ motion induced by patient respiration. These images, which manifest organs of interest at different breathing phases, can be obtained from a respiratory correlated CT imaging (FORD et al. 2003) or a slow rotating cone-beam CT (SONKE et al. 2003). Clinical application of this measurement has been focused on lung cancer treatment; however, it has no limit to be extended to the other cancer treatment where patient respiration effect is significant, such as liver cancer treatment. In principle, both types of sequential imaging are 4D CT imaging, although the terminology of 4D CT imaging has been specifically used to describe the 3D sequential CT images induced by patient respiration. Commonly, either an onboard or an off-board volumetric imager can be applied to measure patient organ motion for the purpose of offline planning modification; however, onboard imaging is a favorable tool for both online and offline treatment planning modification and adjustment.

25.4
Estimation and Evaluation

It has been discussed that temporal variation of patient/organ shape and position during the whole treatment course could be modeled as a random process of organ subvolume displacement, denoted

as $\left\{ \bar{u}_t(v) \mid t \in \bigcup_{i=1}^{n} T_i \subset T \right\}, \forall v \in V,$ or multiple

subprocesses during each treatment delivery, denoted as $\left\{ \bar{u}_t(v) \mid t \in T_i \right\}, \forall v \in V; i = 1,...,n$. The former has been primarily considered in the design of offline imaging and planning modification as indicated as the outer loop of the adaptive system in Fig. 25.1; the latter, on the other hand, has to be considered additionally in the design of online image-guided adjustment – the inner loop of Fig. 25.1. Although process parameter estimation and treatment evaluation are normally performed in the outer feedback loop, they will be utilized to modify and update the planning and adjustment parameters for both offline and online planning modification and adjustment.

As has been discussed in section 25.2, the key parameters of a temporal variation process are the systematic variation and the standard deviation of random variation of subvolume displacement [see also Eq. (4)]. These parameters, therefore, have to be estimated for either offline or online mechanisms. Moreover, the cumulative dose/volume relationship

of organs of interest and the corresponding biological indexes can also be estimated and evaluated. In addition, effects of dose per fraction on a critical normal organ should also be considered in the cumulative dose evaluation when online image guided hypo-fractionation is implemented, because the effect of temporal variation of organ fraction dose can be significantly enlarged in a hypo-fractionated treatment (YAN and LOCKMAN 2001).

Multiple imaging (or sampling) performed in the early part of treatment has been the common means to estimate the mean $\bar{\mu}_t$ (the systematic variation) and the standard deviation $\bar{\sigma}_t$ (the characteristic of the random variation) of a temporal variation process. Estimation can be performed once using multiple images obtained early in treatment course, in batches, or continuously. In general, imaging schedule in an adaptive radiotherapy system has been selected in a treatment protocol based on a pre-designed strategy of adaptive treatment. In case of the single offline adjustment of patient position during the treatment course, optimal sampling schedule of four to five observations obtained daily in the early treatment course has been suggested (YAN et al. 2000) and proved to be a favorable selection with respect to the criteria of minimal cumulative displacement (BORTFELD et al. 2002); however, so far, there has not been any systematic study to explore the relationship between the imaging schedule and the treatment dose/volume factor. A preliminary study (BIRKNER et al. 2003) demonstrated that there was only a marginal improvement for prostate cancer IMRT, when offline-planning modification is continuously performed compared with a single modification after five measurements. Parameter estimation for a given temporal variation process could be straightforward. Example of such process is the organ motion induced by patient respiration. In this case, the motion can be characterized using 4D CT imaging and fluoroscopy before the treatment if the process is stationary; otherwise, the estimation can be performed multiple times during the treatment course.

25.4.1
Parameter Estimation for a Stationary Temporal Variation Process

Without loss of generality, let $\left\{ \bar{u}_{t_i} \mid t_i \in T; i = 1,...,k \right\}$ be the measurements (the sample size k is commonly small except for the respiratory motion) of subvolume displacement for an organ of interest obtained from k CT image measurements, or displacement of a reference point when radiographic images or fluoroscopy are used for the measurements. The standard unbiased estimations of the constant mean, $\bar{\mu}$, and the constant standard deviation, $\bar{\sigma}$, based on the measurements are

$$\hat{\bar{\mu}} = \frac{1}{k}\sum_{i=1}^{k}\bar{u}_{t_i}; \quad \hat{\bar{\sigma}} = \sqrt{\frac{1}{k-1}\sum_{i=1}^{k}(\bar{u}_{t_i} - \hat{\bar{\mu}})^2}.$$

In addition, the potential residuals between the true and the estimation can also be evaluated based on the standard confidence interval estimation as

$$\left\| \hat{\bar{\mu}} - \bar{\mu} \right\| \le t_{\alpha/2,k-1}\cdot\frac{\bar{\sigma}}{\sqrt{k}}; \quad \bar{\sigma} \le \sqrt{\frac{k-1}{\chi^2_{1-\alpha,k-1}}}\cdot\hat{\bar{\sigma}},$$

where the factor $t_{\alpha/2,k-1}$ has the t-distribution with $k-1$ degrees of freedom and $\chi^2_{1-\alpha,k-1}$ has the χ^2-distribution with $k-1$ degrees of freedom, and both them have the confidence $1-\alpha$.

The other method of estimating the systematic variation is the Wiener filtering (WIENER 1949). Applying the Wiener filtering theory, the optimal estimation of the systematic variation is constructed by minimizing the expectation of the estimation and the truth, such as $Min\, E\left[(\hat{\bar{\mu}}-\bar{\mu})^2\,|\,\bar{u}_{t_1},...,\bar{u}_{t_k},\bar{\Sigma}_{\mu},\bar{M}_{\sigma}\right]$, with conditions of the k measurements, the standard deviation of the individual systematic variations,

$$\bar{\Sigma}_{\bar{\mu}} = \begin{pmatrix} \sigma_{\mu_1} & 0 & 0 \\ 0 & \sigma_{\mu_2} & 0 \\ 0 & 0 & \sigma_{\mu_3} \end{pmatrix},$$

and the root-mean-square of the individual standard deviations of the random variations,

$$\bar{M}_{\bar{\sigma}} = \begin{pmatrix} \mu_{\sigma_1} & 0 & 0 \\ 0 & \mu_{\sigma_2} & 0 \\ 0 & 0 & \mu_{\sigma_3} \end{pmatrix}.$$

Consequently, the estimation of the systematic variation is

$$\hat{\bar{\mu}} = c\cdot\frac{1}{k}\sum_{i=1}^{k}\bar{u}_{t_i}, \quad c = k\cdot\bar{\Sigma}_{\bar{\mu}}\cdot\left(k\cdot\bar{\Sigma}_{\bar{\mu}} + \bar{M}_{\bar{\sigma}}\right)^{-1}.$$

It has been proved that for a temporal variation process, $\bar{\Sigma}_{\bar{\mu}}$ and $\bar{M}_{\bar{\sigma}}$ could have similar values; therefore, the Wiener estimation can be simplified as

$$\hat{\bar{\mu}} = \frac{1}{k+1}\cdot\sum_{i=1}^{k}\bar{u}_{t_i}.$$

In addition to the mean and the standard deviation, knowledge of the probability density $\varphi(v)$ of each subvolume displacement in a temporal variation process could also be useful; however, except for patient respiratory motion that can be determined directly using 4D CT and fluoroscopy (as described in section 25.3.2), the majority of temporal variations can only be practically measured a few times, and using these small numbers of measurements to estimate a probability distribution is most unlikely possible. Therefore, pre-assumed normal distribution has been applied in the clinic. It has been demonstrated that the actual treatment dose in an organ subvolume is most likely determined by its systematic variation and the standard deviation of its random variation. The actual shape of the displacement distribution is less important (YAN and LOCKMAN 2001); therefore, the pre-assumption of the normal distribution, $\hat{\varphi}(v) = N\left(\hat{\bar{\mu}}(v),\,\hat{\bar{\sigma}}^2(v)\right)$, is acceptable in the treatment dose estimation.

Parameter estimation of stationary temporal variation process can be applied for both offline and online feedback. Examples of offline feedback include using multiple radiographic portal imaging to characterize patient setup variation, multiple CT imaging to characterize internal organ motion, and 4D CT/fluoroscopy imaging to characterize respiratory organ motion. Application for online feedback is currently limited to characterize intra-treatment organ motion assessed by multiple portal imaging and portal fluoroscopy. For the online CT image-guided prostate treatment, parameter estimation for intra-treatment variation also depends on patient anatomical conditions. There has been a study (GHILEZAN et al. 2003) that showed that the intra-treatment variation of prostate position was primarily controlled by the rectal filling conditions.

25.4.2 Parameter Estimation for a Non-stationary Temporal Variation Process

Parameters to be estimated in a non-stationary process are similar to those in a stationary process; however, instead of constants, they can be piecewise constants, such as respiration-induced organ motion during lung cancer treatment, or a continuous function of time, such as bladder-filling-induced motion during treatment delivery. It is relatively simple to estimate the process parameters that are piecewise constants. The estimation in each constant period will be performed as same as the one for a stationary process; however, the estimation for a process with parameter as a continue function will be less straightforward. The most common method to estimate a function based on finite number of samples is the least-squares estimation. With pre-selected orthogonal base of functions $\vec{\phi}^{(j)}(t) = \left[\phi_1^{(j)}(t)\quad\phi_2^{(j)}(t)\quad\cdots\quad\phi_m^{(j)}(t)\right]$, i.e.

$\phi_i^{(j)}(t) = t^{i-1}$, $i = 1, 2, ..., m$, the estimations for both the systematic variation and the standard deviation of the random variation are

$$\hat{\mu}_t^{(j)} = \sum_{i=1}^{m} a_i^{(j)} \cdot \phi_i^{(j)}(t); \quad \hat{\sigma}_t^{(j)} = \sqrt{\frac{\sum_{i=1}^{k} (u_{t_i}^{(j)} - \hat{\mu}_{t_i}^{(j)})^2}{k - m}},$$

$j = 1, 2, 3,$ where

$$\begin{bmatrix} a_1^{(j)} \\ \vdots \\ a_m^{(j)} \end{bmatrix} = \left(\Phi^T \cdot \Phi \right)^{-1} \cdot \Phi^T \cdot \begin{bmatrix} u_{t_1}^{(j)} \\ \vdots \\ u_{t_k}^{(j)} \end{bmatrix}; \quad \Phi = \begin{bmatrix} \vec{\phi}^{(j)}(t_1) \\ \vdots \\ \vec{\phi}^{(j)}(t_k) \end{bmatrix}.$$

As the extension of the Wiener filter, the Kalman filter has also been applied to estimate the systematic variation for a non-stationary process assuming that the systematic variation is a linear function of time (Yan et al. 1995; Lof et al. 1998; Keller et al. 2004). In addition, similar to the description in the previous section, the probability density of each subvolume displacement can be estimated as $\hat{\varphi}_t(v)$ for a respiratory motion or a normal distribution $\hat{\varphi}_t(v) = N(\hat{\vec{\mu}}_t(v), \hat{\sigma}_t^2(v))$.

Applications of parameter estimation for a non-stationary process have been few. One study (Ford et al. 2002) attempted to determine the reproducibility of patient breathing-induced organ motion. It revealed that the mean of patient respiration-induced organ motion could considerably vary during the course of NSCLC due to treatment and patient related factors; therefore, multiple measurements of 4D CT and fluoroscopy, i.e., once a week, may be necessary to manage adaptive treatment of lung cancer. Regarding parameter estimation for a continuous function, one study (Heidi et al. 2004) has been performed to estimate bladder expansion and potential standard deviation for the online image-guided bladder cancer treatment, where bladder subvolume position was modeled as a linear function of time.

25.4.3
Estimation of Cumulative Dose

Including temporal variation in cumulative dose estimation can be performed using the knowledge of subvolume displacement distribution. At present, the construction is performed assuming time invariant spatial dose distribution calculated from the treatment planning. It implies that the dose distribution remains constant spatially regardless the changes of patient anatomy; however, this assumption can only be acceptable if spatial dose variation induced by the changes of patient body shape and tissue density is insignificant, i.e., during the prostate cancer treatment; otherwise, the dose has to be recalculated using each new feedback image. Of course, this can only be performed when CT image feedback is applied.

Cumulative dose for each subvolume in the organ of interest can be evaluated as

$$D(v) = \sum_{i=1}^{n} d_{t_i}(v) \quad ; \quad d_{t_i}(v) = \int_{t \in T_i} d_t(\vec{x}_r(v) + \vec{\xi}_t(v)) \cdot dt$$

or

$$D(v) = \sum_{i=1}^{n} d_{t_i}(v) \cdot \frac{\frac{\beta}{\alpha} + d_{t_i}(v)}{\frac{\beta}{\alpha} + d_s}$$ with considering the biological effect of dose per fraction, where d_s is the standard fraction dose 1.8 or 2 Gy. This dose expression is a very general and can be simplified based on the attributes of a temporal variation.

For a stationary process with the time invariance dose distribution, the cumulative dose in an organ of interest can be estimated directly utilizing the probability density of subvolume displacement $\hat{\varphi}(v)$ and the planned dose distribution d_p as

$$\hat{D}(v) = n \cdot \int_{R^3} d_p(\vec{x}_r(v) + \vec{\xi}(v)) \cdot \hat{\varphi}(\vec{\xi}) \cdot d\vec{\xi} \qquad (5)$$

if the planned dose per treatment fraction is fixed. When an offline planning modification is performed at the $k+1$ treatment delivery, based on the previous k image measurements, then the cumulative dose can be estimated by considering the treatments which have been delivered (Birkner et al. 2003), such that

$$\hat{D}(v) = \sum_{t=1}^{k} d_p(\vec{x}_t(v)) + (n-k) \cdot \int_{R^3} d_p(\vec{x}_r(v) + \vec{\xi}(v)) \cdot \hat{\varphi}(\vec{\xi}) \cdot d\vec{\xi}.$$

The estimation can also be performed using the mean and the standard deviation of a temporal variation (Yan and Lockman 2001), such that

$$\hat{D}(v) = n \cdot d_p(\vec{x}_r) + \hat{\vec{\mu}}^T \cdot n \cdot \nabla d_p[\vec{x}_r, \vec{x}_r + \hat{\vec{\mu}}] + \frac{n}{2} \cdot \hat{\vec{\sigma}}^T \cdot \frac{\partial^2 d_p(\vec{x}_r + \hat{\vec{\mu}})}{\partial \vec{x}^2} \cdot \hat{\vec{\sigma}}$$
$$(6)$$

where $\nabla d_p[\vec{x}_r, \vec{x}_r + \hat{\vec{\mu}}]$ is the mean dose gradient in the interval $[\vec{x}_r, \vec{x}_r + \hat{\vec{\mu}}]$, and $\frac{\partial^2 d_p(\vec{x}_r + \hat{\vec{\mu}})}{\partial \vec{x}^2}$ is the dose curvature at point $\vec{x}_r + \hat{\vec{\mu}}$.

Equation (6) provides a very important structure on the parameter design of adaptive planning and adjustment (discussed in the next section).

Using the estimated dose, radiotherapy dose response parameters, such as the EUD, NTCP, and TCP, can be evaluated using the common methods that have been discussed elsewhere.

25.5
Design of Adaptive Planning and Adjustment

Design of adaptive planning and adjustment contains computation and rules to select planning and adjustment parameters, and to update the schedule of imaging, delivery, and adjustment. Ideally, imaging/verification, estimation/evaluation, and planning/adjustment should be performed with the identical sampling rates, and the planning/adjustment parameters should be selected in such way that the adaptive radiotherapy system can be completely optimized; however, this is most unlikely possible when clinical practice is considered. Only a few possibilities have been investigated and are discussed here.

25.5.1
Design Objectives

Objectives in the design of adaptive planning/adjustment are commonly specified in an adaptive treatment protocol. The objectives are (a) to improve treatment accuracy by reducing the systematic variation, (b) to reduce the treated volume and improve dose distribution by reducing the systematic variation and compensating for patient specific random variation, (c) to reduce the treated volume and improve dose distribution by reducing the both systematic and random variations, and (d) to additionally improve treatment efficacy by alternating daily dose per fraction and number of fractions. Clearly, an objective has to be selected based on expected treatment goals and available technologies. The first two can be implemented using an offline feedback technique. Conversely, online image guided adjustment or planning modification has to be implemented to achieve the objectives (c) or (d). Most of offline techniques have implemented the replanning and adjustment once during the treatment course, except for the case when a large residual appeared in the estimation. On the other hand, most of online techniques have aimed to adjust patient treatment position only by moving the couch and/or beam aperture; therefore, it is also important to implement a hybrid technique, where offline planning is performed to modify the ongoing treatment plan in certain time intervals (e.g., weekly) during online daily adjustment process.

25.5.2
Adaptive Planning and Adjustment Parameters

Given organs of interest, the target and surrounding critical normal structures, the aim of an adaptive treatment planning is to design and modify treatment dose distribution in response to the temporal variations observed in the previous treatments. Considering the dose expressed in Eq. (6), four factors play the key roles on treatment quality and can be considered in the adaptive treatment planning and adjustment design; these are two patient/organ-geometry related factors, the systematic variation $\bar{\mu}$ and the standard deviation of the random variation $\bar{\sigma}$ for each subvolume in the organs of interest, and two patient dose-distribution-related factors, the dose gradient ∇d_t and the dose curvature $\dfrac{\partial^2 d_t}{\partial \bar{x}^2}$ at each spatial point in the region of interest. Theoretically, any treatment planning and adjustment parameter, which can control these factors, can be selected to modify and improve the treatment.

Planning and adjustment parameters can be divided into two classes: one contains patient-positioning parameters, such as couch position and rotation, beam angle, and collimator angle, which can be applied to reduce both the systematic and random variations $\{\bar{\mu}, \bar{\sigma}\}$; however, these parameters can only adjust variations induced by rigid body motion and improve position accuracy and precision, but have limits to manage variations induced by organ deformation and cannot improve treatment plan qualities; the other contains dose-modifying parameters, such as target margin, beam aperture, beam weight, and beamlet intensities. These parameters are typically used to adjust dose distribution, thus modifying $\left\{ \nabla d_t, \dfrac{\partial^2 d_t}{\partial \bar{x}^2} \right\}$ in the region of interest to improve ongoing treatment qualities. In addition, prescription dose, dose per fraction, and number of fractions have also been used as parameters for adaptive planning (YAN 2000).

25.5.3
Adaptive Planning and Adjustment Parameter and Schedule: Selection and Modification

In an ideal adaptive radiotherapy system, design of planning and adjustment should have a function of automatically selecting on going planning/adjustment parameters and schedules of imaging, delivery, and adjustment; thus, a new treatment plan can be calculated by including the observed temporal variations and estimation, optimized using the selected parameters and executed with the new schedules. In principle, a set of pre-specified rules and control laws could be used in the design, which match the parameters of temporal variation process to the parameters and schedules of adaptive planning and adjustment.

Basic rules can be created utilizing the discrepancies between the ideal treatment under the ideal condition (i.e., no temporal variation occurs) and the "actual" treatment that includes the temporal variations. The discrepancies can be either the organ volume/dose discrepancy, or the discrepancies of EUD, TCP, and/or NTCP determined from the planned dose, $\{ D(v) \, | \, v \in V_i, \, i = 1, ..., l \}$, in organs of interest, V_i, calculated without considering temporal variations vs those determined from the estimated dose distribution $\{ \hat{D}(v) \, | \, v \in V_i, \, i = 1, ..., l \}$ constructed from a treatment plan created using pre-specified planning/adjustment parameters and including the estimation of temporal variations. A set of predefined tolerances $\{ \delta_V, \, \delta_D, \, \delta_{EUD}, \, \delta_{TCP}, \, \delta_{NTCP} \}$ is then used to test whether or not the discrepancies, $V(\Delta D \leq \delta_D) \leq \delta_V$, $\Delta EUD \leq \delta_{EUD}$, $\Delta TCP \leq \delta_{TCP}$, and/or $\Delta NTCP \leq \delta_{NTCP}$, hold within the predefined ranges. In addition, these tolerances can also be utilized to evaluate and rank the potential treatment quality with respect to different groups of planning/adjustment parameters and adjustment methodology (offline or online). Depended on the variation type (rigid or non-rigid) and the objectives of planning/adjustment, the planning and adjustment parameters could be selected as (a) couch position/rotation, beam angle, and/or collimator angle (online or offline position adjustment for a rigid body motion), (b) target margin, beam aperture, beam weight (intensities), and/or prescription dose (online or offline planning modification), and (c) beamlet intensity, prescription dose, and/or dose per fraction plus number of fractions (online planning modification). Contrarily, a subset of patients, who have insignificant temporal variation, can also be identified; therefore, no re-planning and adjustment are necessary for this subset.

There have been limited studies on utilizing control laws to automatically modify the planning and adjustment parameters. A decision rule (BEL et al. 1993) has been proposed and applied for the offline adjustment of systematic variation induced by daily patient setup. This decision rule is constructed by assuming the statistical knowledge of patient setup variation, and automatically schedules the setup adjustment based on the estimated systematic variation, and pre-designed "action levels." The other method to control the offline planning and adjustment has been "no action level" but including estimated residuals in the target margin design, and primarily single modification after four to five consecutive observations (YAN et al. 2000; BIRKNER et al. 2003). An early investigation (LOF et al. 1998) on the adaptive planning has modeled the cumulative dose and beamlet intensities (control parameters) recursively using a linear system, and created a quadratic objective from the prescribed doses and the estimated doses. Based on the optimal control theory (BRYSON and HO 1975), the intensity fluence adjustment therefore follows a standard linear feedback law – a linear function of the dose discrepancy in organs of interest. Intuitively, the beamlet intensities in the treatment should be adjusted proportionally to the estimated dose discrepancy. Similar methodology has been also proposed for adaptive optimization using the tomotherapy delivery machine (WU et al. 2002). In addition to beam intensity fluence, a control law has also been proposed to manipulate the prescription dose per treatment fraction and the total number of treatment fractions in an online image-guided process (YAN 2000). This control law utilizes temporal variation of dose/volume of critical normal organs to select the most effective dose of the fraction and the total number of fractions.

Most problems in adaptive radiotherapy are easily described but hardly solved. Compared with a direct 4D inverse planning after k number of observations, control laws derived from an ideal system model commonly provide only limited roles in the clinical implementation. For the clinical practice, most temporal variations of patient/organ shape and position can be described using stationary random processes, and therefore the control mechanism is straightforward. Applying one or few planning modifications, the systematic variation can be maximally eliminated and thus patient treatment can be significantly improved in an offline adjustment process. Residuals are commonly inverse proportional to the frequency of the verification, estimation, and adjustment. These residuals could be significantly large to diminish the anticipated gain of adaptive treatment for certain patients; therefore, a decision rule should be applied to modify the schedule of imaging and adjustment if these patients are identified during the treatment course.

Sampling rates of imaging/verification, estimation/evaluation, and adaptive planning/adjustment should be scheduled to match the rate of the aimed temporal variations. Mismatch results in significant downgrading of the expected treatment quality; therefore, before selecting objectives in an adaptive treatment design, specifically for an online adjustment process, one should ensure that appropriate sampling rates of imaging/verification, estimation/ evaluation, and adaptive planning/adjustment could be implemented.

25.6
Adaptive Planning and Adjustment

Adaptive planning and adjustment are implemented with the pre-design parameters. The adaptive planning is often performed including the temporal variations in the planning dose calculation; therefore, it has also been called "4D treatment planning." There have been two methods to perform a 4D planning. The first (indirect method) does not directly include the temporal variation in the planning dose calculation. Instead, it constructs the PTV and margins of organs at risk based on the characteristics of patient specific temporal variations and a generic planned dose distribution, and then performs a conventional conformal or inverse planning accordingly. The second (direct method) performs treatment planning by directly including the temporal variations in the dose calculation as has been discussed in section 25.4.3. The adaptive treatment plan designed with the expected dose distribution can best compensate for the temporal variations. Consequently, pre-designed target margin is either unnecessary or used only to compensate for the residuals of the estimation.

25.6.1
Indirect Method

Planning technique in the indirect method is primarily the same as the conventional one except for the definition of the planning target volume and the margins of organs at risk. For a rigid body motion without significant rotation, the patient specific target margin in each direction j after k observations can be constructed by considering the residuals of the estimations of the systematic and random variations (Yan et al. 2000), such that

$$m^{(j)}(c_j) = t_{\alpha/2,k-1} \cdot \frac{\hat{\sigma}^{(j)}}{\sqrt{k}} + c_j \cdot \sqrt{\frac{k-1}{\chi^2_{1-\alpha,k-1}}} \cdot \hat{\sigma}^{(j)},$$

where $t_{\alpha/2,k-1}$ and $\chi^2_{1-\alpha,k-1}$ have been defined in section 25.4.1. The factor c_j is determined by ensuring that the potential dose reduction in the target with the corresponding margin is less than a pre-defined dose tolerance δ, such that

$$\Delta D(c_j) = n \cdot \int_{-\infty}^{\infty} \left[\begin{array}{l} d_p\left(x_r^{(j)}\right) \\ - d_p\left(x_r^{(j)} + \xi^{(j)} - m^{(j)}(c_j)\right) \end{array} \right] \cdot \hat{\varphi}(\xi^{(j)}) \cdot d\xi^{(j)} \le \delta,$$

where $d_p(x_r^{(j)})$ is the planning dose around the CTV edge $x_r^{(j)}$ on the j axis. $\hat{\varphi}(\xi^{(j)})$ is the estimated probability distribution with the mean $t_{\alpha/2,k-1} \cdot \dfrac{\hat{\sigma}^{(j)}}{\sqrt{k}}$ (the residual of the systematic variation) and the standard deviation $\sqrt{\dfrac{k-1}{\chi^2_{1-\alpha,k-1}}} \cdot \hat{\sigma}^{(j)}$.

It is clear that the calculation of dose discrepancy here is approximated assuming the spatial invariance of planning dose distribution. In addition, this evaluation can also be approximated using the dose gradient and curvature around the CTV edge as indicated by Eq. (6), such that

$$\Delta D(c_j) \approx n \cdot \hat{\sigma}^{(j)}$$

$$\cdot \left(\begin{array}{l} \dfrac{t_{\alpha/2,k-1}}{\sqrt{k}} \cdot \dfrac{\partial d_p}{\partial x^{(j)}}[\tau, \pi] \\ + \dfrac{\hat{\sigma}^{(j)}}{2} \cdot \dfrac{\partial^2 d_p(\pi)}{\partial x^{(j)2}} \end{array} \right)_{\tau = x_r^{(j)} - m^{(j)}(c_j);\ \pi = x_r^{(j)} - m^{(j)}(c_j) + t_{\alpha/2,k-1} \frac{\hat{\sigma}^{(j)}}{\sqrt{k}}} \le \delta.$$

This method can be further extended to construct CTV-to-PTV margin that compensates for much broad type of variations including organ deformation. Let ∂CTV represent the boundary of CTV, then the 3D margin can be constructed using the vector normal to target surface at each boundary point $v \in \partial CTV$, such that

$$\bar{m}(c,v) = t_{\alpha/2,k-1} \cdot \frac{\hat{\sigma}(v)}{\sqrt{k}} + c \cdot \sqrt{\frac{k-1}{\chi^2_{1-\alpha,k-1}}} \cdot \hat{\sigma}(v).$$

Similarly, the c is determined such that the following inequality

$$\Delta D(c,v) =$$

$$n \cdot \int_{R^3} \left[\begin{array}{l} d_p\left(\bar{x}_r(v)\right) \\ - d_p\left(\bar{x}_r(v) + \bar{\xi}(v) - \bar{m}(c,v)\right) \end{array} \right] \cdot \hat{\varphi}(\bar{\xi}) \cdot d\bar{\xi} \le \delta$$

holds; therefore, the patient-specific PTV can be formed by creating a new surface with the vectors $\bar{m}(c,v), \forall v \in \partial CTV$.

Typical adaptive planning/adjustment with using the indirect method includes (a) using imaging measurements to perform the estimation, (b) adjusting patient position or beam aperture to correct the estimated systematic variation, (c) constructing a patient specific PTV, and (d) performing a conventional treatment planning or inverse planning. Since patient-specific PTV construction is also dependent on the planning dose distribution, primarily the dose gradient and curvature around the neighborhood of the CTV edge, control parameters, which can directly adjust the dose gradient and curvature, are important for the adaptive treatment planning.

25.6.2
Direct Method

In the direct method of 4D planning, temporal variations are directly included in the planning dose calculation. Consequently, dose distribution in the neighborhood of each subvolume of organs of interest can be designed by selecting beam aperture or modulating beam intensity fluence to effectively compensate for the systematic and the random variations estimated from previous observations; therefore, parameters of temporal variation and dose distribution are automatically included in the treatment planning optimization. However, because the position of each subvolume in a 4D planning is a distribution function rather than static, it considerably increases the time and complicity of the dose calculation.

Most 4D adaptive planning have been completed using an inverse planning engine that searches the optimal beam intensity fluence based on the objective function calculated using the estimated dose discussed in section 25.4.3. Objective function and search algorithm commonly remain the same as those in the conventional inverse planning, but dose computation or estimation is much time-consuming, specifically when including feedback volumetric CT images in the computation becomes necessary. Some simplifications on dose computation have been applied to reduce the calculation time. One study (BIRKNER et al. 2003) has directly included the samples of organ subvolume displacement in the dose estimation during the inverse planning iteration, such that for each subvolume, v, the expected dose after k treatment delivery is computed as

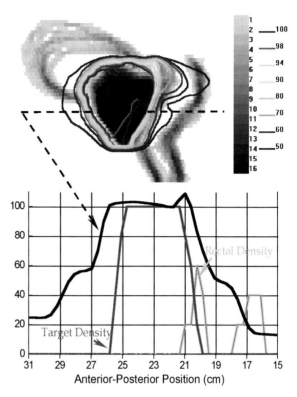

Fig. 25.4 Dose distribution (*colored isodose lines*) calculated from a 4D inverse planning of prostate cancer superimposed on the organ occupancy density (*gray area*)

$$\hat{D}(v, \Phi) = \sum_{t=1}^{k} d_t\left(\bar{x}_t(v)\right) + \frac{(n-k)}{k^2}$$
$$\cdot \sum_{i,j=1}^{k} d\left(\bar{x}_r(v) + \bar{u}_i(v) + \bar{w}_j(v), \Phi\right) \quad (7)$$

where $\bar{u}_j(v)$ is the sample of subvolume displacement induced by internal organ motion, and $\bar{w}_j(v)$ is the sample of subvolume displacement induced from patient setup. Φ is the beam-intensity map to be optimized. It has been demonstrated that the optimization result converges after five observations in an offline adaptive process for prostate cancer radiotherapy. Figure 25.4 shows a typical dose distribution from the 4D inverse planning, where the dose gradient and curvature in the adjacent region between target and normal structure were designed to best compensate for the variations induced from treatment setup and internal organ motion.

The fundamental difference between the 4D planning in an offline process and the one in an online process is the dose construction in the objective function. Offline 4D planning aims to optimize treatment plan for all remaining treatment. Meanwhile, online

Halverson KJ et al. (1991) Study of treatment variation in the radiotherapy of head and neck tumors using a fiber-optic on-line radiotherapy imaging system. Int J Radiat Oncol Biol Phys 21:1327–1336

Heidi L et al. (2004) A model to predict bladder shapes from changes in bladder and rectal filling. Med Phys 31:1415–1423

Hugo G et al. (2004) A method of portal verification of 4D lung treatment. Proc XIIIIth International Conference on The Use of Computers in Radiotherapy (ICCR), Seoul, Korea

Keller H et al. (2004) Design of adaptive treatment margins for non-negligible measurement uncertainty: application to ultrasound-guided prostate radiation therapy. Phys Med Biol 49:69–86

Liang J et al. (2003) Minimization of target margin by adapting treatment planning to target respiratory motion. Int J Radiat Oncol Biol Phys 57:S233

Lof J et al. (1998) An adaptive control algorithm for optimization of intensity modulated radiotherapy considering uncertainties in beam profiles, patient setup and internal organ motion. Phys Med Biol 43:1605–1628

Lujan AE et al. (1999) A method for incorporating organ motion due to breathing into 3D dose calculation. Med Phys 26:715–720

Marks JE, Haus AG (1976) The effect of immobilization on localization error in the radiotherapy of head and neck cancer. Clin Radiol 27:175–177

McInerney T, Terzopoulos D (1996) Deformable models in medical image analysis: a survey. Med Image Anal 1:91–108

Moerland MA et al. (1994) The influence of respiration induced motion of the kidneys on the accuracy of radiotherapy treatment planning: a magnetic resonance imaging study. Radiol Oncol 30:150–154

Nuyttens JJ et al. (2001) The small bowel position during adjuvant radiation therapy for rectal cancer. Int J Radiat Oncol Biol Phys 51:1271–1280

Nuyttens J et al. (2002) The variability of the clinical target volume for rectal cancer due to internal organ motion during adjuvant treatment. Int J Radiat Oncol Biol Phys 53:497–503

Pluim JPW et al. (2003) Mutual information based registration of medical images: a survey. IEEE Trans Med Imaging 10:1–21

Roeske JC et al. (1995) Evaluation of changes in the size and location of the prostate, seminal vesicles, bladder, and rectum during a course of external beam radiation therapy. Int J Radiat Oncol Biol Phys 33:1321–1329

Ross CS et al. (1990) Analysis of movement of intrathoracic neoplasms using ultrafast computerized tomography. Int J Radiat Oncol Biol Phys 18:671–677

Sonke J et al. (2003) Respiration-correlated cone beam CT: obtaining a four-dimensional data set. Med Phys 30:1415

Van Herk M et al. (2000) The probability of correct target dosage: dose-population histograms for deriving treatment margins in radiotherapy. Int J Radiat Oncol Biol Phys 47:1121–1135

Wiener (1949) Extrapolation, interpolation and smoothing of stationary time series. M.I.T. Press, Cambridge, Massachusetts

Wong E (1983) Introduction to random processes. Springer, Berlin Heidelberg New York

Wu C et al. (2002) Re-optimization in adaptive radiotherapy. Phys Med Biol 47:3181–3195

Yan D (2000) On-line adaptive strategy for dose per fraction design. Proc XIIIth International Conference on The Use of Computers in Radiotherapy. Springer, Berlin Heidelberg New York

Yan D, Lockman D (2001) Organ/patient geometric variation in external beam radiotherapy and its effects. Med Phys 28:593–602

Yan D et al. (1995) A new model for "Accept Or Reject" strategies in on-line and off-line treatment evaluation. Int J Radiat Oncol Biol Phys 31:943–952

Yan D et al. (1999) A model to accumulate the fractionated dose in a deforming organ. Int J Radiat Oncol Biol Phys 44:665–675

Yan D et al. (2000) An off-line strategy for constructing a patient-specific planning target volume for image guided adaptive radiotherapy of prostate cancer. Int J Radiat Oncol Biol Phys 48:289–302

26 Predictive Compensation of Breathing Motion in Lung Cancer Radiosurgery

Achim Schweikard and John R. Adler

CONTENTS

26.1
Introduction

The success of radiosurgical methods for brain tumors suggests that improved methods for high-precision treatment delivery could dramatically improve treatment outcome in radiation therapy.

Radiosurgery has been limited to brain tumors, since stereotactic fixation is difficult to apply to tumors in the chest or the abdomen (ADLER et al. 1999; HAYKIN 1996; LUJAN et al. 1999; RIESNER et al. 2004; SCHWEIKARD et al. 1998; WINSTON and LUTZ 1988). To apply radiosurgical methods to tumors in the chest and abdomen, it is necessary to take into account respiratory motion. Respiratory motion can move the tumor by more than 1 cm. Without compensation for respiratory motion, it is necessary to enlarge the target volume with a safety margin. For small targets, an appropriate safety margin produces a very large increase in treated volume. For a spherical tumor of radius 1 cm, a safety margin of 0.5–1 cm would have to be added to ensure that the tumor remains within the treated volume at all times. The ratio between the radius and volume of a sphere is cubic; thus, a margin

of 1 cm will cause an eightfold increase in treated volume. Furthermore, observed motion ranges (>3 cm) suggest that a 1-cm margin may not be sufficient in all cases. An enlarged margin of 2 cm would result in a 27-fold increase of dose in this example; thus, an accurate method capable of compensating for respiratory motion would be of utmost clinical relevance.

Three problems arise in this context:
1. Hysteresis: the inhalation curve may differ from the exhalation curve.
2. The direction of the internal target motion may not be the same as the direction of the surface motion (e.g., consider the diaphragm, moving in inferior–superior direction, while the abdomen surface moves in anterior-posterior direction).
3. The velocity of the motion may vary along the motion curve.

Our previous work has investigated the design of a new sensor method for tracking respiratory motion (SCHWEIKARD et al. 2000). In this prior work, a method for correlating internal motion to external surface motion was developed. The treatment beam itself is moved by a robotic arm, based on a modified Cyberknife system. Our measurements have shown that an accuracy of 1.5 mm is achievable with this technique.

This prior work does not include predictive compensation of the time lag stemming from image acquisition time, and robot motion lag. In this chapter we describe an integrated system using this new sensor technique, and present first clinical results with the use of the integrated system. We also develop a method for *predictive* tracking, designed to capture the regularity of the breathing pattern, and predict target motion ahead in time, to compensate for the inherent system lag.

A. SCHWEIKARD, PhD
Institute for Robotics and Cognitive Systems, University Luebeck, Ratzeburger Alee 160, 23538 Luebeck, Germany
J. R. ADLER, MD
Department of Neurosurgery, R-205 Stanford University Medical Center, 300 Pasteur Drive, Stanford, CA 94305, USA

26.2
Related Work

Intra-treatment displacements of a target due to respiration have been reported to exceed 3 cm in the abdomen, and 1 cm for the prostate (MORRILL 1996; SONTAG 1996). WEBB (2000) discusses methods for achieving improved dose conformality with robot-based radiation therapy. Specifically, trade-offs between treatment path complexity and conformality are analyzed. The experience reported suggests that very high conformality can be achieved with robotic systems by modulating the intensity of the beam; however, this assumes the ideal situation of a stationary target, unaffected by respiration. This assumption holds only for few anatomic sites, such as the brain, but not for tumors close to the lung or diaphragm. Recent studies suggest that even prostate tumors are subject to respiratory motion to a non-negligible amount.

Conventional radiation therapy with medical linear accelerators (LINAC systems) uses a gantry with two axes of rotation movable under computer control (WEBB 1993). This mechanical construction was designed to deliver radiation from several different directions during a single treatment. It was not designed to track respiratory motion.

Respiratory gating is a technique for addressing the problem of breathing motion with conventional LINAC-based radiation therapy. Gating techniques do not directly compensate for breathing motion, i.e., the therapeutic beam is not moved during activation; instead, the beam is switched off whenever the target is outside a predefined window. One of the disadvantages of gating techniques is the increase in treatment time. A second problem is the inherent inaccuracy of such an approach. One must ensure that the beam activation cycles can have sufficient length for obtaining a stable therapeutic beam.

KUBO and HILL (1996) compare various external sensors (breath temperature sensor, strain gauge, and spirometer) with respect to their suitability for respiratory gating. By measuring breath temperature, it is possible to determine whether the patient is inhaling or exhaling. It is verified by KUBO and HILL (1996) that frequent activation/deactivation of the linear accelerator does not substantially affect the resulting dose distribution; however, the application of such a technique still requires a substantial safety margin for the following reason: the sensor method only yields relative displacements during treatment but does not report and update the exact absolute position of the target during treatment.

TADA et al. (1998) report using an external laser range sensor in connection with a LINAC-based system for respiratory gating. This device is used to switch the beam off whenever the sensor reports that the respiratory cycle is close to maximal inhalation or maximal exhalation.

Typical variations in the respiratory motion patterns of 1–2 cm for the same patient (in pediatrics), and in the duration of a single respiratory cycle of 2–5 s are reported by SONTAG (1996).

In a paper by SCHWEIKARD et al. (2000), we investigated a method for tracking a tumor during treatment. Stereo X-ray imaging is combined with infrared tracking. X-ray imaging is used as an internal sensor, whereas infrared tracking provides information on the motion of the patient surface. While X-ray imaging gives accurate information on the internal target location, it is not possible to obtain real-time motion information from X-ray imaging alone. In contrast, the motion of the patient surface can be tracked with commercial infrared position sensors with high speed. The main idea of our approach is to use a series of images from both sensors (infrared and X-ray) where image acquisition is synchronized. From a series of sensor readings and corresponding time stamps, we can determine a motion pattern. This pattern correlates external to internal motion, and is both patient- and site specific. Previous work has verified that we can accurately infer the placement of an internal target tumor from such motion patterns.

Below we describe an integrated system using the new correlation method (internal vs external fiducials), and extend the work to include predictive tracking. Several new approaches have been investigated in this context. Clinical results demonstrate the high reliability of the correlation model.

26.3
System

26.3.1
Overview

Figures 26.1 and 26.2 show the system components. A robot arm (modified Cyberknife system) moves the therapeutic beam generator (medical linear accelerator generating a 6-mV radiation beam). Five infrared emitters are attached to the chest and the abdomen surface of the patient. An infrared tracking system records the motion of these emitters. A stereo X-ray camera system (X-ray cameras with nearly orthogonal visual axes) records the position of internal gold markers, injected into the vicinity

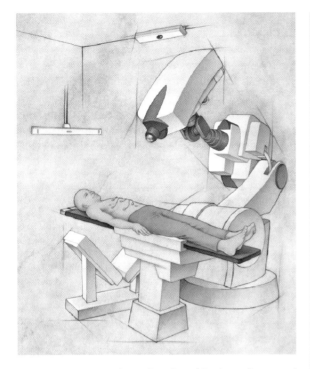

Fig. 26.1 System overview. Infrared tracking is used to record external motion of the patient's abdominal and chest surface. Stereo X-ray imaging is used to record the 3D position of internal markers (gold fiducials) at fixed time intervals during treatment. A robotic arm moves the beam source to actively compensate for respiratory motion

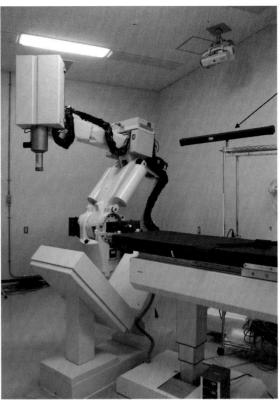

Fig. 26.2 System overview. X-ray sources, camera system, infrared tracking system (*arrow*)

of the target area under tomographic (CT) monitoring. In addition, hardware means for capturing the time points of X-ray image acquisition and infrared sensor acquisition are used to synchronize of the two sensors.

The basic correlation method for computing the position of the internal target has been described in more detail by SCHWEIKARD et al. (2000) and is briefly summarized here. Prior to treatment, small gold fiducial markers are placed in the vicinity of the target organ. Stereo X-ray imaging is used during treatment to determine the precise spatial location of these gold markers. Using stereo X-ray imaging, precise marker positions are established once every 10 s. Limitations on patient radiation exposure and X-ray generator activation times currently prevent the use of shorter intervals; however, even much shorter time intervals would clearly be insufficient for capturing respiratory motion. The typical target velocity under normal breathing is between 5 and 10 mm/s.

External markers (placed on the patient's skin) can be tracked automatically with optical methods

at very high speed. Updated positions can be transmitted to the control computer more than 20 times per second; however, external markers alone cannot adequately reflect internal displacements caused by breathing motion. Large external motion may occur together with very small internal motion, and vice versa. We have observed 2-mm external emitter excursion in combination with 20-mm internal target excursion. Similarly, the target may move much slower than the skin surface. Since neither internal nor external markers alone are sufficient for accurate tracking, X-ray imaging is synchronized with optical tracking of external markers. The external markers are small active infrared emitters (Flashpoint FP 5500, IGT Inc., Boulder, Colo.) attached to the patient's skin (Fig. 26.3). The individual markers are allowed to change their relative placement. The first step during treatment is to compute the exact relationship between internal and external motion, using a series of snapshots showing external and internal markers simultaneously. Each snapshot has a time-stamp which is used to compute the correlation model.

Fig. 26.3 Infrared emitters, attached to the patient's chest, are used as external markers. The position of the emitters is tracked by an infrared tracking system (camera attached to the ceiling of the treatment room)

Previous work has addressed the suitability of this combined X-ray/infrared sensor technique as well as the suitability of the robotic gantry for active tracking. Specifically, the imaging methods used for fiducial detection, stability and robustness of infrared tracking under radiation exposure, accuracy of the correlation method for different subjects, tracking speed, and inherent time lag have been investigated. A systematic lag of 0.1–0.3 s was established for various motion velocities ranging from 3 to 8 mm per second. At a typical speed of 5 mm, this time lag results in a tracking error of 0.5–1.5 mm. This error purely stems from the inherent system lag.

26.4
Clinical Trials

Twelve patients have been treated with this new sensor method. A fully integrated version of the system, including the above prediction scheme has been used in the trials. Figures 26.4–26.6 show the results for representative cases. The figure shows the total error in the correlation model; thus, based on the correla-

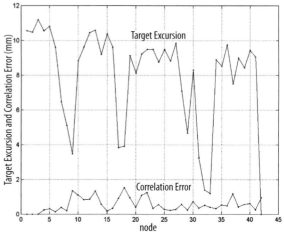

Fig. 26.4 Correlation error and total target excursion. *Top curve:* motion of the target. *Bottom curve:* distance between expected position of the gold markers and actual position. The first four images were used for computing the deformation model; hence, the deviation between model and actual position is zero in the first four nodes

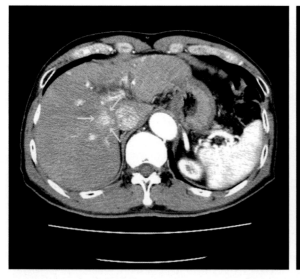
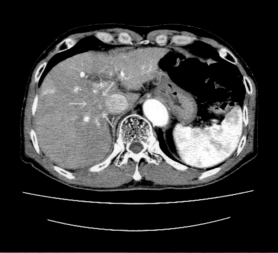

Fig. 26.5 Treatment of a hepatocellular carcinoma with fiducial-based respiration compensation as described above (*left image:* before treatment; *right image:* 3 months after treatment. Total treatment time was 40 min per fraction for three fractions with 80 beam directions and 39 Gy

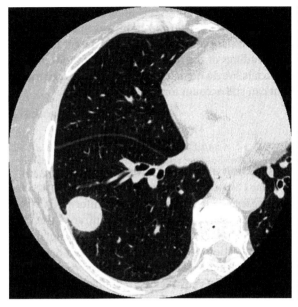
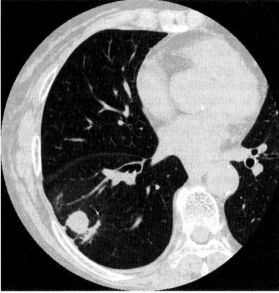

Fig. 26.6 Treatment of an adenocarcinoma with fiducial-based respiration tracking. Treatment with 39 Gy per three fractions. (Courtesy of Osaka University Hospital)

tion model described above, we compute the current position of the target based on the external infrared sensor signal *alone*. At this same time point, we also acquire a pair of X-ray images. We then plot the distance in millimeters from the placement inferred by the correlation model and the actual placement determined from the image. The top curve in Fig. 26.4 shows the total target excursion.

26.5
Extension to Tracking Without Implanted Fiducial Markers

The method described above requires X-ray opaque fiducials to be placed in the vicinity of the target. Stereo X-ray imaging is used to compute the position of these landmarks once every 10 s.

We extend our method in such a way that implanting fiducials is no longer needed. This extension works as described below.

26.5.1
Pre-Treatment Steps

Prior to treatment, two CT scans are taken. One scan shows inhalation and one shows exhalation (Fig. 26.7). From these data a series of intermediate CT scans are computed. These intermediate scans

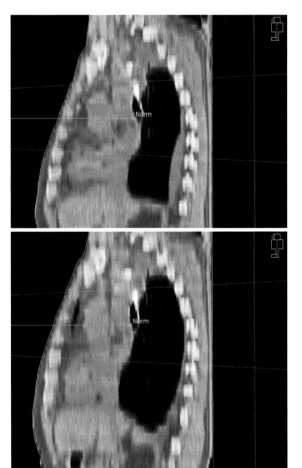

Fig. 26.7 Exhalation/inhalation CT scans of the same subject

particular for intensity-modulated proton techniques (LOMAX et al. 1999; LOMAX et al. 2003a, b; WEBER et al. 2004). The reduction of integral dose with protons is also significant. Although it is true that the clinical relevance of low doses to large volumes is not well known (except perhaps in organs with a parallel or near parallel architecture), there are cases where a reduction in overall normal tissue dose is proven to be relevant, e.g., for pediatric patients (LIN et al. 2000).

27.1.3
Worldwide Proton Therapy Experience

Robert Wilson first proposed the use of protons for radiation therapy at Harvard in 1946 (WILSON 1946).

The first patient was treated in 1954 at Lawrence Berkeley Laboratory (TOBIAS et al. 1958). About 40,000 patients have received proton therapy to date worldwide (see Table 27.1; SISTERSON 2004).

Although the number of patients being treated with protons is steadily increasing, as are the number of facilities operating worldwide, it is still only a small fraction of radiotherapy patients overall that are treated with protons. One reason for this is the cost of a proton facility. Sophisticated proton therapy is presently, and is likely to continue to be, more expensive than sophisticated (i.e., intensity-modulated) X-ray therapy. The construction cost of a current two-gantry proton facility, complete with the equipment, was estimated to be about four times the cost of a two-linac X-ray facility (GOITEIN and JERMANN

Table 27.1 Worldwide proton therapy experience as of July 2004. (From SISTERSON 2004)

Institution	Location	First treatment	Last treatment	No. of patients	Date of total
Berkeley	California, USA	1954	1957	30	
Uppsala	Sweden	1957	1976	73	
Harvard	Massachusetts, USA	1961	2002	9116	
Dubna	Russia	1967	1996	124	
ITEP, Moscow	Russia	1969		3748	June 2004
St. Petersburg	Russia	1975		1145	April 2004
Chiba	Japan	1979		145	April 2002
PMRC(1), Tsukuba	Japan	1983	2000	700	
PSI (72 MeV)	Switzerland	1984		4066	June 2004
Dubna	Russia	1999		191	Nov. 2003
Uppsala	Sweden	1989		418	Jan. 2004
Clatterbridge	UK	1989		1287	Dec. 2003
Loma Linda	California, USA	1990		9282	July 2004
Louvain-la-Neuve	Belgium	1991	1993	21	
Nice	France	1991		2555	April 2004
Orsay	France	1991		2805	Dec. 2003
iThemba LABS	South Africa	1993		446	Dec. 2003
MPRI(1)	Indiana, USA	1993	1999	34	
UCSF – CNL	California, USA	1994		632	June 2004
TRIUMF	Canada	1995		89	Dec. 2003
PSI (200 MeV)	Switzerland	1996		166	Dec. 2003
H.M.I, Berlin	Germany	1998		437	Dec. 2003
NCC, Kashiwa	Japan	1998		270	June 2004
HIBMC, Hyogo	Japan	2001		359	June 2004
PMRC(2), Tsukuba	Japan	2001		492	July 2004
NPTC, MGH	Massachusetts, USA	2001		800	July 2004
INFN-LNS, Catania	Italy	2002		77	June 2004
WERC	Japan	2002		14	Dec. 2003
Shizuoka	Japan	2003		69	July 2004
MPRI(2)	Indiana, USA	2004		21	July 2004
Total				39,612	

2003). According to GOITEIN and JERMANN (2003), the cost of operation of a proton therapy facility is dominated by the business cost (42%, primarily the cost of repaying the presumed loan for facility construction), personnel costs (28%), and the cost of servicing the equipment (21%). For X-ray therapy, the cost of operation was estimated to be dominated by the personnel cost (51%) and the business costs (28%). The ratio of costs of proton vs X-ray therapy per treatment fraction is about 2.4 at present.

27.2
Proton Accelerators

27.2.1
Cyclotron

A cyclotron consists of dipole magnets designed to produce a region of uniform magnetic field. These dipoles are placed with their straight sides parallel but slightly separated. An electric field is produced across the gap by an oscillating voltage. Particles injected into the magnetic field region move on a semicircular path until they reach the gap where they are accelerated. Since the particles gain energy, they will follow a semi-circular path with larger radius before they reach the gap again. In the meantime the direction of the field has reversed and so the particles are accelerated again. As they spiral around, particles gain energy; thus, they trace a larger arc with the consequence that it always takes the same time to reach the gap. The size of the magnets and the strength of the magnetic fields limits the particle energy that can be reached by a cyclotron.

The first cyclotrons being used for proton therapy were initially designed for physics research and were later turned into treatment facilities. Currently, increasingly more cyclotrons are specifically built and dedicated for use in radiation therapy. The envisaged targets in the human body define the specifications for such a cyclotron. The maximum proton beam energy is directly related to the maximum depth in tissue. Low-energy proton beams can only be used for superficial tumors. For example, many cyclotrons currently being used in proton therapy have an energy limit around 70 MeV, which suits them only for treating ocular tumors. In order to be able to treat all common tumors in the human body, the cyclotron has to be able to deliver a beam with energy of up to about 230 MeV, which corresponds to a range in tissue of about 32 cm. However, one has to consider that

the maximum penetration of the proton beam in the patient is reduced if the beam has to be widened by absorbers to reach large field sizes. Field sizes of up to 30×30 cm^2 may be required. While being able to deliver energies that can be used to cover common tumor sizes and locations, the cyclotron has to deliver a dose rate acceptable for treatment, i.e., at least around 2 Gy/min. With respect to the dose rate, one has to keep in mind that the efficiency of a beam delivery system is never 100%. In particular, for double-scattering systems (see section 27.3.1) it can be as low as 20%. Cyclotron intensities can exceed 100 mA, but they can be hardware limited at the ion source to several hundreds of nanoamperes, which corresponds to clinically meaningful dose rates. Higher currents are not safe for treatment because of the short feedback time for machine control. Cyclotrons that are used in a purely clinical environment require a high grade of reliability, low maintenance, and should be easy to operate. In fractionated radiation therapy, machine down times should be minimized because this may result in the necessity to re-calculate daily doses because of tissue repair effects.

Cyclotrons can be either isochronous or synchro-cyclotrons. In addition, they can be superconducting. In an isochronous cyclotron the orbital period is the same for all particles regardless of their energy or radius; thus, the radiofrequency power can operate at a single frequency. Isochronous cyclotrons provide a continuous beam. Since the acceleration of particles in a cyclotron takes usually only tenths of a millisecond, the beam (i.e., via an external injection system) can be turned on and off very quickly. This is an important safety feature. It also allows the beam current to be modified during delivery with very short response times. Both these features are very important when it comes to proton beam-scanning techniques for patient treatment (see section 27.3.2). Superconducting cyclotrons have advantages compared with non-superconducting ones in that they are smaller and not as heavy (BLOSSER et al. 1989).

Figure 27.4 shows a typical example of a cyclotron dedicated for use in radiation therapy. The magnetic field is generated by four sectors with an external yoke diameter of 4.3 m. The system weighs about 200 tons. Since the cyclotron is extracting particles with a fixed energy, an energy selection system is needed in the beamline (Fig. 27.5). The energy selection system consists of a degrader of variable thickness to intercept the proton beam, i.e., a carbon wedge that can be moved in and out of the beam quickly. As a result of the energy degradation, there is an increase in emittance and energy spread, which can be controlled

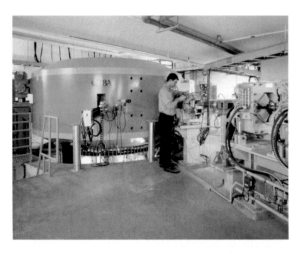

Fig. 27.4 Isochronous cyclotron by Ion Beam Applications, S.A. It extracts protons with an energy of 230 MeV, which corresponds to a range of a ~33.0 cm in water. (Courtesy of Ion Beam Applications, S.A.)

Fig. 27.5 Part of the beamline at the Northeast Proton Therapy Center including the energy selection system to modulate the fixed energy extracted from the cyclotron. (Courtesy of Ion Beam Applications, S.A.)

by slits and magnets. The emittance can be defined as the sum of the phase space areas (or by the area, which encloses the phase space). The phase space of a beam is the distribution of particle position vs momentum direction.

27.2.2
Synchrotron

A synchrotron is a circular accelerator ring. Electromagnetic resonant cavities around the ring accelerate particles during each circulation. Since particles move always on the same radius, the strength of the magnetic field that is used to steer them must be changed with each turn because the particles energy increases. Because of this synchronization of field strength and energy, these accelerators are called synchrotrons. This technique allows the production of proton beams with a variety of energies (unlike the cyclotron which has a fixed extraction energy). A small linear accelerator is often used to pre-accelerate particles before they enter the ring.

One disadvantage of cyclotrons is the inability to change the energy of the extracted particles directly. Energy degradation by material in the beam path leads an increase in energy spread and beam emittance and reduces the efficiency of the system. Another consequence is the need for more shielding because it leads to secondary radiation. In this respect a synchrotron is a more flexible solution. A synchrotron allows beam extraction for any energy. Synchrotrons are, however, much bigger than cyclotrons. A synchrotron delivers a pulsed beam, i.e., it accelerates and extracts protons with a specific repetition rate. Fast extraction delivers the beam after a single turn. This avoids complicated feedback systems; however, for therapeutic applications, slow extraction is needed for machine control reasons. Here the typical extraction pulse is a few seconds. Table 27.2 compares different proton therapy accelerator technologies (COUTRAKON et al. 1999).

Table 27.2 Accelerator technology comparisons for some parameters. (From COUTRAKON et al. 1999)

Type	Synchrotron (rapid cycle)	Synchrotron (slow cycle)	Cyclotron
Energy level selection	Continuous	Continuous	Fixed
Size (diameter; m)	10	6	4
Average power (beam on; kW)	200	370	300
Emittance (RMS unnorm.; μm)	0.2	1–3	10
Repetition rate (Hz)	60	0.5	Continuous
Duty factor (beam-on time)	Pulses	20%	Continuous

27.2.3
Beam Line

The beam has to be transported from the accelerator to the treatment room(s) using magnets for bending, steering, and focusing. In addition, as a safety precaution, detectors monitoring the beam's phase space are located in the beamline. These devices control certain tolerances for beam delivery. Figure 27.5 shows part of the beamline that transports the beam from the cyclotron to different treatment rooms at the Northeast Proton Therapy Center (NPTC). Figure 27.6 shows a

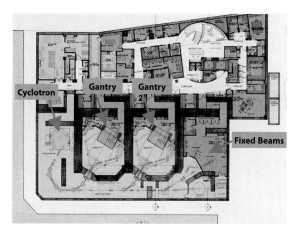

Fig. 27.6 Floor plan of the Northeast Proton Therapy Center. (Courtesy of Ion Beam Applications, S.A.)

typical floor plan of a treatment facility (FLANZ et al. 1995).

Table 27.3 lists the clinical performance specifications of the NPTC (FLANZ et al. 1995), which is based on an isochronous cyclotron and two gantry treatment rooms with double scattering systems as well as one treatment room with horizontal beam lines. These parameters are defined to ensure safe dose delivery taking into account the precise dose deposition characteristics of protons.

27.2.4
Gantry/Fixed Beams

With a fixed horizontal beam line patients can usually be treated only in a seated or near-seated position; however, conformal radiation therapy usually requires multiple beams entering from different directions. In order to irradiate a patient from any desired angle, the treatment head has to be able to rotate. This makes it much easier to position the patient in a reproducible way and similar to the way the patient was positioned during imaging prior to planning and treatment. The ability to deliver beams from various directions is achieved by a gantry system (Fig. 27.7). The beam has to be deflected by magnetic fields in the gantry. Gantries are usually large structures because, firstly protons with therapeutic energies can only be bent

Table 27.3 Clinical specifications of the Northeast Proton Therapy Center. (From FLANZ et al. 1995)

Range in patient	32 g/cm^2 maximum; 3.5 g/cm^2 minimum
Range modulation	Steps of <0.5 g/cm^2
Range adjustment	Steps of <0.1 g/cm^2
Average dose rate	25×25 cm^2 modulated to depth of 32 g/cm^2; dose of 2 Gy in <1 min
Field size	Fixed >40×40 cm^2; gantry 40×30 cm^2
Dose uniformity	2.5%
Effective "source-axis distance"	>3 m
Distal dose fall-off	<0.1 g/cm^2
Lateral penumbra	<2 mm
Time for startup from standby	<30 min
Time for startup from cold system	<2 h
Time for shutdown to standby	<10 min
Time for manual setup in one room	<1 min
Time for automatic setup in one room	<0.5 min
Availability	>95%
Dosimeter reproducibility	1.5% (day); 3% (week)
Time to switch beam to rooms	<1 min
Time to switch energy in one room	<2 s
Radiation levels	ALARA

with large radii, and secondly, beam-monitoring and beam-shaping devices have to be positioned inside the treatment head affecting the size of the nozzle. The nozzle at the NPTC has a length of about 2.5 m, which results in a distance between isocenter and beam entering the gantry of about 3 m. Eccentric gantries are used to reduce the size (PEDRONI et al. 1995). To ensure precise dose delivery the mechanics of the gantry has to be able to keep the isocenter of rotation always within 1 mm under all rotation angles. This requires careful design of the mechanical structure since the overall weight can be several tens of tons.

Treatment nozzles consist of various components for beam shaping and beam monitoring (Fig. 27.8). Beam monitoring ionization chambers detect deviations in beam position, measure the total beam current, and check the beam size and uniformity. Ionization chambers may consist of parallel electrode planes divided into horizontal and vertical strips that allow the quantification of the lateral uniformity of the radiation field. These strips are integrated separately to collect the current in each strip. Such ionization chamber can be used at nozzle entrance, i.e., where the proton beam exits the beam line, to monitors the size of the initial beam spot and the angular distribution of the beam. Beam-shaping devices in the nozzle are scatterers, absorbers, and other patient-specific hardware. The nozzle also has a snout that permits mounting and positioning of a field-specific aperture and compensator along the beam axis. The snout of the nozzle is telescopic to adjust the air gap between the final collimator or compensator and the patient.

 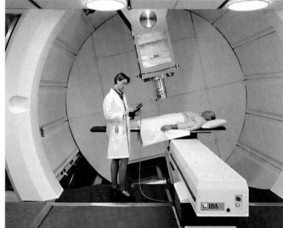

Fig. 27.7 One of the gantries at the Northeast Proton Therapy Center. *Left:* the gantry structure during construction with the steel assembly being visible. *Right:* the gantry treatment room during treatment. The beam delivery nozzle is able to rotate 360° around the movable patient couch. (Courtesy of Ion Beam Applications, S.A.).

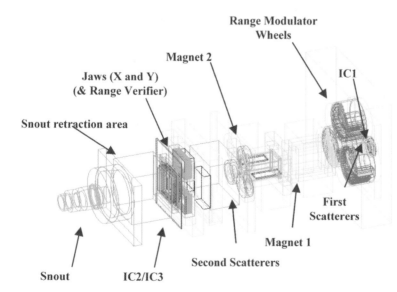

Fig. 27.8 The nozzle at the NPTC (PAGANETTI et al. 2004b). Beam monitoring devices are ionization chambers (IC) and a range verifier (multi-layer Faraday cup). Beam-shaping devices are scattering systems, range modulators, and wobbling magnets. Variable collimators ("jaws") and the snout determine the field size

27.3
Delivery Systems

27.3.1
Broad Beam (Passive Scattering)

27.3.1.1
Scattering System

Field sizes required for tumor treatment range from as small as 1- up to 25-cm diameter. The field aperture has to be covered with a homogeneous particle flux. The narrow beam, with a FWHM of only about 1 cm, incident on the treatment nozzle, must thus be broadened. For small fields a single scattering foil (made of lead) can be used to broaden the beam. For larger field sizes the reduction in proton fluence and the scattering is too big and one turns to a double-scattering system to ensure a uniform, flat lateral dose profile. The double scattering system may contain a first scatterer (set of foils), placed upstream near the nozzle entrance, and a second Gaussian-shaped scatterer placed further downstream (see Fig. 27.8). Double-scattering systems have been described elsewhere (KOEHLER et al. 1977; GRUSELL et al. 1994; PAGANETTI et al. 2004b). The second scatterer may consist of contoured bi-material scatterers. The rationale for a contoured bi-material device is that a high-Z material scatters more with little range loss, whereas a low-Z material scatters less with more loss in range. In order to flatten the field the protons near the field center must be scattered more than the protons further outside the field center. This has to be achieved by maintaining the range modulation (see section 27.3.1.2) across the field. The contoured scatterer is approximately Gaussian shaped. The system creates a broad uniform beam at the final aperture. Key to the optimal solution for beam flattening is not only the achievable beam profile flatness but also to achieve low energy loss in the absorber, thus minimizing the production of secondary radiation.

27.3.1.2
Range Modulator / Ridge Filter

Pristine Bragg peaks are not wide enough to cover most treatment volumes. The incident proton beam forms an SOBP by sequentially penetrating absorbers of variable thickness, e.g., via a range modulator. Each absorber contributes an individual pristine Bragg peak curve to the composite SOBP. A set of pristine peaks is delivered with decreasing depth and with reduced dose until the desired modulation is achieved. Figure 27.9 shows a series of weighted pristine peaks as well as the resulting SOBP when these are superimposed.

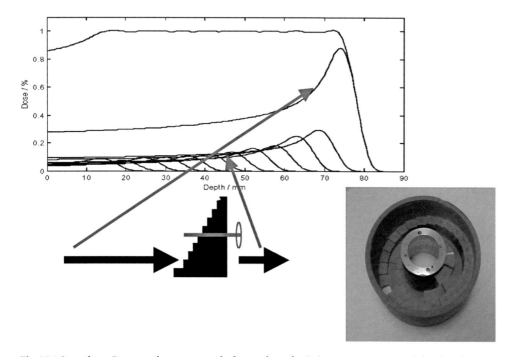

Fig. 27.9 Spread-out Bragg peak as composed of a number of pristine Bragg curves modulated in depth by a set of absorbers of different thickness. The *lower right image* shows a wheel with three different tracks used for different modulation widths.

There are different methods to modulate a field with pristine Bragg curves, e.g., modulation wheels and ridge filters (CHU et al. 1993). A modulator wheel (Fig. 27.9) combines variable thickness absorbers in circular rotating tracks that result in a temporal variation of the beam energy (KOEHLER et al. 1975). Such a wheel typically rotates with about 10 Hz. Modulator wheels are made of a low-Z material (lexan or carbon depending on the designed range interval of a wheel) and a high-Z material (lead). The low-Z material causes slowing down of the beam with little multiple scattering involved and high-Z material is used to adjust the amount of scattering at each depth. Each step segment of the wheel has a specific thickness and covers an angle that represents the weighting of the individual pristine Bragg curves in a SOBP; thus, the angle covered by each step decreases with increasing absorbing power and corresponding decreased range. For small field sizes (i.e., treatment of ocular melanoma) it is sufficient to have modulator wheels made out of plastic material only.

27.3.1.3
Aperture and Compensator

Treatment fields are shaped to a desired target profile using custom milled apertures (Fig. 27.10). Apertures are often made out of brass. It offers the best choice in terms of cost, weight, and production of secondary radiation. The aperture edge, which corresponds to the 50% isodose within a port, is usually defined as the target projection to isocenter plus the 90–50% penumbra plus any setup uncertainties.

The distal part of the dose distribution is shaped according to the desired treatment field using patient-specific milled compensators (Fig. 27.11). Patient-

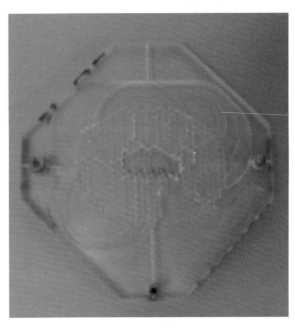

Fig. 27.11 Top view of a range compensator made of plastic. It is used to conform the proton dose distribution to the distal shape of a target. The various depths of the device can be seen through the transparent material.

specific compensators are made out of plastic material and reduce the range of the protons. The maximum required range within a portal, usually defined as the distal 90% of the protons, defines the thinnest point on the compensator. Each part of the compensator controls the range of the protons in its vicinity. The width of the steps can be adjusted to account for uncertainties that may affect the range at various points along the target's cross-sectional profile. Smearing may be included to address patient setup and organ motion uncertainties.

Both aperture and compensator are mounted on a retractable snout on the treatment head. The retractable snout ensures that the air gap between the beam-shaping devices and the patient can always be minimized to reduce effects of scattering in air, which causes softening of the beam penumbra (SISTERSON et al. 1989).

The penumbra varies with treatment depth and beam-line specific hardware settings but a typical

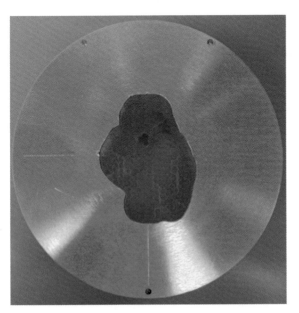

Fig. 27.10 View of a typical brass aperture used to collimate a proton beam. (From BUSSIERE and ADAMS 2003)

value for 16-cm water-equivalent range would be approximately 4.5 mm.

27.3.2
Scanning

Because protons can be deflected magnetically, an alternative to the use of a broad beam is to generate a narrow mono-energetic "pencil" beam and to scan it magnetically across the target volume. Typically, the beam is scanned in a zigzag pattern in the x–y plane perpendicular to the beam direction. This is in close analogy to how a conventional television works (in which, of course, an electron beam is scanned). The depth scan (z) is done by means of energy variation. The method requires neither a collimator nor a compensator.

In practice, it works as follows: one starts with the deepest layer (highest energy) and does one x–y scan. The energy is then reduced, the next layer is painted, and so forth until all 20–30 layers have been delivered. Due to density variations in the patient, the Bragg peaks of one layer are not generally in a plane. Also, it is useful to keep in mind that the distal layers deliver various amounts of dose (depending on the curvature of the distal target surface) to the more proximal regions, such that each layer needs to be intensity modulated in order to generate a uniform target dose. Each layer may be delivered multiple times to reduce delivery errors and uncertainties.

Various modes of particle scanning techniques have been devised, just like different modes of photon IMRT exist:

- Discrete spot scanning: This is a step-and-shoot approach in which the predetermined dose is delivered to a given spot at a static position (constant magnet settings; KANAI et al. 1980). Then the beam is switched off and the magnet settings are changed to target the next spot, dose is delivered to the next spot, and so forth (see Fig. 27.12). This approach is practically implemented at PSI in Switzerland (PEDRONI et al. 1995). There the magnetic scan is performed in one direction only, and the position in the orthogonal direction is changed through a change of the table position. Because the table motion is the slowest motion, it is the last and least often used: first the magnetic scan is performed to create one line of dose (along discrete steps), then the depth is varied by changing the energy, and another line of dose is "drawn" at a more shallow depth. This is repeated until dose is delivered at all relevant depths. Finally, the table

is moved to the next position, and the process is repeated.
- Raster scanning: This method, which is practically realized for heavy ions at the GSI in Darmstadt, Germany (KRAFT 2000), is very similar to discrete spot scanning, but the beam is not switched off while it moves to the next position. Practically, the dose distributions are equivalent for the two methods as long as the scan time from spot to spot is small compared with the treatment time per spot. In general, this is not fulfilled if the scan is done with the treatment table.
- Dynamic spot scanning: Here the beam is scanned fully continuously across the target volume. This method will be used at the NPTC. Intensity (or rather, fluence) modulation can be achieved through a modulation of the output of the source, or the speed of the scan, or both. The combination of the two reduces the required dynamic range of the source output but puts higher demands on the control system.

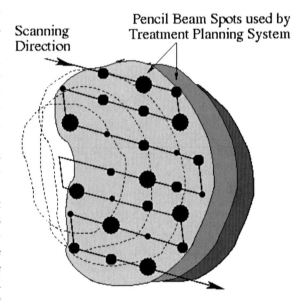

Fig. 27.12 The principle of beam scanning: A narrow pencil beam is scanned across the target volume at various depths. The intensity can be varied from spot to spot, or continuously along the path. (From TROFIMOV and BORTFELD 2003)

One advantage of scanning is that arbitrary shapes of uniform high-dose regions can be achieved with a single beam. With the broad-beam technique, on the other hand, the SOBP is constant across the treatment field and typically delivers some unnecessary amount of dose proximal to the target volume. Another advantage of the scanning approach is that, due to the avoidance of first and second scatterers, the beam has less nuclear interactions outside the patient, and therefore the neutron contamination is smaller (see section 27.5.2). The biggest advantage might be the great flexibility, which can be fully utilized in intensity-modulated proton therapy (IMPT), as we explain below. However, a disadvantage is the technical difficulty to generate very narrow pencil beams that result in an optimal lateral dose fall-off. The scanning approach can also be more sensitive to organ motion than passive scattering (PHILLIPS et al. 1992; BORTFELD et al. 2002).

Another variant of scanning is called "wobbling." Here a relatively broad beam (diameter of the order of 5 cm) is magnetically scanned across the target volume. Because this would result in a broad penumbra, collimators are still required. The main advantage is that field sizes larger than those with passive scattering are easily achievable.

27.4
Proton-Specific Treatment Planning

27.4.1
Broad Beam - Standard Techniques

There are different vendors that offer treatment planning systems for proton-beam therapy. The software packages are either designed for proton therapy planning only or are able to generate plans for conventional photon therapy as well as proton therapy. Different algorithms can be used to calculate dose in the patient. In a broad-beam ray-tracing model the SOBP portion of a depth dose is set to 100% and generic functions are used to describe the proximal build-up as well as the distal fall-off. Lateral penumbra functions are used to form the lateral profiles of beams. With the advent of faster computers more complex and much more accurate pencil-beam models have become the norm replacing the broad-beam model. Pencil-beam algorithms rely on pencil kernels, derived from physical treatment machine data, to model proton range mixing from scatter in the range compensator and patient (GOITEIN and MILLER 1983). The calculated proximal build-up, distal fall-off and lateral penumbra are usually in good agreement with measurements (HONG et al. 1996). In addition to ray-tracing or pencil-beam algorithms there are Monte Carlo dose calculation procedures used mainly in research. They are believed to be more accurate (PAWLICKI and MA 2001) but usually

take too much computing time to be used in routine treatment planning. Monte Carlo dose calculation for proton therapy treatment planning is currently under development (JIANG and PAGANETTI 2004; PAGANETTI et al. 2004a).

As with photons and electrons proton treatments use multiple portals to reduce the overall skin dose to patients. Because proton beams have a sharp distal fall-off, it is possible to aim beams towards critical structures in treatment planning; thus, treatment strategies and treatment options can be different from conventional therapy. The sharp distal dose fall-off of protons (distance from the 90 to the 10% dose level is only a few millimeters) makes it more critical than with photons to understand and limit the uncertainties used in determining the penetration depth required to cover a target. The uncertainties must be incorporated in the treatment planning margins around the target volume. The accuracy of proton-beam delivery may in general allow tighter margins than used conventionally; however, one has to keep in mind that higher accuracy also means that dose delivery is more affected by uncertainties caused by beam delivery, patient setup and immobilization, tissue heterogeneities, and organ motion.

Imaging studies prior to treatment planning and the process of delineating target volumes and structures of interest are identical in proton therapy and in conventional therapy; however, planning and delivery strategies may be different. Typical head and neck cases include four to six non-coplanar fields. During a treatment session between one and three fields are treated. The fields are alternated for subsequent treatment sessions to distribute the daily non-target dose (BUSSIERE and ADAMS 2003).

A dose delivery strategy unique to particle therapy is called "patching." Two fields are combined such that the first field treats a segment of the target avoiding a nearby critical organ with the lateral penumbra. The second field treats the remaining segment, also avoiding the critical organ with the lateral penumbra and matching its distal fall-off 50% dose with the other field's lateral penumbra's 50% dose value. Because of tissue heterogeneity, it can be difficult to obtain a uniform dose along the patch junction; therefore, patch junctions are always selected to be within the target volume. One allows overshoot of the patch field to minimize the low dose region. Using a combination of patch fields with different junctions ensures that the magnitude of the low- and high-dose regions is acceptable. Figure 27.13 shows a typical patch field combination for a skull-base tumor (BUSSIERE and ADAMS 2003).

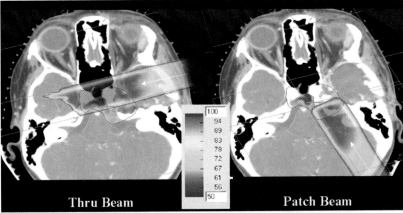

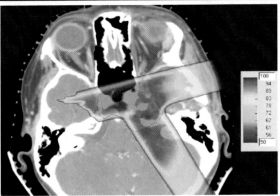

Fig. 27.13 Axial CT image with color-wash dose display resulting from thru-field which irradiates the anterior portion of the target while avoiding the brain stem and patch field which treats the remaining portion of the target while avoiding the brain stem. The *lower figure* shows the combined thru/patch field combination. All doses are given in percent. (From BUSSIERE and ADAMS 2003)

27.4.2
Scanning and IMPT

The recent advances in IMRT have challenged proton therapy. Treatment planning comparison studies have shown that IMRT can produce dose distributions that are comparable with proton distributions in terms of target conformality and dose gradients in the high-dose region (LOMAX et al. 1999). In some cases, IMRT produces even more target-conformal dose distributions in the high-dose region than protons. In the low to medium dose region and in terms of integral dose, protons are, however, always better. Nevertheless, it is somewhat questionable if the higher cost of proton therapy is justifiable if integral dose was the only advantage of protons.

The study mentioned above compares IMRT photon therapy with conventional passive scattering proton techniques. The dose conforming potential of the latter is limited and this has mainly technical reasons. It has nothing to do with the physical dose conformation potential of protons. To fully exploit the physical potential of proton therapy and to permit a fair comparison with IMRT, intensity-modulation techniques have to be introduced into proton therapy as

well. This is then called intensity-modulated proton therapy (IMPT). The name is somewhat misleading because intensity modulation is always required in proton therapy, even for the generation of an SOBP. What we mean by IMPT is a treatment technique that, in analogy with IMRT, delivers intentionally non-uniform dose distributions from each treatment field at a given direction. The desired (generally uniform) dose in the target volume is obtained after superimposing the dose contributions from all fields. The additional degrees of freedom (by not having to produce uniform dose from each direction) can be used to optimize dose distributions in several ways, which we now describe.

The IMPT treatment plans are optimized using an "inverse" treatment planning system, which is similar to inverse planning for photon IMRT (OELFKE and BORTFELD 2001). The main difference is that in IMPT the energy of each pencil beam can be varied in addition to its intensity. This increases the number of degrees of freedom drastically, which increases its dose-shaping potential but also increases the computational and delivery complexity. The calculation can be simplified and the solution can be steered in the desired direction if certain IMPT techniques are pre-selected. LOMAX (1999) has summarized various IMPT techniques whose complexities are between the conventional passive scattering technique and the most general ("3D") technique. One of the simpler techniques is the 2.5D technique, which is actually quite similar to IMRT. It uses poly-energetic SOBP pencil beams, which are individually adapted to the proximal and distal edge of the target volume (Fig. 27.14), such that the dose is constant along the depth of the target volume. The weights (i.e., intensities) of the SOBP pencil beams are modulated across the target volume (symbolized by different colors in Fig. 27.15).

Another technique was devised by DEASY et al. (1997) and is called distal edge tracking (DET). As indicated by its name, it puts Bragg peaks on the distal edge of the target volume only and thereby creates a highly non-uniform dose per treatment field. The

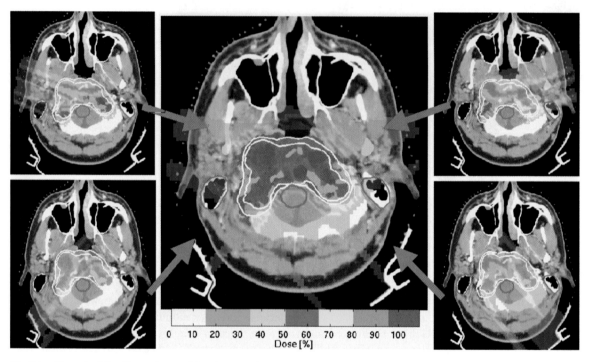

Fig. 27.14 The principle of intensity-modulated proton therapy (IMPT). Non-uniform dose distributions from a number of fields (four in this case) yield the desired (uniform) target dose. (Courtesy of A. Trofimov, Massachusetts General Hospital)

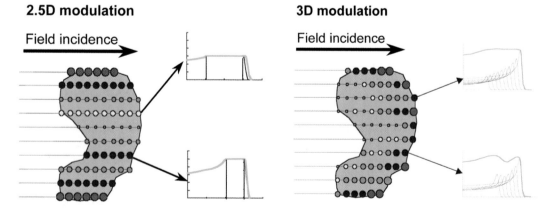

Fig. 27.15 Comparison of the 2.5-D IMPT modulation technique with the most general 3D approach. Different weights are symbolized by different colors. (From LOMAX 1999)

desired uniform dose is obtained by superimposing multiple fields from different directions, in combinations with optimized intensity modulation. The DET technique produces the smallest possible integral dose because every constituent pencil beam delivers the best possible ratio between target dose and dose to the proximal normal structures. Moreover, the DET technique yields very sharp dose gradients because it shapes the dose distribution mainly with the distal edge of the Bragg peak, which is sharper than the lateral fall-off (as long as the energy spectrum is not too wide). However, for obvious reasons DET has greater difficulties in generating a high uniformity of the target dose. Also, it is more sensitive to range uncertainties than, e.g., the 2.5D technique (Fig. 27.15). Furthermore, shaping dose distributions with the distal edge of the Bragg peak with its high LET and RBE may raise some biological flags (see section 27.5.1; WILKENS and OELFKE 2004).

In summary, IMPT treatments can be tailored to yield one of the following advantages:
- Improved dose conformality and steeper dose gradients
- Further reduction of integral dose
- Less sensitivity to range uncertainties and other sources of uncertainty (LOMAX et al. 2001)

A combination of all of these advantages is also possible; however, not all of them may be achievable at the same time to the full extent.

27.5
Biological Effectiveness

27.5.1
Relative Biological Effectiveness

Protons are slightly more biologically effective than photons. In other words, lower dose is required to cause the same biological effect. The relative biological effectiveness (RBE) of protons is defined as the dose of a reference radiation divided by the proton dose to achieve the same biological effect. The fundamental reason for applying a RBE value is to ensure that radiation oncologists can benefit from the large pool of clinical results obtained with photon beams. The RBE adjusted dose is defined as the product of the physical dose and the respective RBE describing the radiosensitivity of the tissue after ion irradiation compared with photon irradiation at a given level of effect. Proton therapy is based on

the use of a single RBE value (equals 1.1 at almost all institutions), which is applied to all proton-beam treatments independent of dose/fraction, position in the SOBP, initial beam energy, or the particular tissue. A generic RBE is only a rough approximation considering experimentally determined RBEs for both in vitro and in vivo systems (PAGANETTI et al. 2002; PAGANETTI 2003). Dependencies of the RBE on various physical and biological properties are disregarded. The RBE of principal concern is that of the critical normal tissue/organ(s) immediately adjacent to or within the treatment volume, i.e., the determinant(s) of NTCP.

Although the fact that a generic RBE cannot be the true RBE for each tissue, dose/fraction, etc., has long been recognized, it was concluded that the magnitude of RBE variation with treatment parameters is small relative to our abilities to determine RBE values (PAGANETTI and GOITEIN 2001; PAGANETTI 2003). The variability of RBE in clinical situations is believed to be within 10–20%. The values for cell survival in vitro indicate a substantial spread between the diverse cell lines. The average value at mid SOBP over all dose levels was shown to be ≈ 1.2 in vitro and ≈ 1.1 in vivo (PAGANETTI et al. 2002). Both in vitro and in vivo data indicated a small but statistically significant increase in RBE for lower doses per fraction. Evaluation of the statistically significant difference in RBE between in vitro and in vivo systems should deal explicitly with the fact that the former uses as the end point the killing of single cells of one cell population (colony formation). The in vivo response reflects the more complex expression of the integrated radiation damage to several tissue systems (cell populations). In addition, the in vivo data refer to various different biological processes (e.g., mutation). The dependency on dose, i.e., increasing RBE with decreasing dose, appears to be far less in vivo compared with the in vitro data. Unfortunately, RBE values from in vivo systems at doses of <4 Gy are quite limited.

The effect of radiation on cells and tissues is a complex and not entirely understood function of the properties of the cell or tissue and the microdosimetric properties of the radiation field. The dependencies of RBE on biological end point and dose are difficult to explain microscopically; however, the LET dependency can be explained based on the concept that ionization density within the sensitive cellular structure (e.g., DNA) increases with LET and that production of non-reparable lesions increases with ionization density. For constant conditions at irradiation a change in response of a cell population to a defined physical dose by different radiation beams is generally accepted as being due to differences in LET. Mean LET is only one parameter which characterizes that microdosimetric structure, and it is only one of several determinants of radiation response. In general, as LET increases, the RBE increases, eventually reaching a maximum and then decreasing (GOODHEAD 1990). Based on these considerations a small increase in the proton RBE across the SOBP and the extension of the penetration of the beam by a few millimeters is expected because of an increasing LET (PAGANETTI et al. 1997; PAGANETTI 1998; PAGANETTI and GOITEIN 2000; WILKENS and OELFKE 2004). Calculations show that LET increases slightly throughout the SOBP and significantly at the terminal end of a SOBP. This measurable increase in RBE over the terminal few millimeters of the SOBP results in an extension of the bio-effective range of the beam of a few millimeters (ROBERTSON et al. 1975; WOUTERS et al. 1996). This needs to be considered in treatment planning, particularly for single-field plans or for an end of range in or close to a critical structure. Clinicians and treatment planners are often reluctant to having the SOBP abutting a critical structure, thus not utilizing one advantage of protons, namely the sharp distal fall-off.

27.5.2
Secondary Radiation

Protons slowing down in matter loose energy not only by coulomb interactions but also by nuclear interactions (LAITANO et al. 1996; MEDIN and ANDREO 1997; PAGANETTI 2002). Nuclear interactions cause secondary radiation. Protons and neutrons are the most important secondary particles from nuclear interactions because they can carry away energy far from the interaction point. Shielding against neutron radiation is therefore important for any proton therapy installation. For example, different combinations of apertures may be used in the treatment head; however, neutron production cannot be avoided. Shielding may reduce the effect of neutrons generated in the scattering system, the aperture, and the compensator, but neutrons are also generated in the patient itself. Nothing can be done to avoid the latter situation. Since the total amount of neutrons produced depends on the amount of material the protons have to penetrate, neutron production can be reduced by extracting to the nozzle the minimum energy needed.

27.6
Patient Positioning and Immobilization Issues, Motion

Proton therapy is, like all highly target-conformal treatment modalities, susceptible to geographical misses. Considerable effort is therefore necessary to position and immobilize the patient. For example, at the NPTC orthogonal X-ray projections are used to detect both translational and rotational positioning errors and correct those errors using a six-axes table within 1 mm or 0.5° (Fig. 27.16). For the most part, positioning and immobilization issues are identical for proton therapy and, for example, IMRT; however, there are a few issues that are specific to protons and other charged particles. They have to do with the simple fact that the range is affected by structures moving in and out of the beam. For example, in prostate treatments the position of the Bragg peak may be significantly altered if parts of the pelvic bone move into the beam, which can happen if on one treatment day the pelvis is rotated compared with the planned position (PHILLIPS et al. 2002). Similar problems can affect treatments in the skull (Fig. 27.16); therefore,

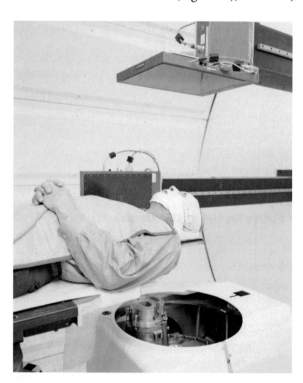

Fig. 27.16 Proton therapy requires, like all highly conformal treatment modalities, a significant effort in patient setup and immobilization. This figure shows the setup at NPTC using orthogonal X-rays (one X-ray source is integrated into the nozzle) and flat-panel detectors. (Courtesy of S. Rosenthal, Massachusetts General Hospital)

in particle therapy it is not only important to ensure that the target volume is always at the same position, but the surrounding structures and especially bony structures should also be at their planned position. The detrimental effect of misalignments can be mitigated to some degree in treatment planning. A common approach in passive scattering proton therapy is to "smear" (thin) the range compensator such that target coverage is ensured even if the position is slightly off; however, this will push the dose into the normal tissues distal to the target volume, and the smearing radius is therefore limited to about 3 mm. Bigger errors cannot be compensated with this method.

Besides alignment errors, proton (and charged particle) therapy is also uniquely affected by internal organ motion, especially in the case of lung tumors. The dose distribution is deformed by the motion of the tumor in the low-density lung tissue. Unless methods such as gating (MINOHARA et al. 2000) are used to "freeze" the motion, this effect must be carefully considered at treatment planning stage. This can be done (MOYERS et al. 2001), but it is fair to say that the proton treatments of lung tumors have not yet fully come of age.

References

Archambeau JO, Slater JD, Slater JM, Tangeman R (1992) Role for proton beam irradiation in treatment of pediatric CNS malignancies. Int J Radiat Oncol Biol Phys 22:287–294

Benk V, Liebsch NJ, Munzenrider JE, Efird J, McManus P, Suit H (1995) Base of skull and cervical spine chordomas in children treated by high-dose irradiation. Int J Radiat Oncol Biol Phys 31:577–558

Blosser H, Bailey J, Burleigh R, Johnson D, Kashy E, Kuo T, Marti F, Vincent J, Zeller A, Blosser E, Blosser G, Maughan R, Power W, Wagner J (1989) Superconducting cyclotron for medical applications. IEEE Trans Magn 25:1746–1754

Bortfeld T, Jokivarsi K, Goitein M, Kung J, Jiang SB (2002) Effects of intra-fraction motion on IMRT dose delivery: statistical analysis and simulation. Phys Med Biol 47:2303–2320

Bussiere MR, Adams JA (2003) Treatment planning for conformal proton radiation therapy. Technol Cancer Res Treat 2:389–399

Chu WT, Ludewigt BA, Renner TR (1993) Instrumentation for treatment of cancer using proton and light-ion beams. Rev Sci Instrum 64:2055–2122

Coutrakon G, Slater JM, Ghebremedhin A (1999) Design considerations for medical proton accelerators. Proc 1999 Particle Accelerator Conference, New York, pp 11–15

Cozzi L, Fogliata A, Lomax A, Bolsi A (2001) A treatment planning comparison of 3D conformal therapy, intensity modulated photon therapy and proton therapy for treatment of advanced head and neck tumours. Radiother Oncol 61:287–297

Deasy JO, Shepard DM, Mackie TR (1997) Distal edge tracking: a proposed delivery method for conformal proton therapy using intensity modulation. In: Leavitt DD, Starkschall G (eds) XIIth International Conference on the Use of Computers in Radiation Therapy. Medical Physics Publishing, Madison, Wisconsin, pp 406–409

Delaney TF, Smith AR, Lomax A, Adams J, Loeffler JS (2003) Proton beam radiation therapy. Cancer Prin Pract Oncol 17:1–10

Flanz J, Durlacher S, Goitein M, Levine A, Reardon P, Smith A (1995) Overview of the MGH-Northeast Proton Therapy Center plans and progress. Nucl Instrum Methods Phys Res B 99:830–834

Fuss M, Hug EB, Schaefer RA, Nevinny-Stickel M, Miller DW, Slater JM, Slater JD (1999) Proton radiation therapy (PRT) for pediatric optic pathway gliomas: comparison with 3D planned conventional photons and a standard photon technique. Int J Radiat Oncol Biol Phys 45:1117–1126

Fuss M, Poljanc K, Miller DW, Archambeau JO, Slater JM, Slater JD, Hug EB (2000) Normal tissue complication probability (NTCP) calculations as a means to compare proton and photon plans and evaluation of clinical appropriateness of calculated values. Int J Cancer 90:351–358

Goitein M, Jermann M (2003) The relative costs of proton and X-ray radiation therapy. Clin Oncol 15:S37–S50

Goitein M, Miller T (1983) Planning proton therapy of the eye. Med Phys 10:275–283

Goodhead DT (1990) Radiation effects in living cells. Can J Phys 68:872–886

Grusell E, Montelius A, Brahme A, Rikner G, Russell K (1994) A general solution to charged particle beam flattening using an optimized dual-scattering-foil technique, with application to proton therapy beams. Phys Med Biol 39:2201–2216

Hara I, Murakami M, Kagawa K, Sugimura K, Kamidono S, Hishikawa Y, Abe M (2004) Experience with conformal proton therapy for early prostate cancer. Am J Clin Oncol 27:323–327

Harsh G, Loeffler JS, Thornton A, Smith A, Bussiere M, Chapman PH (1999) Stereotactic proton radiosurgery. Neurosurg Clin North Am 10:243–256

Hong L, Goitein M, Bucciolini M, Comiskey R, Gottschalk B, Rosenthal S, Serago C, Urie M (1996) A pencil beam algorithm for proton dose calculations. Phys Med Biol 41:1305–1330

Hug EB, Slater JD (1999) Proton radiation therapy for pediatric malignancies: status report. Strahlenther Onkol 175 (Suppl) 2:89–91

Hug EB, Fitzek MM, Liebsch NJ, Munzenrider JE (1995) Locally challenging osteo- and chondrogenic tumors of the axial skeleton: results of combined proton and photon radiation therapy using three-dimensional treatment planning. Int J Radiat Oncol Biol Phys 31:467–476

Hug EB, Devries A, Thornton AF, Munzenrider JE, Pardo FS, Hedley-Whyte ET, Bussiere MR, Ojemann R (2000) Management of atypical and malignant meningiomas: role of high-dose, 3D-conformal radiation therapy. J Neuro-Oncol 48:151–160

Isacsson U, Montelius A, Jung B, Glimelius B (1996) Comparative treatment planning between proton and X-ray therapy in locally advanced rectal cancer. Radiother Oncol 41:263–272

Isacsson U, Hagberg H, Johansson K-A, Montelius A, Jung B, Glimelius B (1997) Potential advantages of protons over conventional radiation beams for paraspinal tumours. Radiother Oncol 45:63–70

Isacsson U, Lennernäs B, Grusell E, Jung B, Montelius A, Glimelius B (1998) Comparative treatment planning between proton and X-ray therapy in esophageal cancer. Int J Radiat Oncol Biol Phys 41:441–450

Jiang H, Paganetti H (2004) Adaptation of GEANT4 to Monte Carlo dose calculations based on CT data. Med Phys 31:2811–2818

Kanai T, Kawachi K, Kumamoto Y, Ogawa H, Yamada T, Matsuzawa H, Inada T (1980) Spot scanning system for proton radiotherapy. Med Phys 7:365–369

Koehler AM, Schneider RJ, Sisterson JM (1975) Range modulators for protons and heavy ions. Nucl Instrum Methods 131:437–440

Koehler AM, Schneider RJ, Sisterson JM (1977) Flattening of proton dose distributions for large-field radiotherapy. Med Phys 4:297–301

Kraft G (2000) Tumortherapy with ion beams. Nucl Instrum Methods Phys Res A 454:1–10

Laitano RF, Rosetti M, Frisoni M (1996) Effects of nuclear interactions on energy and stopping power in proton beam dosimetry. Nucl Instrum Methods A 376:466–476

Lee M, Wynne C, Webb S, Nahum AE, Dearnaley D (1994) A comparison of proton and megavoltage X-ray treatment planning for prostate cancer. Radiother Oncol 33:239–253

Levin CV (1992) Potential for gain in the use of proton beam boost to the para-aortic lymph nodes in carcinoma of the cervix. Int J Radiat Oncol Biol Phys 22:355–359

Lin R, Hug EB, Schaefer RA, Miller DW, Slater JM, Slater JD (2000) Conformal proton radiation therapy of the posterior fossa: a study comparing protons with three-dimensional planned photons in limiting dose to auditory structures. Int J Radiat Oncol Biol Phys 48:1219–1226

Lomax A (1999) Intensity modulation methods for proton radiotherapy. Phys Med Biol 44:185–205

Lomax AJ, Bortfeld T, Goitein G, Debus J, Dykstra C, Tercier P-A, Coucke PA, Mirimanoff RO (1999) A treatment planning inter-comparison of proton and intensity modulated photon radiotherapy. Radiother Oncol 51:257–271

Lomax AJ, Boehringer T, Coray A, Egger E, Goitein G, Grossmann M, Juelke P, Lin S, Pedroni E, Rohrer B, Roser W, Rossi B, Siegenthaler B, Stadelmann O, Stauble H, Vetter C, Wisser L (2001) Intensity modulated proton therapy: a clinical example. Med Phys 28:317–324

Lomax AJ, Cella L, Weber D, Kurtz JM, Miralbell R (2003a) Potential role of intensity-modulated photons and protons in the treatment of the breast and regional nodes. Int J Radiat Oncol Biol Phys 55:785–792

Lomax AJ, Goitein M, Adams J (2003b) Intensity modulation in radiotherapy: photons versus protons in the paranasal sinus. Radiother Oncol 66:11–18

McAllister B, Archambeau JO, Nguyen MC, Slater JD, Loredo L, Schulte R, Alvarez O, Bedros AA, Kaleita T, Moyers M, Miller D, Slater JM (1997) Proton therapy for pediatric cranial tumors: preliminary report on treatment and disease-related morbidities. Int J Radiat Oncol Biol Phys 39:455–460

Medin J, Andreo P (1997) Monte Carlo calculated stopping-power ratios, water/air, for clinical proton dosimetry (50–250 MeV). Phys Med Biol 42:89–105

Minohara S, Kanai T, Endo M, Noda K, Kanazawa M (2000) Respiratory gated irradiation system for heavy-ion radiotherapy. Int J Radiat Oncol Biol Phys 47:1097–1103

Miralbell R, Crowell C, Suit HD (1992) Potential improvement of three dimension treatment planning and proton therapy in the outcome of maxillary sinus cancer. Int J Radiat Oncol Biol Phys 22:305–310

Miralbell R, Lomax A, Russo M (1997) Potential role of proton therapy in the treatment of pediatric medulloblastoma/primitive neuro-ectodermal tumors: spinal theca irradiation. Int J Radiat Oncol Biol Phys 38:805–811

Moyers MF, Miller DW, Bush DA, Slater JD (2001) Methodologies and tools for proton beam design for lung tumors. Int J Radiat Oncol Biol Phys 49:1429–1438

Niemierko A, Urie M, Goitein M (1992) Optimization of 3D radiation therapy with both physical and biological end points and constraints. Int J Radiat Oncol Biol Phys 23:99–108

Nowakowski VA, Castro JR, Petti PL, Collier JM, Daftari I, Ahn D, Gauger G, Gutin P, Linstadt DE, Phillips TL (1992) Charged particle radiotherapy of paraspinal tumors. Int J Radiat Oncol Biol Phys 22:295–303

Oelfke U, Bortfeld T (2001) Inverse planning for photon and proton beams. Med Dosim 26:113–124

Paganetti H (1998) Calculation of the spatial variation of relative biological effectiveness in a therapeutic proton field for eye treatment. Phys Med Biol 43:2147–2157

Paganetti H (2002) Nuclear interactions in proton therapy: dose and relative biological effect distributions originating from primary and secondary particles. Phys Med Biol 47:747–764

Paganetti H (2003) Significance and implementation of RBE variations in proton beam therapy. Technol Cancer Res Treatment 2:413–426

Paganetti H, Goitein M (2000) Radiobiological significance of beam line dependent proton energy distributions in a spread-out Bragg peak. Med Phys 27:1119–1126

Paganetti H, Goitein M (2001) Biophysical modeling of proton radiation effects based on amorphous track models. Int J Radiat Biol 77:911–928

Paganetti H, Olko P, Kobus H, Becker R, Schmitz T, Waligorski MPR, Filges D, Mueller-Gaertner HW (1997) Calculation of RBE for proton beams using biological weighting functions. Int J Radiat Oncol Biol Phys 37:719–729

Paganetti H, Niemierko A, Ancukiewicz M, Gerweck LE, Loeffler JS, Goitein M, Suit HD (2002) Relative biological effectiveness (RBE) values for proton beam therapy. Int J Radiat Oncol Biol Phys 53:407–421

Paganetti H, Jiang H, Adams JA, Chen GT, Rietzel E (2004a) Monte Carlo simulations with time-dependent geometries to investigate organ motion with high temporal resolution. Int J Radiat Oncol Biol Phys 60:942–950

Paganetti H, Jiang H, Lee S-Y, Kooy H (2004b) Accurate Monte Carlo for nozzle design, commissioning, and quality assurance in proton therapy. Med Phys 31:2107–2118

Pawlicki T, Ma C-MC (2001) Monte Carlo simulation for MLC-based intensity-modulated radiotherapy. Med Dosim 26:157–168

Pedroni E, Bacher R, Blattmann H, Boehringer T, Coray A, Lomax A, Lin S, Munkel G, Scheib S, Schneider U, Tourovsky A (1995) The 200-MeV proton therapy project at the Paul Scherrer Institute: conceptual design and practical realization. Med Phys 22:37–53

Phillips MH, Pedroni E, Blattmann H, Boehringer T, Coray A, Scheib S (1992) Effects of respiratory motion on dose uniformity with a charged particle scanning method. Phys Med Biol 37:223–234

Phillips BL, Jiroutek MR, Tracton G, Elfervig M, Muller KE, Chaney EL (2002) Thresholds for human detection of patient setup errors in digitally reconstructed portal images of prostate fields. Int J Radiat Oncol Biol Phys 54:270–277

Robertson JB, Williams JR, Schmidt RA, Little JB, Flynn DF, Suit HD (1975) Radiobiological studies of a high-energy modulated proton beam utilizing cultured mammalian cells. Cancer 35:1664–1677

Rosenberg AE, Nielsen GP, Keel SB, Renard LG, Fitzek MM, Munzenrider JE, Liebsch NJ (1999) Chondrosarcoma of the base of the skull: a clinicopathologic study of 200 cases with emphasis on its distinction from chordoma. Am J Surg Pathol 23:1370–1378

Shioyama Y, Tokuuye K, Okumura T, Kagei K, Sugahara S, Ohara K, Akine Y, Ishikawa S, Satoh H, Sekizawa K (2003) Clinical evaluation of proton radiotherapy for non-small-cell lung cancer. Int J Radiat Oncol Biol Phys 56:7–13

Sisterson JM (2004) Particles Newsletter. http://ptcog.mgh.harvard.edu

Sisterson JM, Urie MM, Koehler AM, Goitein M (1989) Distal penetration of proton beams: the effects of air gaps between compensating bolus and patient. Phys Med Biol 34:1309–1315

Slater JD, Slater JM, Wahlen S (1992a) The potential for proton beam therapy in locally advanced carcinoma of the cervix. Int J Radiat Oncol Biol Phys 22:343–347

Slater JM, Slater JD, Archambeau JO (1992b) Carcinoma of the tonsillar region: potential for use of proton beam therapy. Int J Radiat Oncol Biol Phys 22:311–319

Slater JD, Rossi CJJ, Yonemoto LT, Bush DA, Jabola BR, Levy RP, Grove RI, Preston W, Slater JM (2004) Proton therapy for prostate cancer: the initial Loma Linda University experience. Int J Radiat Oncol Biol Phys 59:348–352

Smit BM (1992) Prospects for proton therapy in carcinoma of the cervix. Int J Radiat Oncol Biol Phys 22:349–353

Tatsuzaki H, Urie MM, Linggood R (1992a) Comparative treatment planning: proton vs X-ray beams against glioblastoma multiforme. Int J Radiat Oncol Biol Phys 22:265–273

Tatsuzaki H, Urie MM, Willett CG (1992b) 3-D comparative study of proton vs X-ray radiation therapy for rectal cancer. Int J Radiat Oncol Biol Phys 22:369–374

Terahara A, Niemierko A, Goitein M, Finkelstein D, Hug E, Liebsch N, O'Farrell D, Lyons S, Munzenrider J (1999) Analysis of the relationship between tumor dose inhomogeneity and local control in patients with skull base chordoma. Int J Radiat Oncol Biol Phys 45:351–358

Thornton A, Fitzek M, Varvares M, Adams J, Rosenthal S, Pollock S, Jackson M, Pilch B, Joseph M (1998) Accelerated hyperfractionated proton/photon irradiation for advanced paranasal sinus cancer. Results of a perspective phase I–II study. Int J Radiat Oncol Biol Phys 42 (Suppl):222

Tobias CA, Lawrence JH, Born JL, McCombs R, Roberts JE, Anger HO, Low-Beer BVA, Huggins C (1958) Pituitary irradiation with high energy proton beams: a preliminary report. Cancer Res 18:121–134

Trofimov A, Bortfeld T (2003) Optimization of beam parameters and treatment planning for intensity modulated proton therapy. Technol Cancer Res Treatment 2:437–444

Wambersie A, Gregoire V, Brucher J-M (1992) Potential clinical gain of proton (and heavy ion) beams for brain tumors in children. Int J Radiat Oncol Biol Phys 22:275–286

Weber DC, Trofimov AV, Delaney TF, Bortfeld T (2004) A treatment plan comparison of intensity modulated photon and proton therapy for paraspinal sarcomas. Int J Radiat Oncol Biol Phys 58:1596–1606

Wenkel E, Thornton AF, Finkelstein D, Adams J, Lyons S, Monte S de la, Ojeman RG, Munzenrider JE (2000) Benign meningioma: partially resected, biopsied, and recurrent intracranial tumors treated with combined proton and photon radiotherapy. Int J Radiat Oncol Biol Phys 48:1363–1370

Wilkens JJ, Oelfke U (2004) A phenomenological model for the relative biological effectiveness in therapeutic proton beams. Phys Med Biol 49:2811–2825

Wilson RR (1946) Radiological use of fast protons. Radiology 47:487–491

Wouters BG, Lam GKY, Oelfke U, Gardey K, Durand RE, Skarsgard LD (1996) RBE Measurement on the 70 MeV Proton Beam at TRIUMF using V79 cells and the high precision cell sorter assay. Radiat Res 146:159–170

Yeboah C, Sandison GA (2002) Optimized treatment planning for prostate cancer comparing IMPT, VHEET and 15 mV IMXT. Phys Med Biol 47:2247-2261

28 Heavy Ion Radiotherapy

Oliver Jäkel

CONTENTS

28.1
Introduction

Within the past two decades not only photon radiotherapy has rapidly developed, but also particle therapy with protons as well as helium and carbon ions has gained increasing interest (JÄKEL et al. 2003). The term "heavy ions" is used here for ions heavier than helium ions. The primary rationale for radiotherapy with heavy charged particles is the sharp increase of dose in a well-defined depth (Bragg peak) and the rapid dose fall-off beyond that maximum

O. JÄKEL, PhD
Abteilung für Medizinische Physik in der Strahlentherapie,
Deutsches Krebsforschungszentrum,
Im Neuenheimer Feld 280, 69120 Heidelberg, Germany

(see Fig. 28.2). The ratio of Bragg peak dose vs dose in the entrance region is larger for heavy ions than for protons. Due to their larger mass, angular and energy straggling of the primary particles is becoming negligible for heavy ions as compared with protons or even helium ions. Heavy ions therefore offer an improved dose conformation as compared with photon and proton radiotherapy with better sparing of normal tissue structures close to the target. In addition, heavy ions exhibit a strong increase of the linear energy transfer (LET) in the Bragg peak as compared with the entrance region. The radiobiological advantage of high LET radiation in tumor therapy is well known from neutron therapy. Unlike in radiotherapy with neutron beams, in heavy ion radiotherapy the high LET region can also be conformed to the tumor.

Between 1977 and 1992, first clinical experiences have been made, especially with helium and neon ions at the Lawrence Berkeley Laboratory and many encouraging results (especially in skull base tumors and paraspinal tumors) were achieved (CASTRO et al. 1994; CASTRO 1997).

The increasing effectiveness of ions with larger charge is shown in Fig. 28.1. Although helium ions are very similar to protons in their biological properties, carbon or neon ions exhibit an increased relative biological effectiveness (RBE) in the Bragg peak as compared with the entrance region (see Fig. 28.1). The RBE ratio (Bragg peak vs entrance region) is highest for carbon ions. For ions heavier than neon, the RBE in the entrance region is even higher than in the Bragg peak (like for argon).

Another disadvantage of very heavy ions for radiotherapy is the increase of nuclear fragmentation processes, which leads to a fragment tail in the depth dose distribution that extends beyond the Bragg peak (see Fig. 28.2).

Currently the availability of heavy ion radiotherapy is limited, as worldwide only three facilities offer carbon ion radiotherapy: two hospital-based facilities in Japan [Heavy Ion Medical Accelerator (HIMAC), Chiba; Hyogo Ion Beam Medical Center (HIBMC), Hyogo] and a physics research facility

pencil beam that is then deflected laterally by two magnetic dipoles to allow a scanning of the beam over the treatment field. Moreover, the energy from a synchrotron can be switched from pulse to pulse in order to adapt the range of the particles in tissue. This way, a target volume can be scanned in three dimensions and the dose distribution can be tailored to any irregular shape without any passive absorbers or patient-specific devices such as compensators or collimators; therefore, the high-dose region can also be conformed to the proximal end of the target volume and the integral dose as well as the volume receiving high LET radiation is minimized.

There is, however, considerable effort necessary in order to monitor on-line the position and intensity of the beam, and to enable a safe and accurate delivery of dose to the patient. Figure 28.4 shows the principle of the active beam delivery system.

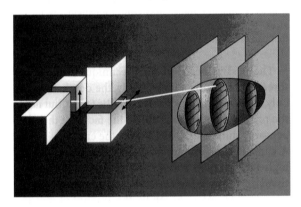

Fig. 28.4. Principle of the active raster scan system used at GSI for carbon ions. A small pencil beam is scanned in vertical and horizontal direction, using two pairs of scanner magnets. By switching the energy of the synchrotron, the position of the Bragg peak can be chosen so that each scanned area is adapted to the extent of the target in depth

28.2.3
Patient Positioning

Due to the high spatial accuracy that is achievable with ion beams, patient fixation and positioning requires special attention. Patient fixation is usually achieved with individually prepared mask systems or whole-body moulds. Highest accuracy during the initial positioning can be achieved by the use of stereotactic methods. Prior to every fraction, the position is verified using X-ray imaging in treatment position. The X-ray images are compared against digitally reconstructed radiographs obtained from the treatment planning CT. Another possibility of position control, which is used at HIMAC, is to do a CT scan of

the patient in treatment position and compare it with reference images used for treatment planning.

To achieve additional freedom, treatment chairs may be used to treat patients in a seated position. In this case, patient movement plays an important role, since the patient tends to relax and move downward in the chair with time. The treatment time therefore has to be minimized and means to control the patient position during therapy are advisable. Additional uncertainties are introduced, if the treatment planning is performed on a conventional horizontal CT scan, with the patient lying on the table. Then the assumption has to be made that organ movement is negligible, when moving the patient from lying to seated position, which is commonly believed to be reasonable for tumors within the cranium. The best way to exclude this uncertainty is certainly to perform the treatment planning CT in seated position, using dedicated vertical CT scanners (KAMADA et al. 1999).

At HIMAC an additional degree of freedom was introduced, by rotating the patient in a mould around its longitudinal axis. Using this method, the problem of organ movement has to be carefully considered and rotation angles are limited to 15° at HIMAC.

28.2.4
Treatment Planning

Since the details of the treatment planning system (TPS) are very much connected to the beam delivery system that is used, TPS for passive and active beam shaping are dealt with separately.

28.2.4.1
TPS for Passive Beam Shaping

Treatment planning systems for the passive systems at Berkeley and HIMAC facilities were developed as research tools within the corresponding institutions, since no commercial products were available at that time (CHEN et al. 1979; ENDO et al. 1996). In the meantime, the HIBMC TPS is available as a fully certified medical product (FOCUS-M by Computerized Medical Systems, St. Louis; and Mitsubishi, Kobe). Furthermore, it is embedded in a modern graphical user interface for 3D conformal radiotherapy planning including all up-to-date available features.

For the passive depth dose shaping system, the depth dose profile is fixed by the modulator hardware throughout the irradiation field and no further optimization is necessary. The modulators were designed in order to achieve a prescribed homogeneous biological effective dose for a single field. The design of the modulators reflects the fixed dependence of the RBE with depth for a certain dose level.

Absorbed Dose Calculation

The algorithms to calculate absorbed dose for a passive beam shaping system are very similar to those used in conventional photon therapy. The beam transport models are relatively simple, as lateral scattering of carbon or even neon ions is very small and the lateral penumbra of the primary beam is preserved almost completely in depth. The modeling of nuclear fragmentation is not a serious problem, because treatment planning can rely on measured depth dose data that include fragmentation. These measurements for the various depth modulators are performed in water and sum up the dose contribution of all fragments. The radiological depth of the beam in tissue is calculated using an empirical relation between X-ray CT numbers and measured particle ranges (MINOHARA et al. 1993; JÄKEL et al. 2001a, b).

Using this procedure, the design of necessary patient-specific devices, such as bolus, compensators, and collimators, can be optimized by computer programs. Similar to photon therapy, only relative values of the absorbed dose may be used, because the absorbed dose scales with the number of monitor units.

Biological Modeling

The situation is more complicated for the calculation of biological effective dose, because the relative biological efficiency (RBE) of an ion beam in tissue depends on the underlying LET spectrum, the cell type, the dose level, and some other quantities. This problem was solved by a number of pragmatic steps and assumptions for passive beam delivery systems (KANAI et al. 1999).

The clinical RBE is replaced by an LET-dependent RBE for in vitro data under well-defined conditions and then linked to clinical data by an empirical factor. At HIMAC, for example, the RBE value for the 10% survival level of human salivary gland (HSG) cells was chosen and then linked to the existing clinical data gained in fast neutron radiotherapy at the corresponding LET level (see also the chapter on "Biological models for treatment planning").

The fractionation scheme and dose per fraction are kept fixed and only one modulator yielding a certain depth dose is used for each treatment field. This is possible, as several fields of a treatment plan are applied on different treatment days; thus, the treatment fields can be considered to be independent and the effective dose values can simply be added up.

Under these conditions, the resulting RBE can be approximated to be only a function of depth. If this function is determined, a corresponding ridge filter can be designed in such a way that the resulting depth dose curve leads to a constant biological effective dose. Consequently, no further biological modeling or optimization is necessary once the ridge filters are designed.

28.2.4.2
TPS for Active Beam Shaping

For an active beam shaping system for ions a research TPS was developed for the GSI facility. The system is a combination of a versatile graphical user interface for radiotherapy planning, called Virtuos (Virtual radiotherapy simulator; BENDL et al. 1993), and a program called TRiP (Treatment planning for particles), which handles all ion-specific tasks (KRÄMER and SCHOLZ 2000; KRÄMER et al. 2000). Virtuos features most tools used in modern radiotherapy planning, whereas TRiP handles the optimization of absorbed as well as biological effective dose and the optimization of the machine control data.

The introduction of a 3D scanning system has some important consequences for the TPS.

A modulator for passive beam shaping is designed to achieve a prescribed homogeneous biological effective dose for a single field. A 3D scanning system, however, can produce nearly arbitrary shapes of the SOBP. The shape of the SOBP therefore has to be optimized separately for every scan point in the irradiation field.

The resulting new demands on the TPS are:
- The beam intensity of every scan point at every energy has to be optimized separately to obtain a homogeneous biological effect.
- As the system is able to apply any complicated inhomogeneous dose distribution, the capability for intensity-modulated radiotherapy with ions should be taken into account.
- All fields of a treatment plan are applied on the same day to avoid uncertainties in the resulting dose due to setup errors.
- The dose per fraction should be variable for every patient.
- The scanner control data (energy, beam position, particle number at every beam spot) have to be optimized for each field of every patient.
- An RBE model has to be implemented, which allows

the calculation of a local RBE at every point in the patient depending on the spectrum of particles at this point. In addition, dedicated quality assurance measures for such a novel TPS have to be developed (Jäkel et al. 2000, Krämer et al. 2003).

Absorbed Dose Calculation

The dose calculation for active beam shaping systems is very similar to the pencil-beam models used for conventional photon therapy and also relies on measured data like for the passive systems. Instead of the measured depth dose data for the SOBPs resulting from the modulators, data for the single energies are needed. If the applied dose is variable, it is necessary to base the calculation of absorbed dose on absolute particle numbers rather than on relative values. For the calculation of absorbed dose, the integral data including all fragments are sufficient.

Before the actual dose calculation starts, the target volume is divided into slices of equal radiological depth. (Here the same empirical methods of range calculation as for passive systems are used.) Each slice then corresponds to the range of ions at a certain energy of the accelerator. The scan positions of the raster scanner are then defined as a quadratic grid for each energy. In the last step, the particle number at each scan point is optimized iteratively until a predefined dose at each point is reached.

Biological Modeling

To fulfill the demands of an active beam delivery on the TPS concerning the biological effectiveness, a more sophisticated biological model is needed. Such a model was developed, e.g., at GSI (Scholz et al. 1997). Its main idea is to transfer known cell survival data for photons to ions, assuming that the difference in biological efficiency arises only from a different pattern of local dose deposition along the primary beam (see also the chapter on "Biological models for treatment planning").

The model takes into account the different energy deposition patterns of different ions and is thus able to model the biological effect resulting from these ions. An important prerequisite for this is, however, the detailed knowledge of the number of fragments produced as well as their energy spectrum. The calculated RBE shows a dependence on the dose level and cell type, if the underlying photon survival data for this respective cell type are known.

The model allows the optimization of a prescribed biological effective dose within the target volume

(Krämer and Scholz 2000; Jäkel et al. 2001b) using the same iterative optimization algorithm as for the absorbed dose. At each iteration step, however, the RBE has to be calculated anew, as it is dependent on the particle number (or dose level). Since this includes the knowledge of the complete spectrum of fragments, the optimization is time-consuming. Again, it has to be pointed out that the dose dependence of the RBE demands the use of absolute dose values during optimization.

28.3
Existing and Planned Facilities

28.3.1
The Bevalac Facility

Pioneering work in the field of radiotherapy with heavy ions was performed at the University of California, Berkeley. The Bevalac provided the scientific and technological basis for many of the current developments in the field of ion radiotherapy. Between 1977 and 1992, 433 patients have been treated here in total with ions heavier than helium. The majority of patients received neon ion treatments, although some patients were also treated with carbon, silicon, and argon beams (Castro and Lawrence 1978; Castro et al. 1994).

At the Bevalac, two treatment rooms, both equipped with a fixed horizontal beam line, were available. The majority of the patients were treated in a sitting position. For treatment planning, a CT scanner was modified to scan patients in a seated position.

The modulation of depth dose was performed using a fixed accelerator energy and passive range modulators, like ridge filters and patient-specific compensators. The lateral width of the beam was produced using a double scattering system and was improved to a beam wobbling system in the 1980s. Patient-specific field collimators were used to confine the beam to the target volume. A magnetic beam scanning system was installed in combination with a passive range modulator, which was used, however, for the treatment of a single patient only.

28.3.2
The HIMAC Facility

The HIMAC started its clinical operation in Chiba, Japan, in 1994. As of February 2002, 1187 patients have been treated with carbon ions (Tsujii et al. 2002). Two redundant synchrotrons deliver carbon ion beams at energies of 290, 350, and 400 MeV/u.

Patients are treated in three different treatment rooms, which are equipped with a vertical beam line, a horizontal beam line, and the third with a vertical and a horizontal beam line, respectively.

For more flexibility, treatment chairs are used additionally to conventional treatment couches. In addition, a tub-like system may be mounted on the treatment table to allow a small angle rotation of the patient around the longitudinal axis. For treatments in seated position, special horizontal CT scanners are available for treatment planning (Kamada et al. 1999).

The beam delivery uses passive range modulation by bar ridge filters to achieve a fixed modulation depth. The lateral field width is obtained by a wobbler system and beam shaping is achieved by patient-specific collimators (Kanai et al. 1999).

28.3.3
The GSI Facility

At the research laboratory GSI in Darmstadt (Germany), a therapy unit began its clinical operation in 1997 (Eickhoff et al. 1999; Debus et al. 2000). As of spring 2004, more than 220 patients have been treated with carbon ions. Only one treatment room is available that is equipped with a treatment couch. An additional treatment chair will come into clinical operation in the near future.

The beam delivery system is completely active and allows 3D scanning of arbitrarily shaped volumes with a spatial resolution of 2 mm in all three directions. Using a magnetic deflection system, the intensity-controlled raster scanner can deliver a monoenergetic pencil beam over an arbitrarily shaped area (Haberer et al. 1993). To do so, a beam of 4–10 mm full-width half-maximum is scanned over a regular grid of points with typically 2- to 3-mm spacing. After completion of a scan, the accelerator energy can be switched from pulse to pulse and another scan can be performed with a different radiological depth. In total, 252 accelerator energies are available.

A feedback loop from the intensity control moves the beam to the next beam spot, when a predefined number of particles is reached. An on-line monitoring of the beam position and a feedback loop to the scanner is used to keep the beam extremely stable at each scan spot.

28.3.4
The HIBMC Facility

The Hyogo Ion Beam Medical Center (HIBMC) started operation with protons in 2001 and with carbon ions in 2002 at Harima Science Garden City, Japan. It is the first facility offering carbon ion and proton treatment at the same facility. As of mid-2002, 28 patients have received carbon ion therapy (Itano et al. 2003). Six therapy rooms are available with seven treatment ports. Three rooms are dedicated to carbon ion beams: one with a vertical beam line; one with a horizontal; and one with a 45° oblique beam line, respectively. Two proton treatment rooms are equipped with commercially designed gantries. The beam delivery system is based on the HIMAC system.

28.3.5
New Facilities

There is currently an increasing interest in heavy ion radiotherapy especially in Europe and Asia (Sisterson 2003). There are proposals for hospital based heavy ion facilities in Lyon (France), Pavia (Italy), Stockholm (Sweden), and Vienna (Austria). Only two facilities (in Italy and Germany), however, have by now reached the phase of construction or call for tenders. In Lanzhou (China) an existing heavy ion research facility is preparing for clinical patient treatments with ions. The therapy facility will be installed within the environment of a research laboratory, similar to the GSI facility. It features only a horizontal beam line.

The Italian project is driven by the CNAO ("Centro Nazionale Adroterapia") and will set up the facility in Pavia near Milan. It will exhibit a synchrotron of about 25 m diameter that is capable of accelerating protons and carbon ions up to energies of 400 MeV/u. It features three treatment rooms equipped with horizontal beam lines, one of them with an additional proton gantry. Beam scanning will be available at all beam lines. The construction of this first phase should start at the beginning of 2004 and the first patient is planned to be treated at the end of 2007. Then, in a second project phase, the construction of two additional treatment rooms equipped with superconducting isocentric ion gantries will take place.

The currently most ambitious and advanced project is the Heavy Ion Therapy accelerator (HIT) which will be installed at the Heidelberg University Hospital (Heeg et al. 2004; Eickhoff et al. 1998). The facility will be equipped with three treatment rooms which will host two horizontal beam lines and one room with a fully rotating isocentric gantry (SPILLER et al. 2000). When completed, this gantry will be the first ion gantry worldwide. The facility will use much of the know-how developed at the GSI facility. The beam delivery in all three treatment rooms, for example, will rely on the active 3D beam scanning method together with the active energy variation of the synchrotron.

One of the biggest technological challenges is the construction of the isocentric ion gantry with an integrated beam scanning system. The scanning magnets will be placed upstream of the last bending magnet in order to reduce the diameter of the gantry to about 13 m. The last bending magnet therefore needs a very large aperture and contributes much to the total weight of the gantry of approximately 600 tons. A very rigid mechanic framework is therefore mandatory to guarantee a sufficient stability of the gantry. The total length of the gantry will be about 20 m.

Furthermore, the design of the synchrotron and beam line will enable the use of carbon ions as well as protons, helium ions, and oxygen ions for radiotherapy. One of the major research goals is thus to evaluate which ion beam modality is best suited for the treatment of a certain type of tumor.

The facility will be completely integrated into the radiological clinic of the Heidelberg University and is designed to treat 1000 patients per year when in full operation. The construction of the facility is scheduled to be finished in 2006 and clinical operation is planned to start at the beginning of 2007.

28.4
Clinical Application of Ion Beams

Considering the physical and biological properties of carbon ions, a potential benefit for carbon ion radiotherapy can be assumed for all tumors with a low α/β ratio and which are surrounded by critical structures. A low α/β ratio has been shown for chordomas, low-grade chondrosarcomas and malignant salivary gland tumors such as adenoid cystic carcinomas and other head and neck tumors. Additional potential indications are bone and soft tissue sarcomas, lung cancer, and prostate cancer. For these tumor entities higher control rates and an improved quality of life can be expected if heavy ion radiotherapy is performed (GRIFFIN et al. 1988; DEBUS et al. 2000).

28.4.1
Clinical Trials at Berkeley

At the time when ion radiotherapy in Berkeley started almost no clinical information was available about the biological effectiveness of ions for various tumors and normal tissues. The studies were basically defined as dose-finding studies. Moreover, most of the patients were irradiated with helium ions, which do not show a significantly increased LET and in total only 433 patients have been treated with heavy ions (mainly with neon). Moreover, the heavy ion treatment was applied mainly in combination with helium ions and/or conventional photon therapy. A total of 299 patients have received a minimum neon ion dose of 10 Gy. Only few patients (e.g., with malignant glioma, pancreatic cancer, and salivary gland tumors) were irradiated solely with heavy ions. For many tumors at the base of skull (SB) pure helium beams were used and neon ions (used only for advanced salivary gland tumors) were avoided because of their large fragmentation tail and the high RBE in normal tissue of the central nervous system.

The 5-year actuarial disease-specific survival (DSS) and local control (LC) rates suggested that helium and neon ions improved the outcome for a number of tumor types given in Table 28.1 as compared with historical results.

For some other tumor types, such as malignant glioma, pancreatic, gastric, esophageal, lung, and advanced or recurrent head and neck cancer, however, the outcome was not significantly better than for low LET irradiation. Also, a large number of patients suf-

Table 28.1. Clinical results from Berkeley (according to CASTRO et al. 1994, 1997; LINSTADT et al. 1991). *DSS* disease-specific survival after 5 years, *LC* local control after 5 years

Indication	DSS (%)	LC (%)	No. of patients	$D_{total}(D_{Ne})$ in Gy
Chondrosarcoma (SB)	83	78	27	65 (–)
Chordoma (SB)	72	63	53	65 (–)
Meningioma (SB)	82	85	27	65 (–)
Advanced salivary gland cancer	50	50	18	65 (>10)
Advanced soft tissue sarcoma	42	59	32	60 (14)
Bone sarcoma	45	59	17	69.6 (16.8)
Locally advanced prostate cancer	90	75	23	76.9 (11.2)
Biliary tract carcinoma	28	44	28	60 (17.8)

Neon ion doses are given in parentheses
Doses refer to median total doses

fered from severe radiation-induced toxicity, which was attributed to the high LET of neon ions and the drawbacks of the passive beam delivery system. A significant improvement in outcome was, however, observed for patients treated between 1987 and 1992, which is attributed to improvements in 3D treatment planning, patient immobilization, and availability of MRI. A review of the clinical data is given by CASTRO (1997), CASTRO et al. (1994), and LINSTADT et al. (1991).

28.4.2
Clinical Trials and Routine Treatments at GSI

Since December 1997, more than 220 patients have been treated with carbon ion radiotherapy at GSI. An overview over the results is given by SCHULZ-ERTNER et al. (2003a). Patients with chordomas (*n*=44) and low-grade chondrosarcomas (*n*=23) of the skull base were treated within a clinical phase-I/II trial with carbon ion radiotherapy only. Median dose was 60 Gy (20 fractions, each 3 Gy). In February 2003 the median follow-up was 20 months. Actuarial 3-year lo-

cal control rate was 100% for chondrosarcomas and 81% for chordomas of the skull base, respectively. Actuarial 3-year overall survival was 91%. Toxicity correlated with radiobiological model estimations. Late toxicity greater than common toxicity criteria (CTC) grade 3 was not observed (SCHULZ-ERTNER et al. 2003a). Local control rates after carbon ion radiotherapy are at least comparable to the results after proton radiotherapy (NOEL et al. 2001), which is currently considered the treatment of choice in chordomas and low-grade chondrosarcomas; therefore, carbon ion radiotherapy is offered at GSI as a routine treatment to patients suffering from these tumors as an alternative to proton radiotherapy since January 2002.

Figure 28.5 shows an example of a treatment plan for patient with a chondrosarcoma close to the brain stem. The excellent dose conformation of the 90% isodose to the target is clearly demonstrated, although only two horizontal treatment fields were used here. The dose sparing of the relevant organs at risk can be seen in the dose volume histogram (Fig. 28.5b).

A clinical phase-I/II study of combined photon radiotherapy plus a carbon ion boost for sacral/spi-

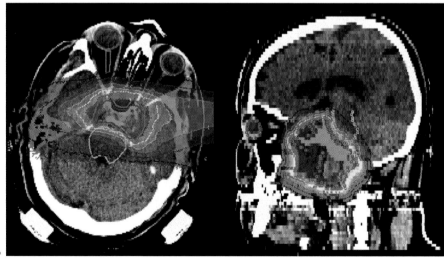

Fig. 28.5a,b. Example of a treatment plan for fully fractionated radiotherapy using two nearly opposing fields of carbon ions, as it was performed at GSI. In **a** the dose distribution at isodose levels of 10, 30, 50, 70, and 90%V is shown, respectively. The 100% dose corresponds to 60 Gy, which was applied in 20 consecutive fractions of 3 Gy. In **b** the dose volume histogram for the target and these organs at risk are shown

nal chordomas and low-grade chondrosarcomas is ongoing. Treatment consists of combined photon IMRT (weekly fractionation 5×1.8 Gy, median dose 50.5 Gy) and a carbon ion boost (weekly fractionation 6×3.0 Gy) to the macroscopic tumor residual after surgery (median total dose was 68.4 Gy). In December 2002 local control was achieved in 8 of 9 patients with cervical spine tumors. Mucositis CTC grade 3 was observed in 3 patients with chordomas of the cervical spine, but none of the patients developed severe late effects to the spinal cord.

Locoregional control was yielded in 7 of 8 patients treated for tumors of the sacrum. Two of 8 patients with sacral chordoma developed distant metastases. Combined photon radiotherapy and carbon ion boost of sacral chordomas is very well tolerated, and no side effects have been observed up to now.

A clinical phase-I/II study for combined photon radiotherapy with a carbon ion boost in locally advanced adenoid cystic carcinomas is ongoing. Therapy consists of combined stereotactically guided photon radiotherapy to the clinical target volume (CTV; CTV dose 54 Gy) and a carbon ion boost to the macroscopic tumor residual (boost dose 18 Gy). A typical treatment plan for this combination therapy is shown in Fig. 28.6.

An interim analysis on 21 patients in December 2002 (median follow-up 14 months) showed an actuarial locoregional control rate of 62% at 3 years; disease-free survival and overall survival were 40 and 75% at 3 years, respectively (Schulz-Ertner et al. 2004). Acute severe side effects CTC grade 3 were observed in

9.5% of the patients, but no radiotherapy-related late effects greater than CTC grade 2 have occurred up to now (Schulz-Ertner et al. 2003b). These results are encouraging, as locoregional control rates are better than in most photon trials and comparable to neutron radiotherapy (Huber et al. 2001; Laramore et al. 1993), however, with minimized toxicity.

28.4.3
Treatments at HIMAC

At HIMAC a number of studies are ongoing using ion radiotherapy for the treatment of tumors of the head and neck, prostate, lung, liver, as well as sarcomas of soft tissue and bone and uterine carcinomas. Between 1994 and February 2002, 1187 patients were treated within clinical trials using carbon ion therapy and performing dose escalation studies. Table 28.2 gives an overview of the most important results (Tsujii et al. 2002). Ion radiotherapy of the prostate was given in combination with hormone therapy and for uterine tumors in combination with conventional photon therapy. The fractionation scheme used is generally 16 fractions in 4 weeks for head and neck tumors as well as for sarcomas of bone and soft tissue. It was significantly shortened for lung cancer (nine fractions in 3 weeks) and liver tumors (12 fractions in 3 weeks) and is being further shortened to four fractions in 1 week for both indications.

The latest results are from dose escalation studies in lung tumors and soft tissue sarcoma.

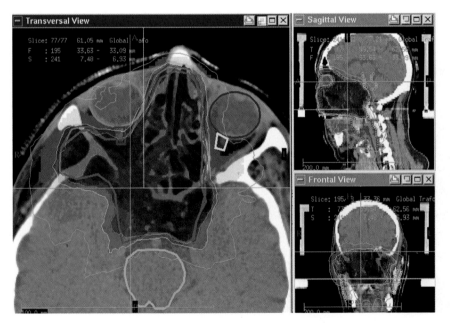

Fig. 28.6. Example of a combination therapy for a patient with adenoid cystic carcinoma, where the larger CTVI (including microscopic spread of the tumor) was treated with photon IMRT in conventional fractionation up to 54 Gy and only the CTVII (defined as GTV plus a margin) was treated with carbon ions in six fractions with a total dose of 18 Gy

Table 28.2. Results from HIMAC (Tsujii et al. 2002). Local control (*LC*) and overall survival (*OS*) are given for 2 and 3 years and refers to the latest completed trial (patient numbers in parentheses).

Indication	LC (2 years; %)	OS (3 years; %)	Patients	$D(C_{12})$ in Gy
Locally advanced head and neck tumors	61	42	170 (134)	52.8–64
Lung cancer (stage I)	100	73	161 (50)	72
Hepatocellular carcinoma (stage T2-N0M0)	83	45	122 (86)	49.5–79.5
Prostate carcinoma (stage T1-3N0M0)	100	97	143 (61)	60–66
Squamous cell carcinoma (uterus; stage T2-4N0M0)	67	36	67 (14)	68.8–72.8
Unresectable bone and soft tissue sarcoma	77	50	95 (64)	52.8–73.6

In two phase-I/II trials for stage-I non-small cell lung cancer (NSCLC), using different fractionation schemes (18 fractions in 6 weeks and 9 fractions in 3 weeks), a dose escalation was performed from 59.4 to 94.5 Gy and from 68.4 to 79.2 Gy, respectively (Miyamoto et al. 2003; Koto et al. 2004). The resulting overall control rates for the 6- and 3-week fractionation were 64 and 84%, respectively. The total recurrence rate was 23.2%.

For unresectable bone and soft tissue sarcomas, a further phase-I/II trial was performed with doses between 52.8 and 73.6 Gy (Kamada et al. 2002), applied in 16 fractions over 4 weeks. The observed overall control rates were 88 and 73% at 1 year and 3 years, respectively.

28.5
Outlook

In the past decade, valuable clinical experience has been gained in heavy ion therapy at HIMAC and GSI. Together with the development of new technologies, especially for beam application and treatment planning, there will certainly be a broader implementation of ions in clinical settings that allow for an optimal exploitation of the physical and biological potential of protons and heavy ions. Among these technologies are inverse treatment planning for particles (Oelfke and Bortfeld 2000, 2001), gating for breath-dependent targets (Minohara et al. 2000; Shirato et al. 2000; Ford et al. 2002), the raster scan system for tumor conform beam application (Haberer et al. 1993), and biological plan optimization for carbon ion radiotherapy (Scholz et al. 1997; Krämer and Scholz 2000; Krämer et al. 2000).

Further research is still required to clarify what indications benefit most from heavy ion therapy and what is the ideal ion species and fractionation scheme. These questions can be answered only in clinical studies performed at dedicated ion facilities such as the HIMAC, HIBMC, or the upcoming Heidelberg facility.

For tumors with proven effectiveness of carbon ion radiotherapy, such as chordomas and low-grade chondrosarcomas of the skull base, clinical phase-III trials are necessary to determine the advantages of carbon ion radiotherapy over other radiotherapy modalities such as modern photon techniques (like IMRT) or proton radiotherapy which is currently considered to be the treatment of choice. Toxicity, quality of life, and socio-economic aspects have to be investigated as study end points besides local control probability.

The encouraging results from HIMAC, especially for bone and soft tissue sarcomas as well as for lung tumors, warrant further clinical investigation of carbon ion radiotherapy in the treatment of these tumors. Clinical phase-III studies will have to be conducted and will not only have to compare different radiotherapy modalities but will have to include a control arm of surgically treated patients.

New immobilization techniques with rigid immobilization devices and pre-treatment control and correction of the patient alignment, together with the development of faster dose calculation and optimization algorithms and better knowledge of the complex biology of carbon ions, will help to enlarge the spectrum of possible indications for carbon ion radiotherapy. The development of a gantry for carbon ion radiotherapy will facilitate the safe treatment of paraspinal tumors by allowing posterior beam directions.

Further investigation is needed in the field of multimodality treatments. There is almost no clinical data available about adjuvant chemotherapy or hormone therapy, although interdisciplinary treatment of patients with tumors that tend to metastasize seems to be warranted. In this connection, major attention has to be drawn to the proper definition of clinical trials for head and neck malignancies, gynecological tumors, and prostate cancer in order to obtain comparability with up-to-date oncological standard therapy regimes.

A combination of different radiotherapy modalities, such as photon IMRT plus carbon ions or protons, might be favorable for a number of indications. Combined radiotherapy offers the possibility for risk-adapted treatment of different target volumes in one patient dependent on the oncological concept based on risk estimation. Furthermore, the combination of carbon ion radiotherapy or protons with photon radiotherapy will guarantee that a higher number of patients will benefit from the advantages of particle radiotherapy despite the still limited availability. Possible indications are head and neck malignancies, malignant primary brain tumors, and prostate cancer. For patients with unfavorable localized prostate cancer a clinical phase-I/II trial of combined photon IMRT with a carbon ion boost, which is a project of the European hadrontherapy network "ENLIGHT," is in preparation (NIKOGHOSYAN et al. 2004).

Due to the biological properties of carbon ion radiotherapy, hypofractionation might be considered for a number of tumor entities. The accelerated fractionation scheme used at GSI for chordomas and low-grade chondrosarcomas of the skull base corresponds to a reduction of treatment time by about 50% compared with conventional photon radiotherapy. First clinical hypofractionation trials for lung cancer are underway at HIMAC and further studies are in preparation. These studies investigate outcome and toxicity of hypofractionation as end points, as well as quality of life and economic aspects. Carbon ion radiotherapy might even turn out to be an optimal treatment option from an economic point of view, as hypofractionation leads to substantial reduction of overall treatment time and hospitalization.

References

Bendl R, Pross J, Schlegel W (1993) VIRTUOS: a program for VIRTUal radiotherapy simulation. In: Lemke HU, Inamura K, Jaffe CC, Felix R (eds) Computer assisted radiology. Proc Int Symp CAR 93. Springer, Berlin Heidelberg New York, pp 676–682

Castro JR (1997) Clinical programs: a review of past and existing hadron protocols. In: Amaldi U, Larrson B, Lemoigne Y (eds) Advances in hadrontherapy. Elsevier, Amsterdam, pp 79–94

Castro J, Lawrence J (1978) Heavy ion radiotherapy. In: Lawrence JH, Budinger TF(eds) Recent advances in nuclear medicine, Grune & Stratton vol 5. New York, pp 119–137

Castro JR, Linstadt DE, Bahary J-P et al. (1994) Experience in charged particle irradiation of tumors of the skull base 1977-1992. Int J Radiat Oncol Biol Phys 29:647–655

Chen G, Singh R, Castro J, Lyman J, Quivey J (1979) Treatment planning for heavy ion radiotherapy. Int J Radiat Oncol Biol Phys 5:1809–1819

Debus J, Haberer T, Schulz-Ertner D et al. (2000) Fractionated carbon ion irradiation of skull base tumours at GSI. First clinical results and future perspectives. Strahlenther Onkol 176:211–216

Eickhoff H, Böhne D, Debus J, Haberer T, Kraft G, Pavlovich M (1998) The proposed accelerator for light ion cancer therapy in Heidelberg. GSI Scientific Report, Darmstadt, Germany, pp 164–165

Eickhoff H, Haberer T, Kraft G, Krause U, Richter M, Steiner R, Debus J (1999) The GSI cancer therapy project. Strahlenther Onkol 175 (Suppl 2):21–24

Endo M, Koyama-Ito H, Minohara S, Miyahara N, Tomura H, Kanai T, Kawachi K, Tsujii H, Morita K (1996) HIPLAN: a heavy ion treatment planning system at HIMAC. J Jpn Soc Ther Radiol Oncol 8:231–238

Ford EC, Mageras GS, Yorke E, Rosenzweig KE, Wagman R, Ling CC (2002) Evaluation of respiratory movement during gated radiotherapy using film and electronic portal imaging. Int J Radiat Oncol Biol 52:522–531

Goldstein LS, Phillips TL, Ross GY (1981) Biological effects of accelerated heavy ions. II. Fractionated irradiation of intestinal crypt cells. Radiat Res 86:542–558

Griffin TW, Wambersie A, Laramore G, Castro J (1988) International clinical trials in radiation oncology. High LET: heavy particle trials. Int J Radiat Oncol Biol Phys 14 (Suppl 1): S83–S92

Haberer T, Becher W, Schardt D, Kraft G (1993) Magnetic scanning system for heavy ion therapy. Nucl Instrum Meth A330:296–305

Heeg P, Eickhoff H, Haberer T (2004) Die Konzeption der Heidelberger Ionentherapieanlage HICAT. Z Med Phys 14:17–24

Huber PE, Debus J, Latz D et al. (2001) Radiotherapy for advanced adenoid cystic carcinoma: neutrons, photons or mixed beam? Radiother Oncol 59:161–167

Itano A, Akagi T, Higashi A et al. (2003) Operation of medical accelerator PATRO at Hyogo Ion Beam Medical Center. Workshop on Accelerator Operation, 10–14 March, Tsukuba, Japan

Jäkel O, Jacob C, Schardt D, Karger CP, Hartmann GH (2001a) Relation between carbon ions ranges and X-ray CT numbers. Med Phys 28:701–703

Jäkel O, Krämer M, Karger CP, Debus J (2001b) Treatment planning for heavy ion radio-therapy: clinical implementation and application. Phys Med Biol 46:1101–1116

Jäkel O, Schulz-Ertner D, Karger CP, Nikoghosyan A, Debus J (2003) Heavy ion therapy: status and perspectives. Technol Cancer Res Treat 2:377–388

Kamada T, Tsujii H, Mizoe J et al. (1999) A horizontal CT system dedicated to heavy-ion beam treatment. Radiother Oncol 50:235–237

Kamada T, Tsujii H, Tsuji H et al. (2002) Efficacy and safety of carbon ion radiotherapy in bone and soft tissue sarcomas. J Clin Oncol 20:4466–4471

Kanai T, Endo M, Minohara S et al. (1999) Biophysical characteristics of HIMAC clinical irradiation system for heavy-ion radiation therapy. Int J Radiat Oncol Biol Phys 44:201–210

Koto M, Miyamoto T, Yamamoto N, Nishimura H, Yamada S, Tsujii H (2004) Local control and recurrence of stage I non-small cell lung cancer after carbon ion radiotherapy. Radiother Oncol 71:147–156

Krämer M, Scholz M (2000) Treatment planning for heavy ion radiotherapy: calculation and optimization of biologically effective dose. Phys Med Biol 45:3319–3330

Krämer M, Jäkel O, Haberer T, Kraft G, Schardt D, Weber U (2000) Treatment planning for heavy ion radiotherapy: physical beam model and dose optimization. Phys Med Biol 45:3299–3317

Laramore GE, Krall JM, Thomas FJ et al. (1993) Fast neutron radiation therapy for locally advanced prostate cancer: final report of an RTOG randomized clinical trial. Am J Clin Oncol 16:164

Linstadt DE, Castro JR, Phillips TL (1991) Neon ion radiotherapy: results of the phase I/II clinical trial. J Radiat Oncol Biol Phys 20:761–769

Minohara S, Kanai T, Endo M, Kawachi K (1993) Effects of object size on a function to convert X-Ray CT numbers into the water equivalent path length of charged particle beam. In: Ando K, Kanai T(eds) Proc Third Workshop on Physical and Biological Research with Heavy Ions, Chiba, Japan, pp 14–15

Minohara S, Kanai T, Endo M, Noda K, Kanazawa M (2000) Respiratory gated irradiation system for heavy-ion radiotherapy. Int J Radiat Oncol Biol Phys 47:1097–1103

Miyamoto T, Yamamoto N, Nishimura H et al. (2003) Carbon ion radiotherapy for stage I non-small cell lung cancer. Radiother Oncol 66:127–140

Nikoghosyan A, Schulz-Ertner D, Didinger B, Jäkel O, Zuna I, Höss A, Wannenmacher M, Debus J (2004) Evaluation of therapeutic potential of heavy ion therapy for patients with locally advanced prostate cancer. Int J Radiat Oncol Biol Phys 58:89–97

Noel G, Habrand JL, Mammar H et al. (2001) Combination of photon and proton radiation therapy for chordomas and chondrosarcomas of the skull base: the Centre de Proton-therapie d'Orsay experience. Int J Radiat Oncol Biol Phys 51:392–398

Oelfke U, Bortfeld T (2000) Intensity modulated radiotherapy with charged particle beams: studies of inverse treatment planning for rotation therapy. Med Phys 27:1246–1257

Oelfke U, Bortfeld T (2001) Inverse planning for photon and proton beams. Med Dosim 26:113–124

Scholz M, Kellerer AM, Kraft-Weyrather W, Kraft G (1997) Computation of cell survival in heavy ion beams for therapy: the model and its approximation. Radiat Environ Biophys 36:59–66

Schulz-Ertner D, Nikoghosyan A, Jäkel O et al. (2003b) Feasibility and toxicity photon and carbon ion radiotherapy for locally advanced adenoid cystic carcinomas. Int. J. Radiat. Oncol. Biol. Phys. 56(2):391–398

Shirato H, Shimizu S, Kunieda T, Kitamura K, van Herk M, Kagei K, Nishioka T, Hashimoto S, Fujita K, Aoyama H, Tsuchiya K, Kudo K, Muyasaka K (2000) Physical aspects of a real-time tumor-tracking system for gated radiotherapy. Int J Radiat Oncol Biol Phys 48:1187–1195

Sisterson J (ed) (2003) Particle newsletter 31, Harvard, Boston, Massachusetts

Spiller P, Boehne D, Dolinskii A et al. (2000) Gantry studies for the proposed heavy ion cancer therapy facility in Heidelberg. Proc 7th European Particle Accelerator Conference (EPAC). Cern, Geneva, pp 2551–2553

Tepper J, Verhey L, Goitein M et al. (1977) In vivo determinations of RBE in a high energy modulated proton beam using normal tissue reactions and fractionated dose schedules. Int J Radiat Oncol Biol Phys 2:1115–1122

Tsujii H, Morita S, Miyamoto T et al. (2002) Experiences of carbon ion radiotherapy at NIRS. In: Kogelnik HD, Sedlmayer F (eds) Progress in radio-oncology, vol 7. Monduzzi Editore, Bologna, pp 393–405

29 Permanent-Implant Brachytherapy in Prostate Cancer

Marco Zaider

CONTENTS

29.1
Introduction

The overall survival benefits of transperineal permanent interstitial implantation among men with early, localized prostate cancer remain uncertain (ALBERTSEN et al. 1998; CHODAK 1998). In favorable-risk patients [prognosis category is defined by pre-treatment prostate-specific antigen (PSA), Gleason score, and tumor stage] this form of treatment is associated with excellent biochemical (generally defined as an undetectable PSA concentration) outcome – close to 90% PSA relapse-free survival at 5 years, a number which compares favorably with outcomes from other forms of local treatment (e.g., external-beam radiotherapy, radical prostatectomy; PESCHEL and COLBERG 2003). This comes, however, with a price tag, namely the risk of (occasionally permanent) rectal and particularly urinary (grade-2 urinary symptoms represent the most common toxicity after prostate brachytherapy) toxicity, as well as sexual dysfunction (JANI and HELLMAN 2003; SANDHU et al. 2000; ZELEFSKY et al. 2000a). For many patients with early, localized prostate cancer the decision to undergo a treatment of questionable

benefit, yet tangibly impacting upon their quality of life (QOL), is understandably difficult. As a result, diminishing the risk of complications, and at the same time maintaining good dosimetric coverage of the tumor, remains the overriding concern in prostate brachytherapy. (Brachytherapy may be less likely to result in impotence in urinary incontinence than other forms of treatment.)

With this in mind, we discuss herein several factors that impact on the quality of the implant. These include:

1. Conventional (dosimetric) treatment planning: Much treatment planning in brachytherapy continues to be performed suboptimally, i.e., by manual (trial-and-error) methods. Applied mathematics has a well-developed repertory of optimization tools (simulated annealing, genetic algorithm, integer programming) that could, and must, be implemented in treatment planning software.

2. *Biological treatment planning* refers to incorporating in treatment planning information on tumor control probability (TCP) and normal-tissue complication probability (NTCP) for all structures of interest. Plan optimization would then be performed based on an overall figure of merit (FM), rather than dosimetrically, as is currently done.

3. Treatment delivery: It is often the case that the insertion of various applicators in or near the intended target results in changes in the target geometry. These changes, and their evolution in time, must be anticipated in the treatment plan – and corrected for. Of particular importance is to make the process of treatment delivery as much as possible operator independent. For instance, in permanent prostate implants there is a recognized variance between the intended radioisotope positions and their actual location in the gland. Imaging tools capable of reconstructing in real time the positions of radioisotopes already inserted could be used to re-optimize the plan.

4. Imaging: Alongside MRI and sextant biopsy, MR spectroscopy (MRS) has been used to identify non-invasively clinically significant tumor An

M. ZAIDER, PhD
Department of Medical Physics, Memorial Sloan-Kettering Cancer Center, 1275 York Avenue, New York, NY 10021, USA

argument can be made, based on TCP models, that in brachytherapy one expects a clinical advantage when the (inevitable) dosimetric hot spots are placed at MRS-positive voxels rather than at random locations in the gland.

The methodology described below reflects, as might be expected, accepted practice at Memorial Sloan-Kettering Cancer Center (MSKCC).

29.2
Treatment Planning

29.2.1
Outline

Pre-planning in prostate brachytherapy (meaning that the plan was generated several days or a week before implantation) has a number of well-recognized deficiencies (NAG et al. 2001):

1. It is difficult to duplicate the patient position in the operating room to match the patient position during the pre-planning simulation.
2. Patient geometry can change over time, e.g., urine in the bladder or feces in the rectum may swell the prostate.
3. A pre-treatment plan may prove impossible to implement due to the needle site being blocked by the pubic symphysis.

At MSKCC the intraoperative treatment-planning sequence is as follows: The patient is placed in the extended lithotomy position and a urinary catheter is inserted. The ultrasound probe is positioned in the rectum and needles are inserted along the periphery of the prostate using a perineal template as a guide. Axial images of the prostate are acquired at 0.5 cm spacing (from base to apex), transferred to the treatment planning system using a PC-based video capture system, and calibrated. For each ultrasound image, prostate and urethra contours as well as the anterior position of the rectal wall are entered. Needle positions are identified on the ultrasound images and correlated with the US template locations. The contours, dose reference points, needle coordinates, and data describing the isotope/activity available along with predetermined dose constraints and their respective weights serve as input for the dose-optimization algorithm.

Dose optimization is achieved using a genetic algorithm (GA) as described below. Isodose distributions for each axial cut and dose-volume histograms

(DVH) for the structures of interest (prostate, urethra, and rectum) are calculated and inspected in the operating room. The ^{125}I or ^{103}Pd radioisotopes are individually loaded using a Mick applicator. The entire planning process from the contouring of images to the generation of the radioisotope loading pattern requires approximately 10–15 min.

About 4 h after the implant, post-implantation CT scans are obtained. For dosimetric evaluation of the implant the prostate, urethra, and rectum are contoured on each axial CT image, and the parameters recommended by the American Brachytherapy Society (NAG et al. 2001) are calculated and compared with the original plan.

29.2.2
Dose Prescription

Dose prescription in prostate brachytherapy makes use of the concept of minimum peripheral dose (mPD), which is defined as the largest-dose isodose surface that completely surrounds the clinical target. This approach has been endorsed by the American Brachytherapy Society (NAG et al. 2001). The mPD is very sensitive to minor variations in the planned dosimetry because it is always given by the lowest dose delivered to any volume of the target – however small. (If 99.9% of the target is covered by the prescription dose and the dose delivered to the remaining 0.1% of the volume is, for example, 10% of the prescription dose, it is this latter number that decides the value of the mPD.) This problem may be avoided by relaxing the requirement that the prescription isodose surface cover 100% of the target; instead, the so-called D_x method can be adopted, which means that a dose of at least D_x is required to be delivered to x% of the target volume. A typical value is D_{95} (D_{100} is, of course, the same as mPD).

The dose constraints for prostate implants are as follows (ZELEFSKY et al. 2000b): mPD to target 144 Gy for ^{125}I and 140 Gy for ^{103}Pd, and the urethra and the rectum are to receive no more than 120% and 78% of the prescription dose, respectively. Typically, we use radioisotopes with air kerma strength of 0.4 U (^{125}I) and 3 U (^{103}Pd).

29.2.3
Dose Calculation Checks

It is common practice at MSKCC to have an independent review of the dose calculation. The review is performed in the operating room prior to radioisotope implantation by a physicist different from the one that calculated the plan. The review determines that:
1. The computer software has been used appropriately
2. Prescription dose is correct
3. User-supplied input data (e.g., isotope, source activity) are accurate
4. Output data look "reasonable"
5. The number of radioisotopes calculated with nomograms (see below) is within 10% of the number used in the plan

Nomograms for permanent volume implants are tools designed to predict the total air kerma strength, S_K, necessary to deliver a stated dose to a known target volume. More precisely, they stipulate the total source strength, S_K, such that the isodose surface corresponding to the prescription dose has a stated volume. The volume nomograms given here (COHEN et al. 2002) were obtained in connection with permanent prostate implants and thus, strictly speaking, the volume in question refers to the geometry of this particular organ.

For a ^{125}I *permanent* implant and a prescription dose of 144 Gy:

$$\frac{S_k}{U} = \begin{cases} 5.709 \left(\dfrac{d_{avg}}{cm}\right) & d_{avg} \leq 3 \ cm \\[2ex] 1.524 \left(\dfrac{d_{avg}}{cm}\right)^{2.2} & d_{avg} > 3 \ cm \end{cases} \quad (1)$$

For a ^{103}Pd *permanent* implant and a prescription dose of 140 Gy:

$$\frac{S_k}{U} = \begin{cases} 29.41 \left(\dfrac{d_{avg}}{cm}\right) & d_{avg} \leq 3 \ cm \\[2ex] 5.395 \left(\dfrac{d_{ave}}{cm}\right)^{2.56} & d_{avg} > 3 \ cm \end{cases} \quad (2)$$

The quantity d_{avg} refers to the average of three orthogonal dimensions of the volume to be implanted. In a typical application of these equations d_{avg} is evaluated and then the total number of radioactive sources needed is calculated by dividing the total required source strength, S_K, by the single-seed strength.

29.2.4
Treatment-Plan Optimization

Potential source positions are localized with respect to a template that is placed in a fixed position relative to the treatment region (the prostate gland). The template, shown in Fig. 29.1, has a rectangular pattern of holes. Needles are inserted through the template grid and radioisotopes are placed along each needle at positions (typically in multiples of 0.5 cm) determined by the treatment plan. A series of parallel ultrasound (US) images is taken through the prostate and firmware in the ultrasound unit overlays a grid of dots onto these images that correspond to the template holes. The grid coordinates on the template and the distance of the US image away from the template uniquely identify the 3D coordinates of each potential radioisotope position relative to the gland anatomy. If inserted needles deviate from the initial grid coordinates, the planning system has provisions for taking this into account in dosimetric calculations.

In the treatment-planning code used at MSKCC optimized radioisotope locations are determined using a genetic algorithm engine (SILVERN 1998; LEE et al. 1999). The GA is a formalism that imitates the principle of "natural selection" (CHAMBERS 1995; GREFENSTETTE et al. 1987; MAN et al. 1999; ZALZALA and FLEMING 1997). A "chromosome" is defined as a sequence of binary data, e.g., for prostate brachytherapy each chromosome represents a specific radioisotope-loading pattern. A collection of chromosomes is referred to as a "population." A symbol in a chromosome represents a location where a source can be placed in the prostate. A 0/1 bit indicates the absence (or presence) of a radioactive seed at that particular location. Using this encoding scheme, each chromo-

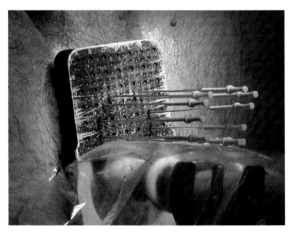

Fig. 29.1. Template used in a prostate-permanent implant

some completely defines an individual treatment plan. Every chromosome in the population is evaluated by an *objective function*, which calculates the doses to all of the recorded optimization points and evaluates how many of these points fall within specified dose tolerances. The larger the number of points, the higher the score the chromosome receives.

For instance (Silvern 1998), let PD be the prescription dose; the *prostate score* is the number of (uniformity) points that satisfy PD≤D≤1.6PD, the *urethral score* is the number of points in the urethra for which D≤1.2PD, and the *rectal score* sums up all points in the rectum for which D≤0.78PD. From this, a *raw score* is obtained:

Raw score = 5* (prostate score) + 35* (rectal score) + 50* (urethral score) (3)

and a *final score* (used in the optimization algorithm) which is a linear function of the raw score [=A*(raw score)+B].

Higher-scoring chromosomes are selectively favored to act as parent chromosomes to produce the next generation. Lower-scoring chromosomes, however, still have a chance of being selected to act as parents. This helps keep the population diversified. The parameters of the objective function include a set of predetermined dose constraints and their respective relative weight factors (e.g., 0.2 for the prostate, 10 for the urethra, 7 for the rectum). Formally, the following steps are followed:

1. Selected parent chromosomes are paired off and sections of these parent chromosomes are interchanged to yield two new "baby" chromosomes (this mimics biological crossover).

2. The newly formed progeny chromosomes are mutated. Mutation is performed by randomly inverting several bits (a 1 is changed to a 0, and vice versa). The mutation probability of each symbol is of the order of 1%. This mutation probability was found to keep the population diversified while not excessively destroying high-quality bit combinations.

3. Next, the chromosomes are again evaluated by the objective function. A new set of parent chromosomes is selected from the population favoring higher-scoring individuals and a new "pass" is started.

The process is repeated for a sufficient number of generations (typically 6000) to repeatedly and reliably yield an adequate treatment plan in a reasonable amount of time.

Other optimization algorithms used in prostate brachytherapy are simulated annealing (Sloboda 1992; Pouliot et al. 1996) and branch-and-bound (Gallagher and Lee 1997; Lee et al. 1999; Zaider et al. 2000).

29.2.5
Example

Figure 29.2 shows a typical 2D ultrasound image with prostate and urethra contours and needle positions. Also shown are isodose contours and seed positions as determined by the GA code. The determination that the treatment plan achieves the required mPD or D_{90} can be made with the aid of dose-volume histograms (DVH), which plot as a function of dose,

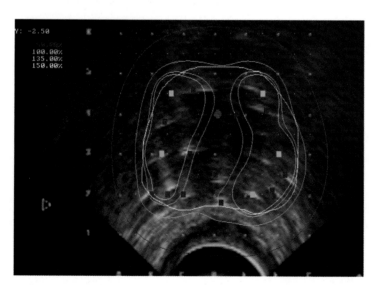

Fig. 29.2. Dose prescription for a permanent prostate implant. In this ultrasound image the *white line* delineates the prostate and the 100% isodose line (mPD=144 Gy) is shown in *green. Green dots* indicate seeds and *red dots* show unused seed locations along needles

D, the probability that a randomly selected voxel volume receives a dose of at least D (this is, of course, the cumulative probability distribution of dose in the target volume). An example is given in Fig. 29.3, which shows the DVH as planned for an LDR treatment of the prostate.

29.3
Functional Imaging

Available imaging modalities [computerized tomography (CT), magnetic resonance imaging (MRI), positron emission tomography (PET)] provide the radiation oncologist with increasingly accurate information on tumor extent and treatment response. Equally important is the ability to detect and treat selectively (e.g., by dose escalation) only those subvolumes of the target that contain *clinically significant* cancer. In radiobiological terms this means tumor cells that are radioresistant, fast proliferating, or both. This methodology, known as targeted radiotherapy, derives from the notion that it is the behavior of these cells that ultimately determines the success or failure of a particular modality of disease management.

Magnetic resonance spectroscopy is based on the observation that in regions of cancer the choline/citrate ratio is elevated (WEFER et al. 2000; SCHEIDLER et al. 1999; KURHANEWICZ et al. 1996; KURHANEWICZ et al. 1995; KURHANEWICZ et al. 2000a; KURHANEWICZ, et al. 2000b). On biochemical grounds this ratio is expected to reflect an increased rate of cell prolifera-

tion, although no direct proof exists yet. As applied to prostate cancer MRS information can be used to reduce treatment morbidity by reducing uncertainties in the location of the tumor, and also to assess tumor aggressiveness (WEFER et al. 2000). [It may be useful to compare (see Table 29.1) the specificity (S_p) and sensitivity (S_e) of MRS with other modalities of cancer detection.] (The Gleason score, a measure of the degree of differentiation of cancer cells, is a strong predictor of how quickly the cancer is growing and also on the likelihood of metastatic spread. There is limited evidence of a correlation between choline levels and histological grade.)

Table 29.1 Data on specificity, sensitivity, positive predictive value, and negative predictive value for the sextant localization of prostate cancer by different techniques. (From WEFER et al. 2000b)

Modality	S_e (%)	S_p (%)
Biopsy	50	82
MRI	67	69
MRS	76	57
MRI+MRS	56	82

Biological imaging is currently in use at MSKCC, in a planning system that employs MRS information to escalate the dose at prostate volumes identified as having increased choline/citrate ratios (ZAIDER et al. 2000). The prostate as a whole is treated to 100–150% of the prescription dose (144 Gy for ^{125}I radioisotopes); however, MRS-positive regions are prescribed 200% of the prescription dose with no upper limit while keeping the urethral dose under 120% (Figs. 29.4, 29.5). A key element in the implementation of this technique is the ability to map MRS-positive volumes (obtained in a gland deformed by the transrectal coil) to the ultrasound images that are used for planning. This topic is discussed by MIZOWAKI et al. (2002).

29.4
Biological Optimization

The notion that a radiation treatment plan should be evaluated in terms of its biological consequences – tumor control probability (TCP) and normal-tissue complication probability (NTCP) – rather than dosimetrically is well understood, and efforts in this direction have been made repeatedly (HAWORTH et al. 2004a, b; D'SOUZA et al. 2004). Typically, a plan

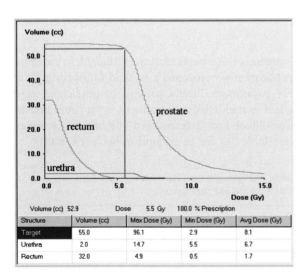

Fig. 29.3. Dose-volume histogram for an LDR treatment plan (prescription dose: 144 Gy)

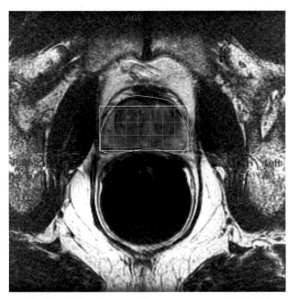

Fig. 29.4. Prostate contours are outlined in MRI pictures on which a grid, representing the MRS voxels, is overlaid. Voxels "suspicious of cancer" are marked (*red numbers 1* and *2*)

is evaluated in terms of a so-called figure of merit (FM) which quantifies the extent to which TCP is maximized while maintaining an acceptable NTCP. Although there is no a priori functional form for the FM, it must satisfy certain logical desiderata, e.g., that FM is monotonically increasing with increasing TCP and monotonically decreasing with increasing NTCP. The simplest such equation is (WANG and LI 2003):

$$FM = TCP \prod_{i=1}^{n}(1-NTCP_i) \qquad (4)$$

where n is the number of organs at risk. In this formulation $FM\varepsilon[0,1]$ and, quite obviously, for a perfect plan, FM=1. As both TCP and NTCP increase with absorbed dose, a realistic plan is a tradeoff between escalating the former while lessening the latter.

Equation (4) may be too simplistic. As observed by us in a recent publication (AMOLS et al. 1997), the concept of optimized treatment plan must ultimately reflect the patient's (or, as a surrogate, the physician's) willingness to accept treatment morbidity in exchange for increased probability of cure. An equation that offers more flexibility in this regard is (AMOLS et al. 1997):

$$FM = [1-(1-TCP)^a]^b \prod_{i=1}^{n}(1-NTCP^{c_i})^{d_i} \qquad (5)$$

where a, b, c_i, and d_i are positive adjustable constants which may be specific to each disease site, modality of treatment, and individual patient and/or physician.

MRS Image

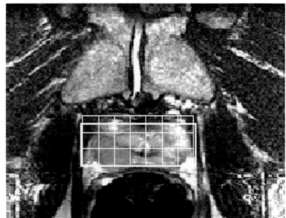

a

MRS Optimized Implant

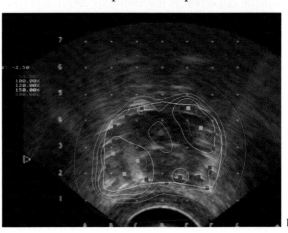

b

Fig. 29.5. An MRS-positive voxels identified in the MRI image (*left panel, yellow*) are treated to a larger dose (*right panel*) in the optimized plan. (Courtesy of M. Zelefsky, MSKCC)

Equation (4) is a particular case of Eq. (5). In the same publication we proposed a method for obtaining the FM parameters. Briefly, patients or physicians were asked to rank triads of treatment plans (from best to worse) based on their assumed TCP and NTCP values (see Table 29.2 for an example of two such triads).

Table 29.2 Examples of TCP/NTCP triads used for obtaining the FM parameters

Triad 1			Triad 2		
	NTCP	TCP		NTCP	TCP
A	0.01	0.40	A	0.05	0.45
B	0.5	0.60	B	0.20	0.70
C	0.10	0.80	C	0.40	0.95

Based on these answers, parameters a, b, c_i, and d_i were sought that reproduce as best as feasible (in the least-squares sense) the desired rankings of the FM (see Eq. (5); the solution to this problem is not unique and the (arbitrary) condition that the parameters thus obtained be closest to a=b=c=d=1 was further imposed).

To apply this type of optimization explicit equations for the FM, TCP, and NTCP are necessary. In the following, equations relevant to prostate plans – as described by us in a recent publication (ZAIDER et al. 2005) – are given.

FM Parameters

Four sets of FM parameters are shown in Table 29.3 (AMOLS et al. 1997). One of them (set 3) represents a physician, set 1 stands for a patient, and sets 2 and 4 correspond to medical physicists at our institution. The two organs at risk considered here are the urethra and the rectum wall. One may use the same parameters, c and d, for both of them.

Table 29.3 Parameters used for calculating the figure of merit for four different respondents

Set	A	B	C	D
1	0.08	0.463	0.757	1.206
2	0.121	0.420	0.490	1.20
3	0.670	3.6×10^{-3}	2.58	1.01
4	0.047	0.186	1.91	1.13

Tumor Control Probability

The TCP may be evaluated using the following equation (ZAIDER and MINERBO 2000):

$$TCP(t) = \left[1 - \frac{S(t)e^{(b-d)t}}{1 + bS(t)e^{(b-d)t} \int_0^t \frac{dt'}{S(t')e^{(b-d)t'}}} \right]^n \tag{6}$$

In Eq. (6) S(t) is the survival probability (proliferation processes excluded) at time t of the n clonogenic tumor cells present at the time treatment started ($t=0$), and b and d are, respectively, the birth and (radiation-independent) death rates of these cells. Equivalently, $b=0.693/T_{pot}$ and d/b is the *cell loss factor*, ϕ, of the tumor. In this expression t refers to any time during or after the treatment. Typically, one would take for t the end of the treatment period or (better) the expected remaining lifespan of the patient. For a permanent implant this distinction is irrelevant and one can take for t any suitably large value, e.g., ten ^{125}I half-lives. The survival probability, S(t), can be expressed in terms of the linear-quadratic equation:

$$S(D) = e^{-\alpha D(t) - q(t)\beta[D(t)]^2} \tag{7}$$

where α and β are adjustable parameters. The function q(t) describes the effect of sublethal damage recovery and can be calculated as follows (ZAIDER et al. 2005):

$$q(t) = \frac{2(\lambda t)^2}{(\mu t)^2 (1 - \lambda^2/\mu^2)(1 - e^{-\lambda t})^2} \left[e^{-(\lambda+\mu)t} + \mu t \left(\frac{1 - e^{-2\lambda t}}{2\lambda t} \right) - \frac{1 + e^{-2\lambda t}}{2} \right] \tag{8}$$

Here, λ is the radioactive decay constant of the isotope used in implantation and μ is the fractional rate of sublethal damage repair.

Urethral Toxicity

Here we are concerned with the probability of unresolved grade-2 (or higher) urethral toxicity. In terms of DU_{20} (meaning that 20% of the urethral volume is treated to a dose of at least DU_{20}) the probability of toxicity at 12 months, $P_{tox,12}$, as a function of dose, was found to be given by (ZAIDER et al. 2005):

$$NTCP_{LDR} = P_{tox,12}(DU_{20}) = \frac{1}{1 + \exp[-(\gamma + \delta\, DU_{20})]} \tag{9}$$

when $\gamma = -2.60 \pm 0.50$ and $\delta = 0.0066 \pm 0.0016$ Gy^{-1}.

Rectal Toxicity in Fractionated Treatments

MACKAY et al. (1997) proposed to describe the probability of late injury to the rectum in terms of a logistic model:

$$NTCP = \left\{ 1 - \prod_i \left[1 - \left(\frac{1}{1 + (D_{50}/D_i)^k} \right)^s \right]^{\frac{v_i}{V}} \right\}^{1/s} \tag{10}$$

Here V is the total volume of the rectal wall and v_i is a rectal wall subvolume containing cells which, upon treatment, receive the same dose (D_i). D_{50}, s, and k are empirically determined parameters. For severe late reactions MACKAY et al. (1997) recommend: k=10.24, s=0.75, and D_{50}=80 Gy – this latter being normalized to a schedule of 2 Gy per fraction. To obtain NTCP for prostate implant (a LDR procedure) one may replace D_{50}/D by the quotient of the corresponding isoeffective doses (IED; ZAIDER et al. 2005):

$$IED = -\frac{1}{\alpha}\log\left[\frac{S(t)e^{(b-d)t}}{1+bS(t)e^{(b-d)t}\int_0^t \frac{dt'}{S(t')e^{(b-d)t'}}}\right] \quad (11)$$

In the absence of cell proliferation (when b=d=0):

$$IED(D) = D\left[1+\frac{\beta}{\alpha}q(t)D\right] = -\log(S)/\alpha \quad (12)$$

an equation commonly taken to represent biologically equivalent dose.

Parameters for Eqs. (6–12) are given by ZAIDER et al. (2005).

29.5
Some Unanswered Questions

Individualized therapy? The typical patient for prostate implant has low-grade disease, low prostate volume (40 cm^3 or less), and at least 10 years of life expectancy. Current data indicate that as many as 85% of all prostate cancers diagnosed through PSA screening would have resulted in death before the cancer would have become symptomatic (McGREGOR et al. 1998). The proportion of overdiagnosis among brachytherapy-treated patients is certainly higher yet, as most of these patients have low-grade tumors (stage T1–T2a, PSA level of 10 ng/ml or less, and Gleason score of 6 or lower). This view is reinforced by an analysis reported by ABBAS and SCARDINO (1997) which indicates that of all patients with histological evidence of prostate cancer only 1 in 14 will have this malignancy as cause of death. The question is then: Who could actually benefit from a treatment that is not only costly but has considerable side effects? As well, among those who do need to undergo treatment it will be essential to be able to establish (preferably non-invasively) the presence and spatial extent of clinically relevant cancer and thus further avoid unnecessary morbidity. MRSI, an approach still under investigation, would be a step in this direction. None of the biomolecular markers currently available are able to distinguish between indolent and life-threatening cancer.

Regarding the curative potential of brachytherapy:

Is real-time planning achievable? Because of unavoidable inaccuracy in radioisotope placement in the prostate gland, there remain potentially significant discrepancies (e.g., cold spots) between the intraoperative plan and its implementation. The solution to this problem consists of developing techniques for automatically identifying, in real time, the locations of radioisotopes already implanted and periodically re-optimize the plan according to the actual distribution of radioisotopes. Techniques for achieving this goal have been described by several groups (TODOR et al. 2002a, b; TODOR et al. 2003; ZHANG et al. 2004; TUBIC et al. 2001), but they are not yet at the stage of clinical use. Real-time planning may also obviate the need for post-implant evaluation.

Do we need 4D planning? Although the dose in a permanent implant is delivered over an extended period of time, the treatment plan is based on the geometry of the target at the time of implantation. Changes in prostate volume (edema shrinkage) as well as seed migration lead to disagreement between the planned and the actual dose delivered to the prostate. Four-dimensional planning refers to a method of planning which incorporates temporal changes in the target-seed configuration during dose delivery. The decay of edema has been quantified by WATERMAN et al. (1997). A planning method making use of the concept of *effective planning volume* has been described by LEE and ZAIDER (2001).

Combined external-beam radiation therapy (EBRT) and LDR implant. Although brachytherapy and EBRT appear to be equally effective treatments for patients with early-stage prostate cancer, for patients with intermediate- and high-risk prognostic features, dose escalation is needed to achieve optimal tumor control. An approach for delivering escalated doses to the prostate is to combine brachytherapy with external beam radiotherapy (CRITZ et al. 2000). When EBRT is combined with an interstitial permanent implant boost, one is in the position to take advantage of either the synergism between the biological effects of the two modalities or the geometry of their respective dose distributions. The fused dose distributions of two different modes of radiotherapy (EBRT and LDR implant) could be best described in terms of their biological equivalent dose distributions applying accepted radiobiological principles (see Eq. (11); FOWLER 1989; LEE et al. 1995; KING et al. 2000; BRAHME et al. 2001; BRAHME 2001). The feasibility of this combined therapy is discussed in a recent paper (ZAIDER et al. 2005).

References

Abbas F, Scardino PT (1997) The natural history of clinical prostate carcinoma. Cancer 80:827–833

Albertsen PC, Hanley JA, Gleason DF, Barry MJ (1998) Competing risk analysis of men aged 55 to 74 years at diagnosis managed conservatively for clinically localized prostate cancer. J Am Med Assoc 280:975–980

Amols HI, Zaider M, Hayes MK, Schiff PB (1997) Physician/patient-driven risk assignment in radiation oncology: Reality or fancy? Int J Radiat Oncol Biol Phys 38:455–461

Brahme A (2001) Individualizing cancer treatment: biological optimization models in treatment planning and delivery. Int J Radiat Oncol Biol Phys 49:327–337

Brahme A, Nilsson J, Belkic D (2001) Biologically optimized radiation therapy. Acta Oncol 40:725–734

Chambers L (1995) Practical handbook of genetic algorithms. CRC Press, Boca Raton, Florida

Chodak GW (1998) Comparing treatments for localized prostate cancer-persisting uncertainty. J Am Med Assoc 280:1008–1010

Cohen GN, Amols HI, Zelefsky MJ, Zaider M (2002) The Anderson nomograms for permanent interstitial prostate implants: a briefing for practitioners. Int J Radiat Oncol Biol Phys 53:504–511

Critz FA, Williams WH, Levinson AK, Benton JB, Holladay CT, Schnell FJ (2000) Simultaneous irradiation for prostate cancer: intermediate results with modern techniques. J Urol 164:738–741

D'Souza WD, Thames HD, Kuban DA (2004) Dose-volume conundrum for response of prostate cancer to brachytherapy: summary of dosimetric measures and their relationship to tumor control probability. Int J Radiat Oncol Biol Phys 58:1540–1548

Fowler JF (1989) The linear-quadratic formula and progress in fractionated radiotherapy. Br J Radiol 62:679–694

Gallagher RJ, Lee EK (1997) Mixed integer programming optimization models for brachytherapy treatment planning. Proc MIA Annual Fall Symposium, pp 278–282

Grefenstette JJ, American Association for Artificial Intelligence, Bolt BaNi, Naval Research Laboratory (U.S.) (1987) Genetic algorithms and their applications. Proc Second International Conference on Genetic Algorithms, 28–31 July 1987, Massachusetts Institute of Technology, Cambridge, Massachusetts. Erlbaum, Hillsdale, New Jersey

Haworth A, Ebert M, Waterhouse D, Joseph D, Duchesne G (2004a) Assessment of I-125 prostate implants by tumor bioeffect. Int J Radiat Oncol Biol Phys 59:1405–1413

Haworth A, Ebert M, Waterhouse D, Joseph D, Duchesne G (2004b) Prostate implant evaluation using tumour control probability: the effect of input parameters. Phys Med Biol 49:3649–3664

Jani AB, Hellman S (2003) Early prostate cancer: clinical decision-making. Lancet 361:1045–1053

King CR, DiPetrillo TA, Wazer DE (2000) Optimal radiotherapy for prostate cancer: predictions for conventional external beam, IMRT, and brachytherapy from radiobiologic models. Int J Radiat Oncol Biol Phys 46:165–172

Kurhanewicz J, Vigneron DB, Nelson SJ, Hricak H, MacDonald JM, Konety B, Narayan P (1995) Ctrate as an in-vivo marker to discriminate prostate-cancer from benign prostatic hyperplasia and normal prostate peripheral zone: detection via localized proton spectroscopy. Urology 45:459–466

Kurhanewicz J, Vigneron DB, Hricak H, Narayan P, Carroll P, Nelson SJ (1996) Three-dimensional H-1 MR spectroscopic imaging of the in situ human prostate with high (0.24–0.1 cm³) spatial resolution. Radiology 198:795–805

Kurhanewicz J, Vigneron DB, Males RG, Swanson MG, Yu KK, Hricak H (2000a) The prostate: MR imaging and spectroscopy – present and future. Radiol Clin North Am 38:115

Kurhanewicz J, Vigneron DB, Nelson SJ (2000b) Three-dimensional magnetic resonance spectroscopic imaging of brain and prostate cancer. Neoplasia 2:166–189

Lee EK, Zaider M (2001) On the determination of an effective planning volume for permanent prostate implants. Int J Radiat Oncol Biol Phys 49:1197–1206

Lee EK, Gallagher RJ, Silvern D, Wuu CS, Zaider M (1999) Treatment planning for brachytherapy: an integer programming model, two computational approaches and experiments with permanent prostate implant planning. Phys Med Biol 44:145–165

Lee SP, Leu MY, Smathers JB, McBride WH, Parker RG, Withers HR (1995) Biologically effective dose distribution based on the linear-quadratic model and its clinical relevance. Int J Radiat Oncol Biol Phys 33:375–389

MacKay RI, Hendry JH, Moore CJ, Williams PC, Read G (1997) Predicting late rectal complications following prostate conformal radiotherapy using biologically effective doses and normalized dose-surface histograms. Br J Radiol 70:517–526

Man KF, Tang KS, Kwong S (1999) Genetic algorithms concepts and designs. Springer, Berlin Heidelberg New York

McGregor M, Hanley JA, Boivin JF, Mclean RG (1998) Screening for prostate cancer: estimating the magnitude of over-detection. Can Med Assoc J 159:1368–1372

Mizowaki T, Cohen G, Fung AYC, Zaider M (2002) Towards integrating functional imaging in the treatment of prostate cancer with radiation: the registration of the MR spectroscopy imaging to ultrasound/CT images and its implementation in treatment planning. Int J Radiat Oncol Biol Phys 54:1558–1564

Nag S, Shasha D, Janjan N, Petersen I, Zaider M (2001) The American Brachytherapy Society recommendations for brachytherapy of soft tissue sarcomas. Int J Radiat Oncol Biol Phys 49:1033–1043

Peschel RE, Colberg JW (2003) Surgery, brachytherapy, and external-beam radiotherapy for early prostate cancer. Lancet Oncol 4:233–241

Pouliot J, Tremblay D, Roy J, Filice S (1996) Optimization of permanent I-125 prostate implants using fast simulated annealing. Int J Radiat Oncol Biol Phys 36:711–720

Sandhu AS, Zelefsky MJ, Lee HJ, Lombardi D, Fuks Z, Leibel SA (2000) Long-term urinary toxicity after 3-dimensional conformal radiotherapy for prostate cancer in patients with prior history of transurethral resection. Int J Radiat Oncol Biol Phys 48:643–647

Scheidler J, Hricak H, Vigneron DB, Yu KK, Sokolov DL, Huang LR, Zaloudek CJ, Nelson SJ, Carroll PR, Kurhanewicz J (1999) Prostate cancer: localization with three-dimensional proton MR spectroscopic imaging – clinicopathologic study. Radiology 213:473–480

Silvern DA (1998) Automated OR prostate brachytherapy treatment planning using genetic optimization. Thesis, Columbia University, New York

Sloboda RS (1992) Optimization of brachytherapy dose distribution by simulated annealing. Med Phys 19:964

Todor DA, Cohen GN, Amols HI, Zaider M (2002a) Operator-free, film-based 3D seed reconstruction in brachytherapy. Phys Med Biol 47:2031–2048

Todor DA, Cohen GN, Amols HI, Zaider M (2002b) Operator-free, film-based 3D seed reconstruction in brachytherapy. Phys Med Biol 47:2031–2048

Todor DA, Zaider M, Cohen GN, Worman MF, Zelefsky MJ (2003) Intraoperative dynamic dosimetry for prostate implants. Phys Med Biol 48:1153–1171

Tubic D, Zaccarin A, Pouliot J, Beaulieu L (2001) Automated seed detection and three-dimensional reconstruction. I. Seed localization from fluoroscopic images or radiographs. Med Phys 28:2265–2271

Wang JZ, Li XA (2003) Evaluation of external beam radiotherapy and brachytherapy for localized prostate cancer using equivalent uniform dose. Med Phys 30:34–40

Waterman FM, Yue N, Corn BW, Dicker AP (1997) Edema associated with I-125 or Pd-103 prostate brachytherapy and its effect on post-implant dosimetry: an analysis based on serial CT acquisition. Int J Radiat Oncol Biol Phys 39:220

Wefer AE, Hricak H, Vigneron DB, Coakley FV, Lu Y, Wefer J, Mueller-Lisse U, Carroll PR, Kurhanewicz J (2000) Sextant localization of prostate cancer: comparison of sextant biopsy, magnetic resonance imaging and magnetic resonance spectroscopic imaging with step section histology. J Urol 164:400–404

Zaider M, Minerbo GN (2000) Tumour control probability: a formulation applicable to any temporal protocol of dose delivery. Phys Med Biol 45:279–293

Zaider M, Zelefsky MJ, Lee EK, Zakian KL, Amols HI, Dyke J, Cohen G, Hu Y, Endi AK, Chui C, Koutcher JA (2000) Treatment planning for prostate implants using magnetic-resonance spectroscopy imaging. Int J Radiat Oncol Biol Phys 47:1085–1096

Zaider M, Zelefsky M, Cohen AM, Chui C, Yorke ED, Ben-Porat L, Happersett MA (2005) Methodology for biologically based treatment planning for combined low dose-rate (permanent implant) and high dose-rate (fractionated) treatment of prostate cancer. Int J Radiat Oncol Biol Phys 61:702–713

Zalzala AMS, Fleming PJ (1997) Genetic algorithms in engineering systems. Institution of Electrical Engineers, London

Zelefsky MJ, Hollister T, Raben A, Matthews S, Wallner KE (2000a) Five-year biochemical outcome and toxicity with transperineal CT-planned permanent I-125 prostate implantation for patients with localized prostate cancer. Int J Radiat Oncol Biol Phys 47:1261–1266

Zelefsky MJ, Cohen G, Zakian KL, Dyke J, Koutcher JA, Hricak H, Schwartz L, Zaider M (2000b) Intraoperative conformal optimization for transperineal prostate implantation using magnetic resonance spectroscopic imaging. Cancer J 6:249–255

Zhang M, Zaider M, Worman M, Cohen G (2004) On the question of 3D seed reconstruction in prostate brachytherapy: the determination of X-ray source and film locations. Phys Med Biol 49:N335–N345

30 Vascular Brachytherapy

Boris Pokrajac, Erich Minar, Christian Kirisits, and Richard Pötter

CONTENTS

30.1
Introduction

Percutaneous transluminal angioplasty (PTA) and percutaneous transluminal coronary angioplasty (PTCA) result in immediate restitution of blood flow, but its major limitation, restenosis, compromises the outcome within 12 months in 30–70% of patients. The arterial re-narrowing or restenosis occurs in 90% of patients mostly up to 9 months after intervention (Kuntz and Baim 1993).

Despite new advances in endovascular techniques, such as atherectomy, laser angioplasty, thrombolysis, coated stents and pharmacological agents (anticoagulants, angiotensin converting enzyme, statins and calcium channel blocker), the problem of restenosis remains unsolved (Tardif et al. 1997). The appearance of endovascular brachytherapy (EVBT) gave hope for final solution of this problem. The EVBT had been proven to reduce significantly restenosis after angioplasty in many experimental coronary studies (Weintraub 1994; Wiedermann et al. 1994). The

B. Pokrajac, MD; C. Kirisits, DSc; R. Pötter, MD
Department of Radiotherapy and Radiobiology, Vienna General Hospital, Medical University of Vienna, Waehringer Guertel 18–20, 1090 Vienna, Austria
E. Minar, MD
Department of Angiology, Vienna General Hospital, Medical University of Vienna, Waehringer Guertel 18–20, 1090 Vienna, Austria

clinical and experimental studies with both gamma and beta radioactive sources delivered by catheter-based systems have demonstrated strong evidence of neointimal hyperplasia reduction and positive arterial remodeling. Positive experience with clinical application of EVBT was gained in several single-center (Condado et al. 1997; King et al. 1998; Minar et al. 1998) and multicenter studies (Greiner et al. 2003; Waksman et al. 2001).

Since Dotter and Judkins in 1964 performed the first successful percutaneous revascularization of superficial femoropopliteal artery (SFA), this therapy has become an important treatment modality for Rutherford stage 3–6 of peripheral vascular disease. Andreas Grüntzig introduced in 1977 the angioplasty for stenotic coronary arteries. Since then, angioplasty has become widely used with approximately 500,000 interventions in the U.S. per year and approximately 150,000 interventions annually in Germany.

30.2
Technique and Indications for Vascular Brachytherapy

The EVBT device for peripheral arteries consists of a remote-controlled afterloader unit, which delivers gamma radiation with an iridium-192 source. This device is already well known and widely available in high-dose-rate (HDR) brachytherapy of cancer. The positioning of the source is computer controlled and the patient is during the treatment audio-visually monitored from outside the irradiation room (Pötter and Van Limbergen 2002).

The afterloader is operated by programming the different source positions and dwelling times, thus enabling the isodose optimization. The treatment planning system is attached to the afterloader unit. For the design of devices for EVBT, the dimensions of sources and catheters have to be small to allow treatment thin arteries (4–8 mm arterial lumen of SFA). An individual adaptation of the active source length

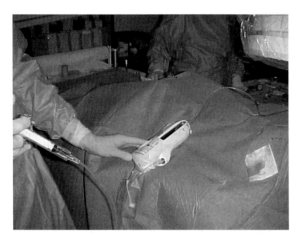

Fig. 30.3 Performing of brachytherapy with strontium-90 source (Beta-Cath system, Novoste, Norcross, Ga., USA)

30.4
Clinical Evidence

Concerning peripheral arteries, important knowledge about brachytherapy for restenosis prophylaxis has been obtained (Teirstein 2000). All performed studies have been single center with the exception of PARIS feasibility study and Vienna-3 trial. The PARIS-pilot study (WAKSMAN et al. 2001), although with relatively small patient numbers (*n*=35), confirmed the positive results of previous trials. The angiographic binary restenosis at 6 months was 17.2% and clinical restenosis at 12 months was 13.3%.

Possible prevention of restenosis using brachytherapy was first investigated and reported for femoropopliteal arteries by the Frankfurt group in the early 1990s based on the well-documented antiprolif-

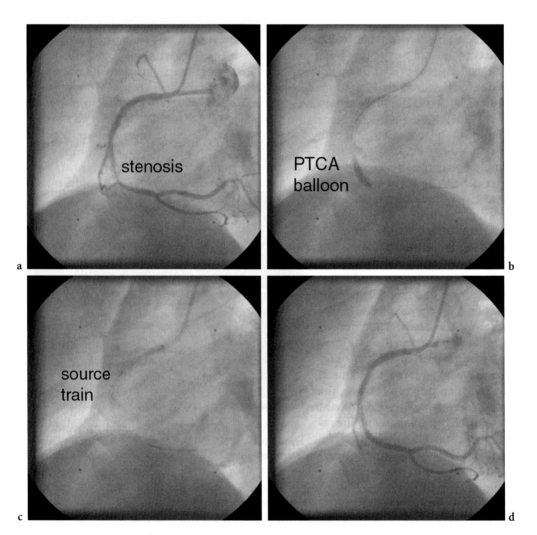

Fig. 30.4 a High-grade in-stent restenosis of right coronary artery. **b** Dilatation balloon (Ø 3.0 mm, length 30 mm). **c** Brachytherapy with strontium-90 source (ASL 60 mm). **d** Completely patent artery 6 months after initial treatment. *PTCA* percutaneous transluminal coronary angioplasty

erative effect of low dose of ionizing radiation in the prevention of keloids. Excellent results from a clinical phase-I/II trial in a limited number of patients (*n*=28) were reported, even after a long follow-up period (6 years), with a binary restenosis rate of 23% (BÖTTCHER et al. 1994; SCHOPOHL et al. 1996).

Vienna-2 (POKRAJAC et al. 2000) was a prospective, randomized trial comparing PTA vs PTA + brachytherapy for non-stented lesions. A total of 113 patients with claudication or critical limb ischemia having de-novo lesions of ≥5 cm, i.e., recurrent lesions after former PTA of any length were included. In this study stenting was not allowed. The mean PTA length was 16.7 cm in brachytherapy cohort vs 14.8 cm in placebo cohort. A well-balanced patient distribution was achieved for the criteria used for stratification ("de-novo stenosis vs recurrence after former PTA," "stenosis vs occlusion," "claudication vs critical limb ischemia," and "diabetes vs non-diabetes"). The dose of 12 Gray (Gy) was the same as in Frankfurt study and has been targeted at 3-mm distance from the iridium source. Primary end point was femoropopliteal patency after 6 months. Crude restenosis rate at 6 months was in the placebo arm 54 vs 28% in brachytherapy arm. Actuarial estimate of the patency rate was at 6 months 45 vs 72%, respectively. The cumulative patency rates at 12 months were 64% in brachytherapy group vs 35% in placebo group. The comparison of restenosis rates for the different subgroups with risk factors (recurrence after former PTA, occlusion and PTA length >10 cm) showed significant decrease of the restenosis rate, if brachytherapy was added. Significant reduction was not achieved in diabetics.

The Vienna-4 study (WOLFRAM et al. 2001) using brachytherapy after PTA + stenting because of unsatisfactory angioplasty (>30% residual stenosis; multiple dissections) was performed with a dose of 14 Gy prescribed 2 mm from the centering catheter surface. The antiplatelet therapy with clopidogrel was given for 1 month 75 mg/die. Binary restenosis of 30% was observed after 6 months follow-up in 33 patients.

The Vienna-3 multicenter trial (POKRAJAC et al. 2003) compared PTA vs PTA followed by brachytherapy using centering radiation delivery catheter without stenting in de-novo and restenotic lesions. The binary restenosis rate at 12 months was 23.4% in the brachytherapy and 53.3% in the placebo group (*p*<0.05). The cumulative patency rates after 24 months were 77% in the brachytherapy and 39% in the placebo group (*p*<0.001). Late thrombosis was not seen.

The Vienna-5 study has investigated angioplasty plus stenting vs angioplasty plus stenting and additional brachytherapy.

There have been several trials in coronary brachytherapy since the mid-1990s. Many of them were double-blind randomized prospective trials, preceded by feasibility studies and followed by registry trials. Results reported thus far refer to far more than 1000 patients and focus mainly on the overall therapeutic effect of vascular brachytherapy.

The double-blind randomized trials based on iridium-192 brachytherapy (manual loading) all have revealed for lesion lengths <50 mm a significant angiographic reduction in restenosis rate at 6 months follow-up; the most outstanding of them are the following:

• SCRIPPS I (TEIRSTEIN et al. 1997) single center: restenotic native coronary arteries or saphenous venous grafts (SVG) in 55 patients from 54 to 17%, reduction rate of 68%
• GAMMA I (LEON et al. 2001) multicenter: in-stent restenosis of coronary arteries in 252 patients from 56 to 34% WRIST (single center): in-stent restenosis of coronary arteries and SVG in 130 patients from 58 to 19%, reduction rate of 67%

The clinical outcome was also significantly improved; however, the difference was mostly somewhat less pronounced. Clinical outcome was measured by major adverse cardiac events (MACE), which includes the need for target lesion and vessel revascularization and myocardial infarction.

For intravascular brachytherapy based on beta sources the results reported in the early experience have been less favorable. The first Geneva study proved feasibility using the Schneider-Sauerwein-Boston Scientific System (yttrium-90, centering balloons) but showed no apparent effectiveness in de-novo lesions (POPOWSKI et al. 1996). The recent multicenter Geneva dose-finding study (VERIN et al. 2001) revealed a significant reduction of restenosis rate from 9, 12, 15 to 18 Gy from 27.5, 17.5, 15.8 to 8.3% (de-novo lesions, 183 patients). The START trial (SUNTHARALINGAM et al. 2002) was the first to show a significant reduction in angiographic and clinical restenosis rate based on beta radiation with the Novoste system (strontium-90) from 45 to 29% and from 24 to 16%, respectively. Interesting in this analysis is that, for the first time in coronary brachytherapy, a more systematic analysis for the different lengths has been performed. Looking at the vessel segment, which received the therapeutic dose, the results in this study are even more favorable. In the INHIBIT (WAKSMAN et al. 2002) trial (in-stent restenosis, 332 patients, Guidant system, phosphorus-32) restenosis rate was dropped from 52% in placebo cohort to 26% in brachytherapy cohort for the whole analyzed segment.

The restenosis at the stent edge represents one of the major problems at present, in particular in coronary beta brachytherapy. Also the radial dose distribution represents a major problem due to the pathological eccentricity of the vessel. Again this issue has more impact on beta brachytherapy devices.

Another important issue which has recently been reported from longer follow-up of the early trials (SCRIPPS 1, WRIST, VIENNA-2) is the occurrence of "late events"; these are restenoses occurring in the second and third year after brachytherapy, which reduce the therapeutic benefit. As such, late events have not been observed in the control arms of these trials. Nevertheless, the results of EVBT to reduce restenosis are still significant after long-term follow-up (TEIRSTEIN et al. 2000; WAKSMAN et al. 2004).

30.5
Side Effects of Brachytherapy

Late stent thrombosis has been reported as sudden thrombotic occlusion within the stent taking place approximately 2–6 months following brachytherapy (COSTA et al. 1999; MINAR et al. 2000a, b). It occurs after stenting and brachytherapy with a frequency going up to 15%. It is rarely seen if no stent is implanted. The underlying mechanism seems to be the delayed re-endothelialization to cover the stent struts. If antiplatelet treatment is given for at least 12 months (e.g., clopidrogel), this serious side effect seems to be minimized and only rarely observed (up to 3%). The aneurysm formation has been associated with irradiation from first experience with coronary brachytherapy in 1996 (CONDADO et al. 1997). This question has still not been clarified. As radiation can induce malignancies, this is of course a serious issue. The importance of this problem is difficult to assess, as patients need long follow-up; however, if many patients with a benign disease (neointimal hyperplasia) are going to be treated with brachytherapy, this issue certainly needs to be taken into consideration (PÖTTER and VAN LIMBERGEN 2002).

30.6
Other Treatment Options

Interventional procedures, such as directional atherectomy, thrombolysis, and rotational ablation, have shown some advantages when compared with plain balloon angioplasty. All these techniques failed to decrease restenosis significantly. Stenting is associated with low arterial patency especially if the long segment is treated. For short lesions an improvement with additional stenting could not be obtained either. The role of sono- and photodynamic therapy has rarely been investigated; thus, no definitive conclusion can be drawn.

Drug-eluting stents covered with rapamycin and paclitaxel are tested in phase-III trials and promises further drop of restenosis. Recently, published results of two randomized studies have shown a significant lowering of restenosis especially in coronary de-novo lesions. In a small study ($n=36$) for femoropopliteal arteries, the results were very promising after implantation of rapamycin-coated stent (DUDA et al. 2002). Binary restenosis was 0 compared with 22% in the control arm. No late thrombosis was observed. Larger trial was conducted in coronary arteries with very good results (MORICE et al. 2002). The restenosis rate of about 6–8% was observed; however, rates up to 35% were observed in diabetic patients and in small-vessel diameters (<3.0 mm). These very promising results should be confirmed in larger clinical trials with brachytherapy.

References

Böttcher HD, Schopohl B, Liermann D, Kollath J, Adamietz IA (1994) Endovascular irradiation – a new method to avoid recurrent stenosis after stent implantation in peripheral arteries: technique and preliminary results. Int J Radiat Oncol Biol Phys 29:183–186

Condado JA, Waksman R, Gurdiel O et al. (1997) Long-term angiographic and clinical outcome after percutaneous transluminal coronary angioplasty and intracoronary radiation therapy in humans. Circulation 96:727–732

Costa MA, Sabate M, van der Giessen WJ et al. (1999) Late coronary occlusion after intracoronary brachytherapy. Circulation 100:789–792

Duda SH, Pusich B, Richter G et al. (2002) Sirolimus-eluting stents for the treatment of obstructive superficial femoral artery disease: six-month results. Circulation 106:1505–1509

Greiner RH, Mahler F, Jäger K, Schneider E, Gallino A (2003) Swiss trials of peripheral endovascular brachytherapy for restenosis prevention after PTA. Radiother Oncol 66 (Suppl 1):S6 (Abstract)

Leon MB, Teirstein PS, Moses JW, Tripuraneni P, Lansky AJ, Jani S, Wong SC, Fish D, Ellis S, Holmes DR, Keriakes D, Kuntz RE (2001) Localized intracoronary gamma-radiation therapy to inhibit the recurrence of restenosis after stenting. N Engl J Med 344:250–256

King SB, Williams DO, Chougule P et al. (1998) Endovascular beta-radiation to reduce restenosis after coronary balloon angioplasty. Results of the Beta Energy Restenosis Trial (BERT). Circulation 97:2025–2030

Kirisits C, Georg D, Wexberg P et al. (2002) Determination and application of the reference isodose length (RIL) for commercial endovascular brachytherapy devices. Radiother Oncol 64:309–315

Kuntz RE, Baim DS (1993) Defining coronary restenosis: newer clinical and angiographic paradigms. Circulation 88:1310–1323

Minar E, Pokrajac B, Ahmadi R et al. (1998) Brachytherapy for prophylaxis of restenosis after long-segment femoropopliteal angioplasty: pilot study. Radiology 208:173–179

Minar E, Pokrajac B, Maca T et al. (2000a) Endovascular brachytherapy for prophylaxis of restenosis after femoro-popliteal angioplasty: results of a prospective, randomized study. Circulation 102:2694–2699

Minar E, Wolfram R, Pokrajac B (2000b) Endovascular brachytherapy and late thombotic occlusion. Circulation 102: el175–el176

Morice MC, Serruys PW, Sousa JE, Fajadet J, Ban Hayashi E, Perin M, Colombo A, Schuler G, Barragan P, Guagliumi G, Molnar F, Falotico R, RAVEL Study Group (2002) A randomized comparison of a sirolimus-eluting stent with a standard stent for coronary revascularization. N Engl J Med 346:1773–1780

Pokrajac B, Pötter R, Maca T, Fellner C, Mittlböck M, Ahmadi R, Seitz W, Minar E (2000) Intra-arterial 192Ir HDR brachytherapy for prophylaxis of restenosis after femoropopliteal percutaneous transluminal angioplasty: the prospective randomized Vienna-2 trial radiotherapy parameters and risk factors analysis. Int J Radiat Oncol Biol Phys 48:923–931

Pokrajac B, Cejna M, Kettenbach J et al. (2001) Intraluminal 192Ir brachytherapy following transjugular intrahepatic portosystemic shunt revision: long terms results and radiotherapy parameters. Cardiovasc Radiat Med 2:133–137

Pokrajac B, Schmid R, Pötter R, Wolfram R, Kirisits C, Mendel H, Kopp M, Minar E (2003) Endovascular brachytherapy prevents restenosis after femoropopliteal angioplasty: results of the Vienna-3 multicenter study. Int J Radiat Oncol Biol Phys 57 (Suppl 2):S250 (Abstract)

Popowski Y, Verin V, Urban P (1996) Endovascular ß-radiation after percutaneous transluminal coronary balloon angioplasty. Int J Radiat Oncol Biol Phys 36:841–845

Pötter R, Van Limbergen E (2002) Endovascular brachytherapy: In: Gerbaulet A, Pötter R, Mazeron JJ, Meertens H, Van Limbergen E (eds) The GEC ESTRO handbook of brachytherapy. ACCO, Leuven, Belgium

Pötter R, Van Limbergen E, Dries W, Popowski Y, Coen V, Fellner C, Georg D, Kirisits C, Levendag P, Marijnissen H, Marsiglia H, Mazeron JJ, Pokrajac B, Scalliet P, Tamburini V, EVA GEC ESTRO Working Group (2001) Recommendations for prescribing, recording, and reporting endovascular brachytherapy. Radiother Oncol 59:339–360

Rubin P, Williams JP, Riggs PN et al. (1998) Cellular and molecular mechanisms of radiation inhibition of restenosis. Part I. Role of the macrophage and platelet-derived growth factor. Int J Radiat Oncol Biol Phys 40:929–941

Schmid R, Kirisits C, Syeda B et al. (2004) Evaluation of geographic miss in endovascular brachytherapy of the coronaries. A new method of assessment. Recommendations for determining the planning target length to avoid geographic miss. The Vienna experience.(Evaluation of geographic miss in endovascular brachytherapy). Radiother Oncol 71(3):311–8

Schopohl B, Liermann D, Jülling-Pohlit L et al. (1996) 192 Ir endovascular brachytherapy for avoidance of intimal hyperplasia after percutaneous transluminal angioplasty and stent implantation in peripheral vessels. Six years of experience. Int J Radiat Oncol Biol Phys 36:835–840

Stoeteknuel-Friedli S, Do DD, Briel C von et al. (2002) Endovascular brachytherapy for prevention of recurrent renal in-stent restenosis. J Endovasc Ther 9:350–353

Suntharalingam M, Laskey W, Lansky AJ et al. (2002) Clinical and angiographic outcomes after use of strontium-90/ yttrium-90 beta radiation for the treatment of in-stent restenosis: results from the stents and radiation therapy 40 (START 40) registry. Int J Radiat Oncol Biol Phys 52:1075–1082

Tardif JC, Cote G, Lesperance J et al. (1997) Probucol and multivitamins in the prevention of restenosis after coronary angioplasty. N Engl J Med 337:365–372

Teirstein P (2000) Fulfilling the promises of percutaneous angioplasty. Circulation 102:2674–2676

Teirstein PS, Massullo V, Jani S, Popma JJ, Mintz GS, Russo RJ, Schatz RA, Guarneri EM, Steuterman S, Morris NB, Leon MB, Tripuraneni P (1997) Catheter-based radiotherapy to inhibit restenosis after coronary stenting. N Engl J Med 336:1697–1703

Teirstein PS, Massullo V, Jani S et al. (2000) Three-year clinical and angiographic follow-up after intracoronary radiation: results of a randomized clinical trial. Circulation 101:360–365

Verin V, Popowski Y, de Bruyne B, Baumgart D, Sauerwein W, Lins M, Kovacs G, Thomas M, Calman F, Disco C, Serruys PW, Wijns W (2001) Dose-Finding Study Group. Endoluminal beta-radiation therapy for the prevention of coronary restenosis after balloon angioplasty. N Engl J Med 344:243–249

Waksman R, Laird JR, Jurkovitz CT et al. (2001) Intravascular radiation therapy after balloon angioplasty of narrowed femoropopliteal arteries to prevent restenosis: results of the Paris feasibility clinical trial. J Vasc Interv Radiol 12:915–921

Waksman R, Raizner AE, Yeung AC et al. (2002) Use of localized intracoronary beta radiation in treatment of in-stent restenosis: the INHIBIT randomized controlled trial. Lancet 359:551–557

Waksman R, Ajani AE, White RL et al. (2004) Five-year follow-up after intracoronary gamma radiation therapy for in-stent restenosis. Circulation 109:340–344

Weintraub WS, and the LRT Study Group (1994) Lack of effect of lovastatin on restenosis after coronary angioplasty. N Engl J Med 331:1331–1337

Wiedermann JG, Marboe C, Schwartz A (1994) Intracoronary irradiation reduces restenosis after balloon angioplasty in a porcine model. J Am Coll Cardiol 23:1491–1498

Wolfram R, Pokrajac B, Ahmadi R et al. (2001) Endovascular brachytherapy for prophylaxis of restenosis after long-segment femoropopliteal stenting: pilot study. Radiology 220:724–729

31 Partial Breast Brachytherapy After Conservative Surgery for Early Breast Cancer: Techniques and Results

Yazid Belkacémi, Jean-Michel Hannoun-Lévi, and Eric Lartigau

CONTENTS

31.1 Introduction

Post-operative radiation therapy (RT) has been shown in a number of randomized trials to reduce the risk of breast cancer recurrence following breast-conserving surgery (Clark 1992; Forrest 1996; Liljegren 1994; Fisher 2002; Veronesi 2002).

Breast-conserving surgery, adjuvant postoperative RT, and systemic therapy has become standard treatment for increasing numbers of women with early breast cancer. Although breast-conserving therapy has become an accepted alternative for early-stage breast cancer patients, a number of studies show that only 10–70% of patients qualifying for brachytherapy actually receive it in North America. In an

effort to overcome this problem, a number of centers have attempted an accelerated regimen using interstitial brachytherapy as the sole radiation modality (Clarke 1994; Fentiman 1991; Kuske and Bolton 1995; Perera 1995, 1997; Vicini 1997, 1999). This treatment approach has the potential to make brachytherapy more attractive to many patients, and poses fewer logistic problems. Moreover, in some patients older than 65 years with or without co-morbidities, the long duration of conventional fractionated RT may significantly alter the quality of life and the treatment observance. Ballard-Barbash et al. (1996) found among a cohort 18704 patients aged 65 years or older that the receipt of postoperative RT declined substantially with age, irrespective of co-morbidity and disease stage. For the age groups 65–69 and 80 years or older, the use of irradiation felt from 77 to 24% in women without co-morbid conditions and 50–12% with two or more co-morbid conditions. For elderly women, with the low risk of local recurrence, accelerated partial breast irradiation during 6 weeks (one fraction per week) seems to be also a reasonable alternative to compare in the future to post-operative standard whole-breast radiation therapy (WBRT; Hannoun-Lévi 2003).

After conservative surgery, WBRT is the standard of care. The RT schedule consists of delivering 45–50 Gy to the whole breast, in 4.5–5 weeks, followed by a boost dose of 10–20 Gy to the tumor bed. Brachytherapy is used in this setting as one of the boost techniques. The rational for elective treatment of the whole breast after conservative surgery stems from the findings of pathological studies of mastectomy specimens in women with localized breast cancer and from preoperative radiological imaging of patients selected for conservative surgery, which demonstrated the existence of unsuspected foci of carcinoma in 16–37% of women (Rosen 1975; Orel 1995; Moon 2002). This finding initially led to the notion that breast-conserving therapy was an inappropriate treatment option for women with early-stage disease; however, the type and location of local recurrences in studies of standard breast-conserving

Y. Belkacémi, MD, PhD; E. Lartigau, MD, PhD
Department of Radiation Oncology, Centre Oscar Lambret, University of Lille II, 3 rue Frédéric Combemale, 59020 Lille, France
J.-M. Hannoun-Lévi, MD, PhD
Department of Radiation Oncology, Centre Antoine Lacassagne, 33 avenue de Valombrose, 06189 Nice Cedex, France

therapy have suggested that these areas of clinically occult carcinoma were either of limited clinical significance or were encountered less frequently when patients were more carefully selected and evaluated to rule out multicentric disease. The lack of clinically significant benefit in elective treatment of the whole breast is clearly demonstrated in randomized trials comparing patients treated with conservative surgery plus RT to those treated with conservative surgery alone. Numerous studies have reported that the local recurrences occur within and surrounding the primary tumor site (65–85%). For this reason and in order to decrease the treatment duration and increase quality of life of patients, partial breast irradiation (PBI) using interstitial brachytherapy technique, balloon implantation (MammoSite), intra-operative RT (IORT), or hypofractionated 3D conformal irradiation of the lumpectomy cavity have been developed for local treatment after conservative surgery. In selected patients with favorable prognostic factors, PBI may be considered as an alternative of postoperative management of early breast cancer. It is currently compared with WBRT in an NSABP phase-III randomized trial.

In this chapter we describe brachytherapy techniques and results in the setting of PBI as a sole postoperative treatment for early breast cancer.

31.2
Interstitial Brachytherapy in Partial Breast Irradiation

31.2.1
Technique

Interstitial brachytherapy implantation technique is dependent on the system used but has to conform to ICRU no. 58 recommendations (ICRU 1997). We describe in this chapter the Paris system (Dutreix and Marinello 1987).

Interstitial brachytherapy using Paris system criteria is applied either with low or high dose rate. The Paris system is based on three principle recommendations:
1. Radioactive wires have to be parallel and their center has to be in the same plan called central plan (Fig. 31.1a).
2. The reference KERMA rate (lineic activity) has to be uniform all along each radioactive wire and identical for all sources.
3. Radioactive wires have to be equidistant. For applications performed with at least two plans,

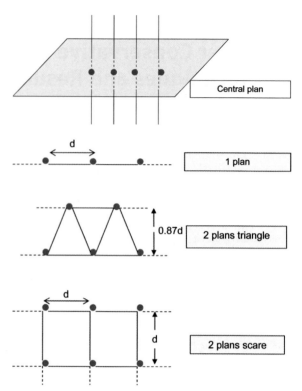

Fig. 31.1a,b. Paris system recommendations

the line intersections with the central plan, have to be placed according to the top of the equilateral triangles or scares (Fig. 31.1b). In case of triangle application, the spacing between the different plans is equal to the spacing between radioactive wires multiplied by 0.87.

In order to avoid acute and late skin side effects, the application has to respect at least 5-mm margin between the iridium-192 wires and the skin surface (Fig. 31.2).

Depending of the location of the tumor in the breast, the plastic tubes or needles have to be placed in order to obtain a reference isodose (85% of the basal dose) encompassing the clinical target volume (CTV). If the tumor bed is distant from the skin or the chest wall, then it is possible to use at least two plans with two to four needles each, in order to cover properly the CTV with sufficient margins (at least 2 cm). In contrast, when the tumor bed is close to the skin, tumor removal includes the skin and allows to apply specific recommendations. In some cases, the tumor may be located at the periphery of the breast (i.e., upper or inner quadrants), which cannot allow the use of two plans due to the small thickness of the target volume (Figs. 31.3, 31.4).

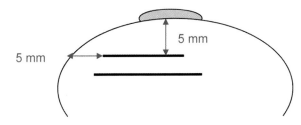

Fig. 31.2. Security margin between iridium-192 sources and the skin to decrease the risk of acute and late cutaneous side effects

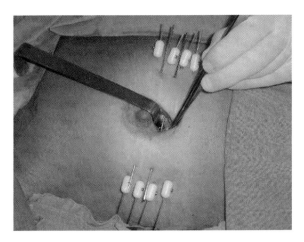

Fig. 31.3. Interstitial low-dose-rate brachytherapy applied according to ICRU report no. 58

When vectors are inserted, templates are used in order to keep constant the geometry of the implant. In PBI setting delivering 45–50 Gy and because the material will be maintained during 8–15 days, templates have to conserve the good position of the vectors but avoid being too tight for the breast.

31.2.2
Partial Breast Irradiation Using Low-Dose-Rate Interstitial Brachytherapy

Low-dose rate (LDR) interstitial brachytherapy is the first technique used for PBI. It could be used as a sole postoperative treatment after conservative surgery or as salvage treatment after local recurrence in patients who refused radical mastectomy.

The implantation of needles or plastic tubes is generally performed during the surgical time. The CTV has to be defined by the surgeon and the radiation oncologist, and must be at least set at 2 cm safety margin around lumpectomy cavity. Radioactive source loading (^{192}iridium, Ir-192) has to be performed at least 5 days after surgery (and never after 6 weeks from the surgery; VICINI 1999). This delay not only allows good healing but also allows benefit of a precise histological report. The dose rate is generally around 0.60 Gy/h in order to complete treatment in 3–4 days. For LDR brachytherapy implants, activity of the Ir-192 sources should be ordered to provide a minimum dose rate of approximately 0.45 Gy/h, with

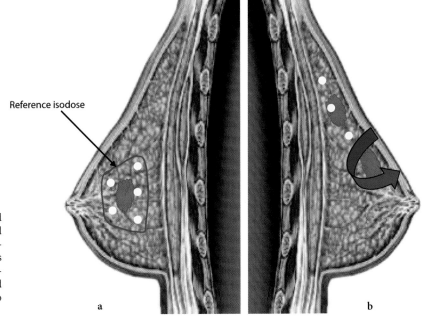

Fig. 31.4. a Tumor bed is located at distance from the skin and the chest wall (implantation using two plans). b Tumor bed is located close to the skin (implantation using one plan) and the skin face to the tumor has to be removed.

Reference isodose

a   b

Table 31.1. Results of partial breast irradiation using low-dose rate brachytherapy. *LR* local recurrence, *DR* dose rate

Reference	Number	Follow-up (months)	DR (Gy/h)	Dose (Gy)	LR (%)	Cosmetic result: good/excellent (%)
JEWELL et al. (1987)	107	52	–	–	2	–
CIONINI et al. (1993)	90	27	–	50–60	4.4	–
FENTIMAN et al. (1996)	27	72	0.40	55	37	83
KUSKE and BOLTON (1995)	26	20	–	45	0	78
KING et al. (2000)	41	75	–	45	2	75
KRISHNAN et al. (2001)	24	47	0.67	20–25	0	100
VICINI et al. (2003)	120	36	0.52	50	1	98

an acceptable range of 0.30–0.70 cGy/h. With a lower dose rate, the risk of local recurrence risk is significantly increased (DEORE 1993).

With interstitial implants (LDR or HDR), it is mandatory to respect a 5-mm minimum distance from the skin to avoid telangiectasia or lasting pigmentation of the skin. Given this last point, all series have shown very good cosmetic results (good to excellent: 80–100% of the patients; KUSKE 1994; FENTIMAN 1996; VICINI 2001; KRISHNAN 2001; WAZER 2002; POLGAR 2002).

The results in terms of local recurrence reported in the literature varies from 0 to 37% with often a short median follow-up (JEWELL 1987; CIONINI 1993; FENTIMAN 1991; FENTIMAN 1996; VICINI 1999; KING 2000; VICINI 2001; KRISHNAN 2001) but good or excellent cosmetic results (Table 31.2). Except for Fentiman et al. (1996), the 5-year probability of local recurrence did not exceed 5%. In this phase-II trial of PBI using LDR brachytherapy has included 27 patients with a follow-up of 6 years. The rate of local recurrence was 37% (10 of 27 patients). The authors concluded that

PBI with "continuous iridium 192 implant delivering 55 Gy in 5 days is not an effective means of achieving local control in patients with operable breast cancer." It is likely that these results are related to patient selection rather than the PBI technique. Indeed, 30% of the patients who developed local recurrences were treated for large tumors (>40 mm); 70% of them had axillary nodal metastases confirmed histologically, 60% had positives margins, 50% had tumor necrosis, and 50% had extensive intraductal component. These risk factors are known to result in a high local recurrence rate.

Conversely, the Beaumont hospital experience is based on more selected patients. One hundred twenty patients with a mean age of 65 years treated for small tumors (mean size 11 mm), without intraductal component (in 95% of the cases) and clear margin >10 mm (in 75% of the cases), received LDR brachytherapy (0.52Gy/h) delivering 50 Gy in 96 h. After 3 years of follow-up, local relapse rate was 1%. The authors did not find any statistical difference when they compared these results with a control

Table 31.2. Second conservative treatment of breast cancer local recurrence using partial breast irradiation with brachytherapy alone or surgery associated with brachytherapy or external beam radiation therapy. *TTT* treatment, *LR* local recurrence, *OS* overall survival, *G3-4 LTC* grade 3-4 of long-term complications, *EBRT* external beam radiation therapy, *PM* partial mastectomy, *d* diameter; *HDR BT* high-dose-rate brachytherapy, *LDR BT* low-dose-rate brachytherapy, *PDR BT* pulse-dose-rate brachytherapy, *NA* data not available

Reference	Number	Primary TTT	Dose of primary TTT (Gy)	LR TTT	Dose for LR TTT (Gy)	Percentage of second LR	5-year OS (%)	G3-4 LTC (%)
MAULARD et al. (1995)	15	EBRT or	45+20	PM/LD RBT (d: 24 mm)	30	26.6	61	13
	23	PM/EBRT	45+20		30×2	17.4	50	15
RESCH et al. (2002)	9	EBRT or	50+10	LDR BT (d: 39 mm)	60	0	NA	0
	8	PM/EBRT	50+10		19+26.6	29.4	NA	0
GUIX et al. (2003)	41	PM/EBRT	50+25	PM/PDR BT	30 (12/5)	10	NA	0
HANNOUN-LEVI et al. (2004)	24	PM/EBRT	50+10	PM/EBRT/PDR BT, PM/HDR BT	30	43.8	95.1	7
	45	PM/EBRT/ BT	50+10	PM/LDR BT, PM/ LDR BT	45	5.3	97.2	11

group of 907 matched patients who received WBRT (VICINI 2001, 2003). The Oschner Clinic group from Los Angeles has also reported results of 51 patients treated between 1992 and 1993. Among the 25 patients who received LDR brachytherapy none had relapsed. In this study, selection criteria were also restrictive to small tumors (<12 mm) without intraductal component (KING 2000).

Recently, CHEN et al. (2003) reported the late toxicity and cosmetic results after a median follow-up of 5.3 years in 199 patients treated with interstitial brachytherapy using either high or LDR. The majority of toxicities were grade 1 at all time intervals. Fat necrosis and telangiectasia increased with the passage of time, although fat necrosis remained an infrequent occurrence. The majority of fat necroses were asymptomatic and detected mammographically. At 5 years, 79 patients were available for toxicity and cosmetic evaluation. Fibrosis and fat necrosis grade ≥2 were observed in 4 and 11%, respectively. There was no toxicity grade 3 and good to excellent results were observed in 99% of the patients.

The other setting in which LDR interstitial brachytherapy may be indicated is as salvage treatment of patients treated conservatively for local recurrence (LR). The standard treatment of LR is total mastectomy and over the years there have been few attempts to modify this approach, although the occurrence of LR appears more and more as a major and detrimental prognostic factor for systemic relapse (COWEN 2000; VICINI 2003). For patients who experience LR after initial conservative radiosurgical procedure and who refuse salvage mastectomy, second conservative breast treatments using conservative surgery and PBI is an option which has been proposed by several teams (Table 31.2; MAULARD 1995; POLGAR 2002; RESCH 2002; GUIX 2003; HANNOUN-LÉVI 2004). After salvage lumpectomy followed by 45 Gy delivered through a LDR interstitial brachytherapy technique, we reported a 94.7% 5-year second local control rate associated with 11% of grade-3 to grade-4 late complications (HANNOUN-LÉVI 2004).

31.2.3
Partial Breast Irradiation Using High-Dose-Rate Interstitial Brachytherapy

The HDR brachytherapy presents some advantages over LDR brachytherapy such as total radioprotection for the medical staff, the possibility to use dose-optimization algorithms and no need of hospitalization during the treatment (KUSKE et al. 1994; PERERA 1997; VICINI 2001; WAZER 2002; POLGAR 2002); however, with HDR brachytherapy technique, the dose per fraction should be no more than 4 Gy to minimize the risk of fat necrosis especially when the tumor is located very close to the skin (WAZER 2001). In HDR brachytherapy, the implantation technique of vector material is hardly different from the technique used in LDR brachytherapy. The positioning of the plastic tubes is also performed during the surgical time, whereas the treatment itself may be delayed. Most authors have adopted a protocol using two fractions a day with doses per fraction ranging between 3.4 and 5.2 Gy (spaced by at least 6 h) for a total dose of 32–37.2 Gy delivered over 4–5 days (Table 31.3). The HDR fraction sizes are different leading to different biological equivalent doses. For instance, concerning late reactive tissues with an α/β ratio of 3 Gy, and assuming a dose per fraction (d/f) of 3.4 Gy, a total dose (TD) of 34 Gy would be equivalent to 43.5 Gy using conventional fractionation. Likewise, a d/f of 5.2 Gy and a TD of 36.4 Gy would be equivalent to 59.7 Gy. Equivalent doses for tumor control would also differ, depending on the assumed α/β ratio. Despite those different schedules, local control results were comparable ranging between 0 and 4.4% with a median follow-up still too short (18–57 months). Cosmetic results were good to excellent in more than two-thirds of the patients (KUSKE 1994; PERERA 1997; VICINI 2001; POLGAR 2002). WAZER et al. (2001) described a high rate of fat necrosis (58%) after a median follow-up of 24 months, depending on the dose rate. This complication was most likely caused by a combination of surgical and radiation parameters. Indeed, the risk of

Table 31.3. Results of partial breast irradiation with two fractions per day using high-dose rate brachytherapy. LR local recurrence

Reference	Number	Follow-up (months)	Dose/fraction (Gy)	Total dose (Gy)	LR (%)	Cosmetic results good/excellent (%)
PERERA et al. (1995)	39	20	3.72	37.20	2.6	–
KUSKE and BOLTON (1995)	26	20	4	32	0	67
VICINI et al. (1999)	45	18	4	32	0	100
WAZER et al. (2002)	32	33	3.4	34	4	88
POLGAR et al. (2002)	83	57	5.20	36.40	4.4	99

grade-3 or grade-4 toxicity was associated exclusively with the volume of the implant as reflected in the number of source dwell positions and volume of tissue exposed to fractional doses of 34 Gy (100%), 51 Gy (150%), and 68 Gy (200%). The authors postulated that complication risk may be more closely related to the volume of tissue encompassed within 150% isodose shell. This is suggested by the fact that the lower incidence of complications are observed when the median values are limited like in the William Beaumont (median 26.4 cm^3, range 15.5–38.8 cm^3) and in the Tufts/Brown (median 37 cm^3, range 14–144 cm^3) studies.

Recently, SHAH et al. (2003) reported toxicity and cosmetic outcome of 141 patients treated for early breast cancer with HDR brachytherapy delivering 32–34 Gy over 4 or 5 days to the lumpectomy cavity with a 2-cm margin. During treatment 15% of the patients experienced a grade-1 skin reaction. The most common toxicity was pockmarks (i.e., discoloration of the skin at the entry and exit sites of the brachytherapy catheters), which became less noticeable with each subsequent follow-up visit. At 12 months, 98% of the patients had an excellent or good cosmetic outcome.

31.3
Brachytherapy Using MammoSite Balloon Device for Partial Breast Irradiation

Although the principle of brachytherapy as a sole radiation modality may be adopted for selected patients, the optimal technique to deliver radiation has to be investigated more thoroughly. Interstitial brachytherapy seems to be less trivial to deliver (FENTIMAN 1996; VICINI 1999; KESTIN 2000) and is not available and therefore not adopted in many centers. To make brachytherapy more accessible, a new breast treatment device (MammoSite RTS; Proxima Therapeutics, Alpharetta, Ga.) has recently been developed (Fig. 31.5). This device delivers a high dose rate at the center of an inflated balloon that is placed intra-operatively (or days following surgery) into the lumpectomy cavity.

In this section we describe the surgical procedure for MammoSite insertion, dosimetric aspects, and results.

31.3.1
MammoSite Device Insertion Procedure After Conservative Surgery

The MammoSite device consists of a catheter with a silicone balloon and a shaft approximately 6 mm in

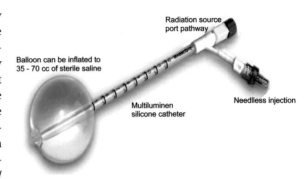

Fig. 31.5. MammoSite radiation therapy system

diameter and 15 mm in length. The shaft contains a small inflation channel and a second larger channel for treatment after passage of the HDR source. An injection port is attached to the inflation channel, and a luer fitting is attached to the inflation channel. An adapter is provided separately to connect with any brand of remote afterloading device (Fig. 31.5). In the initial phase-II studies demonstrating reliability and safety of the device, all patients had a CT scan before initiating treatment and also just before its removal at the end of treatment. Ultrasound or computed tomography were also used to determine the critical parameters of treatment, namely balloon diameter, symmetry, conformity, and its proximity to the chest wall and skin. Before and during treatment, standard orthogonal X-rays are useful for treatment planning and verification of balloon integrity.

The optimal technique and timing for the implantation have not been well defined. It is possible to insert the device peroperatively or utilizing percutaneous ultrasound-guided technique. The disadvantage of the intra-operative implantation is the high rates of disqualified patients (29%) for postoperative histology issues. In the postoperative ultrasound-guided percutaneous placement the disadvantage is related to the unfavorable balloon positioning requiring catheter removal. ZANNIS et al. (2003) reported among 23 patients implanted utilizing percutaneous ultrasound-guided technique only 2 (9%) of secondary device removal.

We recently reported the different steps of the surgical procedure of lumpectomy and intra-operative implantation of the MammoSite (BELKACÉMI 2003). The procedure is presented in Fig. 31.6.

After sentinel node detection and removal, the MammoSite applicator implantation is preceded by a standard lumpectomy. The wide excision creates a lumpectomy cavity around which clear margins

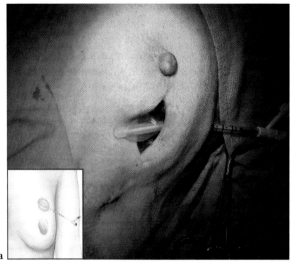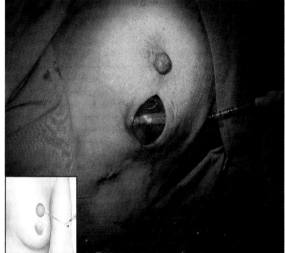

Fig. 31.6a,b. Surgical procedure and intra-operative MammoSite RTS balloon implantation

have to be >5 mm. The MammoSite is placed either through a separate stab wound or through the primary incision site. In the first case, a supplied trocar is used to create a pathway from the skin to the cavity. The uninflated balloon is advanced into the cavity through the trocar pathway. The MammoSite is then inflated with a saline solution (composed of 90% saline and 10% contrast product) and positioned in the tissue to receive RT. The balloon can contain 34–70 ml allowing a "sphere" diameter of 4–5.89 cm. The MammoSite is covered by at least 1 cm of skin thickness. The distance between balloon and skin surface is confirmed by the CT scan measurement and is of paramount importance for skin toxicity and cosmetic results with a minimal limit of 5 mm (DICKLER 2003). A CT scan is generally done at least 48 h after surgery. The transversal CT slices and multiplan reconstructed images allows:

- An evaluation of the conformity of the balloon to the cavity.
- A 3D measurements of the balloon diameters.
- Definition of the exact distance from balloon surface to skin surface and chest wall.
- A check of balloon symmetry and deformation by its contact with the chest wall. In this case a decrease of the injected volume can clear the deformation.

31.3.2
Dosimetric Aspects

The dosimetric aspects and procedure is described by EDMUNDSON et al. (2002). In summary, standard orthogonal radiographs are used for treatment planning. The X-ray markers are used to determine the distance of the balloon center from the remote afterloader. The first step of the procedure is to construct "a center finder" by scribing two lines in a Mylar sheet, intersecting at an approximately 60° angle. A third line bisecting this angle is also scribed. With this tool taped to a light box, an X-ray can be maneuvered over it, such that the two outer lines are tangent to the balloon at some point. The central line is therefore guaranteed to pass through the balloon center and a pencil mark is made on the film following this line. By rotating the film and repeating this procedure several marks are made, and their intersections represent the balloon center. By comparing this point with the marks on the dummy ribbon, the distance to the balloon center can be determined.

For 3D planning treatment, the device balloon is contoured in a 3D planning system. The planning target volume (PTV) is generally determined in the following fashion: an expansion of the balloon in 3D is performed, and the expansion is limited to breast tissue only (i.e., chest wall and skin surface are used as limiting structures). The balloon itself is then removed from this volume, so that volumes reported are limited to the tissue volumes only. Dose-volume histograms of this region are constructed and can be compared with those reported for interstitial treatments. In the MammoSite procedure one of the critical treatment planning parameters is the distance from the balloon surface to the skin, because this directly influences skin dose, and therefore acute and late toxicities, and also cosmetic results (EDMUNDSON 2002; CHEN 2003).

A minimum distance of 5 mm is set for the majority of the protocols in order to keep the skin dose at any point less than 150% of the prescribed dose, which is generally delivered at 1 cm around the balloon; however, the minimum skin distance is not always easily determined from the transverse CT slices alone and is impossible to determine from plane films. In the initial clinical experience with MammoSite, KEISCH et al. (2003) found that the skin–balloon surface distance and balloon-cavity conformance was the main factors limiting the initial use of the device. Ineligibility for poor balloon conformance (i.e., air- or fluid-filled gaps between balloon and cavity) and for limited skin–balloon surface was observed in 7 of 11 and 4 of 11 patients, respectively. The poor conformance may induce significant modification of the absolute tissue volumes receiving 100% (V340), 150% (V510), and 200% (V680) of the total dose of 34 Gy. There is a suggestion in a recent publication (WAZER 2001) that these values may be associated with development of toxicities such as fat necrosis.

The homogeneity index (DHI) as first described by WU et al. (1988) describes the fraction of the volume of the 100% isodose surface (i.e., treated volume) that receives less than 150% of the prescribed dose. According to the initial US experience reported by EDMUNDSON et al. (2002), the DHI with MammoSite is lower than interstitial brachytherapy, but the first device is more reproducible and produced better PTV coverage (EDMUNDSON et al. 2002). Moreover, the conformance of the balloon to the cavity wall is essential to assure appropriate dosimetry. Regarding the 9 of 34 (27%) patients requiring removal in the open setting, the reasons included mostly technical issues, such as conformance, as well as pathological issues such as nodal positivity.

Another particularity described with MammoSite is that the balloon is not perfectly symmetrical; thus, isodose distribution exhibits anisotropy because of self-absorption along the source axis. This anisotropy can be used to advantage by orienting the balloon normal to the skin when somewhat reduced margins between the balloon and skin are encountered.

Recently, DICKLER et al. (2003) suggested that the volume of normal breast tissue treated by MammoSite device is comparable to other methods of interstitial brachytherapy which treat 1–2 cm margin of tissue around the excision cavity. In their study of 21 patients, they compared two techniques using either a single-prescription point, single dwell position optimization, or a six-prescription point, multiple dwell position technique; in the latter, the six points were placed 1 cm from the balloon. Four of them were placed in a plan transverse to the balloon axis perpendicular to the catheter axis and two points along the axis of the catheter; the last were used to compensate for the decreased dose coverage due to anisotropy dose distribution of the source. Compared with the single-prescription-point optimization method, the six-prescription-point method provided better dose coverage. The mean V90 and V100 (percentage of volume receiving 90 and 100% of the prescribed dose) were 97.2 and 88.9% for six points vs 89.5 and 77.6% for one point; however, the six-point optimization method resulted in treatment that was less uniform.

31.3.3
Acute Toxicity

Acute toxicity depends on several parameters. For the skin, the main factor is the balloon–skin surface distance. In the initial US experience on 43 patients, the tumor was located in the upper outer and upper mid-breast region in 60% of all treated patients and the balloon–skin surface distance ranged between 5 and 6 mm in 19%, 7 and 9 mm in 33%, and >10 mm in 49% of patients. Patients experienced only mild-to-moderate side effects, including skin erythema (57%), catheter site drainage (52%), breast pain (43%), ecchymosis (31%), breast edema (15%), and dry desquamation (13%). Overall, that study demonstrated that the device was safe and well tolerated leading to U.S. Food and Drug Administration clearance on 6 May 2002 as an applicator suitable for delivery of radiation therapy to the surgical margins of lumpectomy cavity (KEISCH 2003b).

As compared with interstitial brachytherapy, fat necrosis is not so frequently observed with MammoSite. This may be related to association between clinically evident fat necrosis and volumes receiving 200, 150, and 100% of the prescribed dose (WAZER 2001). The MammoSite device never exceeds these dose-volume cut-offs as shown by EDMUNDSON et al. (2002)

31.3.4
Late Effects, Cosmetic Results, and Local Control

The follow-up is generally higher in the HDR interstitial brachytherapy series than in MammoSite studies. After HDR interstitial brachytherapy, the breast pain, edema, and infections diminish in frequency over the time. Breast fibrosis, hyperpigmentation, and hypopigmentation increase until the 2-year mark and then stabilize.

Fat necrosis and telangiectasia increases with the passage of time, without, however, any impact on cosmetic results. Good to excellent cosmetic result is reported in 95–99% of patients after HDR interstitial brachytherapy (CHEN 2003).

In the few days following brachytherapy using MammoSite device, the most common radiation effect is limited to mild or moderate erythema without desquamation. In addition, other less common, but significant, events which may be related to the device, such as moist desquamation, abscess, or seroma requiring drainage, are possible. In the EDMUNDSON et al. (2002) study, the relatively high rate of acute side effects decreased the good to excellent cosmetic results in the early evaluation done at 1 month. Using Harvard scale, only 88% of the patients had experienced good to excellent results. The authors did not, however, determine precisely the rates of excellent and good results separately, and also did not give the opportunity to the women themselves to evaluate their cosmetic outcome in addition to the evaluation of the physician.

To date, there is only one report on the cosmetic results of patients who received HDR MammoSite brachytherapy delivering 34 Gy in ten fractions with a median follow-up of 21 months (KEISCH 2003b). Among the 28 patients who were followed >1.5 years, 82% had good to excellent cosmetic results. Balloon-to-skin spacing >7 mm was associated with better cosmetic results (78% when the distance was 5–7 mm vs 93% for a distance >7 mm; p=0.045). Two patients experienced asymptomatic fat necrosis. Between 1 and 33 months of follow-up, none of the patients had adverse sequelae requiring surgical or local recurrence.

On the other hand, there is actually no study comparing MammoSite brachytherapy technique to the standard WBRT. In the future NSABP-RTOG phase-III randomized trial, all PBI brachytherapy techniques and 3D-conformal technique can be used to include 6300 patients. VICINI et al. (2003) reported recently a single institution of matched-pair analysis of patients with early-stage breast cancer treated postoperatively by brachytherapy or WBRT. To compare the rate of local recurrence in a comparable group of patients treated with WBRT, each of the 199 limited-field RT patients was matched with one WBRT patient at William Beaumont Hospital. Match criteria included tumor size, lymph node status, age, margins of excision, estrogen receptor status, and the use of tamoxifen therapy. The authors concluded that limited-field RT administered to the region of the tumor bed has 5-year local control rates comparable to those of WBRT in selected patients.

31.4
Conclusion

Local recurrences after conservative surgery and WBRT are most likely to occur in the immediate vicinity of the lumpectomy site. This fact has prompted the investigation of new approach of limited-field RT. Brachytherapy using either low or high dose rates delivering the total dose during a few days after surgery is advocated by several teams. While with interstitial brachytherapy the first results at 5 years are promising, the results with the MammoSite balloon device are still immature with a relatively short follow-up. The balloon catheter applicator has been developed in North America because of the theoretical disadvantages reported after the standard catheter-based interstitial brachytherapy. In the U.S. very few clinicians are familiar with the technique: many patients and health care find the placement, appearance, and the numerous puncture sites disturbing. If a simpler, safer, and quicker technique for the delivery of radiation could be offered to patients with early-stage breast cancer, such an approach could theoretically increase the breast-conserving therapy option to more women and improve their quality of life.

Accelerated PBI is logistically simpler and a more practical method for breast-conserving therapy, but it has to be demonstrated in randomized phase-III trials that it is at least equivalent to WBRT before its routine use.

References

Ballard-Barbash R, Potosky AL, Harlan LC et al. (1996) Factors associated with surgical and radiation therapy for early stage breast cancer in older women. J Natl Cancer Inst 88:716–726

Belkacémi Y, Chauvet MP, Giard S et al. (2003) Partial breast irradiation: high-dose rate peroperative brachytherapy technique using MammoSite. Cancer Radiother 7 (Suppl 1):129s–136s

Chen P, Vicini F, Kestin L et al. (2003) Long-term cosmetic results and toxicity with accelerated partial breast irradiation (APBI) utilizing interstitial brachytherapy. Int Radiat Oncol Biol Phys 57 (Suppl 2):1081, pp S309

Cionini L, Pacini P, Marzano S et al. (1993) Exclusive brachytherapy after conservative surgery in cancer of the breast. Proc Lyon Chir (Abstr), no. 89, pp 128

Clark RM, McCulloch PB, Levine MN et al. (1992) Randomised clinical trial to assess the effectiveness of breast irradiation following lumpectomy and axillary dissection for node negative breast cancer. J Natl Cancer Inst 84:683–689

Clarke DH, Vicini FA, Jacobs H et al. (1994) High dose rate brachytherapy for breast cancer. In: Nag S (ed) High dose

Verification and QA

32 3D Quality Assurance Systems

Bernhard Rhein and Peter Häring

32.1
Overview

With implementation of intensity-modulated radiotherapy (IMRT) conformation of high-dose to complex-shaped target volumes is reachable while having the potential to prevent absorbed dose below the tolerance limit in organs at risk. In 3D conformal radiotherapy (3D CRT) without intensity modulation monitor unit (MU) calculation computed by treatment planning software usually can be verified by hand calculation using basic dosimetry data such as percentage depth dose curve (PDD), tissue phantom ratio (TPR), total scatter factors ($S_{c,p}$) and the knowledge of the treatment plan beam parameters and geometry. In IMRT simple MU verification is not possible as a result of intensity modulation inside the beam; therefore, most IMRT sites perform a pre-treatment dosimetric verification procedure based on

B. Rhein, Dipl. Ing.
P. Häring, Dipl. Ing.
Department of Medical Physics in Radiotherapy, Deutsches Krebsforschungszentrum, Im Neuenheimer Feld 280, 69120 Heidelberg, Germany

either a single beam or a total plan, using ionisation chambers, TLDs, MOSFET detectors, radiochromic or radiographic films or electronic portal imaging devices (camera-based, amorphous silicon or liquid ionization chamber arrays) together with special designed IMRT verification phantoms. As the first step prior to the verification measurement, the patient's IMRT plan or single beams are transferred inside a phantom starting a dose recalculation to consider the phantom geometry. Meanwhile homemade and commercial software tools are available to correlate and evaluate computed and measured data sets more or less automatically.

The time investment in pre-treatment verification is important with regard to when IMRT should be implemented not only in research centres or university hospitals; thus, the real question is not, Do we need an IMRT verification, but instead, how to do it quickly and effectively and what are the necessary tools?

Dosimetric IMRT verification is only one part of the IMRT quality assurance (QA) process consisting of the commissioning and evaluation of the inverse planning software with the dose calculation algorithm, the data transfer to the verify and record system (V&R) and the IMRT delivery at the linear accelerator.

After getting experienced, several institutions divided IMRT QA into a patient-specific QA part and an MLC/linac-specific QA part. When going this way MU (patient-specific) verification can be done by single- or multiple-point dose measurements inside a phantom or by independent dose calculation methods using the Clarkson method, simplified kernel or Monte Carlo algorithms (Ma CM et al. 2000; Xing et al. 2000; Yang et al. 2003; Zhu et al. 2003); however, independent dose or MU calculation will not detect errors in the IMRT delivery process; therefore, independent MU calculations as well as the measurement of single dose points need to be part of a rigorous control of the dose delivery system such as quality assurance test of MLC leaf-positioning accuracy and reproducibility, leakage measurements, leaf speed control, MU to dose linearity, etc.

Herein an overview about IMRT QA, patient- and the linac-specific QA, as well as the dosimetric verification procedure in Heidelberg, is presented.

32.2
IMRT QA

The IMRT QA can be subdivided into three parts: firstly, the commissioning and validation of the dose calculation algorithm; secondly, the data transfer from the planning system to linac verify and record system; and thirdly, the delivery process itself.

32.2.1
Basic Dosimetry for Dose Algorithm Production

In step-and-shoot IMRT (segmented multi-leaf modulation, SMLM) very small field sizes down to $0.5{\times}0.5$ cm^2 are possible. This depends on the width of the leaves and the resolution of the fluence matrix. In small-field dosimetry the use of spatial high-resolution detectors is essential when total scatter factors (Fig. 32.1a) and profile steepness have to be measured correctly. For a $1{\times}1$-cm^2 field the total scatter factor measured with the 0.125-cm^3 chamber and the diamond detector deviates around 25% because of the chamber volume-averaging effect. Figure 32.1b illustrates dose profiles through a 6-MV photon 1-cm slit field measured with detectors (PTW) of different sensitive volumes (diamond detector/0.003 cm^3, a PinPoint chamber/0.015 cm^3, a semiflex ion chamber/0.125 cm^3 and a Farmer-type ion chamber/0.6 cm^3). The profiles were scanned with the detector's highest resolution

orientation. The measured 80–20% penumbras are 3.2, 3.9, 4.8 and 5.3 mm, respectively. LAUB and WONG (2003) reported discrepancies inside an IMRT plan of up to 10% between dose computation and dosimetric verification if dose profiles measured with a 0.125-cm^3 chamber were used for dose-kernel production.

Extensive recommendations for commissioning dosimetry and dose-computation algorithm validation are given in the report of the IMRT subcommittee of the AAPM radiation therapy committee published by EZZEL et al. (2003).

32.2.2
RTP File Transfer QA

The IMRT file transfer from the inverse treatment planning system via network to the linac's V&R system needs to be checked separately. For example, the RTP file coming from the inverse planning system should be imported in the V&R system at the linac. Afterwards, the imported RTP file should be exported again. The original RTP file from the inverse planning system and the exported RTP file from the V&R system must be identical.

32.2.3
IMRT Dosimetric Pre-treatment Verification

In general, dosimetric IMRT pre-treatment verification means the plan transfer of the patient's original IMRT plan inside a phantom (hybrid phantom), dose recalculation to consider the phantom geometry and measurement of the delivered dose distribution inside the phantom. This procedure verifies the whole IMRT QA

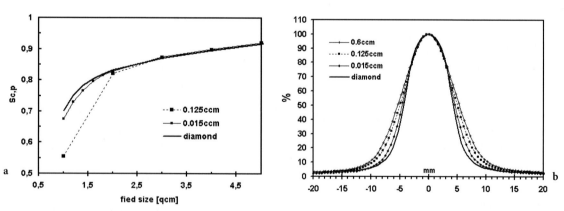

Fig. 32.1. a Total scatter factors $S_{c,p}$ of a 6-MV photon beam in 5 cm water measured with detectors of different sensitive volumes. **b** Dose profile of a 6-MV photon beam with a field size of $1{\times}5$ cm^2 at a depth of 5 cm in water measured with detectors of different sensitive volumes

chain, namely the dose algorithm, the RTP file transfer to the V&R system and the dose delivery at the linac.

The IMRT pre-treatment verification can be divided into two types: single beam verification and total plan verification. In single beam verification the IMRT plan is divided into the single beams transferred into and recalculated inside a solid water slab phantom at a certain depth. If electronic portal imaging devices (EPIDs) are used, the dose will be recalculated at the location of the EPID at a certain build-up depth. Single beam verification usually is applied in dynamic multi-leaf modulation (DMLM), whereas in SMLM many institutions prefer total plan verification because of the complex structure of the single beam fluence modulation. In SMLM the fluence is subdivided by a sequencer in a number of discrete fluence steps; therefore, single beam dosimetry inside a phantom is difficult because non-electronic equilibrium and high-dose gradient regions occur at many parts of the single beam dose distribution.

32.2.3.1
Single Beam Verification

Electronic Portal Imaging Devices

When EPID systems became available they were generally used for checking patient positioning before treatment by doing single or double exposures in order to replace radiographic films. Although image quality of these first-generation devices was poor, several institutions tried to adapt EPIDs for dosimetric measurements and for QA of the multi-leaf collimator (LIJUN et al. 1998). As the present EPIDs are much better concerning image resolution, contrast, image acquisition speed, image distortion and mechanical and dosimetric stability, they also found their way to IMRT verification. In principle, there are two possible ways to do measurements for treatment verification, both based on measuring the fluence or dose delivered by the treatment device. The first alternative is to measure fluence maps during a verification run without the patient. Several papers show that raw data can be processed to indicate either fluence or dose. Figure 32.2 shows a comparison between predicted and measured absolute dose with the SRI-100 camera-based EPID (Electa, Philips) for a 10-MV beam (PASMA et al. 1999). The system was calibrated against an ion chamber inside a polystyrene phantom at a source chamber distance of 160 cm.

Other devices, for example, a liquid-filled EPID and amorphous silicon flat-panel EPIDs, were investigated in single beam IMRT pre-treatment verification by VAN ESCH et al. 2001, PARTRIDGE et al. 2000,

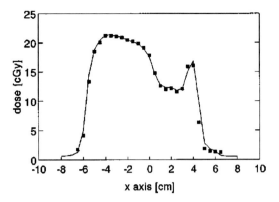

Fig. 32.2. Predicted (*lines*) and with electronic portal imaging device-measured (*squares*) absolute dose profiles for a 25-MV intensity-modulated beam for a prostate cancer patient. (From PASMA et al. 1999)

Vieira et al. 2003; WARKENTIN et al. 2003 and ZEIDAN et al. 2004.

As an alternative to these described EPIDs, specially designed verification detectors have been presented for single beam verification such as the [beam imaging system (BIS), Scanditronix Wellhöfer], a 27×27 air-filled ionisation chamber 2D array (PTW) or a 2D diode array (JURSINIC and NELMS 2003).

The second alternative is one of the actual developments in IMRT verification: the reconstruction of the delivered dose inside the patient from measured transmitted fluence through the patient. As above, the EPIDs are used to measure fluence respective exit dose but now with the patient in place. Since the patient absorbs most of the dose and therefore is a great source of scattered photons and electrons, the raw data form the EPID can show big differences compared with the fluence delivered by the treatment machine. Data have to undergo a scatter correction, and absorption from the patient has to be calculated. As these parameters, scatter and absorption, rely strongly on the patient and organ movement, this information can be acquired by reconstructing a megavoltage (MV) CT scan with the EPID before the IMRT is delivered to the patient (PARTRIDGE et al. 2002).

Film Dosimetry

Film dosimetry still is the most distributed dose integrating 2D dosimetry method. The radiographic film Kodak X-OMAT XV2 has been reported by MARTENS et al. (2002) to be a suitable detector to characterize single IM beams. Here field doses do not exceed 1 Gy resulting in a maximum optical density of 2.5 making a rescaling of the calculated monitor unnecessary for single beam verification.

In total beam verification the maximum dose per fraction can be up to 2.5 Gy, especially in IMRT plans with an integrated boost exceeding the dynamic range of this radiographic film; therefore, a new Kodak film with extended dose range EDR2 is available with an optical density of about 1.5–1.7 at an absorbed dose of 2 Gy.

The energy-dependent response of the EDR2 and the XV2 films remain within 3% as long as the field size does not increase above 15×15 cm^2 (DOGAN et al. 2002; ESTHAPPAN et al. 2002; OLCH 2002; ZHU XR et al. 2002).

With increasing field size the mean energy inside the field will decrease because of the higher fraction of scattered low-energy photons and therefore the film response will increase. The 6 MV (Siemens PRIMUS) photon spectrum in water at 5 cm depth was calculated using the MC code BEAMnrc at the field sizes 2×2, 10×10 and 20×20 cm^2. The main difference in photon fluence is detectable between 100 and 400 keV while the ratio of the mass energy absorption coefficients increases dramatically below 100 keV with a maximum at 60 keV (Fig. 32.3).

The field-size-dependent response of the EDR2 films was examined experimentally. Optical densities were evaluated after the films were exposed with 2 Gy at field sizes between 2×2 and 20×20 cm^2. In IMRT the field segments (SMLM) are in the range between 1×1 and 10×10 cm^2. In DMLM the width of the moving field aperture is of the same order. If film calibration conditions are defined near the measurement conditions, e.g. at 5×5cm^2 in a certain phantom depth, the energy- or field-size dependent EDR2 response remains within ±2% (Fig. 32.4). In comparison, the variation of the K$_Q$ values for PTW M31002 varies only 1 per mille between the field size of 2×2 and 20×20 cm^2. The K$_Q$ values were computed according to the IAEA dosimetry protocol TRS 398.

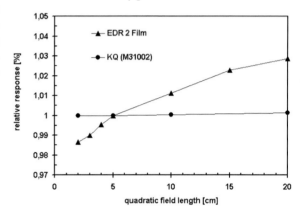

Fig. 32.4. Response of Kodak EDR 2 film as a function of field size (*triangles*). In comparison, the variation of the KQ values for a PTW M31002 0.125-cm^3 ion chamber is illustrated (*dots*).

BUCCIOLINI et al. (2004) and RHEIN et al. (2002) published film calibration methods minimizing the effect of field-size dependency in film dosimetry.

The above-mentioned accuracy in film dosimetry is only achievable if the film processing as well as film scanner is included in a continuous QA procedure.

Radiochromic films show nearly no energy dependency (MUENCH et al. 1991), but they have a low response and need up to 50 Gy to produce a reasonable optical density. It is advantageous that they need no film processing, especially in times where more clinics go filmless, but on the other hand, they are very expensive.

32.2.3.2
Total Beam Verification

In total beam verification the complete IMRT plan is transferred from the patient CT data set into a phantom (hybrid phantom) and the dose is recalculated to

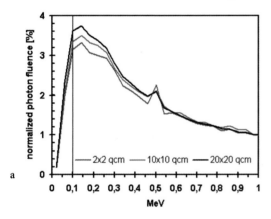

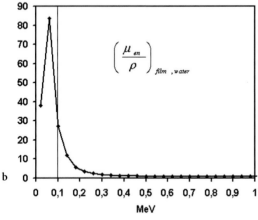

Fig. 32.3. a Siemens PRIMUS 6-MV photon fluence spectra calculated with the MC code BEAMnrc at the field sizes 2×2, 10×10 and 20×20 cm^2 at 5 cm depth in water. **b** Energy-dependent ratio of the mass energy absorption coefficients film to water

consider the phantom geometry. Especially in SMLM mode this procedure is advantageous compared with single beam verification because dosimetry can be done now inside the more or less homogeneous high-dose region of the total plan dose distribution. Since the original patient RTP file is used, this procedure verifies the whole IMRT QA chain, namely the dose algorithm, the file transfer to the V&R system and the dose delivery at the linac.

32.2.3.3
Verification Phantoms

In single beam verification mainly solid water slab phantoms are used together with films and/or ionisation chambers while in total IMRT plan verification increasingly more special phantoms have been developed. There are all kinds of shapes, close to a patient's outline, cylinders or just cubic phantoms, mainly made of solid water, in use. Different phantoms have the ability to be equipped with films, ionisation chambers, TLDs and even with bang gel and ferrous sulphate gel inserts (GUM et al. 2002; LOW et al. 1998a; MA CM et al. 2003; PALIWAL et al. 2000; RHEIN et al. 2002; RICHARDSON et al. 2003; TSAI et al. 1998). Many of them are commercially available from various dosimetry companies.

Two film phantoms and two ionisation chamber phantoms used for total IMRT plan (SMLM with a Siemens PRIMUS linac) verification at the German Cancer Research Centre (DKFZ) are demonstrated below.

Figure 32.5a shows a phantom built for verifications in the head and neck. It is a cylindrical solid water phantom (diameter 20 cm, height 20 cm) that is adaptable to the stereo-tactical system. For the body region a slightly modified solid water slab phantom ($30 \times 30 \times 20$ cm^3) is shown in Fig. 32.5b. Both phantoms can be used to perform film measurements as well as point measurements with a diamond detector or a small volume ion chamber along the central axis. With both phantoms it is possible to mark the coordinate system onto the films with a guided needle for data correlation and evaluation with IMRT software described later.

Fig. 32.5. a Solid water head and neck phantom for film or ion chamber dosimetry. **b** Solid water slab phantom for film or ion chamber dosimetry in the body region. **c** Solid water ion chamber matrix phantom for simultaneous measurements with up to 12 ion chambers. **d** Lucite cylinder phantom for dosimetry with the liquid ionisation chamber array LA48

With the same outline as the head and neck phantom, an ion chamber solid water matrix phantom (diameter 20 cm, height 20 cm) is shown in Fig. 32.5c. The phantom has boreholes in a 5×5 matrix, each hole separated by 2 cm. Up to five calibrated small-volume ionisation chambers (PTW M31002 / 0.125 cm^3 or M31014 / 0.015 cm^3) are placed simultaneously inside this phantom and connected via a multi-detector box to the PTW MULTIDOS electrometer. Chamber alignment in the z axis within 1-mm accuracy can be done easily with a special straight edge. Holes not used are closed with solid water plugs.

Another ion chamber phantom made of lucite (diameter 20 cm, height 20 cm) is shown in Fig. 32.6d. This phantom was also built for multiple point measurements with the PTW liquid ionization chamber array LA48. Calibrated to indicate the absorbed dose in water, this setup is able to simultaneously measure point doses every 8 mm along the central axis reducing verification effort compared with single point measurements. Both ion chamber phantoms are also adaptable to the stereo-tactical systems at DKFZ. These phantoms are commercialized by PTW but also other dosimetry companies offer similar IMRT verification phantoms.

There is an ongoing discussion on the question if more patient realistic anthropomorphic phantoms should be preferred instead of homogeneous solid water phantoms. Most of the treatment planning systems use pencil-beam kernel algorithms for dose calculation in combination with water-equivalent pathlength concept to consider tissue heterogeneities. They are not able to do accurate dose calculation at tissue interface regions of large density differences with non-electronic equilibrium as is done in the nasopharyngeal region, for example. It is noted that the problem of accurate dose calculation and measurement in regions without electronic equilibrium is a general problem itself and is not only related to IMRT.

32.2.3.4
Methods for Comparison of 2D Dose Distributions

As mentioned previously, DKFZ is equipped with PTW dosimetry hardware and software; therefore, the following figures are screenshots from their verification software. Similar software tools are also commercially available from other dosimetry vendors. Additionally, many institutions made own homemade IMRT verification software programs.

Unlike point dose measurements, where a geometric correlation and evaluation of single dose point is simple, 2D measuring devices as radiographic or radiochromic films or EPIDs overextend the user by the number of information acquired. Special IMRT verification software is necessary to read the calculated dose cube from different treatment planning systems and measured film or EPID dose distributions in order to correlate and to evaluate both data sets. Standard evaluation tools are the overlay of absolute or relative isodoses and profiles. The upper left side in Fig. 32.6 show one slice of the dose distribution of a paraspinal tumour treated with 6-MV photons in a five-beam SMLM configuration transferred and recalculated inside the described head and neck phantom. The lower left part shows the corresponding film dose distribution. On the upper right side absolute dose profiles (blue: calculated; green: film measurement) are illustrated. Superposed dose profiles and isodoses (Fig. 32.7) give a good impression about dose accuracy and spatial translation.

Another comparison method is to evaluate different matrices by pixel-by-pixel subtraction of two data sets (upper left part in Fig. 32.8). This method is less helpful if inhomogeneous dose distributions, such as those generated by IMRT, have to be examined. Setup errors of the phantom and the detector, uncertainties of the dose calculation algorithms and, of course, failures of the delivery system itself make it almost impossible to define and find tolerance criteria to decide whether a verification has failed or not. Several papers have been published on this problem. The gamma index method was proposed and evaluated by Low et al. (1998a, b) and Low and DEMPSY (2003) and is widely used in IMRT verification software tools. The index is calculated based on a dose difference and distance to agreement criteria using measured dose data as reference. With Δd as the acceptable distance to agreement, e.g. 3 mm, ΔD as dose difference criteria, e.g. 3%, $r_{(\vec{r}e,\vec{r}r)}$ as distance and $\delta_{(\vec{r}e,\vec{r}r)}$ as dose difference of the reference $\vec{r}r$ to examination point $\vec{r}e$ the total gamma value set $\Gamma_{(\vec{r}e,\vec{r}r)}$ is:

$$\Gamma_{(\vec{r}e,\vec{r}r)} = \sqrt{\frac{r^2_{(\vec{r}e,\vec{r}r)}}{\Delta d^2} + \frac{\delta^2_{(\vec{r}e,\vec{r}r)}}{\Delta D^2}}$$

This equation describes an ellipsoid. The minimum of the gamma value set is the gamma value of the point under examination.

$$\gamma_{(\vec{r}r)} = \min\left\{\Gamma_{(\vec{r}e,\vec{r}r)}\right\} \forall \{\vec{r}e\}$$

The minimum of the gamma matrix resulting in the calculated and measured data is 1 if the point under evaluation has been found on the surface of the ellipsoid, or <1 if the point is inside the ellipsoid. If $\gamma \leq 1$, the criteria for plan acceptance are fulfilled.

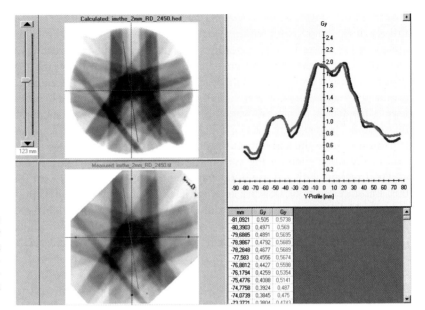

Fig. 32.6. *Upper left:* computed dose distribution of a paraspinal tumour transferred into the head and neck phantom. *Lower left:* corresponding irradiated film. *Upper right:* superposed dose profiles (*green:* film measurement; *blue:* computed)

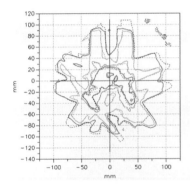

Fig. 32.7. Superposed relative 30, 50, 80 and 95% isodose lines. The *dashed lines* correspond to the film measurement normalized to 1.92 Gy, and the *solid lines* correspond to the dose computation normalized to 1.94 Gy

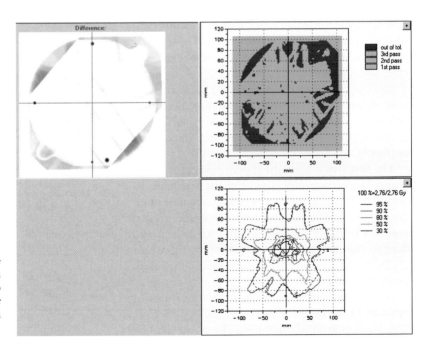

Fig. 32.8. *Upper left:* difference matrix; *upper right:* γ distribution (*red:* γ>1, *green:* γ≤1) according to Huskens et al. (2002). *Lower right:* same isodose superposition as in Fig. 32.7

Displayed in a colour scale, green $\gamma \leq 1$ and red $\gamma > 1$, this result can help to find regions where deviations are above the limits. In most cases in which regions with $\gamma > 1$ are present, the superposed isodose lines, dose profiles and even the absolute dose comparison are in good agreement. Moreover, an experienced user is needed to do a final evaluation on basis of this method (upper right part in Fig. 32.8). Here it is helpful to have the superposed isodose information at the same time on the screen to be able to assess if regions with $\gamma > 1$ are in relevant regions or not. Enhancements of Low's γ concept have been published by Bakai et al. (2003) and Depuydt et al. (2002).

32.3
IMRT Pre-treatment Verification at DKFZ

The DKFZ is equipped with two 6-MV Siemens PRIMUS linear accelerators performing step-and-shoot IMRT. Actually, only 20% of dosimetric IMRT verifications are accomplished with the film dosimetry method, basically at paraspinal tumours where the verification of the spatial dose distribution with respect to the spinal cord is more important than the accuracy of the absolute delivered dose. Again with well-defined film calibration conditions with respect to IMRT delivery the additional error in absolute dose measurement with EDR2 films is within ±2% compared with ion chamber dosimetry. The whole film verification process requires about 1.5 h per patient including phantom setup, film processing, densitometry, and data evaluation.

At all other tumour sites IMRT QA has been separated into a patient specific verification part using up to five ion chambers (PTW M310002 0.125 cm^3) simultaneously inside the matrix phantom and a MLC or linac-based QA part. The ion chamber interference among each other was investigated and found to be negligible.

The first step is the patient plan transfer with the original segment monitor units into the matrix phantom and recalculating the resulting absolute dose distribution inside the phantom. Since the original patient RTP file is imported in the V&R system and used to deliver the dose at the phantom, the whole chain, dose algorithm, RTP file transfer and the dose delivery at the linac are verified.

We use the matrix phantom for all locations of tumours in the head and neck and the body region. The decision as to where the ion chambers shall be located inside the phantom can easily be done with the PTW VERISOFT tool shown in Fig. 32.9 exemplified on a prostate case in the isocentre plane. The blue circles show the 25 possible ion chamber locations inside the phantom. In this case the yellow numbered positions 1 to 5 indicate the chosen ion chamber positions. When moving the mouse pointer onto a circle the mean dose inside this circle is indicated. The size of the circles can be adjusted in an initialisation file corresponding to the size of the used ion chambers. In Table 32.1 for the ion chamber positions 1 to 5 the computed and the measured dose as well as the deviations are listed.

We try to do IMRT verifications only once per week in one verification cycle to reduce effort for the phantom setup at the linac. With this philosophy around five IMRT dosimetric verifications can be done within 1 h without any further data processing.

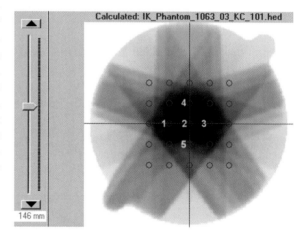

Fig. 32.9. Computed dose distribution of a five-beam, 6-MV prostate IMRT plan. *Blue circles:* Superposition of the 5×5 possible ion chamber positions inside the matrix phantom. *Numbers1–5:* chosen chamber positions for verification

Table 32.1. Comparison of measured dose with ion chambers vs dose computation of the prostate case given in Fig. 32.9

Position	Computed dose (Gy)	Measured dose (Gy)	Deviation (%)
1	2.450	2.437	+0.5
2	2.687	2.652	+1.3
3	2.509	2.487	+0.88
4	2.339	2.345	−0.26
5	2.018	1.983	+1.8

32.4
Linac-Based QA

Since patient-specific QA has been reduced to measure only a few dose points inside the high-dose region, MLC and linac QA has to be performed more carefully. The same is valid if independent MU calculation has been performed.

The major linear accelerator vendors all have MLC integrated into the linac head showing different design and leaf resolution. Double-focussed moving on a circle, or linear-moving single-focussed with rounded leaf faces, can be found; these, as well as several accessory mini-MLCs, are capable of delivering IMRT in dynamic DMLM or static SMLM mode. Depending on the IMRT delivery mode and the MLC design, some QA procedures are more important than others. A good overview is given by EZZEL et al. (2003).

The main focus in the following section is on the accuracy of the field size, the leaf position, leaf speed, leakage, and rotational accuracy of the gantry, the couch, the collimator and the alignment of the room lasers.

32.3.1
MLC Field-Size and Leaf-Speed Tests

The first question we have to answer is: How accurate do we need to be? If we set a minimum dosimetric error resulting from leaf-positioning uncertainties of 5%, which we want to achieve in any situation, we can easily develop the maximum positioning error by a simple measurement; therefore, the total scatter

factor function for a 6-MV photon beam (Siemens PRIMUS) was measured with a diamond detector in 5 cm water depth at an SSD of 95 cm as shown in Fig. 32.10a. The in-plane jaws were set to be constant at 5-cm field length to prevent uncertainties coming from the jaws.

The dashed curves in Fig. 11b indicate the ±5% dosimetric uncertainty. Hereby the left part of the error bar of the abscissa can be defined as the required field size plus or minus tolerance. For a 1-cm MLC slit field the field size tolerance is ±1 mm and ±0.2 mm for a 0.5-cm slit field (see Fig. 32.10b).

In DMLM the size of a moving aperture has an important influence on the absolute dose. In the worst case scenario a broad field is irradiated using a small aperture. In this scenario small errors in the size of the moving field or a wrong leaf speed can cause a large dose error. In DMLM therefore special tests where the leaves are moving in high speed, in low speed and with different speed are essential.

32.4.2
MLC Leaf-Position Accuracy Tests

Directly correlated with the previously described tests are special leaf-position tests. A widespread test is the so-called garden fence test. In DMLM a film is exposed in step-and-shoot mode by moving a 1- or 2-mm MLC strip, e.g. in 2-cm intervals over the film. In SMLM the width of the strips should be equal to or nearby the minimal permitted MLC field size (Fig. 32.11a, b). Thereby not only the leaf positions but also the dosimetric field sizes for all

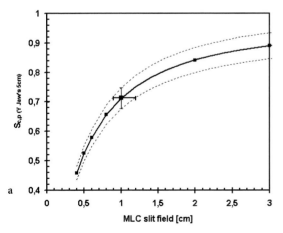
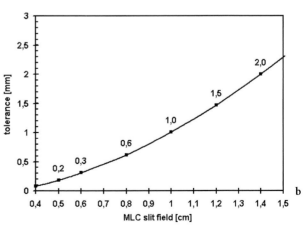

Fig. 32.10. a *Solid curve:* Total scatter function (6-MV photons) measured with a diamond detector for variable MLC field sizes. *Dashed curves:* ±5% uncertainties. **b** Mechanical MLC field-size accuracy to fulfil the requirement of ±5% dosimetric accuracy

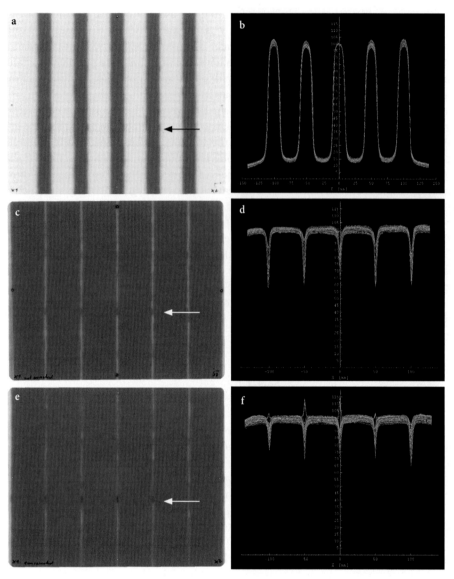

Fig. 32.11. Leaf position quality assurance tests for a double-focussed MLC. **a** Garden fence test with 5×2-cm MLC stripes. **c** Film irradiated with 5-cm abutting MLC strips at the same position. **e** Same situation as in **c** but with a field overlapping of 1 mm of two adjacent field components. **b, d, f** Scanned dose profiles. The *arrows* in **a, c** and **e** indicate a leaf out of tolerance

leaves can be estimated in relation to the room lasers marked by the pinholes on the film. Another leaf-position test is shown in Fig. 32.11c, d where a film was irradiated with 5-cm abutting MLC strips at the same position with a double-focussed MLC. The film shows under-dose lines around 40% of the maximum field dose which are caused by the effect of incomplete compensation for the penumbra caused by the finite size of radiation source, scattering effects and incomplete mechanical focussing of the leaves. Since this under dosage is also strongly effected by the leaf positions, this test is highly sensitive for position inaccuracies as indicated by the red arrows in Figure 32.11a, c and e, where one leaf of the right leaf bank is out of tolerance.

The effect of under-dosed match lines cannot be solved by MLC calibration because, as shown in Fig. 32.12 for example, for large head and neck cases single segments can have leaves opened at both sides while the leaves in between remain closed. The leaves have to be calibrated in a way that these situations produce no increased leakage. The film on the right side was exposed with leaves closed every 2 cm with 100 MU at 6-MV photons.

These under-dose lines look unappealing but may not be too serious. It is a controversial question as to whether or not this under-dose is of clinical importance. The absolute value of the under dose is up to 40% of a single intensity level. This is 8% of the maximum intensity in cases of a five-step intensity map and as a rough estimation 1% of a beam arrangement

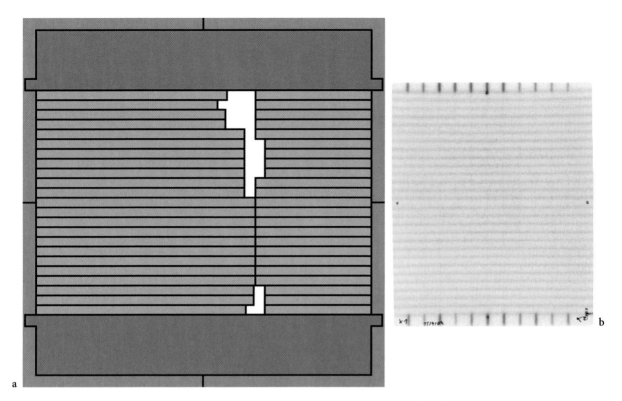

Fig. 32.12. a Typical segment for a head and neck case treated in segmented multi-leaf modulation mode. **b** Leakage test for closed leaves for a double-focussed MLC. The film was irradiated with closed leaves every 2 cm. The outer-leaf pairs remained open at 4 mm to define the in-plane jaws to 30 cm.

of seven beams. Furthermore, there are smoothing effects such as gravity, gantry sag and patient positioning uncertainties from fraction to fraction. It is possible to reduce the under-dose line effect by means of a field overlapping of two adjacent field components of about 1 mm as demonstrated in Fig. 32.11e, f.

32.4.3
Linac Isocentre and Room Laser Adjustment

Regardless of whether SMLM or DMLM mode is employed to form intensity-modulated beams, both are precise methods to deliver complex and highly conformal dose distributions; therefore, all other equipment, such as room lasers, gantry, couch and collimator movements, should fulfil the same quality standards as known from radiosurgical treatment techniques. Requirements should be defined according to the single isocentric movements of the gantry, the collimator and the treatment as depicted below. The diameter of the inner circle of the three stars (gantry, collimator and couch) should not be much bigger than 1 mm (Fig. 32.13).

The same care should be taken for the adjustment of the room lasers. TREUER et al. (2000) pub-

lished a method irradiating a small film box in three different orientations with a multiple convergent arc technique with a small circular tungsten collimator. Figure 32.14a illustrates such a film box mounted on the treatment couch in vertical orientation and a tungsten collimator with a 1-mm cone. The nine non-coplanar arcs (140°) technique used in Heidelberg for radiosurgical treatments is shown in Fig. 32.14b. This technique produces a sharp circular dose distribution on the films inside the cassette. The distances between the pinholes and the centre of the distribution determined with film densitometry indicate the deviation from the ideal laser position. The corresponding irradiated films in vertical and horizontal orientation together with the densitometric evaluation are shown in Fig. 32.15. The measured deviation between the centre of the optical density and the pinhole indicating the room laser were determined as δx: 0.1 mm, δy: 0.5 mm and δz: 0.2 mm resulting in the total spatial deviation of

$$\vec{r} = \sqrt{\delta x^2 + \delta y^2 + \delta z^2} = 0.55\ mm$$

The tolerance limit should be ≤1 mm.

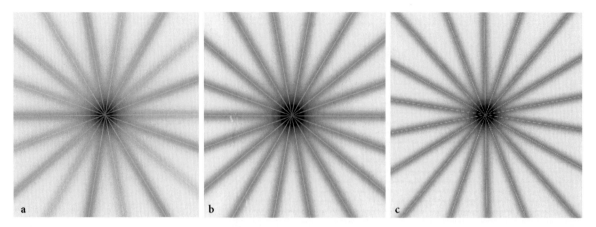

Fig. 32.13. Isocentric star shots of **a** the gantry, **b** the treatment couch and **c** the collimator. The films were exposed with field width of 4 mm.

Fig. 32.14. a Film box used for room laser quality assurance mounted in vertical orientation at the treatment couch. **b** Nine non-coplanar arc technique for radiosurgical treatments with circular tungsten collimators

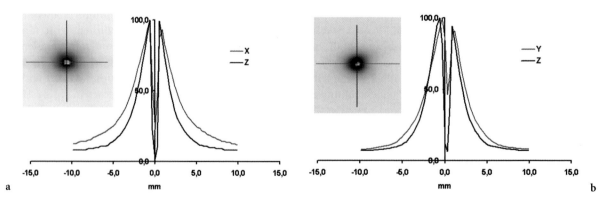

Fig. 32.15a,b. Quality assurance of room laser adjustment. **a** Extracted dose profiles through the pinhole along the stereotactic x,z coordinates. The film box was mounted in horizontal orientation. **b** Extracted dose profiles through the pinhole along the stereotactic y,z coordinates. The filmbox was mounted in vertical orientation.

References

Bakai A, Alber M, Nüsslin F (2003) A revision of the γ-evaluation concept for the comparison of dose distributions. Phys Med Biol 48:3543–3553

Bucciolini M, Buonamici FB, Casati M (2004) Verification of IMRT fields by film dosimetry. Med Phys 31:161–168

Depuydt T, Van Esch A, Huyskens DP (2002) A quantitative evaluation of IMRT dose distributions: refinement and clinical assessment of the gamma evaluation. Radiother Oncol 62:309–319

Dogan N, Leybovich LB, Sehti A (2002) Comparative evaluation of Kodak EDR2 and XV2 films for verification of intensity modulated radiation therapy. Phys Med Biol 47:4121–4130

Esthappan J, Mustic S, Harms WB, Dempsey JF, Low DA (2002) Dosimetry of therapeutic photon beams using an extended dose range film. Med Phys 29:2438–2445

Ezzel GA, Galvin JM, Low DA, Palta JR, Rosen I, Sharpe MB, Xia P, Xiao Y, Xing L, Yu C (2003) Guidance document on delivery, treatment planning, and clinical implementation of IMRT: report of the IMRT subcommittee of the AAPM radiation therapy committee. Med Phys 30:2089–2115

Gum F, Scherer J, Bogner L, Rhein B, Bock M (2002) Verification of IMRT treatment plans using an inhomogeneous anthropomorphic Fricke gel phantom and 3D magnetic resonance dosimetry. Phys Med Biol 47:N67–N77

Jursinic PA, Nelms BE (2003) A 2-D diode array and analysis software for verification of intensity modulated radiation therapy delivery. Med Phys 30:870–879

Laub WU, Wong T (2003) The volume effect of detectors in the dosimetry of small fields used in IMRT. Med Phys 30:341–347

Lijun M, Boyer AL, Findley DO, Geis PB, Mok E (1998) Application of a video-optical beam imaging system for quality assurance of medical accelerators. Phys Med Biol 43:3649–3659

Low DA, Dempsy JF (2003) Evaluation of the gamma dose distribution comparison method. Med Phys 30:2455–2464

Low DA, Gerber RL, Mutic S, Purdy JA (1998a) Phantoms for IMRT dose distribution measurement and treatment verification. Int J Radiat Oncol Biol Phys 40:1231–1235

Low DA, Harms WB, Mutic S, Purdy JA (1998b) A technique for the quantitative evaluation of dose distributions. Med Phys 25:656–661

Ma CM, Pawlicki T, Jiang SB, Li JS, Deng J, Mok E, Kapur A, Xing L, Ma L, Boyer AL (2000) Monte Carlo verification of IMRT dose distributions from a commercial treatment planning optimization system. Phys Med Biol 45:2483–2495

Ma CM, Jiang, SB, Pawlicki T, Chen Y, Li JS, Deng J, Boyer AL (2003) A quality assurance phantom for IMRT dose verification. Phys Med Biol 48:561–572

Martens C, De Wagter C, De Neve W (2002) The value of radiographic film for the characterization of intensity-modulated beams. Phys Med Biol 47:2221–2234

Muench PJ, Meigooni AS, Nath R, McLaughlin WL (1991) Photon energy dependence of the sensitivity of radiochromic film and comparison with silver halide film and LIF TLDs used for brachtherapy dosimetry. Med Phys 4:769–775

Olch AJ (2002) Dosimetric performance of an enhanced dose range radiographic film for intensity-modulated radiation therapy quality assurance. Med Phys 29 2159–2168

Paliwal B, Tomé W, Richardson S, Mackie TR (2000) A spiral phantom for IMRT and tomotherapy treatment delivery verification. Med Phys 27:2503–2507

Partridge M, Saymond-Taylor JRN, Evans PM (2000) IMRT verification with a camera based electronic portal imaging system. Phys Med Biol 45:183–196

Partridge M, Ebert M, Hesse BM (2002) IMRT verification by three-dimensional dose reconstruction from portal beam measurements. Med Phys 29:1847–1858

Pasma KL, Dirkx MLP, Kroonvijk M, Visser AG, Heijmen BJM (1999) Dosimetric verification of intensity modulated beams produced with dynamic multileaf collimation using an electronic portal imaging device. Med Phys 26:2373–2378

Rhein B, Häring P, Debus J, Schlegel W (2002) Dosimetrische Verfikation von IMRT- Gesamtplänen am Deutschen Krebsforschungszentrum Heidelberg. Z Med Phys 12:122–132

Richardson SL, Tomé WA, Orton NP, McNutt TR, Paliwal BR (2003) IMRT delivery verification using a spiral phantom. Med Phys 30:2553–2558

Treuer H, Hoevels M, Luyken K, Gierich A, Kocher M, Müller R-P, Sturm V (2000) On isocentre adjustment and quality control in linear accelerator based radiosurgery with circular collimators and room lasers. Phys Med Biol:45:2331–2342

Tsai JS, Wazer DE, Ling MN, Wu JK, Fargundes M, DiPetrillo T, Kramer B, Koistinen M, Engler MJ (1998) Dosimetric verification of the dynamic intensity modulated radiation of 92 patients. Int J Radiat Oncol Biol Phys 40:1213–1230

Van Esch A, Vanstraelen B, Verstraete J, Kutcher G, Huyskens D (2001) Pre-treatment dosimetric verification by means of a liquid filled electronic portal imaging device during dynamic delivery of intensity modulated treatment fields. Radiother Oncol 60:181–190

Vieira SC, Dirkx MLP, Pasma KL, Heijmen BJM (2003) Dosimetric verification of X-ray fields with steep dose gradients using an electronic portal imaging device. Phys Med Biol 48:157–166

Warkentin B, Steciv S, Rathee S, Fallone BG (2003) Dosimetric IMRT verification with a flat-panel EPID. Med Phys 30:3143–3155

Xing L, Chen J, Luxton G, Li GJ, Boyer AL (2000) Monitor unit calculation for an intensity modulated photon field by a simple scatter summation algorithm. Phys Med Biol 45:1–7

Yang Y, Xing L, Li JG, Palta J, Chen Y, Luxton G, Boyer A (2003) Independent dosimetric calculation with inclusion of head scatter and MLC transmission for IMRT. Med Phys 30:2937–2947

Zeidan OA, Li JG, Ranade M, Stell AM, Dempsey JF (2004) Verification of step and shoot IMRT delivery using a fast video-based electronical portal imaging device. Med Phys 31:463–476

Zhu J, Yin FF, Kim JH (2003) Point dose verification for intensity modulated radiosurgery using Clarkson's method. Med Phys 30:2218–2221

Zhu XR, Jursinic PA, Grimm DF, Lopez F, Rownd JJ, Gillin MT(2002) Evaluation of Kodak EDR2 film for dose verification of intensity modulated radiation therapy delivered by a static multileaf collimator. Med Phys 29:1687–1692

33 Quality Management in Radiotherapy

Guenther H. Hartmann

CONTENTS

G. H. HARTMANN, PhD
Department of Medical Physics in Radiotherapy, Deutsches Krebsforschungszentrum, Im Neuenheimer Feld 280, 69120 Heidelberg, Germany

33.1
Introduction

It is anticipated that radiotherapy will have an increasing role in treating cancer patients; however, all recent developments in radiotherapy technology and the improved performance of modern equipment cannot be fully exploited unless a high degree of accuracy and reliability in dose delivery is reached. To meet this requirement, sustained efforts have to be applied to all areas of the radiotherapy process.

It is a characteristic feature of modern radiotherapy that this process is a multi-disciplinary process involving complex equipment and procedures for the delivery of treatment; therefore, it is extremely important that (a) the radiation therapist cooperate with specialists in the various disciplines in a close and effective manner, and (b) the various procedures (related to the patient and that related to the technical aspects of radiotherapy) be subjected to careful quality control. The establishment and use of a comprehensive quality system (QS) is an adequate measure to meet all these requirements.

A number of organizations, such as WHO, AAPM, ESTRO, and IPEM, and also other publications have given background discussion and recommendations on the structure and management of a QS in radiotherapy (AMERICAN ASSOCIATION of PHYSICS in MEDICINE 1994a; MAYLES et al. 1999; LEER et al. 1995, 1998; McKENZIE et al. 2000; THWAITES et al. 1995; VAN DYK et al. 1993; WORLD HEALTH ORGANIZATION 1988). Historically, actions summarized under the term "quality assurance" (QA) has long been carried out in many areas of radiotherapy, particularly in the more readily defined physical and technical aspects of equipment, dosimetry, and treatment delivery. Presently, however, it is generally appreciated that the concept of quality in radiotherapy is broader than a restricted definition of technical maintenance and quality control of equipment and treatment delivery, and instead that it should encompass a comprehensive approach to all activities in the radiotherapy department; thus, it has clinical, physical, and administrative components. Herein the term quality management system (QMS) is used to denote this comprehensive approach including organizational aspects. It is favored because it is recognized that partial organization of only some of the key steps in the radiotherapy process is not sufficient to guarantee to patients that each individual will receive the best available care for their disease (Leer et al. 1998). It is also an most important characteristic of a good QMS that it provide its various instructions in a formal and written scheme. Only this characteristic can ensure that all important aspects of quality are defined, documented, understood, and put into practice.

The next sections start with a discussion on why a radiotherapy center should introduce a quality system. It provides departmental heads with arguments to convince themselves of the need to initiate a quality project in their own department. A section follows that provides a basic framework for a comprehensive and formal quality management program. The next section deals with a quality system for external radiotherapy equipment, since in particular the functional performance of equipment ultimately influences the geometrical and dosimetric accuracy of the applied dose to the patient. Quality issues related to brachytherapy are not explicitly addressed since in general QA practices are less rigorously defined than in external beam therapy. The final section summarizes various clinical aspects. They refer to those areas which link together the work of radiation oncologists, radiation therapists, and medical physicists.

33.2
Objectives of a QMS in Radiotherapy

33.2.1
Maintaining the Required Accuracy

An assessment of clinical requirements in radiotherapy indicates that a high accuracy is necessary to arrive at a control rate as high as possible, and at the same time maintaining complication rates within acceptable levels. Requirements of accuracy and precision as applied in a radiotherapy context can be found in various publications, as well as discussions of dosimetric and geometric uncertainty requirements, e.g., Dutreix (1999), Mijnheer et al. (1987), Dobbs and Thwaites (1999), and Van Dyk and Purdy (1999).

The clinical requirements for accuracy are based on evidence from dose-response curves for tumor

control probability (TCP) and normal tissue complication probability (NTCP), as shown in Fig. 33.1; both of these need careful consideration in designing radiotherapy treatments for good clinical outcome.

The steepness of a given TCP or NTCP curve against dose defines the change in response expected for a given change in delivered dose; thus, uncertainties in delivered dose translate into either reductions in TCP or increases in NTCP, both of which worsen the clinical outcome. The accuracy requirements are defined by the most critical curves, i.e., very steeply responding tumors and normal tissues.

From a consideration of the available evidence on clinical data, various recommendations have been made about required accuracy in radiotherapy:

- The International Commission on Radiological Units and Measurements report 24 (1976) reviewed TCP data and concluded that an uncertainty of 5% is required in the delivery of absorbed dose to the target volume. This has been widely quoted as a standard; however, it was not stated explicitly what confidence level this represented. It is generally interpreted as 1.5 or 2 times the standard deviation (SD) and this assumption has been broadly supported by more recent assessments. For example, Mijnheer et al. (1987), considering NTCP, and Brahme et al. (1988), considering the effect of dose variations on TCP, recommend a standard uncertainty of 3–3.5%. In general, the

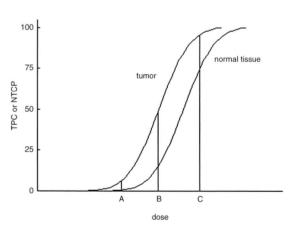

Fig. 33.1 Tumor control probability (*TCP*) and the probability of normal tissue complication (*NTCP*) as a function of radiation dose, in a hypothetical case. If normal tissue injury is to be avoided altogether, the radiation dose cannot exceed *A*; the *TCP* is low. By accepting a certain probability of normal tissue injury, the radiation dose can be increased to *B*, and the *TCP* is significantly improved; however, the question always is: Which level of *NTCP* is acceptable? A further increase in the radiation dose above *C* results in an increased complication rate with very little improvement in TCP

smallest of these numbers might be applicable to the simplest situations, with the minimum number of parameters involved, whereas the larger figure (3.5%) is more realistic for practical clinical radiotherapy when more complex treatment situations and patient factors are considered.

- Geometric uncertainty relative to target volumes or organs at risk also lead to dose problems, either underdosing of the required volume (decreasing the TCP) or overdosing of nearby structures (increasing the NTCP). Consideration of these effects has led to recommendations on spatial uncertainty of between 5 and 10 mm (at the 95% confidence level). The figure of 5 mm is generally applied to the overall equipment-related mechanical/geometric problems, whereas larger figures (typically 8 or 10 mm) are used to indicate overall spatial accuracy including representative contributions for problems related to the patient and to clinical setup. The latter factors obviously depend on the site involved, the method of immobilization, and the treatment techniques employed.

Thus, the recommended accuracy on dose delivery is generally 5–7% on the 95% confidence level, depending on the factors intended to be included. On spatial accuracy, figures of 5–10 mm (95% confidence level) are usually given, depending on the factors intended to be included. These are general requirements for routine clinical practice. In some specific applications, such as stereotactic radiotherapy, IMRT, or radiotherapy with ion beams, better accuracy might be demanded which consequently has an impact on increased QA efforts.

These recommendations are for the end point of the radiotherapy process, i.e., for the entire treatment as delivered to the patient; therefore, on each of the steps that contribute to the final accuracy correspondingly smaller values are required, such that when all are combined the overall accuracy is met. The aim of a QMS program is to maintain each step within an acceptable tolerance; therefore, if a more complex treatment technique, more stages, sub-stages, or other parameters and factors are involved, a correspondingly more complex QMS is required.

33.2.2
Minimizing the Risk of Errors and Accidents

Human error will always occur in any organization and any activity. In radiotherapy, there is the potential failure to control the initial disease which, when it is malignant, is eventually lethal to the patient or there is the risk to normal tissue from increased exposure to radiation; thus, in radiotherapy an accident or a misadministration is significant, if it results in either an underdose or an overdose. It is another important objective of a QMS to minimize the number of occurrences of "errors" and to identify them at the earliest possible opportunity, thereby minimizing their consequences.

When is a difference between prescribed and delivered dose considered to be at the level of an accident or a misadministration in external beam radiotherapy?

- From the general aim for an accuracy approaching 5% (95% confidence level), about twice this seems to be an accepted limit for the definition of an accidental exposure, i.e., a 10% difference.
- For example, in several jurisdictions, levels are set for reporting to regulatory authorities, if equipment malfunctions are discovered which would lead to a 10% difference in a whole treatment or 20% in a single fraction.
- In addition, from clinical observations of outcome and of normal tissue reactions, there is good evidence that differences of 10% in dose are detectable in normal clinical practice. Additional dose applied incidentally outside the proposed target volume may lead to increased complications.

Review of Accidents
The International Atomic Energy Agency (2000a) has recently analyzed a series of accidental exposures in radiotherapy to draw lessons in methods for prevention of such occurrences. A classification of causes for accidents in external beam therapy, as shown in Tables 33.1 and 33.2, was given. For more details the reader is referred to the original IAEA publication.

Problems with Equipment
Table 33.1 refers to accidents related to problems with equipment, such as calibration of the beam output. Such events involve all patients treated with the beam until the problem is discovered. The most important events were those resulting in an error in the determination of dose rate, and therefore wrong irradiation times for patients treated under these conditions. In the three worst cases, 115, 207, and 426 patients were involved, with dose deviations of up to 60% and many deaths. In addition, there were two major accidents related to maintenance of accelerators, one of them involving 27 patients (several of whom died as a direct result of radiation exposure).

Table 33.1 Events related to problems with equipment

Equipment	Step at which the event occurred	No. of events
Radiation measurement system (including ionization chambers and electrometer)	Calibration of reference system	1
	Intercomparison with secondary system	1
	Routine use	3
Treatment machine	Commissioning (acceptance)	6
	Calibration (annual)	3
	Constancy check (daily, weekly)	1
	Malfunction of machine	7
	Incorrect use	6
Simulator	Malfunction	3
Treatment planning system	Commissioning and input of basic data	5
	Routine use	6

Table 33.2 Events involving individual patients

Process	Step at which the event occurred	No. of events
Prescription	Miscommunication of prescription	6
	Error in use of images	3
Treatment planning	Documentation	1
	Calculation of treatment time or monitor units	3
	Incorrect use of treatment planning system	4
Execution of treatment	Patient identification	2
	Documentation of patient setup	9
	Incorrect operation of treatment machine	2
	Final review at completion of treatment	1

Events Involving Individual Patients

Some of the accidents reviewed occurred due to different conventions for the incorporation of different diagnostic imaging modalities that are used in treatment planning, including those from simulators, computer tomography scanners, nuclear medicine, ultrasound, etc. Also, accidents were caused by incomplete or inaccurate documentation of patient charts for all aspects of treatment planning, including isodose curves, use of wedges, special blocking and placement, and orientation of beams, since these factors are used to calculate beam-on times for all fields. There were cases of treatments to the wrong person or the wrong anatomical site, and of incorrect tumor dose and overdose to normal tissues because of ineffective institutional protocol for patient identification, such as a photograph of the patient, unique patient hospital number, ID bracelet, and verbal confirmation of patient identification.

These incidents are representative of typical causes. Recording, categorizing, and analyzing differences in delivered and prescribed doses in radiotherapy can be carried out at many levels. The above lists give one example for the relatively small number of events reported, where large differences are involved, i.e., misadministrations.

Other evaluations have been reported from the results of in vivo dosimetry programs or other audits of radiotherapy practice, where smaller deviations, or "near misses," have been analyzed. Similar lists of causes with similar relative frequencies have been observed. In any wide-ranging analysis of such events, at whatever level, a number of general observations can be made:

1. Errors may occur at any stage of the process and by every staff group involved. Particularly critical areas are interfaces between staff groups, or between processes, where information is passed across the interface.

2. Most of the immediate causes of accidental exposure are also related to the lack of an adequate QS program or a failure in its application.

33.2.3
Further Arguments

Sometimes the implementation of a QMS is criticized of imposing too much additional bureaucratic effort that is not really helpful. LEER et al. (28) have given further arguments for the practical advantage of a QMS.

Continuing Quality Improvement

A QMS provides the framework for implementing the organizational structure, responsibilities, procedures, and resources. It must be formal, however, at the same time it must be flexible. In particular, it must allow changes to be incorporated. If such changes are based on a systematic mechanism, such as regular audits and assessments as part of the QMS, a quality system has the potential to ensure continuing quality improvement.

Increase of Efficiency

A QMS provides a tool for a good control of all documents involved in the process of radiotherapy. The feature of allowing documents to be traced is the base to determine any discrepancy between aims and actual performance. By feeding the results of the quality analysis back into the process, this mechanism provides the means to increase the efficiency.

Chance for a Cultural Change

A QMS is most effectively introduced by involving peoples at all levels, especially the people who do the actual work. For example, when introducing a new technique it has a distinct advantage if the personnel, who are to use the new technique, offer to contribute to the written QA protocol, rather than have this task imposed on unwilling personnel. This approach may gradually result in a cultural change.

Raised Morale of Personnel

Associated with the importance which a quality system places on training and developing personnel is the raised morale among personnel who feel, correctly, that their individual needs are being given a high priority. This, of course, feeds back into the system as a benefit. Since a quality system ensures that good communication lines are established in the organization, the misunderstandings which can arise between relatively isolated groups of personnel are less likely to occur, and there is a general awareness of being part of a well-run organization, which, itself, contributes to the raised morale.

Reduction of the Chance of Litigation

As a good QMS reduces the likelihood of errors, it will also reduce the likelihood of litigation. In a seemingly informed society litigation is continuing to gain prominence. So even when treatments are given according to protocol, patients increasingly use litiga-tion, for instance when getting normal yet not anticipated side effects. Such litigation often demands that detailed documentation be reviewed sometimes many years after the treatment was completed. A QMS provides a method that the documentation is readily available, providing not only good defense against such litigation but also an increased credibility that will help to defend the institution in the face of such litigation.

QMS as a Management Tool

Good management requires good tools. A QMS provides the required framework to practically accomplish the management task; thus, good management is not left to chance.

Increase of Competitiveness

A QMS may increase the competitiveness of a radiotherapy centers. In particular, when using website information increasingly offered by radiotherapy centers for comparing efficiency, the successful implementation of a QMS may be an advantage. While radiotherapy centers should aspire to a quality system for the positive benefits which it brings, there can be no doubt that to be surrounded by neighboring centers who run quality systems can be a motivating force.

33.3 Comprehensive Quality Management Program

33.3.1 Introduction

This section is an attempt to provide a basic framework for comprehensive quality management program. The document of ISO 9000 is normally referred to as a standard for a quality system. Its general principles and structures, however, must be translated into language appropriate to radiotherapy and covering the complete radiotherapy process. This process can be simply summarized as shown in Fig. 33.2.

This again underlines that quality management should include not only technical aspects of the treatment with ionizing radiation, but also quality aspects from the patient perspective. Those aspects are predominantly summarized in the last section clinical aspects.

For the purpose of QM and a QM program in radiotherapy, the following series of tasks should be accomplished:

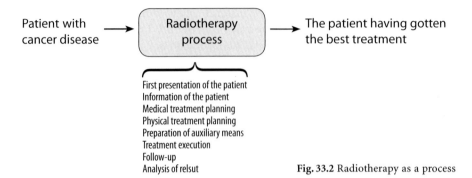

First presentation of the patient
Information of the patient
Medical treatment planning
Physical treatment planning
Preparation of auxiliary means
Treatment execution
Follow-up
Analysis of relsut

Fig. 33.2 Radiotherapy as a process

1. Creation of a QM project team
2. Structuring the quality system
3. Formulation of the department's policy
4. Description (or definition) of department structure
5. Consideration of other requirements
6. Development of a quality manual
7. Implementation
8. Quality audit and continuous quality improving

33.3.2
QM Project Team

In order to assure that all relevant aspects of quality, a QM project team should be formed that represents the many disciplines within radiation oncology. This committee should include a member each of the areas of medicine, physics, and treatment. It is important that these members be appointed by the department chairman and that the committee be given the authority to perform its tasks. If it is possible, a short training of the members of the project team by professionals in quality management should be a starting activity. The committee should directly report to the head of the department on quality issues.

33.3.3
Structure of the Quality System

When introducing a quality system it is useful to differentiate between three hierarchical levels:
- Level 1 reflects (a) the definition of its objectives, (b) the strategies developed to meet these objectives, and (c) the responsibilities and the structure of supervision of all functions having an impact on quality the quality management. These items are summarized under the conception of "policy of the department."
- Level 2 refers to all procedures for which a formal organization is required. In this context procedure

is a document containing information on a definition of the scope of the procedure (what it covers and is about), of the respective responsibilities of those involved (who is responsible for doing what, who is in charge of which areas), and the outline of the practical actions to be undertaken (what is to be done).
- Finally, Level 3 work instructions explain in detail, for each separate area of practical action contained in a Level-2 document, how this is to be implemented at the practical level (how to do it).

33.3.4
Policy

The implementation of a QA system is successful only when it is clear what the position and the aims of the department are; therefore, the head of department has to define the position and basic goals of the department. How can this be accomplished? For a starting point, there are several sources available. As an example, in Germany several associations provide a series of documents and committees as outlined in Fig. 33.3.

In the United States, the Inter-Society Council for Radiation Oncology (ISCRO), the Joint Commission on the Accreditation of Healthcare Organizations (JCAHO) the America College of Radiology (ACR), and the American College of Medical Physics (ACMP) have issued documents and requirements. More practical details are outlined in the report of the AMERICAN ASSOCIATION OF PHYSICS IN MEDICINE Radiation Therapy Committee Task Group 40: Comprehensive QA for Radiation Oncology (1994a).

The objectives should then be laid down in a formal document, also called the policy manual, with a long-term vision on the desired development and the general consequences for budget, infrastructure, and staffing. Answering the following questions can be helpful to write one's own policy manual (LEER 1998):

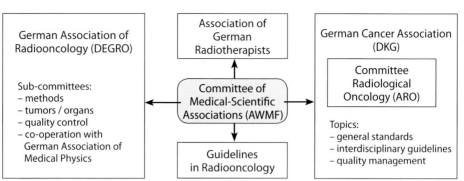

Fig. 33.3 Network of German documents steered by the Committee of Medical–Scientific Associations ("Arbeitsgemeinschaft der Wissenschaftlichen Medizinischen Fachgesellschaften")

- What is the expected number of cancer patients with morbidity and referral pattern?
- What is the expected number of cancer patients who are candidates for radiotherapy?
- What is the expected evolution of these numbers in the next 5–10 years?
- How is the relationship to other radiotherapy departments in the neighborhood?
- Are there agreements on the referral of patients for special treatment, e.g., stereotactic radiotherapy or hyperthermia to the department? Should such agreements be made?
- Are there agreements for the referral of patients with special types of cancer whose multidisciplinary treatment should be centralized in the hospital, e.g., cancer in children, bone tumors, or whethersuch agreements should be made?
- Are there official relations with other departments with respect to training, scientific work, etc.?
- Which consequences of these points are expected for the budget, infrastructure, and staffing?

33.3.5
Responsibilities

The next item of the policy manual is the description and/or definition of responsibilities. Clearly defined responsibility and accountability is of utmost importance for a comprehensive QM system. Without a clear channel of accountability, some components of the quality program will be missed or it might be assumed that someone else is responsible for a specific component. Ultimately, the head each department will be responsible for the overall departmental policy, for all quality matters and for the implementation and maintenance of a documented quality system; however, it must be clearly stated that the quality committee acts on its own authority in his area.

Responsibility and accountability is also involved in the organizational relationship (1995); therefore, it is useful to generate an image of the structure of the department, so that each individual clearly understands his or her own position in that structure, as well as that of others. In addition, the links and relationships between different individuals and staff groups should be clear. These relationships can be: (a) hierarchical, along direct management lines; (b) functional, describing an advisory relationship; or (c) operational, between members of a team working on a particular project. The role and responsibilities of each individual should be clearly defined, with particular attention given to adjoining or overlapping areas of responsibility or to areas of multidisciplinary cooperation. This ensures that there are no omissions due to misinterpretation of who should carry out particular tasks. Communication between different individuals and groups should be clearly structured, with records of meetings and decisions.

33.3.6
Requirements for Quality

All requirements for the quality system must be described and/or defined in the policy manual; thus, all needs can be explicitly expressed or translated into a set of quantitatively or qualitatively stated requirements for the characteristics of a process, procedure, or organization, in order to enable its realization and examination. Special attention must be given to include obligations resulting from laws and other jurisdictional regulation. As an example, the regulations for so-called medical products for which laws have been harmonized throughout Europe have a distinct impact on the quality control of radiotherapy equipment.

33.3.7
Quality Manual

The quality manual is the key document to practically implement a quality program. This usually progresses through four consecutive periods: a preparatory phase; a development phase; an implementation phase; and a consolidation phase. The steps described in the subsections above are the main components of the preparatory phase. The next following phase, the development phase, deals with describing (or defining) policy and structures, preparing procedures, and preparing work instructions. This is the point where the quality manual enters. It has a dual purpose: external and internal. Externally, to collaborators in other departments, in management and in other institutions, it helps to indicate that the department is strongly concerned with quality. Internally, it provides the department with a framework for further development of quality and for improvements of existing or new procedures. Basically, the quality manual will answer two questions for each item: What is the standard required (as a particular example: What tolerances are required on treatment unit positioning precision?), and how to meet this requirement?

Very often generally agreed upon or recommended standards are already in existence for a number of areas, particularly concerning the more well-defined and more easily measured technical aspects, e.g., beam calibration, beam dosimetry, mechanical stability of equipment, performance of treatment planning systems, patient positioning, dose delivery, etc. In other areas, agreed upon standards still need to be developed. Finally, some standards will be internal to a given department (e.g., waiting time before treatment) and will have to be developed locally. It is sensible to begin with standards set at levels which are not too strict but which can be met in the majority of cases. Setting standards which cannot be met routinely is useless and kills the credibility of the quality system. When deviations from these initial levels become exceptional, then a more stringent set can be introduced as a new target to work to.

Since a complete quality manual will address a large number of procedures, experience has shown that the best way to avoid omitting any important feature of a procedure is to formalize the process of describing procedures. A possible approach is given in the next section which explicitly deals with the quality program of radiotherapy equipment.

Finally, it is recommended that everything should not be considered at once. Some aspects need to be addressed first, logically those where the need for improvement has been identified as urgent, whereas others can wait to be dealt with later because they are less critical. Setting up priorities is thus an important step in writing a useful quality manual.

33.3.8
The Treatment Planning Process as a Guide to Structure the Quality Manual

Treatment planning is a process that begins with patient data acquisition and continues through physical treatment planning, plan implementation, and treatment verification. Sometimes the picture of the radiotherapy chain as shown in Fig. 33.4 is used to describe this sequence of various procedures.

The treatment planning procedures and related QA procedures as listed in Table 33.3 are taken from the report of AMERICAN ASSOCIATION OF PHYSICS IN MEDICINE Radiation Therapy Committee Task Group 40 (1994a). This list is a useful guide to structure the quality manual in its content.

33.3.9
Implementation

The next step to be performed is the phase of implementation which covers a training phase and a vali-

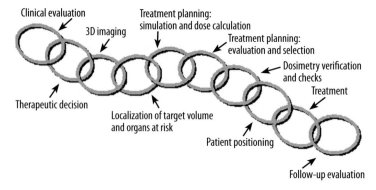

Fig. 33.4 Chain of procedures involved in the radiotherapy process also referred to as the "radiotherapy chain"

Table 33.3 Treatment planning process and related quality assurance (QA) procedures. (From AAPM Radiation Therapy Committee Task Group 1994). TPS treatment planning system

Process	Related QA procedure
Positioning and immobilization	Port films, laser alignment
Simulation	Simulator QA including image and mechanical integrity
Patient data acquisition (CT, MR, manual contouring)	CT, MR, QA including image and mechanical integrity
Data transfer to treatment planning	QA of entire data-transfer process, including digitizers, digital data transfer
Definition of target volumes	Peer review (patient planning conference, chart rounds)
Aperture design	Independent check of delivery, peer review
Computation of dose distribution	Machine data from commissioning and QA of treatment machine; accuracy and QA of TPS
Plan evaluation	Peer review of plan (chart rounds), independent check by radiation oncology physicist
Prescription	Written, signed, and dated
Computation of monitor units	TPS QA, independent check
Production of blocks, beam modifiers (MLC)	QA for block cutting and compensator systems, MLCs, port film review
Plan implementation	Review of setup by treatment planning beam, chart review
Patient QA	Treatment plan review, chart review after new or modified field, weekly chart review, port film review, in vivo dosimetry for unusual fields, critical organ doses, status check, follow-up

dation phase. At this point the following obstacle is frequently met: it is the initial belief that a quality system is too bureaucratic and rigid. Indeed, there is an element of loss of comfort which must be taken into account and the level of bureaucracy will increase in the department, essentially to cope with the demands on document control. It might also appear that the whole system is too stringent and inappropriate for the size of the department and the level of activities; therefore, a subtle equilibrium needs to be found between a quality manual which is too loose and one which is too stringent. A careful design and implementation of the local quality system can contain this to a level where the benefits gained are recognized to be worth this cost. The best way to convince the personnel of the advantage of a quality system is to involve the personnel as soon as possible in the development of procedures and in the implementation of the quality system.

From a practical point of view, the following recommendations may be helpful:
- Training sessions should be introduced that are short but repeated.
- They should initially be oriented towards very practical subjects.
- They should be concerned with routine daily work before addressing wider principles.

- The aims, objectives, and rationale for the approach should be explicitly explained to the personnel at any level.
- Time for reaction and discussion, including criticism and proposals for alteration, should be allowed and provided.
- Comments should be taken into account.

The objective of validation phase is to test the new procedures, as to their appropriateness and feasibility. Advice from the users is important and must be used as a source of improvement. Any suggested improvements at this stage must be carried out rapidly. Care should be taken that any inefficient or unaccepted procedures not be left unchanged, as they will act as sources of generalized de-motivation towards, and criticism of, the whole approach. The head of department must be involved and openly interested during this phase.

33.3.10
Quality Audit and Continuous Quality Improvement

33.3.10.1
Definition

Quality audit is a systematic and independent examination to determine whether or not quality activities and results comply with planned arrangements, and whether or not the arrangements are implemented effectively and are suitable to achieve the stated objectives. Any result of non-compliance found during the process of quality audit then must be feed back into quality manual. The important point is: the quality manual must be kept flexible in such a way that it allows the possibilities of changes and improvement based on regular audits and assessments.

A detailed discussion of the structure and operation of various types of quality audit is given in papers by ESTRO (LEER et al. 1998), IPEM (SHAW 1996), INTERNATIONAL ATOMIC ENERGY AGENCY (1999), and McKENZIE et al. (2000).

Parameters of quality audits are given below. Quality audits:

- Are performed by personnel not directly responsible for the areas being audited, preferably in cooperative discussion with the responsible personnel.
- Evaluate the need for improvement or corrective action.
- Should not be confused with a surveillance or inspection.
- Can be conducted for internal or external purposes.
- Can be applied at any level of a QA program.
- Must be against pre-determined standards, linked to those that the QA program is trying to achieve.
- Should require action if those standards are not met.
- Should be regular and form part of a quality feedback loop to improve quality.

33.3.10.2
Practical Quality Audit Modalities

- Postal audit with mailed dosimeters (usually TLD): these are generally organized by SSDL or agencies, such as the IAEA, Radiological Physics Center (RPC) in the U.S., ESTRO (EQUAL), national societies, national quality networks, etc. They can be applied at various levels in the clinical dosimetry chain and can include procedural audit by using a questionnaire.
- Quality audit visits can audit practical aspects in detail, limited only by time. They can audit procedural aspects by questioning staff and by inspection of procedures and records.

Finally, possible contents of a quality audit visit are given:

- Check infrastructure, e.g., equipment, personnel, patient load, existence of policies and procedures, QA program in place, quality improvement program in place, radiation protection program in place, data and records, etc.
- Check documentation, e.g., content of policies and procedures, QA program structure and management, patient dosimetry procedures, simulation procedures, patient positioning, immobilization and treatment delivery procedures, equipment acceptance and commissioning records, dosimetry system records, machine and treatment planning data, QC program content, tolerances and frequencies, QC and QA records of results and actions, preventive maintenance program records and actions, patient data records, follow-up and outcome analysis, etc.
- Carry out check measurements of beam calibration, field size dependence, electron cone factors, depth dose, electron gap corrections, wedge transmissions (with field size), tray etc., factors, mechanical characteristics, patient dosimetry, dosimetry equipment comparison, temperature and pressure measurement comparison, etc.
- Carry out check measurements on other equipment, such as simulator, CT scanner, etc.
- Assess treatment planning data and procedures. Measure some planned distributions in phantoms.

33.4
Quality System for Radiotherapy Equipment

33.4.1
General

This is the system that makes sure that the technical or instrumental devices and equipment involved in the process of radiotherapy meet all specifications and requirements. It mainly includes inspection and verification according to an established quality protocol. To this end, it is extremely important to give the greatest attention to the elaboration of this quality protocol (which is also part of the quality manual introduced above) before starting any other actions.

The following text concentrates on the major items and systems of radiotherapy equipment. Again, there are many sets of national and international recommendations and protocols covering requirements on quality for various radiotherapy equipment items, e.g., from INTERNATIONAL ELECTROTECHNICAL COMMISSION (1989a,b), AMERICAN ASSOCIATION OF PHYSICS IN MEDICINE (1984, 1994a), IPEM (SHAW 1996; MAYLES et al. 1999), that should be consulted. These texts give recommended tests, test frequencies, and tolerances; some give test methods (MAYLES et al. 1999; INTERNATIONAL

ELECTROTECHNICAL COMMISSION 1989a); other sources give practical advice on QA and QC tests for many items of equipment (VAN DYK et al. 1993).

33.4.2
Terminology

It has already been mentioned that in particular those actions carried out in the more readily defined physical and technical aspects of equipment, dosimetry, and treatment delivery often has been summarized under the term "quality assurance" (QA). Since many documents are still using the term in this sense, it is also used herein.

In Table 33.4 further expressions are given which are frequently used in a QA-specific language and which are used in subsequent subsections.

33.4.3
QA Protocol

33.4.3.1
Structure of Equipment-Specific QA Program

Every radiotherapy unit is involved in the implementation of new equipment. The normal case is to purchase equipment. While the purchase process varies from one institution to another, there are certain steps common to any institution. In Table 33.5, QA actions accompanying the implementation are listed according to the sequence on which they have to be done. A general QA program for equipment typically includes actions as listed in Table 33.5.

Many protocols and procedures of levels 2 and 3 normally may be already available in the departments. In any case, they have to be rewritten according to a uniform format. Table 33.6 gives an example of a uniform format.

33.4.3.2
Clinical Needs Assessment and
Initial Specification for Purchase

The decision to purchase a particular technology or the implementation of a particular technique or procedure is based on clinical needs and/or the improvement in efficiency of treatment. They should be clearly analyzed. The answer to the questions listed in Table 33.7 may serve that purpose.

In preparation for procurement of equipment, a detailed specification document must be prepared:

- This should set out the essential aspects of the equipment operation, facilities, performance, service, etc., as required by the customer.

Table 33.4 Expressions frequently used in a QA-specific language

Term	Aim or meaning
Quality characteristic	One of several characteristic determining the quality; it may be quantitative or qualitative
Safety characteristic	A specific type of a quality characteristic referring to safety
Performance characteristic	A quality characteristic referring to the performance (of equipment)
Specific test characteristic	A quantitative performance characteristic that is determined during the acceptance test; for determination a test procedure including conditions must be defined
Baseline value	Value of a specific characteristic obtained during commissioning und used for subsequent periodic tests as reference value
Tolerance level	Performance to within the tolerance level gives acceptable accuracy in any situation
Action level	Performance outside the action level is unacceptable and demands action to remedy the situation

Table 33.5 Quality assurance actions accompanying implementation according to sequence

Subsequent QA actions for a specific equipment	Aim or test characteristic
Clinical needs assessment	
Initial specification and purchase process	Expression of all specification data in acceptable units of measure
Acceptance testing	Compliance with specifications
Commissioning for clinical use, including calibration where applicable	Establishment of baseline performance values
Periodic QA tests	Monitoring the reference performance values
Additional quality control tests after any significant repair, intervention, or adjustment, or when there is any indication of changes in performance	Same as with periodic tests
Planned preventive maintenance program	

Table 33.6 A uniform format for protocols

Scope and objective	What is the aim of a specific test?
Responsibility:	Who is responsible to perform the test and to document the results?
Documentation:	Where is the documentation stored?
Method:	Definition of functional performance characteristics and/or specific test characteristics
	Description of test procedures and test conditions
	Acceptance and commissioning tests
	Determination of baseline values
	Periodic tests
	Determination of numerical values for reference
	Values, tolerance, and action levels

Table 33.7 Questions to which the answer is helpful to assess clinical needs

Which patients will be affected by this technology?
What is the likely number of patients per year?
Number of procedures or fractions per year?
Will the new procedure provide cost savings over old techniques?
Would it be better to refer patients to a specialist institution?
Is the infrastructure available to handle the technology?
Will the technology enhance the academic program?
What is the organizational risk in implementation of this technology?
What is the cost impact?
What maintenance is required?

- A multi-disciplinary team from the department should be involved in contributing to the specification, including input from radiotherapy physicists, radiation oncologists, radiotherapy technologists, engineering technicians, etc. It would generally be expected that liaison between the department and the suppliers would be by a radiotherapy physicist.
- In response to the specifications, the various interested suppliers should indicate how the equipment they offer will meet the specifications; if there are any areas that cannot be met or if there are any limiting conditions under which specified requirements can or cannot be met, etc.
- Decisions on procurement should be made by a multi-disciplinary team, comparing specifications as well as considering costs and other factors.

33.4.3.3
Acceptance Tests

Acceptance of equipment is the process in which the supplier demonstrates the baseline performance of the equipment to the satisfaction of the customer, i.e., that the specifications contained in the purchase order are fulfilled and that the environment is free of radiation and electrical hazards to staff and patients. The tests are performed in the presence of a manufacturer's representative. Upon satisfactory completion of the acceptance tests, the physicist signs a document certifying that these conditions have been met. When the physicist accepts the unit, the final payment is made for the unit, ownership of the unit is transferred to the institution, and the warranty period begins. These conditions place a heavy responsibility on the physicist in correct performance of these tests.

Acceptance tests may be divided into three groups: (a) safety checks; (b) mechanical checks; and (c) dosimetry measurements.

33.4.3.3.1
Safety Checks

Acceptance tests begin with safety checks to assure a safe environment for staff and public.

Interlocks, warning lights, patient monitoring equipment

The initial safety checks should verify that all interlocks are functioning properly. These interlock checks should include the door interlock, all radiation beam-off interlocks, all motion-disable interlocks, and all emergency-off interlocks.

The door interlock prevents irradiation from occurring when the door to the treatment room is open. The radiation beam-off interlocks halt irradiation but they do not halt the motion of the treatment unit or patient treatment couch. The motion-disable interlocks halt motion of the treatment unit and patient treatment couch but they do not stop machine irradiation. Emergency-off interlocks typically disable power to the motors that drive treatment unit and treatment couch motions and power to some of the radiation producing elements of the treatment unit. The idea is to prevent both collisions between the treatment unit and personnel, patients or other equipment and to halt undesirable irradiation.

The medical physicist must verify the proper function of all these interlocks and assure that all personnel operating the equipment have a clear understanding of each. After verifying that all interlocks and emergency-off switches are operational, all warning lights should be checked. Next, the proper functioning of the patient monitoring audio-video equipment can be verified. The audio-video equipment is often useful for monitoring equipment or gauges during the acceptance testing and commissioning involving radiation measurements.

Radiation Survey

After completion of the interlock checks, the medical physicist should perform a radiation survey in all areas outside the treatment room. For cobalt units and linear accelerators operated below 10 MeV, a photon survey is required. For linear accelerators operated above 10 MeV, the physicist must survey for neutrons in addition to photons. The survey should be conducted using the highest energy photon beam. To assure meaningful results the physicist should perform a preliminary calibration of the highest energy photon beam before conducting the radiation survey. Photon measurements will require both a Geiger counter and an ionization chamber survey meter. Neutron measurements will require a neutron survey meter.

Collimator and Head Leakage

Shielding surrounds the target on a linear accelerator or the source on a cobalt-60 unit. Most regulations require this shielding to limit the leakage radiation to a 0.1% of the useful beam at one meter from the source. The adequacy of this shielding must be verified during acceptance testing.

33.4.3.3.2
Mechanical Checks

The mechanical checks establish the precision and accuracy of the mechanical motions of the treatment unit and patient treatment couch. Items to be checked are listed in Table 33.8.

Table 33.8 Mechanical check items

Collimator axis of rotation
Photon collimator jaw motion
Congruence of light and radiation field
Gantry axis of rotation
Patient treatment couch axis of rotation
Radiation isocenter
Optical distance indicator
Gantry angle indicators
Collimator field size indicators
Patient treatment couch motions

Dosimetry measurements

Dosimetry measurements establish that the central axis percentage depth doses and off-axis characteristics of clinical beams meet the specifications. The

characteristics of the monitor ionization chamber of a linear accelerator or a timer of a cobalt-60 unit are also determined. Dosimetric items to be checked are listed in Table 33.9. More details are given by PODGORSAK (2003).

Table 33.9 Dosimetric check items

Photon energy
Photon beam uniformity
Photon penumbra
Electron energy
Electron beam "bremsstrahlung" contamination
Electron beam uniformity
Electron penumbra
Monitor characteristics
Arc therapy

33.4.4
Commissioning

Following acceptance of equipment, a full characterization of its performance over the whole range of possible operation must be undertaken. This is generally referred to as commissioning. Another definition is that commissioning is the process of preparing procedures, protocols, instructions, data, etc., for clinical service. Clinical use can only begin when the physicist responsible for commissioning is satisfied that all aspects have been completed and that the equipment and any necessary data, etc., are safe to use on patients. Depending on the type of equipment, acceptance and commissioning may partially overlap.

Another useful recommendation is that acceptance and commissioning together will establish the baseline-recorded standards of performance to which all future performance and periodic tests will be referred. Where appropriate, commissioning will incorporate calibration to agreed upon protocols and standards. For critical parts of commissioning, such as calibration, an independent second check is recommended.

Commissioning includes a series of tasks that generally should consist of the following:
- Acquiring all radiation beam data required for treatment
- Organizing this data into a dosimetry data book
- Entering this data into a computerized treatment planning system (TPS)
- Developing all dosimetry, treatment planning, and treatment procedures
- Verifying the accuracy of these procedures

- Establishing quality control tests and procedures
- Training all personnel

The completion of all the tasks associated with placing a treatment unit into clinical service can be estimated to require from 1.5 to 3 weeks per energy following completion of the acceptance tests. The time will depend on machine reliability, amount of data measurement, sophistication of treatments planned, and experience of the physicist.

The next section provides a short description of measurements which are typically done for photon beams.

33.4.4.1
Acquisition of Photon Beam Data

Beam Calibration

The basic output calibration of a clinical radiation beam, by virtue of a direct measurement of dose or dose rate in water under specific reference conditions, is also referred to as reference dosimetry. For many reasons, the dose should be determined in terms of absorbed dose to water under well-defined reference conditions. Calibrated air-filled ionization chambers placed at a reference point in water should be used. Calibration means that the chamber has a calibration coefficient traceable to a standards dosimetry laboratory.

Before such a chamber is used in radiotherapy machine output calibration, the user must identify and follow a dosimetry protocol (code of practice) appropriate for the given radiation beam. A dosimetry protocol provides the formalism and the data to relate a calibration of a chamber at a standards laboratory to the measurement of absorbed dose to water under reference conditions in the clinical beam. Dosimetry protocols are generally issued by national or regional organizations, such as the AMERICAN ASSOCIATION OF PHYSICS IN MEDICINE (1999; North America), INSTITUTE OF PHYSICS AND ENGINEERING IN MEDICINE AND BIOLOGY (1996; UK), DEUTSCHES INSTITUT FÜR NORMUNG (1997; Germany), NEDERLANDSE COMMISSIE VOOR STRALINGSDOSIMETRIE; The Netherlands and Belgium, and NORDIC ASSOCIATION OF CLINICAL PHYSICS (1980; Scandinavia), or by international bodies such as the INTERNATIONAL ATOMIC ENERGY AGENCY (2000b) and the INTERNATIONAL COMMISSION ON RADIOLOGICAL UNITS AND MEASUREMENTS (2001). This procedure ensures a high level of consistency in dose determination among different radiotherapy clinics in a given country and also between one country and another.

Beam calibration according to a dosimetry protocol is usually performed at a reference depth, z_{ref}, and under well-defined reference conditions. It is recommended that this result obtained under reference condition is generally used for any normalization of other relative dosimetric data such as percentage depth dose or output factors in order to generate a consistent data base.

Central Axis Percentage Depth Doses

Central axis depth-dose distributions should be measured in a water phantom. Ionization chambers, and in particular plane-parallel ionization chambers, are generally recommended. Since the stopping-power ratios and perturbation effects can be assumed to a reasonable accuracy to be independent of depth for a given beam quality and field size, relative ionization distributions can be used as relative distributions of absorbed dose, at least for depths at and beyond the depth of dose maximum. It is recommended that a chamber be placed at the desired depth with its reference point. For plane-parallel chambers this point is on the inner surface of the entrance window, at the center of the window. For cylindrical chambers it is on the central axis at the center of the cavity volume; however, if a cylindrical ionization chamber is used, the displacement effect of the chamber type must be taken into account. This requires that the complete depth-ionization distribution be shifted towards the surface a distance equal to 0.6 r_{cyl}, where r_{cyl} is the cavity radius of the cylindrical ionization chamber.

Central axis percentage depth-dose values should be measured over the range of field sizes from 4×4 to 40×40 cm^2. Increments between field sizes should be no greater than 5 cm but are typically 2 cm. Measurements should be made to a depth of 35 or 40 cm. Field sizes smaller than 4×4 cm^2 require special attention.

To make accurate measurements in the build-up region, extrapolation chambers, or well-guarded fixed separation plane-parallel chambers should be used. Attention should be paid to the use of certain solid-state detectors (some types of diodes and diamond detectors) to measure depth-dose distributions. Only a solid-state detector whose response has been regularly verified against a reference detector (ionization chamber) should be selected for these measurements.

Output Factors

The radiation output, in cGy/MU for a linear accelerator and cGy/min for a cobalt unit, increases with an increase in collimator opening or field size. The output factor is determined as the ratio of corrected dosimeter readings measured under a given set of non-reference conditions to that measured under reference conditions, typically a 10×10-cm^2 field. Special attention should be given to the uniformity of the radiation fluence over the chamber cavity. This is especially important for field sizes smaller than 5×5 cm. Some accelerators have very pronounced V-shaped photon beam profiles which usually vary with depth and field size. For large detectors it may be difficult to accurately correct for this variation. Thimble chambers with large cavity length and plane-parallel chambers with large collecting electrodes should therefore be avoided in situations where the beams have pronounced V-shaped profiles.

Output factors are usually given as a function of equivalent square fields. This approach assumes the output for rectangular fields is equal to the output of its equivalent square field. This assumption must be verified by measuring the output for a number of rectangular fields at their z_{max}. If the outputs of rectangular fields vary from the output of their equivalent square field by more than 2%, it may be necessary to have a table or graph of output factors for each rectangular field. The output of a rectangular field may also depend on whether the upper or lower jaw forms the long side of the field. This effect is sometimes referred to as the collimator exchange effect and should be investigated as part of the commissioning process.

There are two other collimator-dependent quantities: the collimator scatter factor and the phantom scatter factor. The collimator scatter factor is measured "in air" with a build-up cap large enough to provide electronic equilibrium. Typically, these values are normalized to a 10×10-cm^2 field.

A problem arises for small high-energy photon field sizes as the size of the build-up cap approaches or exceeds the size of the field. In this case the collimator scatter correction factor may be determined by placing the ionization chamber at an extended shaded-surface display (SSD) but with the field defined at the nominal SSD. With the chamber at 200 cm the collimator scatter correction factor can be measured for fields with dimensions down to 4×4 cm^2 at 100 cm. These relative measurements should all be performed under the same conditions.

As the output factor is the product of the collimator scatter correction factor and the phantom scatter correction factor, the phantom scatter correction factor may be found by dividing the output factor by the collimator scatter correction factor.

Blocking Tray Factors

High-density shielding blocks are normally supported on a plastic tray to correctly position them within the radiation field can be used to protect normal critical structures within the irradiated area. This tray attenuates the radiation beam. The amount of beam attenuation provided by the tray must be known to calculate the dose received by the patient. The attenuation for solid trays is easily measured by placing an ionization chamber on the central axis of the beam at reference depth in phantom in a 10×10-cm^2 field. The ratio of the ionization chamber signal with the tray in the beam to the signal without the tray is the blocking tray transmission factor.

Multileaf Collimators

On most current treatment machines multileaf collimators (MLC) are finding widespread application. One must differentiate between add-on MLCs and those MLC systems replacing at least one set of conventional jaws. For add-on MLCs, the central axis percentage depth doses, the penumbra, and the output factor should be measured for a set of MLC-defined fields.

Leakage through the MLC consists of transmission through the leaves and leakage between the leaves. Leakage between the leaves is easily demonstrated by exposing a film placed perpendicularly to the collimator axis of rotation with the leaves fully closed. Leakage through the leaves can be determined by comparing the umbra region of transverse beam profiles for fields defined by the MLC to fields defined by the collimator jaws.

Central axis Wedge Transmission Factors

In wedged photon beams the radiation intensity varies strongly in the direction of the wedge. For output measurements in such beams the detector dimension in the wedge direction should be as small as possible. A small thimble chamber aligned with its axis perpendicular to the wedge direction is recommended. The coincidence of the central axes of the beam, the collimator, and the wedge should be ensured prior to making the output measurements.

The central axis wedge transmission factor is the ratio of the dose (or dosimeter reading) at a specified depth on the central axis of a specified field size with the wedge in the beam to the dose for the same conditions without the wedge in the beam. Central axis wedge transmission factors determined for one field size at one depth are frequently used to calculate beam-on time or monitor unit settings for all wedged fields and depths; however, the central axis wedge transmission factors may be a function of both depth and field size. Moreover, the field size variation may depend not only on the width of the field along the gradient of the wedge but also on the length of the field. In other words, the central axis wedge transmission factor for a given wedge for a 10×10-cm² field may differ from the central axis wedge transmission factor for a 10×20-cm² field even when the 10 cm is along the wedge gradient in both cases. These dependencies require measuring central axis percentage depth doses with the wedge in the beam for the range of field sizes. The dose with the wedge in the beam can then be related to the calibrated dose rate by measuring the central axis wedge transmission factor at one depth for each field size.

Dynamic Wedge

Linear accelerators can be equipped with a so-called dynamic wedge technique by moving one of the independent collimator jaws while the opposite jaw remains stationary during irradiation. Clinical implementation of dynamic wedges requires measurement of central axis percentage depth doses, central axis wedge transmission factors, and transverse beam profiles of the dynamic wedges. These measurements are generally more complicated due to the modulation of the photon fluence during the delivery of the radiation field.

The central axis percentage depth dose may be measured by integrating the dose at each point during the entire irradiation of the dynamic wedge field. The central axis wedge transmission factors are determined by taking the ratio of the collected ionization at a specified depth for the dynamic wedge field to the collected ionization at the same specified depth for the open field with the same collimator and monitor unit settings.

It is important to note that the central axis wedge transmission factors for dynamic wedges may have much larger field-size dependence than physical wedges and the field size dependence for dynamic wedges may not be asymptotic. Some manufacturer's implementations of the dynamic wedge technique show a significant change in the trend of the central axis wedge transmission factor as the field width changes between 9.5 and 10 cm. This change in the central axis wedge transmission ratio has been demonstrated to approach 20%. This characteristic should be carefully investigated on each machine. Dynamic wedge transverse beam profiles can be measured with a detector array or an integrating dosimeter such as radiochromic film. When a detector array is used, the sensitivity of each detector must be determined.

Transverse Beam Profiles/Off-Axis Energy Changes

Transverse beam profiles are measured to characterize the dose at points off the central axis. Frequently, off-axis data are normalized to the dose on the central axis at the same depth. These data are referred to as off-axis ratios. Off-axis ratios are combined with central axis data to generate isodose curves.

The number of profiles and the depths at which these profiles are measured depends on the requirements of the TPS. Some systems require these profiles at a few equally spaced depths, others require several profiles at specified depths, and some require only one off-axis profile for the largest field size measured "in-air" with a build-up cap. Transverse beam profiles should be measured in addition to those on which the beam model was determined to verify the accuracy of the TPS algorithms. Of course, these profiles should be measured for both open and wedged fields. The profiles of the wedged field can then be combined with the central axis percentage depth dose values for wedged fields to generate wedged isodose curves. Any change in wedge factor with depth is then included in the isodose curves.

Other Parameters

Knowledge of other dosimetric parameters is useful. Examples are the entrance dose, the dose at interfaces at small air cavities, the exit dose at the surface of the patient, the dose at bone–tissue interfaces, or that between a metallic prosthesis and tissue. Another parameter is the exact virtual source position. For recommended measurements the reader is again referred to PODGORSAK (2003).

Acquisition of Electron Beam Data Measurements

Many recommendations as given in the photon beam section also apply to electron beams. Special attention has to be given to ionization chamber measurements because of the relation between measured charge and absorbed dose that is energy dependent. For instance, the energy decreases with increasing depth; therefore, a depth ionization curve is not equivalent to a depth dose curve. Further details cannot be addressed; instead, the reader is referred to modern dosimetry protocols such as the International Code of Practice issued by the INTERNATIONAL ATOMIC ENERGY AGENCY (2000b).

The items which have to be determined in the commissioning process of electron beams are as follows:
- Central axis percentage depth dose
- Output factors
- Secondary collimators
- Metal cut-outs

- Skin collimation
- Transverse beam profiles
- Virtual source position

Treatment Planning System

A computerized TPS is an essential tool in the process of a radiotherapeutic treatment of cancer patients. After the installation of a TPS in a hospital, acceptance testing and commissioning of the system is required, i.e., a comprehensive series of operational tests has to be performed before using the TPS for treating patients. These tests, which should partly be performed by the vendor and partly by the user, do not only serve to ensure the safe use of the system in a specific clinic, but also help the user in appreciating the possibilities of the system. The major aspect of the acceptance and commissioning of the system is to test its fundamental performance and gain an understanding of the algorithms used for the dose prediction. This provides the knowledge of the limitations of the system and a considerable part of this understanding should be gained by comparison with experimental measurement in phantoms for test cases of varying complexity. Some information on this should also be obtainable from the manufacturer, from the literature, and from user's groups. In the past some irradiation accidents happened with patients undergoing radiation therapy, which were related to the misuse of a TPS and/or to a lack of understanding of how the TPS works. In many of these accidents, a single cause could not be identified but usually there was a combination of factors contributing to the occurrence of the accident. The most prominent factors were deficiencies in education and training, and a lack of QA procedures.

Over recent years, increased attention has been paid to QA of TPS by various national and international organizations. Examples include Van Dyk et al. (1999), Shaw (1996), Swiss Society for Radiobiology and Medical Physics (1997), Fraass et al. (1998), Mayles et al. (1999), International Atomic Energy Agency (2004), and Nederlandse Commissie voor Stralingsdosimetrie (2003). These reports provide recommendations for specific aspects of QA of a TPS, such as anatomy description, beam description, and dose calculations. More practical recommendations for commissioning and QA of a TPS are provided by the new ESTRO booklet "Quality Assurance of Treatment Planning Systems – Practical Examples for External Photon Beams" (European Society for Therapeutic Radiology and Oncology 2004). These protocols should be consulted for more details. The exact requirements depend on the level of complexity of the system and of the treatment planning techniques used clinically.

As a rule, any uncertainty concerning the operation or output of a TPS should be tested by comparing the performance of the TPS to measurements in suitable phantoms. Generic tolerances of 2% have often been quoted for isodose distributions where dose gradients are not steep and 2 mm where dose gradients are steep. These values may typically be applied to single-field or single-source isodose distributions; however, these will not necessarily be applicable in less simple situations. A similar generic tolerance of 2% is often quoted in monitor-unit calculations, which again may need careful consideration in complex situations. Discussion of the acceptable tolerances for different situations is given, for example, by Van Dyk and Purdy (1999), and Venselaar and Welleweerd (2001).

33.4.5
Periodic Tests

33.4.5.1
Test Program

It is essential that the performance of treatment equipment remains consistent within accepted tolerances throughout its clinical life, as patient treatments will be planned and delivered on the basis of performance measurements at acceptance and commissioning; therefore, an ongoing quality test program of periodic performance checks must started immediately after commissioning. If these measurements identify departures from expected performance, corrective actions are required. A formalized program to check equipment should specify the following:

- Object of test to be performed and parameters to be tested
- Specific equipment used to perform the tests
- Measurement condition
- Frequency of the tests
- Staff group or individual performing the tests, as well as the individual supervising and responsible for the standards of the tests and for actions which may be necessary if problems are identified
- Expected results
- Tolerance and/or action levels
- The actions required when the tolerance levels are exceeded

No one program is necessarily suitable in all circumstances and may need tailoring to the specific equipment and the departmental situation. For example, programs must be flexible for additional testing whenever it seems necessary, following repair, observed equipment behavior, or indications of problems from the periodic tests. Also, frequencies may need to be adjusted in the light of experience with a given machine. Staff resources available to undertake the program may limit what can be checked, which may have an effect on the structure of the test program. Tests should be designed to provide the required information as rapidly as possible with minimal time and equipment. Often customized devices are very useful to make tests easier. Examples of periodic tests are given in more detail for linear accelerators and for a TPS in the subsections below.

34,4.5.2
Uncertainties, Tolerances, and Action Levels

In relevant documents, such as the "Guide to the Expression of Uncertainty in Measurement" (INTERNATIONAL ORGANIZATION FOR STANDARDIZATION 1993), uncertainties are classified according to the method of evaluation as either *type A*, meaning that they have been assessed by statistical analysis of series of observations, or *type B*, meaning that they have been assessed by other means. In the latter case the uncertainty is evaluated by scientific judgment based on all available information on the possible variability of the measured quantity. In previous textbooks and still in common practice, uncertainties are frequently described as random (*a posteriori*) or systematic (*a priori*). Use of the type-A and type-B classification is given preference. Irrespective of how uncertainties are assessed, the uncertainties at different steps are usually combined in quadrature to estimate overall values.

For communicating uncertainties in an unmistakable way it is helpful to introduce the "standard uncertainty" which is that uncertainty expressed as a standard deviation. Uncertainties are also related to confidence levels. To associate a specific level of confidence with the range of uncertainty requires explicit or implicit assumptions regarding the probability distribution of the measurement result. In many situations there are good reasons to justify the assumption of a Gaussian distribution. In this case the standard uncertainty is associated with a confidence level of 68%, whereas an uncertainty expressed by two standard deviations is associated with a 95% confidence level. In radiotherapy, this 95% confidence level is normally the goal.

If a measurement is performed appropriately, it is expected to give the best estimate of the particular measured parameter; however, this will have an associated uncertainty, dependent on the measurement technique. Therefore, the tolerance set for the parameter must take into account the uncertainty of the measurement technique employed. If the measurement uncertainty is greater than the tolerance or action level set, then random variations in the measurement will lead to unnecessary intervention, increased downtime of equipment, and inefficient use of staff time. On the other hand, tolerances should be set with the aim of achieving the overall uncertainties desired in radiotherapy, which are of the order of 2.5–3.5% standard uncertainty. Variances of individual parameters can be combined in quadrature for combined factors and this can now be used to determine specific tolerance limits for individual parameters.

Action levels are related to tolerances. For example, action levels are often set at approximately twice the tolerance level, although some critical parameters may require tolerance and action levels to be set much closer to each other or even at the same value. What happens if a measurement falls between tolerance and action level? For practical reasons some flexibility in monitoring and adjustment is needed. For example, if a measurement on the constancy of dose/MU indicates a result between the tolerance and action levels, then it may be permissible to allow clinical use to continue until this is confirmed by measurements the next day before taking any further action. If repeated measurements remain consistently between tolerance and action levels, adjustment is required. Any measurement at any time outside the action level requires immediate investigation and, if confirmed, rectification.

Different sets of recommendations may use very different approaches to set tolerance levels and/or action levels, and this should be borne in mind in comparing values from different sources; in some, the term tolerance level is used to indicate values that in others may be closer to action levels, i.e., some workers use the term tolerance to indicate levels at which adjustment or correction is necessary.

34.4.5.3
Periodic Tests for Linear Accelerators

As an example for periodic tests, a typical test program for linear accelerators as suggested by the AMERICAN ASSOCIATION OF PHYSICS IN MEDICINE (1994a) is listed in Table 33.10. The experimental techniques for performing these tests are not described. Techniques

Table 33.10 Quality assurance of medical accelerators

Frequency	Procedure	Action level	Frequency	Procedure	Action level
Daily:	Dosimetry		Annually:	Dosimetry	
	X-ray output constancy	3%		X-ray/electron output calibration constancy	2%
	Electron output constancy	3%		Field size dependence of X-ray Output constancy	2%
	Mechanical			Output factor constancy for electron applicators	2%
	Localizing lasers	2 mm		Central axis parameter constancy	2%
	Resistance indicator	2 mm		Off-axis factor constancy	2%
	Safety			Transmission factor constancy for all treatment accessories	2%
	Door interlock	Functional			
	Audiovisual monitor	Functional		Wedge transmission factor constancy	2%
Monthly:	Dosimetry			Monitor chamber linearity	1%
	X-ray output constancy	2%		X-ray output constancy vs gantry angle	2%
	Electron output constancy	2%		Electron output constancy vs gantry angle	2%
	Backup monitor constancy	2%			
	X-ray central axis dosimetry parameter constancy	2%		Off-axis factor constancy vs gantry angle	2%
	Electron central axis dosimetry parameter constancy	2 mm at ther. depth		Arc mode	Manufacturer's specifications
	X-ray beam flatness constancy	2%		Safety interlocks: follow manufacturer's test procedures	Functional
	Electron beam flatness constancy	3%		Mechanical checks	
	X-ray and electron symmetry	3%		Collimator rotation isocenter	2-mm diameter
	Safety interlocks			Gantry rotation isocenter	2-mm diameter
	Emergency off switches	Functional		Couch rotation isocenter	2-mm diameter
	Wedge, electron cone interlocks	Functional		Coincidence of collimator, gantry	2-mm diameter
	Mechanical checks			Couch axes with isocenter	2-mm diameter
	Light/radiation field coincidence	2 mm or 1% on a side		Coincidence of radiation and mechanical isocenter	2-mm diameter
	Gantry/collimator angle indicators	1°		Tabletop sag	2 mm
	Wedge position	2 mm or 2% change in transmission factor		Vertical travel of table	2 mm
	Tray position	2 mm			
	Applicator position	2 mm			
	Field size indicators	2 mm			
	Cross-hair centering	2-mm diameter			
	Treatment couch position indicators	2 mm/1°			
	Latching of wedges, blocking tray	Functional			
	Jaw symmetry	2 mm			
	Field light intensity	Functional			

as well as programs for other equipment, such as for cobalt-60 teletherapy machines, treatment simulators, CT scanners, or test equipment, can be adopted from a number of publications (AMERICAN ASSOCIATION OF PHYSICS IN MEDICINE 1994a, b; WIZENBERG 1982; AMERICAN NATIONAL STANDARDS INSTITUTE 1974, 1978). In Germany various periodic tests (so-called constancy tests) are explicitly described in part 5 of the respective German Standards (DIN).

Some comments have to be made to Table 33.10:
- The level values listed in the third column should be interpreted to mean real action levels.
- For constancy, percent values are ± the deviation of the parameter with respect its nominal value; distance values are referenced to the isocenter or nominal SSD.
- The electron output needs not to be checked daily for all electron energies, but all energies are to be checked at least twice weekly.
- Jaw difference is defined as difference in distance of each jaw from the isocenter, determined at the isocenter distance.
- Most wedge transmission factors are field-size and depth dependent.
- The action level values for radiation output are 3% and 2% for daily and monthly checks, respectively. The 2% value for the monthly checks is given because these are normally performed with calibrated ionization chambers. The daily check, however, may also be performed by any device, which has at least the precision adequate to verify compliance with a tolerance value of 3%.

34.4.5.4
Periodic Tests for TPS

For safety and security reasons, periodic quality control of particular parts of a TPS is also an important aspect of the QA process of a TPS. In this subsection tests are discussed that should be performed periodically at specified time intervals. For a software device, such as a TPS, one must be concerned about data files, integrity of the software executables, failures or problems in hardware peripherals and general system configuration, as well as the process that uses the software. The main aims of a routine periodic QA program for the TPS include the following:
- Confirm the integrity and security of the TP data files that contain the external beam and brachytherapy information used in dose and MU calculations.
- Verify the correct functioning and accuracy of peripheral devices used for data input, including

the digitizer tablet, CT, MR, video digitizer, simulator control system, and devices for obtaining mechanical simulator contours. One must separately consider the devices themselves and the networks, tape drives, software, transfer programs, and other components, which are involved in the transfer of the information from the device to the TPS.
- Check the integrity of the actual TPS software.
- Confirm the function and accuracy of output devices and software, including printers, plotters, automated transfer processes, connections to computer-controlled block cutters, and/or compensator makers, etc.

The AAPM Radiation Therapy Committee Task Group 53 (FRAASS et al. 1998) has given recommendations on periodic tests of a TPS as specified in Table 33.11. Commercial manufacturers often make their own recommendations regarding ongoing QA of their planning systems. Each radiation oncology physicist should review all the recommendations and develop a program of periodic testing that will match the planning system characteristics and its user base. The frequency of testing of each specific feature of the RTP system should depend on how that feature is used in the clinic and how critical that feature is from a safety point of view.
A series of reviews and training sessions is recommended to be included as part of the periodic QA program.

33.5
Clinical Aspects of QA

According to report of AAPM Radiation Therapy Committee Task Group 40 (AMERICAN ASSOCIATION OF PHYSICS IN MEDICINE 1994a), the method of peer review is particularly helpful in establishing a QA program for the clinical aspects. The three important components of such peer review are: new planning conference; chart review; and film/image review. Although already thoroughly discussed in the report, the essential steps to perform such peer reviews are briefly described in the following subsection.

33.5.1
New Patient Planning Conference

New patient planning conference should be attended by radiation oncologists, radiation therapists, and medical physicists. In some countries, such as the

Table 33.11 Periodic RTP process QA checks

Recommended frequency	Item	Comments/details
Daily	Error log	Review report log listing system failures, error messages, hardware malfunctions, and other problems; triage list and remedy any serious problems that occur during the day
	Change log	Keep log of hardware/software changes
Weekly	Digitizer	Review digitizer accuracy
	Hard-copy output	Review all hard-copy output, including scaling for plotter and other graphics-type output
	Computer files	Verify integrity of all RTP system data files and executables using checksums or other simple software checks; checking software should be provided by the vendor
	Review clinical planning	Review clinical treatment planning activity; discuss errors, problems, complications, difficulties; resolve problems
Monthly	CT data input into RTP system	Review the CT data within the planning system for geometrical accuracy, CT number consistency (also dependent on the QA and use of scanner), and derived electron density
	Problem review	Review all RTP problems (both for RTP system and clinical treatment planning) and prioritize problems to be resolved
	Review of RTP system	Review current configuration and status of all RTP system software, hardware, and data files
Annually	Dose calculations	Annual checks; review acceptability of agreement between measured and calculated doses for each beam/source
	Data and I/O devices	Review functioning and accuracy of digitizer tablet, video/laser digitizer, CT input, MR input, printers, plotters, and other imaging output devices
	Critical software tools	Review BEV/DRR generation and plot accuracy, CT geometry, density conversions, DVH calculations, other critical tools, machine-specific conversions, data files, and other critical data
Variable	Beam parameterization	Checks and/or recommissioning may be required due to machine changes or problems
	Software changes, including operating system	Checks and/or recommissioning may be required due to changes in the RTP software, any support/additional software such as image transfer software, or the operating system

U.S., a further differentiation between a medical physicist and a dosimetrist is being made. The aim of the conference is:

- Presentation of:
  - Pertinent medical history
  - Physical and
  - Treatment strategy
- Discussion of:
  - Time required for planning and preparation accessory devices and blocks
  - Prescribed dose
  - Critical organ doses
  - Patient positioning
  - Possible field arrangements
- Possibly further discussion of the need for:
  - Special point dose calculations (critical organs)
  - In vivo (or special in phantom) dosimetry

- Designated points where the cumulative dose is required (overlapping fields, previous radiotherapy)

Interaction of all participants at the planning conference is helpful in resolving technical issues, which otherwise could potentially lead to delays and errors.

33.5.2
Chart Review

The radiation chart accompanies the patient during the entire process of radiotherapy. A number of individuals review the various entries in the chart. They should address the following items:

- Patient identification
- Initial physical evaluation of patient and pertinent

clinical information
- Treatment planning
- Signed and witnessed consent form
- Treatment execution
- Clinical assessment during treatment
- QA checklists

The authors of the report recommend that:
- Charts be reviewed
  - At least weekly
  - Before the third fraction following the start or a field modification
  - At the completion of treatment
- The review be signed and dated by the reviewer
- The QA team oversee the implementation of a program which defines
  - Which items are to be reviewed
  - Who is to review them
  - When are they to be reviewed
  - The definition of minor and major errors
  - What actions are to be taken, and by whom, in the event of errors
- All errors be reviewed and discussed by the QA team including Documentation of Consequences
- A random sample of charts be audited at intervals prescribed by the QA team

33.5.3
Film/Image Review

The review of films and/or digital images mainly refers to those obtained during the localization procedure using portal imaging and verification imaging. Whereas a portal image is obtained by exposing the film (or the image detector) to only a small fraction of the daily treatment dose, verification images are single-exposure images that record the delivery of the entire dose for each fraction from each field.

It is recommended that all initial portal images be reviewed, signed, and dated by the radiation oncologist before the first treatment for curative or special palliative irradiations, and before the second fraction for palliative irradiations.

A general recommendation on how often ongoing portal and verification images should be produced and reviewed cannot be given. There are several studies showing that clinically significant localization changes (errors) frequently may occur, and that this frequency can be reduced by increasing the number of ongoing portal imaging; however, time and cost are still considerable obstacles to introduce at a daily

imaging procedure, and it would be worthwhile to always arrive at an well-controlled localization. New concepts, such as adaptive radiotherapy, are currently being developed to overcome this problem of resources. The AAPM report recommends that portal or verification images of all fields be obtained at least once a week.

References

American Association of Physics in Medicine (1984) Physical aspects of quality assurance in radiation therapy. Report series no. 13. American Institute of Physics, New York

American Association of Physics in Medicine (1994a) Comprehensive QA for radiation oncology. Report of AAPM Radiation Therapy Committee Task Group 40. Med Phys 21:581–618

American Association of Physics in Medicine (1994b) Code of practice for accelerators. Report of AAPM Radiation Therapy Committee Task Group 45 Med Phys 21:1093–1121

American Association of Physics in Medicine (1999) Task Group 51. Protocol for clinical reference dosimetry of high-energy photon and electron beams. Med Phys 26:1847–1870

American National Standards Institute (1974) Guidelines for maintaining Co-60 and Cs-137 teletherapy equipment. Report no. 449. American National Standards Institute, Washington DC

American National Standards Institute (1978) Procedures for periodic inspection of Co-60 and Cs-137 teletherapy equipment. Report no. 449.1. American National Standards Institute

Brahme A, Chavaudra J, Landberg T, McCullough E, Nüsslin F, Rawlinson A, Svensson G, Svensson H (1988) Accuracy requirements and quality assurance of external beam therapy with photons and electrons. Acta Oncol 27 (Suppl 1)

Deutsches Institut für Normung (1997) Dosismessverfahren nach der Sondenmethode für Photonen- und Elektronenstrahlung, Teil 2: Ionisationsdosimetrie. Deutsche Norm DIN 6800-2, Deutsches Institut für Normung, Berlin

Dobbs H, Thwaites DI (1999) Quality assurance and its conceptual framework. Institute of Physics and Engineering in Medicine, York, UK, Chap. 1

Dutreix A (1984) When and how can we improve precision in radiotherapy? Radiother Oncol 2:275–292

European Society for Therapeutic Radiology and Oncology (2004) Quality assurance of treatment planning systems: practical examples for external photon beams. ESTRO, Brussels, booklet no. X

Fraass B, Doppke K, Hunt M, Kutcher G, Starkschall G, Stern R, Van Dyk J (1998) American Association of Physicists in Medicine Radiation Therapy Committee Task Group 53: Quality assurance for clinical radiotherapy treatment planning. Med Phys 25:1773–1836

Institute of Physics and Engineering in Medicine and Biology (1996) The IPEMB code of practice for the determination of absorbed dose for X-rays below 300 kV generating potential (0.035 mm Al–4 mm Cu; 10–300 kV generating potential). Phys Med Biol 41:2605–2625

International Atomic Energy Agency (1999) Standardized quality audit procedures for on-site dosimetry visits to radiotherapy hospitals. IAEA DMRP-199907-IU. International Atomic Energy Agency, Vienna

International Atomic Energy Agency (2000a) Lessons learned from accidental exposures in radiotherapy. Safety Reports Series no. 17. International Atomic Energy Agency, Vienna

International Atomic Energy Agency (2000b) Absorbed dose determination in external beam radiotherapy: an international code of practice for dosimetry based on standards of absorbed dose to water. Technical report series no. 398. International Atomic Energy Agency, Vienna

International Atomic Energy Agency (2004) Commissioning and quality assurance of computerized treatment planning. Technical Report Series no. 430. International Atomic Energy Agency, Vienna

International Commission on Radiological Units and Measuremenets (1976) Determination of dose in a patient irradiated by beams of X or gamma rays in radiotherapy procedures. ICRU report 24. ICRU, Bethesda, Maryland

International Commission on Radiological Units and Measurements (2001) Dosimetry of high-energy photon beams based on standards of absorbed dose to water. ICRU report 64. J ICRU 1

International Electrotechnical Commission (1989a) Medical electrical equipment: medical electron accelerators – functional performance characteristics. IEC 976. International Electrotechnical Commission, Geneva, Switzerland

International Electrotechnical Commission (1989b) Medical electrical equipment: medical electron accelerators in the range 1 MeV to 50 MeV – guidelines for performance characteristics. IEC 977. International Electrotechnical Commission, Geneva, Switzerland

International Organisation for Standardization (1993) Guide to the expression of uncertainty in measurement. International Organisation for Standardization, Geneva, Switzerland

Leer JW, Corver R, Kraus JJ, Togt JC, Buruma OJ (1995) A quality assurance system based on ISO standards: experience in a radiotherapy department. Radiother Oncol 35:75–81

Leer JW, McKenzie AL, Scalliet P, Thwaites DI (1998) Practical guidelines for the implementation of a quality system in radiotherapy. European Society for Therapeutic Radiology and Oncolgy. Physics for Clinical Radiotherapy, booklet no. 4. ESTRO, Brussels

Mayles WPM, Lake R, McKenzie A, Macaulay EM, Morgan HM, Jordan TJ, Powley SK (eds) (1999) Physics aspects of quality control in radiotherapy. IPEM report 81. Institute of Physics and Engineering in Medicine, York, UK

McKenzie A, Kehoe T, Thwaites DI (2000) Quality assurance in radiotherapy physics. In: Williams JR, Thwaites DI (eds) Radiotherapy physics in practice. Oxford Medical Publishing, Oxford

Mijnheer B, Battermann J, Wambersie A (1987) What degree of accuracy is required and can be achieved in photon and neutron therapy. Radiother Oncol 8:237–252

Nederlandse Commissie voor Stralingsdosimetrie (1997) Dosimetry of low and medium energy X-rays, a code of practice for use in radiotherapy and radiobiology, report NCS-10. NCS, Delft, The Netherlands

Nederlandse Commissie voor Stralingsdosimetrie (2003) Quality assurance of 3-D treatment planning systems; practical guidelines for acceptance testing, commissioning, and periodic quality control of radiation therapy treatment planning systems. The Netherlands Commission on Radiation Dosimetry. Delft, The Netherlands

Nordic Association of Clinical Physics (1980) Procedures in external radiation therapy dosimetry with electron and photon beams with maximum energies between 1 and 50 MeV. Acta Radiol Oncol 19:55–79

Podgorsak EB (ed) (2003) Review of radiation oncology physics: a handbook for teachers and students. Educational Reports Series. International Atomic Energy Agency, Vienna

Shaw JE (ed) (1996) A guide to commissioning and quality control of treatment planning systems. IPEM report 68. Institute of Physics and Engineering in Medicine, York, UK

SSRPM Report 7 (1997) Quality control of treatment planning systems for teletherapy. Swiss Society for Radiobiology and Medical Physics: ISBN 3-908125-23-5

Thwaites D, Scalliet P, Leer JW, Overgaard J (1995) Quality assurance in radiotherapy. European Society for Therapeutic Radiology and Oncology Advisory Report to the Commission of the European Union for the Europe Against Cancer Programme. Radiother Oncol 35:61–73

Van Dyk J (ed) (1999) Radiation oncolgy overview. In: The modern technology for radiation oncology: a compendium for medical physicists and radiation oncologists. Medical Physics Publishing, Madison, Wisconsin

Van Dyk J, Purdy J (1999) Clinical implementation of technology and the quality assurance process. In: Van Dyk J (ed) The modern technology for radiation oncology: a compendium for medical physicists and radiation oncologists. Medical Physics Publishing, Madison, Wisconsin, pp 19–52

Van Dyk J, Barnett RB, Cygler JE, Shragge PC (1993) Commissioning and quality assurance of treatment planning computers. Int J Radiat Oncol Biol Phys 26:261–273

Venselaar JLM, Welleweerd J (2001) Application of a test package in an intercomparison of the performance of treatment planning systems used in a clinical setting. Radiother Oncol 60:203–213

Wizenberg MJ (ed) (1982) Quality assurance in radiation therapy: a manual for technologists. American College of Radiology, Philadelphia

World Health Organization (1988) Quality assurance in radiotherapy. WHO, Geneva, Switzerland

Subject Index

List of Contributors

JOHN R. ADLER, PhD
Professor, Department of Neurosurgery
R-205 Stanford University Medical Center
300 Pasteur Drive
Stanford, CA 94305-5327
USA

N. AGAZARYAN, PhD
Department of Radiation Oncology, Suite B265
David Geffen School of Medicine at UCLA
University of California Los Angeles
200 UCLA Medical Plaza
Los Angeles, CA 90096-6951
USA

DIMOS BALTAS, PhD
Professor, Department of Medical Physics
Klinikum Offenbach
Starkenburgring 66
63069 Offenbach am Main
Germany

YAZID BELKACÉMI, MD, PhD
Department of Radiation Oncology
Centre Oscar Lambret
3 rue Frédéric Combemale
59020 Lille
and University of Lille II
France

ROLF BENDL, PhD
Abteilung Medizinische Physik
in der Strahlentherapie
Deutsches Krebsforschungszentrum
Im Neuenheimer Feld 280
69120 Heidelberg
Germany

THOMAS BORTFELD, PhD
Professor, Department of Radiation Oncology
Massachusetts General Hospital
30 Fruit Street
Boston, MA 02114
USA

LIONEL G. BOUCHET, PhD
Vice President
Technology Development
ZMed, Inc., Unit B-1
200 Butterfield Drive
Ashland, MA 01721
USA

GEORGE T. Y. CHEN, MD
Professor, Department of Radiation Oncology
Massachusetts General Hospital
30 Fruit Street.
Boston, MA 02114
USA

KARSTEN EILERTSEN, PhD
Department of Medical Physics
The Norwegian Radium Hospital
University of Oslo
Montebello
Ullernchauséen 70
0310 Oslo
Norway

MATTHIAS FIPPEL, PhD
Privatdozent, Universitätsklinik für Radioonkologie
Medizinische Physik
Universitätsklinikum Tübingen
Hoppe-Seyler-Straße 3
72076 Tübingen
Germany

K.H. GROSSER, PhD
Radiologische Klinik der Universität Heidelberg
Abteilung Strahlentherapie
Im Neuenheimer Feld 400
69120 Heidelberg
Germany

ANCA-LIGIA GROSU, MD
Privatdozent, Department of Radiation Oncology
Klinikum rechts der Isar
Technical University Munich
Ismaningerstrasse 22
81675 München
Germany

PETER HÄRING
Dipl.-Ing. (FH), Department Medical Physics in Radiotherapy
Deutsches Krebsforschungszentrum
Im Neuenheimer Feld 280
69120 Heidelberg
Germany

JEAN-MICHEL HANNOUN-LÉVI, MD, PhD
Department of Radiation Oncology
Centre Antoine Lacassagne
33 avenue de Valombrose
06189 Nice Cedex
France

GUENTHER H. HARTMANN, PhD
Professor, Department of Medical Physics
in Radiotherapy
Deutsches Krebsforschungszentrum
Im Neuenheimer Feld 280
69120 Heidelberg
Germany

KLAUS K. HERFARTH, MD
Privatdozent, Department of Radiation Oncology
University of Heidelberg
Im Neuenheimer Feld 400
69120 Heidelberg
Germany

JÜRGEN HESSER, PhD
Professor, Department of ICM
Universitäten Mannheim und Heidelberg
B6, 23–29, C
68131 Mannheim
Germany

RALF HINDERER, PhD
Abteilung Medizinische Physik
in der Strahlentherapie
Deutsches Krebsforschungszentrum
Im Neuenheimer Feld 280
69120 Heidelberg
Germany

G. D. HUGO, PhD
Department of Radiation Oncology
William Beaumont Hospital
3601 W. Thirteen Mile Road
Royal Oak, MI 48073
USA

OLIVER JÄKEL, PhD
Privatdozent, Abteilung Medizinische Physik
in der Strahlentherapie
Deutsches Krebsforschungszentrum
Im Neuenheimer Feld 280
69120 Heidelberg
Germany

MARC KACHELRIESS, PhD
Professor, Institute of Medical Physics (IMP)
University Erlangen-Nürnberg
Henkestrasse 91
91052 Erlangen
Germany

CHRISTIAN P. KARGER, PhD
Privatdozent, Abteilung Medizinische Physik
in der Strahlentherapie
Deutsches Krebsforschungszentrum
Im Neuenheimer Feld 280
69120 Heidelberg
Germany

MARC L. KESSLER, PhD
Professor, Department of Radiation Oncology
The University of Michigan Medical School
1500 East Medical Center Drive, Box 0010
Ann Arbor, MI 48109-0010
USA

CHRISTIAN KIRISITS, DSc
Department of Radiotherapy and Radiobiology
Vienna General Hospital
Medical University of Vienna
Waehringer Guertel 18–20
1090 Vienna
Austria

P. KNESCHAUREK, PhD
Professor, Department of Radiation Oncology
Klinikum rechts der Isar der
Technischen Universität München
Ismaninger Straße 22
81675 München
Germany

S. KRIMINSKI, PhD
Department of Radiation Oncology, Suite B265
David Geffen School of Medicine at UCLA
University of California Los Angeles
200 UCLA Medical Plaza
Los Angeles, CA 90096-6951
USA

ERIC LARTIGAU, MD, PhD
Professor, Chief Department of Radiation Oncology
Centre Oscar Lambret
3 rue Frédéric Combemale
59020 Lille
and University of Lille II
France

SANFORD L. MEEKS, PhD
Professor, Department of Radiation Oncology
Room W189Z-GH
University of Iowa
200 Hawkins Drive
Iowa City, IA, 52242
USA

ERICH MINAR, MD
Department of Angiology
Vienna General Hospital
Medical University of Vienna
Waehringer Guertel 18–20
1090 Vienna
Austria

MICHAEL MOLLS, MD
Professor, Department of Radiation Oncology
Klinikum rechts der Isar
Technical University Munich
Ismaningerstrasse 22
81675 Munich
Germany

SASA MUTIC, MS
Professor, Washington University School of Medicine
Mallinckrodt Institute of Radiology
Siteman Cancer Center
4921 Parkview Place
St. Louis, MS 63141
USA

Simeon Nill, PhD
Abteilung Medizinische Physik
in der Strahlentherapie
Deutsches Krebsforschungszentrum
Im Neuenheimer Feld 280
69120 Heidelberg
Germany

Uwe Oelfke, PhD
Privat Dozent, Abteilung Medizinische Physik
in der Strahlentherapie
Deutsches Krebsforschungszentrum
Im Neuenheimer Feld 280
69120 Heidelberg
Germany

Dag Rune Olsen, PhD
Professor, Institute for Cancer Research
The Norwegian Radium Hospital
University of Oslo
Montebello
Ullernchauséen 70
0310 Oslo
Norway

Nigel P. Orton, PhD
Medical Physicist
Department of Human Oncology
University of Wisconsin Medical School
CSC K4/B 100
600 Highland Avenue
Madison, WI 53792
USA

Harald Paganetti, PhD
Massachusetts General Hospital
Harvard Medical School
Fruit Street 30
Boston, MA 02114
USA

Andrea Pirzkall, MD
Assistant Adjunct Professor
Departments of Radiation Oncology
Radiology and Neurological Surgery
University of California, San Francisco (UCSF)
505 Parnassus Avenue, L-75, Box 0226
San Francisco, CA 94143
USA

Richard Pötter, MD
Professor, Department of Radiotherapy
and Radiobiology
General Hospital of Vienna
Medical University of Vienna
Waehringer Guertel 18–20
1090 Vienna
Austria

Boris Pokrajac, MD
Department of Radiotherapy and Radiobiology
Vienna General Hospital
Medical University of Vienna
Waehringer Guertel 18–20
1090 Vienna
Austria

Andreas Pommert, PhD
Institut für Medizinische Informatik (IMI)
Universitätsklinikum Hamburg-Eppendorf
Martinistrasse 52
20246 Hamburg
Germany

Bernhard Rhein
Dipl.-Ing. (BA), Department Medical Physics in Radiotherapy
Deutsches Krebsforschungszentrum
Im Neuenheimer Feld 280
69120 Heidelberg
Germany

Mark A. Ritter, MD, PhD
Associate Professor
Department of Human Oncology
University of Wisconsin Medical School
CSC K4/B 100
600 Highland Avenue
Madison, WI 53792
USA

Eike Rietzel, PhD
Siemens Medical Solutions
Particle Therapy
Henkestrasse 127
91052 Erlangen
Germany

Michael Roberson, BS
Department of Radiation Oncology
The University of Michigan Medical School
1500 East Medical Center Drive, Box 0010
Ann Arbor, MI 48109-0010
USA

Georgios Sakas, PhD
Professor, Fraunhofer Institut für
Graphische Datenverarbeitung
Fraunhoferstr. 5
64283 Darmstadt
Germany

Lothar R. Schad, PhD
Professor, Abteilung Medizinische Physik
in der Radiologie
Deutsches Krebsforschungszentrum
Postfach 101949
69009 Heidelberg
Germany

Wolfgang Schlegel, PhD
Professor, Abteilung Medizinische Physik
in der Strahlentherapie
Deutsches Krebsforschungszentrum
Im Neuenheimer Feld 280
69120 Heidelberg
Germany

Christian Scholz, PhD
Siemens Medical Solutions
Oncology Care Systems
Hans-Bunte-Str. 10
69123 Heidelberg
Germany

ACHIM SCHWEIKARD, PhD
Professor, Institute for Robotic and
Cognitive Systems
University Luebeck
Ratzeburger Allee 160
23538 Luebeck
Germany

TIMOTHY D. SOLBERG, PhD
Professor, Medical Physics
Department of Radiation Oncology
University of Nebraska Medical Center
987521 Nebraska Medical Center
Omaha, NE 68198–7521
USA
and
Department of Radiation Oncology, Suite B265
David Geffen School of Medicine at UCLA
University of California Los Angeles
200 UCLA Medical Plaza
Los Angeles, CA 90096-6951
USA

LISA D. SPRAGUE, PhD
Department of Radiation Oncology
Klinikum rechts der Isar
Technical University Munich
Ismaningerstrasse 22
81675 Munich
Germany

DMITRY STSEPANKOU
Department of ICM
Universitäten Mannheim und Heidelberg
B6, 23–29, C
68131 Mannheim
Germany

S. E. TENN, MS
Department of Radiation Oncology, Suite B265
David Geffen School of Medicine at UCLA
University of California Los Angeles
200 UCLA Medical Plaza
Los Angeles, CA 90096-6951
USA

CHRISTIAN THIEKE, MD, PhD
Department of Radiation Oncology
German Cancer Research Center (DKFZ)
Im Neuenheimer Feld 280
69120 Heidelberg
Germany

WOLFGANG A. TOMÉ, PhD
Associate Professor
Department of Human Oncology
University of Wisconsin Medical School
CSC K4/B 100
600 Highland Avenue
Madison, WI 53792
USA

N. M. WINK, MS
Department of Radiation Oncology, Suite B265
David Geffen School of Medicine at UCLA
University of California Los Angeles
200 UCLA Medical Plaza
Los Angeles, CA 90096-6951
USA

DI YAN, D.Sc
Director, Clinical Physics Section
Department of Radiation Oncology
William Beaumont Hospital
Royal Oak, MI 48073-6769
USA

MARCO ZAIDER, PhD
Department of Medical Physics
Memorial Sloan-Kettering Cancer Center
1275 York Avenue
NewYork, NY 10021
USA

NIKOLAOS ZAMBOGLOU, MD, PhD
ProfessorAbteilung Medizinische Physik
Klinikum Offenbach
Starkenburgring 66
63069 Offenbach am Main
Germany

MEDICAL RADIOLOGY   Diagnostic Imaging and Radiation Oncology

Titles in the series already published

 Springer

MEDICAL RADIOLOGY Diagnostic Imaging and Radiation Oncology

Titles in the series already published

 Springer